Lecture Notes in Physics

Edited by J. Ehlers, München, K. Hepp, Zürich, and
H. A. Weidenmüller, Heidelberg
Managing Editor: W. Beiglböck, Heidelberg

38

Dynamical Systems, Theory and Applications

Battelle Seattle 1974 Rencontres

Edited by J. Moser

Springer-Verlag
Berlin Heidelberg GmbH 1975

Editor
Prof. Dr. Jürgen Moser
New York University
Courant Institute
of Mathematical Sciences
251 Mercer Street
New York, N.Y. 10012/USA

Library of Congress Cataloging in Publication Data

Battelle Rencontres, Seattle, 1974.
Dynamic systems.

(Lecture notes in physics ; 38)
Bibliography: p.
Includes index.
1. Differential equations, Nonlinear--Congresses.
2. Statistical mechanics--Congresses. 3. Ergodic
theory--Congresses. I. Moser, Jürgen, 1928-
II. Title. III. Series.
QA371.B35 1974 530.1'5'535 75-14488

ISBN 978-3-540-07171-6 ISBN 978-3-540-37505-0 (eBook)
DOI 10.1007/978-3-540-37505-0

Preface

These Proceedings contains lectures held at the Battelle Rencontres in Seattle. The conference was devoted to the study of the time evolution of dynamical systems as they occur in various fields of application. Here the concept of dynamical systems was understood in the widest sense, encompassing finite particle systems in the classical sense, infinite systems of statistical mechanics and wave propagation phenomena described by nonlinear partial differential equations.

In the spirit of the Battelle Rencontres, it was the aim to bring physicists and mathematicians of various fields together to get a wide angle view and uncover the ideas bridging the various specialized subjects. It became evident, for instance, that there is a close connection between ergodic theory and statistical mechanics. Another topic drawing on various different fields is provided by the striking wave phenomena described in the work of Kruskal, Newell and Flaschka, which can be viewed as the first significant examples of integrable Hamiltonian systems of infinitely many degree freedoms -- a concept which until now had been investigated only for finite dimensional systems. This subject is related to spectral theory and scattering theory in a most surprising way.

While at the meeting in Seattle, these lectures were interspersed, in the Proceedings they are divided into three groups: (I) Statistical Mechanics, (II) Ergodic Theory, and (III) Nonlinear Differential Equations, although this organization is rather arbitrary. There were three main lectures by Kruskal, Lanford and Ornstein, whose papers head the three groups. Some of the following papers are expository in nature while others, like those of Lanford, Mather and McGehee, contain new results which have not yet been published. Of course, at the meeting there were several other lectures held which have not been written up or which have appeared elsewhere.

It is a pleasant task to express on behalf of the participants their thanks to the Battelle Institute for providing not only the financial basis and the housing but also the pleasant surroundings and facilities for this meeting. In particular, the assistance of Drs. Murray Mercier and William Kern was invaluable during the organization of the conference. In the scientific planning, I had the assistance of many participants to whom I wish to express my thanks here, especially to J. Lebowitz, without whom the conference could not have succeeded as it did. The typing was carried out expertly and cheerfully by Constance Engle.

New York, March 1975 Jürgen Moser

Table of Contents

BATTELLE RENCONTRES PARTICIPANTS

ADLER, ROY
IBM, Yorktown Heights, New York

AIZENMAN, MICHAEL
Yeshiva University

BERGMANN, ERICK
Battelle-Geneva Research Center

CHANNELL, PAUL
Lawrence Berkeley Laboratory

CONLEY, CHARLES C.
University of Wisconsin

CURRY, JAMES
University of California at Berkeley

DI LIBERTO, FRANCESCO
Istituto di Fisica Teorica Italy

FENICHEL, NEIL
University of British Columbia Vancouver, B. C., Canada

FLASCHKA, HERMANN
University of Arizona

FORD, JOSEPH
Georgia Institute of Technology

GALLAVOTTI, GIOVANNI
Institute Voor Theoretische Fysica, The Netherlands

GIDAS, BASILIS
University of Washington

GOLDSTEIN, SHELDON
Institute for Advanced Study

HAAG, RUDOLPH
Pinneberg, Germany

HEPP, KLAUS
Eidgenössische Technische Hochschule, Zürich-Hönggerberg Switzerland

KERN, WILLIAM C.
Battelle Memorial Institute

KRUSKAL, MARTIN
The Weizmann Institute of Science

LANFORD, OSCAR E. III
University of California at Berkeley

LEBOWITZ, JOEL L.
Yeshiva University

Lieb, Elliott
Massachusetts Institute of Technology

LIND, DOUGLAS A.
University of California at Berkeley

LIU, CHEN Y.
Battelle Memorial Institute

McGEHEE, RICHARD
University of Minnesota

MATHER, JOHN
Harvard University

MILFORD, FREDERICK J.
Battelle Memorial Institute

MIURA, ROBERT M.
Vanderbilt University

MOSER, JÜRGEN
New York University

NEWELL, ALAN C.
Clarkson College of Technology

ORNSTEIN, DONALD S.
Stanford University

RICHARDSON, RICHARD L.
Battelle Pacific Northwest Laboratory

RUELLE, DAVID
Institut des Hautes Etudes Scientifiques, France

RÜSSMANN, HELMUT
Mainz-Lerchenberg, Germany

SMOLLER, JOEL
The University of Michigan

SMORODINSKY, M.
Tel Aviv University

WEISS, BENJAMIN
Hebrew University Jerusalem, Israel

ZABUSKY, NORMAN J.
Bell Laboratories, Inc.

ZEHNDER, EDWARD
Institute for Advanced Study

TIME EVOLUTION OF LARGE CLASSICAL SYSTEMS

Oscar E. Lanford III *

Department of Mathematics, University of California, Berkeley, California

1. Introduction

Summary. We begin with some very general and elementary remarks about nonequilibrium statistical mechanics. We then establish our notation for discussing finite systems of classical point particles, construct the microcanonical ensemble, and sketch some of the relations between statistical mechanics and ergodic theory.

....

In these lectures, we will be concerned with the time evolution of systems of many point particles moving according to the laws of classical mechanics. The ultimate objective of the investigation of these questions is an understanding of the time-dependent behavior of macroscopic matter analogous to and generalizing the understanding of stationary (equilibrium) behavior provided by equilibrium statistical mechanics. Because we must build on the insights of equilibrium statistical mechanics, we begin with a brief description of that theory.

The striking fact from which equilibrium statistical mechanics begins is the contrast between the enormous complexity of matter on the molecular level and the relative simplicity of its macroscopic behavior. If we idealize the molecules making up matter as classical point particles, then the microscopic states of a piece of matter containing n molecules are parametrized by the points of a 6n-dimensional manifold, the phase space of the system. For one mole of matter, n is about 6×10^{23}. On the other hand, the macroscopic states of matter form a two-dimensional manifold whose points may be labeled, for example, by temperature and density.

* Supported in part by NSF Grant GP-42225.

That is, the temperature and density determine most other observable quantities, such as the pressure, the compressibility, the specific heat, etc.

The formalism of equilibrium statistical mechanics, invented in the second half of the nineteenth century by Boltzmann and Gibbs, provides a procedure for computing macroscopic properties starting from the fundamental physics governing molecules. Rather than trying to look at properties of individual phase points, Boltzmann and Gibbs investigated probability measures of the phase space. (For historical reasons, these probability measures are generally referred to as ensembles.) A particularly important role is played by the canonical ensemble obtained by normalizing the measure

$$\exp\,[-\beta\, H(\underline{x})]\; d\underline{x}$$

where

$\underline{x}$ denotes a phase point (i.e., a point of phase space)

$\underline{x} = (q_1, p_1, \ldots, q_n, p_n)$

$d\underline{x} = dq_1 dp_1 \ldots dq_n dp_n$

$H(\underline{x})$ is the energy of the phase point

$\beta = (kT)^{-1}$, where T is the absolute temperature and k is a constant, called Boltzmann's constant.

The fundamental principle of statistical mechanics is the:

Boltzmann-Gibbs Prescription. The equilibrium value of any quantity at temperature T is the average value of that quantity in the canonical ensemble.

It is remarkable that this simple principle (or a quantum generalization of it) appears to provide a completely correct basis for the prediction of macroscopic behavior.

Although a completely satisfactory mathematical explanation for the success of the Boltzmann-Gibbs prescription has not yet been given, it seems likely that such an explanation will eventually be constructed along the following lines: Macroscopic properties of matter are determined, not by arbitrary functions on the phase space, but by a special class of functions called macroscopic observables. If the number of particles is very large, the values of these macroscopic observables are determined almost exactly by the energy and density, except perhaps on a set of very small measure. Such statements can only become exact in the limit as the number of particles becomes infinite with the density held fixed (and finite), i.e., in the so-called thermodynamic limit. For a discussion of the extent to which the above statements can be shown to be correct, see [10].

By comparison to equilibrium statistical mechanics, the theory of nonequilibrium phenomena is in a very primitive state. There is no central unifying principle like the Boltzmann-Gibbs prescription from which all quantities can in principle be computed. To illustrate the sort of questions we would like to be able to answer, let us examine a simple prototype problem. Consider a container divided in half by an internal partition:

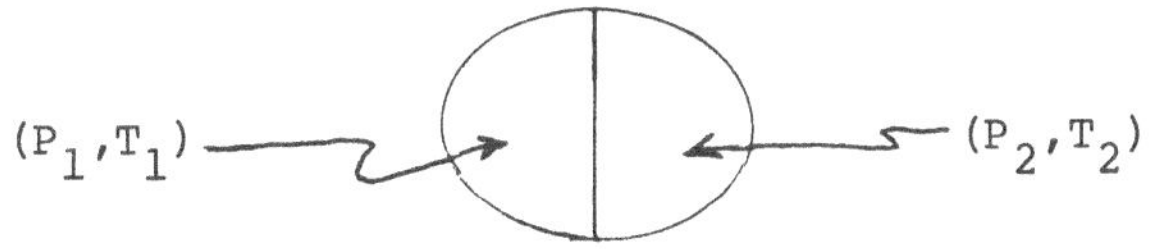

In the two halves we put different quantities of gas, each separately in thermodynamic equilibrium but with different temperatures and pressures. At time zero, we remove the partition and allow the system to evolve; a satisfactory theory of nonequilibrium statistical mechanics should predict what will happen.

In practice, we know at least schematically what will happen. The system eventually arranges itself to be in equilibrium throughout the container. The process by which it reaches equilibrium may be very complicated, perhaps involving shock waves, turbulence, etc., but this process is governed to an excellent approximation by a set of hydrodynamic equations which are coupled nonlinear partial differential equations including

(a) a heat flow equation

(b) equations for viscous compressible fluid flow

One of the fundamental problems of nonequilibrium statistical mechanics is the derivation of these hydrodynamic equations from the underlying molecular dynamics. On a more practical level, the theory should give a prescription for computing the "transport coefficients" (heat conductivity, viscosity, etc.) appearing in these equations.

In any attempt to derive hydrodynamic equations, it is necessary to come to terms with the fact that the hydrodynamic equations are irreversible and dissipative while the underlying molecular dynamics is reversible. The irreversibility of the macroscopic description reflects, for example, the empirical facts that heat always flows from a warm body to a cooler one, never the reverse, and that viscosity (friction) damps out fluid motion. Reversibility (or time-reversal invariance) of the molecular dynamics means that, if $(q_1(t),\ldots,q_n(t))$ is a solution of the equations of motion, so is $(q_1(-t),\ldots,q_n(-t))$. This implies that, for every solution in which heat flows from a warm region to a cooler one, there is another equally good solution for which exactly the reverse happens. Thus, at the very least, the macroscopic hydrodynamic description cannot follow from the molecular dynamics alone; some additional element must be added.

The additional element which seems to work is the introduction of probabilistic considerations. That is, instead of trying to show that hydrodynamic equations always hold, one instead tries to show that they hold with probability very close to one. This approach must be applied with a considerable amount of care as the time development of probability measures is just as reversible as that of the phase points themselves. What appears to be true is something like the following: For a large class of "reasonable" initial probability distributions, if the number of particles is very large and if one looks only at appropriately chosen "macroscopic observables", hydrodynamic equations hold with probability very close to one. The task of formulating such a result precisely, to say nothing of proving it, is far from being accomplished. For one attempt in this direction, see the derivation of the Boltzmann equation in the final chapter of these notes.

We now turn from these generalities to the precise formulation of the microscopic physics from which we are supposed to begin. We will be talking extensively about systems of point particles and will need some streamlined notation for describing these systems. Let us consider a system of n particles confined to a region Λ of physical space $\mathbb{R}^\nu$.* The phase space of such a system is $(\Lambda \times \mathbb{R}^\nu)^n$, which is $2n\nu$ dimensional. To specify a point of this phase space, we must give n positions $q_1,\ldots,q_n$ (each of which belongs to Λ) and n momenta $p_1,\ldots,p_n$ belonging to $\mathbb{R}^\nu$. We will write $\underline{q}$ for $(q_1,\ldots,q_n)$; $\underline{p}$ for $(p_1,\ldots,p_n)$; x_i for $(q_i,p_i) \in \Lambda \times \mathbb{R}^\nu$ (the one-particle phase space); and $\underline{x}$ for $(x_1,x_2,\ldots,x_n) = (q_1,p_1; \ldots; q_n,p_n)$. The energy corresponding to the phase-point $\underline{x}$ is

*We are of course mostly interested in $\nu = 3$, but do not commit ourselves to this value since there is occasionally something interesting to be learned about the dependence of various results on ν.

$$H(\underline{x}) = T(\underline{p}) + U(\underline{q})$$

$$T(\underline{p}) = \frac{1}{2m} \sum_i p_i^{\,2} \; ; \qquad\qquad U(\underline{q}) = \sum_{i<j} \Phi(q_i - q_j) \; ,$$

where Φ is the interparticle (pair) potential. The equations of motion are

$$\frac{dq_i}{dt} = \frac{p_i}{m} = \frac{\partial H}{\partial p_i} \; ; \qquad\qquad \frac{dp_i}{dt} = \sum_{j \neq i} F(q_i - q_j) = - \frac{\partial H}{\partial q_i} \; ;$$

here, $F(q) = -\operatorname{grad} \Phi(q)$ is the interparticle force.

In the above brief discussion, we have not dealt with the problem of keeping the particles confined to Λ. We will generally ignore this problem, but we mention briefly here a few ways to deal with it:

1. We may add to the potential energy an external potential $\sum_i v(q_i)$ where $v(q)$ is small (together with its first derivatives, at least) in the interior of Λ but where $v(q)$ goes to ∞ as q approaches the boundary of Λ.
2. We may take Λ to be a ν-dimensional torus, i.e., a ν-dimensional rectangular parallelepiped "with periodic boundary conditions".
3. We may assume that Λ has a smooth boundary and make particles undergo elastic reflection when they strike the boundary.

Any one of these approaches will work; we will generally leave it to the reader to work out the necessary details.

We will let T^t denote the solution mappings to the equations of motion, i.e., if $\underline{x}(t)$ is a solution curve, then $T^t\underline{x}(0) = \underline{x}(t)$ for any t. Because the equations of motion have Hamiltonian form, we have:

(1) (Conservation of energy): $H \circ T^t = H$

(2) (Liouville's Theorem): The Lebesgue measure $d\underline{x} = dq_1 \ldots dq_n \, dp_1 \ldots dp_n$ is invariant under the flow T^t.

Let us now consider a value of E for the energy which is greater than the minimum value of H, and let

$$\Sigma(E) = \{\underline{x}: H(\underline{x}) = E\} ;$$

$\Sigma(E)$ is called the <u>energy surface</u>. Since $\Sigma(E)$ is carried into itself by the flow T^t, we would like to find a way to "restrict" Lebesgue measure to $\Sigma(E)$ in order to get an invariant measure. We of course cannot do this in a completely elementary way since the Lebesgue measure of the "surface" $\Sigma(E)$ is zero. If E is a regular value of H (i.e., if $\Sigma(E)$ contains no points $\underline{x}_0$ where $(\text{grad } H)(\underline{x}_0) = 0$, then it is an easy exercise to "extract a factor of dE from $d\underline{x}$" leaving a differential of degree $2n\nu - 1$ which gives an invariant measure on $\Sigma(E)$. It is interesting to note, however, that by using the special form of H we can avoid any possible problems about singular values of H. The point is as follows: We have $H(\underline{q},\underline{p}) = \frac{1}{2m} \cdot \underline{p}^2 + U(\underline{q})$; by introducing polar coordinates for the momentum integral, we get

$$d\underline{x} = d\underline{q}\, d\underline{p} = d\underline{q} \cdot |\underline{p}|^{n\nu-1}\, d|\underline{p}|\, d\omega(\underline{p}) ,$$

where $d\omega(\underline{p})$ denotes the angular integration. On the other hand, regarding the energy as a function of $\underline{p}$ with $\underline{q}$ held fixed, we get $dE = |\underline{p}|\,d|\underline{p}| \,/\, m$, and hence

$$d\underline{x} = d\underline{q}\ \frac{|\underline{p}|^{n\nu-2}}{m}\ dE\ d\omega(\underline{p})$$

$$= d\underline{q}\ \frac{[2m(E-U(\underline{q}))]^{n\nu/2\ -1}}{m}\ dE\ d\omega(\underline{p}) .$$

This is slightly formal, but it is not hard to verify that it gives

$$\int f(\underline{x})\, d\underline{x} = \int dE \int_{U(\underline{q})\leq E} d\underline{q}\ \frac{[2m(E-U(\underline{q}))]^{n\nu/2\ -1}}{m} \int_{|\underline{p}| = [2m(E-U(\underline{q}))]^{1/2}} d\omega(\underline{p})\ f(\underline{q},\underline{p})$$

where f is for example any continuous function of compact support on $(\Lambda \times \mathbb{R}^{\nu})^{n}$. Thus, for example,

$$\lim_{\Delta E \downarrow 0} \frac{1}{\Delta E} \int_{E \le H(\underline{x}) \le E+\Delta E} f(\underline{x})\, dx = \int_{U(\underline{q}) \le E} d\underline{q}\ \frac{[2m(E-U(\underline{q}))]^{n\nu-1}}{m} \int_{|\underline{p}|=\sqrt{2m(E-U(q))}} d\omega(\underline{p})\ f(\underline{q},\underline{p})$$

for all such functions f, provided that $n\nu \geq 2$. We may take the right-hand side of this equation as defining, via the Riesz Representation Theorem, a measure on $(\Lambda \times \mathbb{R}^{\nu})^{n}$ which is

(1) concentrated on $\Sigma(E)$ (assuming U is continuous)

(2) invariant under T^{t} since it is a limit of invariant measures.

We will refer to this measure as Lebesgue measure cut down to the energy surface. If Λ has finite volume and U is bounded below, this measure is finite (integrate the constant function 1), and hence may be normalized to form a probability measure. This probability measure is called to microcanonical probability measure or microcanonical ensemble. Its physical interpretation is obvious; it is the probability distribution on $\Sigma(E)$ which assigns "equal weight" to phase points with energy E.

Before going on, let us discuss briefly an interesting system which does not quite fit into the above framework but which is in some sense a limiting case of it. This is the system of hard spheres of diameter d which move freely except for undergoing elastic collisions. What happens in a binary collision is completely determined by the requirements of conservation of (kinetic) energy and linear momentum together with the condition that momentum transferred in collisions is orthogonal to the plane of contact. (What happens in collisions of more than two particles is less clear, but such collisions are supposed to be sufficiently improbable to be neglected). For some purposes, it is convenient to think of this system as given by a two-body potential Φ:

$$\begin{aligned}\Phi(q_1-q_2) &= 0 \quad \text{if} \quad |q_1-q_2| \geq d \\ &= \infty \quad \text{if} \quad |q_1-q_2| < d\,.\end{aligned}$$

The phase space is $\{(q_1,\ldots,q_n,p_1,\ldots,p_n) \in (\Lambda \times \mathbb{R}^\nu)^n\colon |q_i-q_j| \geq d$ for all $i \neq j\}$. The solution mappings preserve energy, and (although this is not entirely obvious) Liouville's Theorem holds. We can again cut down to the energy surface to form the microcanonical ensemble. (In this case, the energy surface is just the product of the sphere $\sum_i \frac{p_i^2}{2m} = E$ with the position space $\{(q_1,\ldots,q_n) \in \Lambda^n\colon |q_i-q_j| \geq d$ for $i \neq j\}$.)

Although the microcanonical ensemble is certainly a sensible starting point for the development of statistical mechanics, it is natural to ask for some justification of its use in computing equilibrium values of observables. In the 1880's, Boltzmann proposed such a justification which has been enormously fruitful for later developments as it led to the beginnings of ergodic theory. It is a little hard to sort out what Boltzmann actually said from what others have attributed to him, and the account which follows is more properly viewed as myth than as history. The reader interested in the correct history is referred to Boltzmann's lectures [1] and to [6] and [7].

Boltzmann's argument went something like this: Start with a more or less general function f on the energy surface. Although $f \circ T^t$ may vary with time, experience shows that it eventually settles down to an equilibrium value. This equilibrium value may conveniently be written as a time average:

$$\bar{f}(\underline{x}) = \lim_{\tau\to\infty} \frac{1}{\tau} \int_0^\tau dt\, f(T^t\,\underline{x})\,.$$

By the invariance of the microcanonical measure μ,

$$\int \bar{f}\, d\mu = \lim_{\tau\to\infty} \frac{1}{\tau} \int_0^{\tau} dt \int d\mu\; f\circ T^t = \int f\, d\mu \;.$$

If, now, we know that $\bar{f}(\underline{x})$ is constant on the energy surface, the constant must be $\int \bar{f}\, d\mu$ or $\int f\, d\mu$, i.e., the equilibrium value of f is the same as the microcanonical average of f.

Boltzmann tried to complete the justification by arguing that the phase point $T^t\,\underline{x}$ should wander freely over the whole energy surface and thus that $\bar{f}$ must be constant since it is constant along orbits. His argument on this point was, however, not very clear, and forty years of confusion ensued. The picture was considerably clarified by Birkhoff's Pointwise Ergodic Theorem and by the introduction of the notion of metric transitivity or <u>ergodicity</u>. The Birkhoff theorem asserts at least that the time average $\bar{f}(\underline{x})$ exists almost everywhere -- a fact which Boltzmann took for granted. The definition of ergodicity, while it really only served to put the hard problems somewhere else, provided the machinery for making Boltzmann's argument precise. The definition is as follows: We say that a system (X,μ,T^t) consisting of a space, a measure on it, and a group of measure-preserving transformations, is <u>ergodic</u> if the only invariant measurable sets are either null sets or complements of null sets. A very elementary argument shows that this is the same thing as saying that the only invariant functions are equal almost everywhere to constants. Since the time average

$$\bar{f}(\underline{x}) = \lim_{T\to\infty} \frac{1}{T} \int_0^{T} dt\; f(T^t\underline{x}) \;,$$

whose existence is assured by the Birkhoff theorem, is invariant under the flow, it must be equal almost everywhere to a constant, and this constant must be given by $\int f\, d\mu$. In other words, using the standard terminology of the physics literature, time averages are equal to ensemble averages. This, then, provides an airtight justification for the use of the microcanonical ensemble provided:

(a) one can prove that systems of physical interest are ergodic on their energy surfaces, and

(b) one is willing to ignore sets of measure zero with respect to the microcanonical ensemble.

Ergodicity is only the first of a long list of desirable properties a dynamical system can have. Many of these stronger conditions have physical interpretations some of which are discussed in [9]. We will not give these interpretations except in a single particularly interesting case -- that of the mixing property. Recall that a dynamical system is <u>mixing</u> if for all measurable sets A,B,

$$\lim_{|t|\to\infty} \mu(T^t A \cap B) = \mu(A)\,\mu(B)$$

or, equivalently (assuming $\mu(B) \neq 0$)

$$\lim_{|t|\to\infty} \frac{\mu(T^t A \cap B)}{\mu(B)} = \mu(A) ,$$

i.e., for any B of non-zero measure and any A, the fraction of B occupied by $T^t A$ is, for large t, very nearly the same as the fraction of the whole space occupied by $T^t A$. In one more reformulation: the system is mixing if, for each measurable A, the set $T^t A$ becomes uniformly spread out over the whole space as $t \to \infty$. This property evidently implies ergodicity, and examples show that it is strictly stronger than ergodicity. A straightforward approximation argument shows that if mixing holds then

$$\lim_{|t|\to\infty} \int f \circ T^t \cdot g \, d\mu = \int f \, d\mu \cdot \int g \, d\mu$$

for all integrable f and bounded measurable g. Now suppose we consider a probability measure $d\rho$ which is absolutely continuous with respect to μ, i.e.,

$$d\rho = f(\underline{x})\ d\mu \ ,$$

where $f \geq 0$ and $\int f\ d\mu = 1$. If we let the measure evolve with the flow, we obtain a one-parameter family of measures

$$d\rho^t = f \circ T^{-t}\ d\mu \ .$$

The mixing property now implies

$$\lim_{|t|\to\infty} d\rho^t = d\mu$$

in the sense that

$$\lim_{|t|\to\infty} \int g\ d\rho^t = \int g\ d\mu$$

for all bounded measurable g. Thus we may summarize:

(a) ergodicity means that the only <u>invariant</u> measure absolutely continuous with respect to μ is μ itself

(b) mixing means that any measure absolutely continuous with respect to μ converges to μ when followed forward or backward in time.

In physical terms, mixing means that any statistical state converges to the "equilibrium" microcanonical ensemble provided it is absolutely continuous with respect to it. Note, incidentally, that this offers one way out of the irreversibility paradox -- although $d\rho$ can be reconstructed (in principle) from $d\rho^t$ for any finite t, the reconstruction becomes more and more delicate as t becomes large and eventually becomes completely impossible when $|t| \to \infty$. On the other hand, the mixing property does not distinguish between going forward in time and going backward, i.e., $d\rho^t$ converges to $d\mu$ either as $t \to \infty$ or as $t \to -\infty$.

Having discussed the usefulness of ergodic theory in understanding the approach to equilibrium, we next point out some of its limitations. To fix ideas, let us return to our model problem of a container of gas initially divided in half by an internal wall, the two halves initially containing different quantities of gas in thermal equilibrium at different temperatures. At time zero, we remove the internal wall and watch how the system develops by measuring the fraction of particles $\phi(\underline{x},t)$ in the left-hand half of the container as a function of time. As our notation indicates, $\phi(\underline{x},t)$ depends on the initial phase point $\underline{x}$ as well as on t. If we knew that the microcanonical ensemble on the appropriate energy surface was ergodic, we could conclude that

$$\lim_{\tau\to\infty} \frac{1}{\tau} \int_0^\tau dt\ \phi(\underline{x},t) = \frac{1}{2}$$

for almost all $\underline{x}$. If we knew in addition that the microcanonical ensemble was mixing, we would also know that

$$\lim_{t\to\infty} \bar{\phi}(t) = \frac{1}{2}$$

where $\bar{\phi}(t)$ denotes the mean value of $\phi(\underline{x},t)$ with respect to some appropriate ensemble describing the allowed initial phase-points. These results, while helpful, do not deal with many of the important aspects of the behavior of the system. Specifically:

(1) They offer no help in the task of computing $\bar{\phi}(t)$.

(2) They do not imply the important fact that, for all but a very small set of initial phase points $\underline{x}$, $\phi(\underline{x},t)$ stays very near to $\bar{\phi}(t)$ over quite a long interval of times.

Point (2), which depends on the fact that the system is composed of a very large number of particles, becomes a precise statement only in the limit as the number of particles goes to infinity.

Ergodic theory, in fact, implies the apparently counter-intuitive fact that $\phi(\underline{x},t)$ <u>cannot</u> approach 1/2 as $t \to \infty$ (except perhaps for a null set of $\underline{x}$'s). This follows from a simple argument due to Poincaré: If A is any measurable subset of the energy surface, then the set of $\underline{x}$'s in A such that $T^t\underline{x}$ is outside A for all sufficiently large t is a set of measure zero. (This result, incidentally, holds even if the system is not ergodic.) If, for example, we start the system out with all particles in the left-hand half and if we let A be the subset of the energy surface composed of phase points with all particles in the left-hand half, then those $\underline{x}$'s in A such that

$$\lim_{t\to\infty} \phi(\underline{x},t) = \frac{1}{2}$$

cannot have $T^t\underline{x}$ in A for arbitrarily large values of t and hence must form a null set.

The above remark, although striking, is entirely irrelevant from a practical point of view, for reasons which again have to do with the large number of particles in systems of interest. The point is that the set A has extremely small measure. For example, a cubic millimeter of gas at a pressure of 1 atmosphere and a temperature of 0° C. contains about 2.7×10^{16} molecules -- a comparatively small number as such things go. If the particles do not interact, the probability that they are all in the left-hand half of the volume is $(1/2)^n \simeq 10^{-8\times 10^{15}}$. For interacting molecules with reasonably realistic potentials the probability is not too far from this. If the system is ergodic, then the fraction of the time it spends in A is given by the probability of A, and since this is so very small, we can expect to have to wait a totally unreasonable length of time before it departs from equilibrium to return to A.

References to Part 1

The two classic works on statistical mechanics, both still quite readable, are

[1] Boltzmann, L., Vorlesungen über Gastheorie, Leipzig: J. A. Barth, 1896-98, translated by S. G. Brush as Lectures on Gas Theory, Berkeley: University of California Press, 1964.

[2] Gibbs, J. W., Elementary Principles in Statistical Mechanics, New Haven: Yale University Press, 1902, reprinted New York: Dover, 1960.

A particularly penetrating analysis and critique of statistical mechanics is given in

[3] Ehrenfest, P and T., Begriffliche Grundlagen der statistischen Auffassung in der Mechanik, in Encyclopädie der mathematischen Wissenschaften, translated by M. J. Moravcsik as The Conceptual Foundations of the Statistical Approach to Mechanics, Ithaca: Cornell University Press, 1959.

For more recent discussion of the foundations of nonequilibrium statistical mechanics, see

[4] Grad, H., Principles of the Kinetic Theory of Gases, in Handbuch der Physik, ed. S. Flügge, Vol. 12, Berlin, Göttingen, Heidelberg: Springer-Verlag, 1958. and

[5] Grad, H., Levels of Description in Statistical Mechanics and Thermodynamics, Delaware Seminar in the Foundations of Physics, ed. M. Bunge, pp. 49-72, New York: Springer, 1967.

For the history of the development of statistical mechanics, see

[6] Brush, S. G., Foundations of Statistical Mechanics 1845-1915, Arch. Hist. Exact Sci. 4, 145-183 (1967). and

[7] Brush, S. G., Proof of the Impossibility of Ergodic Systems: The 1913 Papers of Rosenthal and Plancherel, Transport Theory

and Statistical Physics 1, 287-298 (1971).

A very useful survey of nonequilibrium statistical mechanics from a more practical point of view is given in the second half of

[8] Uhlenbeck, G. E., and Ford, G. W., Lectures in Statistical Mechanics, Providence: American Mathematical Society, 1963.

For a discussion of the relation between modern ergodic theory and statistical mechanics, see

[9] Lebowitz, J. L., and Penrose, O., Modern Ergodic Theory, Physics Today, 26: 2, 23-29 (1973).

For one attempt at a mathematical justification of the Boltzmann-Gibbs prescription, see

[10] Lanford, O.E., III, Entropy and Equilibrium States in Classical Statistical Mechanics, in 1971 Battelle Recontres: Statistical Mechanics and Mathematical Problems, ed. A. Lenard, Lecture Notes in Physics No. 20, Berlin, Heidelberg, New York: Springer-Verlag, 1973, pp. 1-113.

2. States of Infinite Classical Systems

Summary. This chapter is preliminary in character. We construct the phase space for systems of infinitely many particles and mention some elementary facts about measures on it. We then construct the probability measures representing thermodynamic equilibrium for noninteracting particles and outline the process of passing to the infinite system limit of thermodynamic equilibrium states for interacting systems. Finally, we define the notion of Gibbs measures on the infinite system phase space, and state a simplified version of Ruelle's probability estimates for such measures.

....

We have developed above the point of view that much of statistical mechanics, both equilibrium and nonequilibrium, depends in an essential way on the fact that the systems we are dealing with are large. In order to exploit this fact effectively, we want to let the systems we investigate become infinitely large. We emphasize that we are not considering the theory of infinite systems for its own sake so much as for the fact that this is the only precise way of removing inessential complications due to boundary effects, etc., i.e., we regard infinite systems as approximations to large finite systems rather than the reverse.

We will be concerned primarily with the properties of statistical states of the infinite system which we will generally obtain as limits of probability measures on finite system phase spaces. By a statistical state, we mean a probability measure on the infinite system phase space, but this latter concept requires a definition. Let us first define, for any bounded Borel set Λ, $\mathcal{X}(\Lambda)$ to be the union over n of the n-particle phase spaces for systems contained in Λ. That is:

$$\mathfrak{X}(\Lambda) = \bigsqcup_{n=0}^{\infty} (\Lambda \times \mathbb{R}^{\nu})^{n}_{symm} .$$

We have introduced one small twist in that we have replaced $(\Lambda \times \mathbb{R}^{\nu})^{n}$ by $(\Lambda \times \mathbb{R}^{\nu})^{n}_{symm}$, which means the set of nonordered n-tuples of points of $\Lambda \times \mathbb{R}^{\nu}$ or equivalently the quotient of $(\Lambda \times \mathbb{R}^{\nu})^{n}$ by the action of the group of permutations of the factors. This reflects the fact that we want to treat the particles as indistinguishable, and leads to a number of simplifications. For example, if we write $\Lambda = \Lambda_1 \cup \Lambda_2$ where $\Lambda_1 \cap \Lambda_2 = \emptyset$, it is easy to see that there is a natural identification

$$\mathfrak{X}(\Lambda) \cong \mathfrak{X}(\Lambda_1) \times \mathfrak{X}(\Lambda_2)$$

obtained simply by splitting each phase-point into the part describing the particles in Λ_1 and the part describing the particles in Λ_2. If the particles carried labels, this identification would not hold (since it would be necessary to distinguish having particle 1 in Λ_1 and particle 2 in Λ_2 from the reverse arrangement).

We next define the infinite system phase space, denoted by $\mathfrak{X}$ or $\mathfrak{X}(\mathbb{R}^{\nu})$, as the space of all equivalence classes of (possibly finite or even empty) sequences

$$\underline{x} = (x_i) = (q_i, p_i) \quad \text{in} \quad \mathbb{R}^{\nu} \times \mathbb{R}^{\nu}$$

such that, for each bounded set B in $\mathbb{R}^{\nu}$, the number of i's such that $q_i \in B$ is finite. Here, the equivalence classes are formed by grouping together all possible rearrangements of a given sequence. The q_i's and p_i's are of course interpreted as representing the positions and momenta of a finite or countably infinite assembly of particles, and we are requiring that there be only finitely many particles in any bounded region of physical space.

If $\phi(q,p)$ is a function of $\mathbb{R}^{\nu} \times \mathbb{R}^{\nu}$ which vanishes for sufficiently large $|q|$, we can define a function $\sum \phi$ on $\mathfrak{X}$ by

$$(\sum \phi)\ ((q_i,p_i)) = \sum_i \phi(q_i,p_i) \ .$$

Note that only finitely many terms in the sum over i are different from zero, and that $\sum \phi$ really depends only on the equivalence class of the sequence (q_i,p_i). We equip $\mathfrak{X}$ with the weakest topology making the functions $\sum \phi$ continuous for all continuous functions ϕ vanishing for sufficiently large $|q|$. It may be shown that $\mathfrak{X}$, equipped with this topology, is a Polish space, i.e., the topology may be obtained from a metric with respect to which $\mathfrak{X}$ is complete. This means that $\mathfrak{X}$, although not locally compact, has quite a reasonable measure theory, e.g., that every finite Borel measure on $\mathfrak{X}$ is inner regular. We will really not be much concerned with the topology except through the Borel structure it defines.

It is convenient to define also $\mathfrak{X}(M)$ for any Borel set M, by analogy to our definition of $\mathfrak{X} = \mathfrak{X}(\mathbb{R}^\nu)$. If we do this, we get Borel isomorphisms

$$\mathfrak{X}(M_1 \cup M_2) \simeq \mathfrak{X}(M_1) \times \mathfrak{X}(M_2) \quad \text{if} \quad M_1 \cap M_2 = \emptyset \ .$$

Moreover, in $N \subset M$, there is a natural projection

$$\Pi_N^M \colon \mathfrak{X}(M) \to \mathfrak{X}(N)$$

which discards those particles in $M \setminus N$. In particular, we have projections $\Pi_N \colon \mathfrak{X} \to \mathfrak{X}(N)$ for each N. If μ is a Borel measure on $\mathfrak{X}$, then $\Pi_N \mu = \mu_N$ is a Borel measure on $\mathfrak{X}(N)$. If we let N range over the bounded Borel sets in $\mathbb{R}^\nu$, we obtain a family $\{\mu_N\}_{N \text{ bounded}}$ of Borel measures. These measures μ_N are moreover <u>consistent</u>, or a <u>projective system of measures</u>, i.e., if $N \subset M$, then

$$\mu_N = \Pi_N^M \mu_M \ .$$

Conversely, a standard argument in measure theory shows that any consistent family $\{\mu_N\}_{N \text{ bounded}}$ of Borel measures on the $\mathfrak{X}(N)$

comes from a uniquely determined Borel measure μ on $\mathfrak{X}$. Thus, probability measures on $\mathfrak{X}$ (statistical states of the infinite system) may be constructed in a straightforward way from "local data".

As a particularly simple example of such statistical states, let us construct the states describing noninteracting gases. For any bounded Borel set Λ, let dx_Λ denote the (infinite) measure on $\mathfrak{X}(\Lambda)$ given by

$$\frac{1}{n!}\, dq_1 \ldots dq_n\, dp_1 \ldots dp_n \quad \text{on} \quad (\Lambda \times \mathbb{R}^\nu)^n_{\text{symm}}$$

for each n. A simple computation shows that, if $\Lambda = \Lambda_1 \cup \Lambda_2$ with $\Lambda_1 \cap \Lambda_2 = \emptyset$, then under the usual identificationof $\mathfrak{X}(\Lambda)$ with $\mathfrak{X}(\Lambda_1) \times \mathfrak{X}(\Lambda_2)$, dx_Λ becomes $dx_{\Lambda_1} \otimes dx_{\Lambda_2}$. To construct a noninteracting state, we choose an inverse temperature β (a positive number) and a chemical potential μ (a real number which may be adjusted to give any desired value for the density). On each $\mathfrak{X}(\Lambda)$, construct a probability measure

$$\frac{e^{-\frac{\beta}{2m}\sum\limits_{i=1}^{n} p_i^2}\, e^{+\beta\mu n}\, d\underline{x}_\Lambda}{\int e^{-\frac{\beta}{2m}\sum\limits_{i=1}^{n} p_i^2+\beta\mu n}\, d\underline{x}_\Lambda} = \mu^{(0)}_{\beta,\mu,\Lambda}$$

The denominator may easily be computed to be

$$\sum_{n=0}^{\infty} \frac{1}{n!} \left(\int_{\mathbb{R}^\nu} e^{-\frac{\beta}{2m} p^2}\, dp \right)^n \left(e^{\beta\mu}\right)^n \left(V(\Lambda)\right)^n = \exp\left[\left(\frac{2\pi m}{\beta}\right)^{\nu/2} e^{\beta\mu}\, V(\Lambda) \right] .$$

From this formula and the factorizability of $d\underline{x}_\Lambda$ it follows easily that

$$\mu^{(0)}_{\beta,\mu,\Lambda_1 \cup \Lambda_2} \simeq \mu^{(0)}_{\beta,\mu,\Lambda_1} \otimes \mu^{(0)}_{\beta,\mu,\Lambda_2} ,$$

which in particular implies that

$$\prod_{\Lambda_1}^{\Lambda} \mu^{(0)}_{\beta,\mu,\Lambda} = \mu^{(0)}_{\beta,\mu,\Lambda_1}$$

whenever $\Lambda_1 \subset \Lambda$. Thus, the family of measures $(\mu^{(0)}_{\beta,\mu,\Lambda})$ is consistent and hence gives a probability measure on $\mathfrak{X}$, which we will temporarily denote by $\mu^{(0)}_{\beta,\mu}$.

Note a number of elementary features of these measures.

(1) They are invariant under the action of the group of space-translations on $\mathfrak{X}$. More explicitly, if $a \in \mathbb{R}^{\nu}$ and if $\underline{x} = (q_i, p_i) \in \mathfrak{X}$, we define

$$\tau_a \, \underline{x} = (q_i + a,\ p_i) \ .$$

Then each measure $\mu^{(0)}_{\beta,\mu}$ is invariant under the τ_a's.

(2) The momentum distributions are Maxwellian. This means that the momenta of the various particles are all independent of each other and of the positions of the particles, and that each momentum has a Gaussian distribution. More concretely, for each bounded Λ and each n, the measure obtained by restricting $\mu^{(0)}_{\beta,\mu,\Lambda}$ to $(\Lambda \times \mathbb{R}^{\nu})^n_{\text{symm}}$ has the form of the product of a measure on Λ^n_{symm} with the measure

$$e^{-\frac{\beta}{2m}[p_1^2+\ldots+p_n^2]} \, dp_1 \ldots dp_n \left(\frac{\beta}{2\pi m}\right)^{n\nu/2} \quad \text{on} \quad (\mathbb{R}^{\nu})^n \ .$$

The measures $\mu^{(0)}_{\beta,\mu}$ share this property with equilibrium states for interacting systems we will discuss shortly.

(3) If E is any subset of the one-particle phase space $\mathbb{R}^{\nu} \times \mathbb{R}^{\nu}$, we let n_E denote the number of particles in E, regarded as a function on $\mathfrak{X}$. Thus, if $\underline{x} = (q_i, p_i)_{i=1,2,\ldots}$, then

$$n_E(\underline{x}) = \text{number of } i\text{'s such that } (q_i, p_i) \in E \ .$$

n_E is finite almost everywhere if (and only if)

$$\sigma(E) = \int_E e^{\beta\mu} \, e^{-\beta p^2/2m} \left(\frac{\beta}{2\pi m}\right)^{\nu/2} dq \, dp < \infty \ .$$

In this case, n_E has a mean $\sigma(E)$ and a Poisson distribution, i.e.

$$\mu^{(0)}_{\beta,\mu}(\{\underline{x}\colon n_E(x) = j\}) = \frac{\sigma(E)^j}{j!\ e^{\sigma(E)}} \quad \text{for} \quad j = 0,1,2,\ldots .$$

Also, if $E_1,\ldots,E_j$ are disjoint sets in $\mathbb{R}^\nu \times \mathbb{R}^\nu$, the random variables $n_{E_1},\ldots,n_{E_j}$ are <u>independent</u>. Thus, the family or random variables $\{n_E\}_{\sigma(E)<\infty}$ is a generalization of the Poisson process (or, more precisely, of the increments of that process).

(4) We have already remarked that $\mu^{(0)}_{\beta,\mu}$ is invariant under space translations. It is, moreover, invariant under the time-development for the noninteracting gas. That is, if we define

$$T^t(q_i,p_i) = (q_i + t\,\frac{p_i}{m}\,,p_i)$$

each $\mu_{\beta,\mu}$ is invariant under the group of transformations T^t.*
We claim, in fact, that the measure has a much more general invariance property: Let α^t be any group of transformations on $\mathbb{R}^\nu \times \mathbb{R}^\nu$ preserving the measure σ, and "lift" α^t to a group of transformations A^t on $\mathcal{X}$ by letting it act on each particle separately, i.e., put $A^t(x_i) = (\alpha^t x_i)$. Then A^t leaves μ invariant. To see this, we note that $n_E(A^t\underline{x}) = n_{\alpha^{-t}E}(\underline{x})$. Since the joint distributions of the n_E's are determined entirely by σ, and, since σ is invariant under α^t, the joint distributions of the $n_E \circ A^t$'s are identical with the joint distributions of the n_E's. Since the n_E's generate the σ-algebra of all measurable sets, the invariance

---

*A little care is required here. Even if $(q_i,p_i) \in \mathcal{X}$, it is not necessarily the case that the sequence $(q_i + t\,\frac{p_i}{m},p_i)$ has only finitely many particles in each bounded space region, e.g., we may put $p_i/m = -q_i$ so all particles are at the origin at time one. It may be shown, however, that such problems happen only for a set of initial (q_i,p_i) of measure zero.

follows. The ergodic properties of the flow T^t are discussed in the article of Goldstein, Lebowitz, and Aizenmann.

Our next project will be to describe the statistical states which represent equilibrium of infinite interacting systems. This will be considerably more involved than the above discussion of non-interacting equilibrium states and, although it is obviously important in equilibrium statistical mechanics, its interest for the nonequilibrium theory may be puzzling. There are two reasons why we need to consider these states. The first is that "transport coefficients" which enter into macroscopic "hydrodynamic" equations may be computed (in principle) from a knowledge of unequal time equilibrium correlations. For example, the diffusion constant ought to be equal to the limit, as $t \to \infty$, of the equilibrium average of the square of the distance traveled by a typical particle in t units of time divided by t (and where the existence of the limit is highly unobvious). A second reason is a reflection of our ignorance: We will see in the following chapter how to prove the existence of solutions to the equations of motion for a set of initial data with probability one with respect to any equilibrium state, but not for more general initial data.

We can begin as in the noninteracting situation just described, i.e., we choose an inverse temperature β and a chemical potential μ, and for each bounded set Λ we construct a probability measure $_{\Lambda}\mu$ (called the <u>grand canonical probability measure</u>, or <u>grand canonical ensemble</u>) on $\mathcal{X}(\Lambda)$ by

$$_{\Lambda}\mu = \frac{\exp\left[-\beta H(\underline{x}_{\Lambda}) + \beta\mu n(\underline{x}_{\Lambda})\right] d\underline{x}_{\Lambda}}{\Xi(\Lambda,\beta,\mu)}$$

where

$$\Xi(\Lambda,\beta,\mu) = \int d\underline{x}_{\Lambda} \exp\left[-\beta H(\underline{x}_{\Lambda}) + \beta\mu n(\underline{x}_{\Lambda})\right] =$$

$$= \sum_{n=0}^{\infty} \frac{1}{n!} e^{\beta\mu n} \left(\frac{2\pi m}{\beta}\right)^{n\nu} \int_{\Lambda^n} dq_1 \ldots dq_n \exp\left[-\beta U(q_1,\ldots,q_n)\right] .$$

We will assume (at the very least) that the potential energy U is <u>thermodynamically stable</u> i.e., that there is a constant B such that

$$U(q_1,\ldots,q_n) \geq -nB$$

for all n and all $q_1,\ldots,q_n \in \mathbb{R}^\nu$. This condition guarantees that Ξ is finite for all Λ,β,μ.

For the noninteracting system, this was all we needed to do, i.e., the measures ${}_\Lambda\mu$ were consistent and hence defined a probability measure on $\mathfrak{X}$. In the present case, things are more complicated: the ${}_\Lambda\mu$ are not, in general, consistent. To see this, let

$$\Lambda = \Lambda_1 \cup \Lambda_2 \quad \text{with} \quad \Lambda_1 \cap \Lambda_2 = \emptyset ,$$

Then, identifying $\mathfrak{X}(\Lambda)$ with $\mathfrak{X}(\Lambda_1) \times \mathfrak{X}(\Lambda_2)$, we can write

$$\underline{x}_\Lambda = (\underline{x}_{\Lambda_1}, \underline{x}_{\Lambda_2})$$

and

$$H(\underline{x}_\Lambda) = H(\underline{x}_{\Lambda_1}) + H(\underline{x}_{\Lambda_2}) + W(\underline{x}_{\Lambda_1}, \underline{x}_{\Lambda_2})$$

where $W(\underline{x}_{\Lambda_1}, \underline{x}_{\Lambda_2})$ (which depends only on the positions of the particles involved) is the energy of interaction between the particles in Λ_1 and those in Λ_2. More concretely, if $\underline{x}_{\Lambda_1} = (\underline{q}_{\Lambda_1}, \underline{p}_{\Lambda_1})$, and if $\underline{q}_{\Lambda_1} = (q_1',\ldots,q_n')$, and similarly if

$$\underline{q}_{\Lambda_2} = (q_1'',\ldots,q_{n''}'') ,$$

then

$$W(\underline{x}_{\Lambda_1}, \underline{x}_{\Lambda_2}) = \sum_{i'=1}^{n'} \sum_{i''=1}^{n''} \Phi(q_{i'}' - q_{q''}'') .$$

Inserting the above formula for $H(\underline{x}_\Lambda)$ in the formula for ${}_\Lambda\mu$ and integrating over $\underline{x}_{\Lambda_2}$ to project ${}_\Lambda\mu$ onto $\mathfrak{X}(\Lambda_1)$, we obtain

$$\prod\nolimits^{\Lambda}_{\Lambda_1}\, {}_{\Lambda}\mu = \exp\,[-\,\beta H(\underline{x}_{\Lambda_1}) + \beta\mu n_{\Lambda_1}]\; d\underline{x}_{\Lambda_1} \times \Big\{\Xi(\Lambda,\beta,\mu)^{-1}$$

$$\int d\underline{x}_{\Lambda_2}\, \exp\,[-\,\beta H(\underline{x}_{\Lambda_2}) + \beta\mu n(\underline{x}_{\Lambda_2})]\, \exp\,[-\beta W(\underline{x}_{\Lambda_1},\underline{x}_{\Lambda_2})]\Big\} .$$

In order that $\prod^{\Lambda}_{\Lambda_1}\, {}_{\Lambda}\mu$ be equal to ${}_{\Lambda_1}\mu$ it is necessary and sufficient that the term in braces in the above equation be a constant, i.e., not depend on $\underline{x}_{\Lambda_1}$. The explicit dependence of $W(\underline{x}_{\Lambda_1},\underline{x}_{\Lambda_2})$ on $\underline{x}_{\Lambda_1}$ makes this constancy essentially impossible, so the ${}_{\Lambda}\mu$ are not (very often) consistent.

A little reflection indicates, however, that the family $({}_{\Lambda}\mu)$ is not what we really want to use as local data for an infinite-system equilibrium state μ. Instead μ should be determined by the requirement that the measure $\prod^{\mathbb{R}^{\nu}}_{\Lambda_1} \mu\ (\equiv \mu_{\Lambda_1})$ describes the probability distribution on $\mathcal{X}(\Lambda_1)$ induced by ${}_{\Lambda}\mu$ for a much larger Λ containing Λ_1. Thus, we would like to take

$$\mu_{\Lambda_1} = \lim_{\Lambda\to\infty} \prod\nolimits^{\Lambda}_{\Lambda_1} ({}_{\Lambda}\mu)$$

where $\Lambda \to \infty$ indicates the limit along a sequence of regions which increase to fill all of space in some reasonable way. Note that, if the limit exists for all bounded regions Λ_1, the family of limiting measures (μ_{Λ_1}) is consistent, since if $\Lambda_1' \subset \Lambda_1$ we have

$$\mu_{\Lambda_1'} = \lim_{\Lambda\to\infty} \prod\nolimits^{\Lambda}_{\Lambda_1'} ({}_{\Lambda}\mu) = \lim_{\Lambda\to\infty} \prod\nolimits^{\Lambda_1}_{\Lambda_1'} \prod\nolimits^{\Lambda}_{\Lambda_1}\, {}_{\Lambda}\mu = \prod\nolimits^{\Lambda_1}_{\Lambda_1'} \Big(\lim_{\Lambda\to\infty} \prod\nolimits^{\Lambda}_{\Lambda_1}\, {}_{\Lambda}\mu\Big)$$

$$= \prod\nolimits^{\Lambda_1}_{\Lambda_1'} (\mu_{\Lambda_1}) .$$

(We are, evidently, ignoring a number of technical points about the sense in which the measures converge, the continuity of projection maps, etc.)

Now it does not seem to be easy to prove

$$\lim_{\Lambda\to\infty} \prod\nolimits_{\Lambda_1}^{\Lambda} ({}_{\Lambda}\mu)$$

exists with any degree of generality. One can, however, make a compactness argument which shows that a subsequence of Λ's can be chosen such that this limit exists for all bounded Λ_1. (This compactness argument, which will be described later, imposes some further requirements on the interaction.) By passing to different subsequences, one may (in principle) obtain several different limiting consistent families of measures. Each such consistent family gives rise to a measure on $\mathcal{X}$, and a measure on $\mathcal{X}$ obtained in this way will be called an <u>infinite-volume cluster point of finite volume ensembles</u>. The class of states so obtained is a first approximation to our desired notion of equilibrium states for the infinite system. Note that these states all have one simplifying feature in common -- they have Maxwellian momentum distributions. Thus, the only interesting parts of these measures are the probabilities on the space of possible positions of particles.

To make our general definition of equilibrium state, we will need to use the notion of conditional probability. It will in fact be convenient to use a version of this notion which deviates slightly from the standard terminology of mathematical probability theory. Suppose we have a probabilistic system equipped with two sets of "variables". The set of possible values for the first set of variables will be called Y; that for the second set of variables will be called Z. Suppose we have measured the first set of variables and have found a point $y \in Y$, and we want to know what sort of probabilistic predictions we can make about the value of the second variable, assuming that we know the joint distribution as a probability measure μ on $Y \times Z$. By such a probabilistic prediction we mean a probability measure $\mu(dz|y)$ on Z for each

$y \in Y$, which should be connected with the original measure $\mu(dy\,dz)$ by the formal relation

$$\mu(dy\,dz) = \mu(dz|y)\ \mu_Y(dy)$$

where μ_Y means the projection of μ onto Y. More precisely, we define a <u>conditional probability distribution</u> on Z given Y to be a mapping

$$y \to \mu(\cdot|y)$$

from Y to probability measures on Z such that

(i) For each measurable subset A of Z, $y \to \mu(A|y)$ is measurable

(ii) For any bounded measurable function f on $Y \times Z$,

$$\int f\,d\mu = \int_Y \mu_Y(dy) \int_Z \mu(dz|y)\ f(y,z)\ .$$

This definition of conditional probability, although intuitively natural, is not popular with probabilists because some annoying technical restrictions are necessary in order to insure its existence. These problems, however, turn out to be irrelevant for our purposes. Let us note one case in which the existence of the conditional probability is easy to establish: Assume that the measure μ is absolutely continuous with respect to a product measure, i.e., that

$$\mu(dy\,dz) = \mu(y,z)\ dy\,dz\ .$$

In this case, we may take

$$\mu(dz|y) = \frac{\mu(y,z)\ dz}{\int \mu(y,z)\ dz}\ .$$

Now let us return to the problem of characterizing equilibrium states of the infinite system. We first consider the grand canonical ensemble in a region Λ, and let Λ_1 be a smaller region. Let $\Lambda_2 = \Lambda \setminus \Lambda_1$, identify $\ast(\Lambda)$ with $\ast(\Lambda_1) \times \ast(\Lambda_2)$, and write

$\underline{x}_\Lambda = (\underline{x}_{\Lambda_1}, \underline{x}_{\Lambda_2})$. Then we have

$$_\Lambda\mu = \frac{\exp[-\beta H(\underline{x}_{\Lambda_1})-\beta H(\underline{x}_{\Lambda_2})-\beta W(\underline{x}_{\Lambda_1},\underline{x}_{\Lambda_2})+\beta\mu n(\underline{x}_{\Lambda_1})+\beta\mu n(\underline{x}_{\Lambda_2})]\; d\underline{x}_{\Lambda_1}\; d\underline{x}_{\Lambda_2}}{\Xi(\Lambda,\beta,\mu)}.$$

Let us now seek the conditional probability distribution on $\mathcal{X}(\Lambda_1)$ given $\mathcal{X}(\Lambda_2)$. This is easily seen to be

$$(1) \qquad \frac{\exp\,[-\,\beta H(\underline{x}_{\Lambda_1}) - \beta W(\underline{x}_{\Lambda_1},\underline{x}_{\Lambda_2}) + \beta\mu n(\underline{x}_{\Lambda_1})]\; d\underline{x}_{\Lambda_1}}{\Xi(\Lambda_1,\beta,\mu|\underline{x}_{\Lambda_2})}$$

where

$$\Xi(\Lambda_1,\beta,\mu|\underline{x}_{\Lambda_2}) = \int d\underline{x}_{\Lambda_1}\; \exp\,[-\beta H(\underline{x}_{\Lambda_1})-\beta W(\underline{x}_{\Lambda_1},\underline{x}_{\Lambda_2}) + \beta\mu n(\underline{x}_{\Lambda_1})].$$

(In obtaining this formula, we have cancelled a factor of $\exp\,[-\beta H(\underline{x}_{\Lambda_2}) + \beta\mu n(\underline{x}_{\Lambda_2})] \;/\; \Xi(\Lambda,\mu,\beta)$.) The advantage of this formula over previous ones is that the dependence of Λ has nearly disappeared. Thus, for example, if the interparticle potential Φ has finite range (i.e., if $\Phi(q) = 0$ for all sufficiently large q), then $W(\cdot,\underline{x}_{\Lambda_2})$ (and hence the conditional probability) depends on $\underline{x}_{\Lambda_2}$ only through those particles whose distance from Λ_1 is less than the range of the potential. We can therefore, formally at least, pass to the limit as Λ becomes infinitely large. This leads us to the following definition: A probability measure μ on $\mathcal{X}$ is said to be a <u>Gibbs state</u> (for the interaction Φ, the inverse temperature β, and the chemical potential μ) if, for <u>every</u> bounded measurable set Λ_1, the conditional probability on $\mathcal{X}(\Lambda_1)$ given $X(\Lambda_2)$ (where $\Lambda_2 = \mathbb{R}^\nu \setminus \Lambda_1$) is given by (1). (Note: In our original derivation of (1), Λ_2 was supposed to be $\Lambda \setminus \Lambda_1$, where Λ is bounded. Here, we want to use the same formula but with $\Lambda_2 = \mathbb{R}^\nu \setminus \Lambda_1$. This requires a reinterpretation of W and gives rise to some mild technical problems; see below.) The above argument makes it at least plausible that every infinite volume cluster point of finite volume grand-canonical ensembles is a Gibbs state. On the other hand, the definition itself implies that every Gibbs state is a limit of finite-volume ensembles with appropriate (stochastic) "boundary conditions".

There are unfortunately a number of technical adjustments which must be made to the above definition of Gibbs state. We began by assuming that the potential Φ was stable, but it turns out that we actually need a slightly stronger condition, which goes by the name of superstability. Let us divide the position space $\mathbb{R}^{\nu}$ into cubes of unit size centered at the integral points $\mathbb{Z}^{\nu}$. For $\alpha \in \mathbb{Z}^{\nu}$, let $\Delta(\alpha)$ denote the cube centered at α, and for $\underline{x} \in \mathfrak{X}$ let $n_{\alpha}(\underline{x})$ denote the number of particles in $\Delta(\alpha)$. We will say that the potential Φ is superstable if there exist constants A,B with $A > 0$ such that

$$U(\underline{q}) \geq -Bn(\underline{q}) + A \sum_{\alpha} n_{\alpha}^{2}(\underline{q}) \quad \text{for all} \quad n, q_1, \ldots, q_n .$$

Roughly speaking, superstability means that the potential energy per particle goes to infinity if one tries to put too many particles particles into a small region. It is obvious that if $\Phi = \Phi_1 + \Phi_2$ with Φ_1 stable and Φ_2 superstable then Φ is superstable. It is not hard to see that, if $\Phi_2 \geq 0$ and if Φ_2 is bounded away from zero on a neighborhood of zero, then Φ_2 is superstable. Finally, most usable criteria for stability require that Φ grow like const. $\times |q|^{-\nu-\varepsilon}$ for small q, and in this case a small Φ_2 of the indicated sort can be subtracted from Φ without dest ing stability. Using arguments like this, it can be shown that nearly any potential of practical interest (except the zero potential!) is superstable if it is stable.

We also need a condition on the behavior of the potential at large separations. Recall that the definition of Gibbs state involves

$$W(\underline{x}_{\Lambda_1}, \underline{x}_{\Lambda_2})$$

where $\Lambda_2 = \mathbb{R}^\nu \setminus \Lambda_1$. If $\underline{q}_{\Lambda_1} = (q_1^{(1)}, \ldots, q_n^{(1)})$ and $\underline{q}_{\Lambda_2} = (q_1^{(2)}, q_2^{(2)}, \ldots)$, then

$$W(\underline{x}_{\Lambda_1}, \underline{x}_{\Lambda_2}) = \sum_{i=1}^{n} \sum_{j} \Phi(q_i^{(1)} - q_j^{(2)}) \ .$$

The sum evidently makes sense if Φ has finite range, but troubles can be expected to arise if the $q_j^{(2)}$'s are spread out fairly uniformly and if $\Phi(q)$ is not integrable at infinity. It turns out that what is needed to make the proofs work is that Φ satisfy an inequality of the form

$$\Phi(q) \geq -\chi(|q|)$$

for large q, where $\chi(r)$ is nonnegative, nonincreasing, and satisfies

$$\int^{\infty} r^{\nu-1} \chi(r)\, dr < \infty \ .$$

A potential satisfying this condition will be said to be <u>lower regular</u>.

Even requiring that the potential be lower regular does not eliminate all difficulties in the definition of $W(\underline{x}_{\Lambda_1}, \underline{x}_{\Lambda_2})$ since the definition of the space $\mathfrak{X}$ allows the density of particles to grow as rapidly as desired with distance from the origin. To get around this problem, we add to the definition of Gibbs state the requirement that μ assign probability zero to the set of all $\underline{x} \in X$ such that

$$\sup_{r \in \mathbb{Z}_+} \left\{ \frac{1}{r^\nu} \sum_{\alpha : |\alpha| \leq r} n_\alpha^2(\underline{x}) \right\} = \infty \ .$$

On the complementary set there is no problem in defining W and hence the definition of Gibbs state makes sense.

Now suppose that our potential Φ is superstable, let Δ be a unit cube, and let ${}_\Delta\mu$ denote the grand canonical ensemble in Δ. Then, since

$$H(\underline{x}) = \sum_i \frac{p_i^2}{2m} + U(\underline{q}) \geq \sum_{i=1}^{n} \frac{p_i^2}{2m} + An^2 - Bn \ ,$$

we get

$${}_\Delta\mu \leq c_1 \exp\left[- c_2 n^2 - \beta \sum_i \frac{p_i^2}{2m}\right] d\underline{x}_\Delta$$

for some strictly positive constants c_1, c_2. In particular, the probability of finding n particles in Δ decreases like $e^{-\text{const.}\cdot n^2}$ as $n \to \infty$. One of the principal technical tools in this subject is a theorem of Ruelle (see [7]) which says that arbitrary Gibbs states satisfy similar estimates.

Theorem (Ruelle's Probability Estimate for Superstable Potentials). Let Φ be a superstable lower regular potential, let μ be a Gibbs state for Φ, and let Δ be a unit cube in $\mathbb{R}^\nu$. Then there exist constants γ_1, γ_2, with $\gamma_2 > 0$, such that

$$\mu_\Delta(d\underline{x}_\Delta) \leq \gamma_1 \, e^{-\gamma_2 n^2(\underline{x}_\Delta)} \, e^{-\beta \sum_{i=1}^{n} \frac{p_i^2}{2m}} \, d\underline{x}_\Delta \ .$$

Moreover, γ_1, γ_2 may be taken to depend only on Φ, the inverse temperature, and the chemical potential, not on Δ or the particular Gibbs state μ under consideration. The measure μ_Δ may be replaced in the above by

$$\prod\nolimits_\Delta^\Lambda \, {}_\Lambda\mu$$

for any region Λ containing Δ.

The last statement of the theorem contains in particular the compactness needed to construct infinite volume cluster points of finite volume ensembles. For positive potentials Φ, this theorem follows almost trivially from the definitions of Gibbs state and superstability. The proof for general superstable potentials is quite involved.

References to Part 2

The infinite system phase space $\mathfrak{X}$ was introduced in a different form and from a totally different point of view (as a subset of the spectrum of a commutative C^* algebra) in

[1] Ruelle, D., States of Classical Statistical Mechanics, J. Math. Phys, 8 (1967) 1657-1668.

Another development, closer in spirit to the one adopted here, is given in Section 2 of

[2] Lanford, O.E., III, The Classical Mechanics of One-Dimensional Systems of Infinitely Many Particles II. Kinetic Theory, Commun. Math. Phys. 11, (1969), 257-292. Or in

[3] Lenard, A., Correlation Functions and the Uniqueness of the State in Classical Statistical Mechanics, Commun. Math. Phys. 30, (1973) 35-44.

For the theory of integration on Polish spaces, see

[4] Badrikian, A., Seminaire sur les Fonctions Aléatoires Linéaires et les Mésures Cylindriques, Lecture Notes in Mathematics No. 139, Berlin, Heidelberg, New York: Springer-Verlag, 1970, Exposé 8.

The definition of Gibbs state for lattice systems appeared independently in

[5] Dobrushin, R. L., Gibbsian Random Fields for Lattice Systems with Pair Interactions, Funct. Anal. Appl. 2 (1968) 291-301. And

[6] Lanford, O.E., III, and Ruelle, D., Observables at Infinity and States with Short Range Correlations in Statistical Mechanics, Comm. Math. Phys. 13 (1969) 194-215.

The full technicalities involved in the definition for continuous systems with superstable interactions, as well as the proof of

Ruelle's probability estimates, are given in

[7] Ruelle, D., Superstable Interactions in Classical Statistical Mechanics, Commun. Math. Phys. 18 (1970) 127-159.

3. Dynamics of Systems of Infinitely Many Particles: Existence

Summary. We show how to construct solutions to Newton's equations of motion for systems of infinitely many point particles interacting by nonsingular two-body forces which decrease fairly rapidly at infinity. The construction gives solutions for almost all initial phase points with respect to any Gibbs state. The main idea is to exploit the formal fact that a Gibbs state ought to be invariant under the flow given by solutions to the equations of motion; the principal technical tool is a probabilistic *a priori* estimate (Proposition 4) which holds almost everywhere with respect to each Gibbs state.

....

We now come to the problem of constructing the time evolution for a system of infinitely many particles. This means that we want to start with a point $\underline{x} = (q_i, p_i)$ in $\mathcal{X}$ and solve Newton's equations:

$$\frac{dq_i}{dt} = \frac{p_i}{m} ; \qquad \frac{dp_i}{dt} = \sum_{j \neq i} F(q_i - q_j)$$

(where

$$F(q) = - \text{grad}\, \Phi(q))$$

with initial conditions

$$q_i(0) = q_i ; \qquad p_i(0) = p_i .$$

These equations, unfortunately, are rather intractable. It has already been pointed out that even for noninteracting particles (i.e., $F = 0$) the equations do not have solutions for all initial $\underline{x}$'s because the natural procedure for solving them may lead to a situation with infinitely many particles in a bounded region of space. Of course, for noninteracting particles there is no particular problem in allowing infinitely many of them to be present in a bounded region of space, but for interacting particles this can be expected to lead to difficulties in the expression for the force on a given particle. Moreover, although the troublesome initial phase points for the noninteracting situation are easy to identify "by inspection," this need no longer be true for interacting particles, where the equations of motion themselves might, in principle, lead to their own breakdown by building up the large velocities needed to bring in many particles from very far away.

To obtain a completely satisfactory resolution of this difficulty, one ought to proceed by finding a set of initial $\underline{x}$'s for which well-behaved solutions can be shown to exist and which is large enough to describe any reasonable physical situation. This has been done, more or less, for a system of particles moving in one dimension and interacting by smooth, finite range potentials (see Chapter V). For particles moving in several dimensions, the techniques used in this proof fail. There is, however, a partial result in the desired direction. We will show that there exists a subset $\hat{\mathcal{X}}$ of $\mathcal{X}$ such that:

(1) well-behaved global (i.e., all time) solutions to the equations of motion exist for any initial $\underline{x}$ in $\hat{\mathcal{X}}$

(2) $\hat{\mathcal{X}}$ is a set of measure one for every Gibbs state (for the potential defining the interparticle interaction).

This provides us with the machinery needed to investigate time-dependent phenomena in equilibrium states. On the other hand, the theorem we will prove has the peculiar feature that it does not exhibit a single phase point with infinitely many particles which can be shown to have a nontrivial time development.

By way of introduction, we note some properties which hold almost everywhere with respect to every Gibbs state. Let us label the particles in some systematic way, e.g., in order of increasing distance from the origin. The momenta $p_1, p_2, p_3, \ldots$ are then independent and Gaussian. Although the probability that any given $|p_i|$ is large is very small, we can be certain that if we search long enough we will find an arbitrarily large momentum. In other words, the sequence $\{p_i\}$ is unbounded with probability one. It is important for our purposes to know how fast it becomes unbounded. One way to describe the typical behavior is the following:

Remark. $\sup_{i \geq 2} \frac{|p_i|}{\sqrt{\log(i)}} < \infty$ with probability one.

To see this we note that there exist constants c_1, c_2, with $c_2 > 0$, such that

$$\mu\{|p_i| \geq \alpha\} \leq c_1 \exp(-c_2\alpha^2) \quad \text{for all } \alpha .$$

Thus

$$\mu\{|p_i| \geq \lambda \sqrt{\log(i)}\} \leq c_1 \cdot i^{-c_2\lambda^2} ,$$

or

$$\mu\{\sup_i \frac{|p_i|}{\sqrt{\log(i)}} \geq \lambda\} \leq c_1 \sum_i i^{-c_2\lambda^2} ,$$

which tends to zero as λ tends to infinity.

This remark is slightly unnatural in that it appears to depend on the labeling of the particles. It may be shown, however, that

$$\sup_i \frac{|p_i|}{\sqrt{\log_+(q_i)}} < \infty \quad \text{with probability one.}$$

(See Lemma 2.) In this formula, we have introduced the notation:

$$\log_+(q) \equiv \begin{cases} \log(|q|) & \text{if } |q| \geq e \\ 1 & \text{otherwise.} \end{cases}$$

This is merely a convenient way of cutting off the logarithm for small values of its argument.

Let us now apply the above inequality to the noninteracting system. We have $q_i(t) = q_i + \frac{p_i}{m} \cdot t$; $p_i(t) = p_i$, so if

$$\sup_i \frac{|p_i|}{\sqrt{\log_+(q_i)}} < \infty ,$$

then there exists an M (depending on (q_i, p_i)) such that

$$|q_i(t) - q_i| \leq M \sqrt{\log_+(q_i)}$$

for all i and all t between zero and one. This means that particles initially far from the origin travel in unit time only a very small fraction of their initial distance from the origin, i.e., that they remain relatively well localized. Our strategy will be to prove similar estimates for interacting systems and to use the localization so obtained to prevent the development of singularities in the solutions we construct. As in the above trivial case, the localization estimates will hold only almost everywhere with respect to equilibrium states, but in contrast to the above situation it will not be easy to determine explicitly the set on which they hold.

Let us begin by proving a theorem which is of relatively little direct use in itself, since it assumes that solutions have already been constructed, but which displays the main idea of our argument in a very simple form. We will be considering a two-body interaction Φ which is superstable and lower regular (so that we may speak of Gibbs states for Φ); other conditions will be imposed as we go along.

Proposition 1. Let μ be a Gibbs state for Φ, and let T^t be a one-parameter group of mappings of $\ast$ into itself given by solutions of the equations of motion with the interaction Φ, and leaving the measure μ invariant. Then for almost every $\underline{x}$ in $\ast$ there exists a positive number M such that

$$|q_i(t) - q_i| \leq M \log_+(q_i) \quad \text{for all } i \text{ and } 0 \leq t \leq 1$$

(where $(q_i(t), p_i(t)) = T^t\underline{x}$).

(It may be noted that the right-hand side of the above inequality is $M \log_+(q_i)$ rather than $M\sqrt{\log_+(q_i)}$ as we obtained in the absence of interaction. Examination of our argument will show that we could have obtained the estimate with the square root with no more difficulty; we have chosen the above statement only to simplify the formulas.)

Proof: We define a function B on $\mathfrak{X}$ by

$$B((q_i,p_i)) = \sup_i \frac{|p_i|/m}{\log_+(q_i)} .$$

We regard B as everywhere defined, but taking on the value $+\infty$ at some points of $\mathfrak{X}$.

Lemma 2. B is integrable with respect to μ.

We postpone the proof to the end of the chapter. We next define (temporarily)

$$\bar{B}(\underline{x}) = \int_0^1 dt\ B(T^t\underline{x}) ,$$

and we note that

$$\int \bar{B}\, d\mu = \int \mu(d\underline{x}) \int_0^1 dt\ B(T^t\underline{x}) = \int_0^1 dt \int \mu(d\underline{x})\ B(T^t\underline{x})$$

(by Fubini's theorem)

$$= \int_0^1 dt \int \mu(d\underline{x})\ B(\underline{x}) \quad \text{(by the invariance of } \mu \text{ under } T^t\text{)}$$

$$= \int B\, d\mu < \infty .$$

Thus, $\bar{B}$ is μ-integrable, and, in particular, $\bar{B}$ is finite almost everywhere. We now claim that, if $\bar{B}(\underline{x}) < \infty$, then $T^t\underline{x}$ satisfies an inequality of the desired type. To see this note that

$$\int_0^1 dt\ \frac{|dq_i(t)/dt|}{\log_+(q_i(t))} \leq \int_0^1 dt\ \sup_j \frac{|p_j(t)|/m}{\log_+(q_j(t))} = \bar{B}(\underline{x}) \quad \text{for all } i.$$

We now need

Lemma 3. For each $b > 0$, there exists a number $M(b)$ such that any differentiable $\mathbb{R}^\nu$-valued function $q(t)$ satisfying

$$(1) \qquad \int_{t_1}^{t_2} \frac{|dq/dt|}{\log_+(q)}\, dt \le b$$

also satisfies

$$|q(t_2) - q(t_1)| \le M(b) \log_+(q(t_1)) .$$

This lemma, together with preceding observations, immediately implies Proposition 1. The proof of the lemma is elementary calculus. Let $q(t)$ satisfy (1), and let

$$q_{max} = \max_{t_1 \le t \le t_2} |q(t)|$$

Then (replacing the denominator $\log_+(q(t))$ in (1) by $\log_+(q_{max})$),

$$|q(t_1) - q(t_2)| \le b \log_+(q_{max})$$

but also

$$|q(t_1)| \ge q_{max} - b \log_+(q_{max}) .$$

The second inequality implies the existence of $c(b)$ such that

$$\frac{\log_+(q_{max})}{\log_+(q(t_1))} \le c(b) ;$$

by the first inequality, we may then take

$$M(b) = b\, c(b) .$$

(The above proof should, of course, logically be written in the reverse order.)

We next sharpen and generalize Proposition 1. It is convenient, first of all, to make estimates which hold for all times, not just for times between zero and one. This can be done efficiently by redefining

$$\bar{B}(\underline{x}) = \frac{1}{\pi} \int_{-\infty}^{\infty} \frac{dt}{1+t^2} B(T^t \underline{x}) .$$

Then for any positive τ, and any t between $-\tau$ and τ,

$$\left| \int_0^t dt_1 \, B(T^{t_1}\underline{x}) \right| \leq \int_{-\tau}^{\tau} dt_1 \, B(T^{t_1}\underline{x}) \leq \pi(1+\tau^2) \, \bar{B}(\underline{x}) \, ,$$

so Lemma 3 immediately gives

$$|q_i(t) - q_i| \leq M\big(\pi(1 + \tau^2) \, \bar{B}(\underline{x})\big) \log_+(q_i) \quad \text{for all} \quad i$$

provided $|t| \leq \tau$.

Next we observe that the proof of Proposition 1 really does not use the fact that T^t is obtained by solving the equations of motion. What is essential is that T^t leaves μ invariant and that $dq_i(t)/dt = p_i(t)/m$. In fact, all we really need (aside from invariance) is that $|dq_i(t)/dt| \leq |p_i(t)|/m$. Thus, we have actually proved:

Proposition 4. Let μ be a Gibbs state; let T^t be a flow on $\mathcal{X}$ leaving μ invariant such that, if $T^t(q_i, p_i) = (q_i(t), p_i(t))$, then $|dq_i(t)/dt| \leq |p_i(t)|/m$, and let $B(\underline{x})$, $\bar{B}(\underline{x})$ be defined as above. Then

(i) $\int \bar{B} \, d\mu = \int B \, d\mu < \infty$ and

(ii) If $\bar{B}(\underline{x}) < \infty$,

$$|q_i(t) - q_i| \leq M\big(\pi(1+\tau^2) \, \bar{B}(\underline{x})\big) \log_+(q_i)$$

for all i, and all t with $|t| \leq \tau$ (where M is as defined in Lemma 3).

We will use Proposition 4 in the construction of solutions to the equations of motion. In order to make matters as simple as possible, we will first assume that the interparticle potential Φ is once continuously differentiable (so $F(q)$ is continuous) and has finite range. We will note later how to remove the restriction to finite-range interactions.

The idea of the construction is to find approximate solutions to the equations of motion to which Proposition 4 can be applied and then to obtain exact solutions by a compactness argument. The approximate solutions are constructed as follows: For each positive integer s, let Λ_s denote the ball of radius s centered at the origin, and let $T^t_{(s)}$ be the one-parameter group of mappings $\mathcal{X}$ onto itself giving the time-development for the following dynamics:

(a) If q_i is initially outside Λ_s (i.e., if $|q_i| > s$) then q_i and p_i stay constant for all t.

(b) The particles initially inside Λ_s move according to the usual laws of dynamics

$$\frac{dq_i^{(s)}(t)}{dt} = \frac{p_i^{(s)}(t)}{m} \; ; \qquad \frac{dq_i^{(s)}(t)}{dt} = \sum_{j\neq i} F\left(q_i^{(s)}(t) - q_j^{(s)}(t)\right)$$

with elastic reflection at the boundary of Λ_s. The sum over j in $dp_i^{(s)}/dt$ includes those j's corresponding to fixed particles outside Λ_s as well as to the moving particles inside Λ_s, i.e., the particles outside Λ_s exert time-independent forces on the particles inside Λ_s.

<u>Proposition 5</u>. Any Gibbs state μ for the interaction Φ is invariant under any of the flows $T^t_{(s)}$.

<u>Proof</u>: The argument is best expressed in geometric language. For any s, we may decompose

$$\mathcal{X} \simeq \mathcal{X}(\Lambda_s) \times \mathcal{X}(\Lambda_s')$$

where $\Lambda_s' = \mathbb{R}^\nu \setminus \Lambda_s$. In accordance with this decomposition, we may partition $\mathcal{X}$ into "horizontal lines" each of which may be identified with $\mathcal{X}(\Lambda_s)$. Two points $\underline{x}$, $\underline{x}'$ are in the same horizontal line, or <u>fiber</u>, if they have the same projection onto $\mathbb{R}^\nu \setminus \Lambda_s$, i.e., if they are identical outside Λ_s. The fibers are labeled by the possible "vertical coordinates", i.e., the points of $\mathcal{X}(\Lambda_s')$. On the fiber labeled by $\underline{x}_{\Lambda_s'} \in \mathcal{X}(\Lambda_s')$, we may define a Hamiltonian

$$H(\underline{x}_{\Lambda_s} | \underline{x}_{\Lambda'}) = H(\underline{x}_{\Lambda_s}) + W(\underline{x}_{\Lambda_s}, \underline{x}_{\Lambda_s'})$$

where $H(\underline{x}_{\Lambda_s})$ is the usual Hamiltonian on $\mathcal{X}(\Lambda_s)$. Intuitively, a Gibbs state μ has the property that, when restricted to the fiber labeled by $\underline{x}_{\Lambda_s'}$, it gives a measure differing only by normalization from the measure

$$\exp\,[-\beta H(\underline{x}_{\Lambda_s} | \underline{x}_{\Lambda_s'}) + \beta\mu n(\underline{x}_{\Lambda_s})]\, d\underline{x}_{\Lambda_s} .$$

(Of course, this must be interpreted carefully since the μ-measure of each fiber is zero, but the correct interpretation is provided by the definition of conditional probability.) On the other hand, the flow $T^t_{(s)}$ carries each fiber into itself since it does not change the component of the phase point in $\mathcal{X}(\mathbb{R}^\nu \setminus \Lambda_s)$. Moreover, the flow obtained by restricting $T^t_{(s)}$ to the fiber labeled by $\underline{x}_{\Lambda_s'}$ is the Hamiltonian flow given by the Hamiltonian $H(\underline{x}_{\Lambda_s} | \underline{x}_{\Lambda_s'})$ and elastic reflection at the wall of Λ_s. Conservation of energy, Liouville's Theorem, and the definition of Gibbs state imply that the restriction of a Gibbs state to each fiber is invariant under the flow $T^t_{(s)}$ on that fiber and hence that each Gibbs state is invariant under $T^t_{(s)}$.

Now we define

$$\bar{B}_{(s)}(\underline{x}) = \frac{1}{\pi} \int_{-\infty}^{\infty} \frac{dt}{1+t^2} B(T^t_{(s)}\underline{x})$$

$$\bar{B}_\infty(\ ) = \liminf_{s \to \infty} \bar{B}_{(s)}(\underline{x}) \quad .$$

By Proposition 4, $\int \bar{B}_{(s)}\, d\mu = \int B\, d\mu$ independent of s, so Fatou's Lemma gives

$$\int \bar{B}_\infty \, d\mu \leq \liminf_s \int \bar{B}_{(s)}(x) \, d\mu = \int B \, d\mu < \infty .$$

Thus, in particular, $\bar{B}_\infty(\underline{x}) < \infty$ for almost all $\underline{x}$. We now state

Proposition 6. If $\bar{B}_\infty(\underline{x}) < \infty$, there exists a solution to the equations of motion with initial data $\underline{x}$.

Proof: Since $\bar{B}_\infty(\underline{x}) = \liminf_{s \to \infty} \bar{B}_{(s)}(\underline{x})$, the finiteness of $\bar{B}_\infty$ is equivalent to the existence of a finite number b and an increasing sequence s_n of positive integers tending to infinity such that

$$\bar{B}_{(s_n)}(\underline{x}) \leq b$$

for all n. If we write $T^t_{(s)} \underline{x} = (q_i^{(s)}(t), p_i^{(s)}(t))$, we obtain from Lemma 3 that

$$| q_i^{(s_n)}(t) - q_i | \leq M\left(\pi(1+\tau^2)b\right) \log_+(q_i)$$

for all n, all i, and all t with $|t| \leq \tau$. We write

$$F_i^{(s)}(t) = \sum_{j \neq i} F\left(q_i^{(s)}(t) - q_j^{(s)}(t)\right) ,$$

i.e., F_i is the force exerted on the ith particle by all the others. As long as the ith particle does not collide with the wall of Λ_s, we have

$$\frac{d}{dt} p_i^{(s_n)}(t) = F_i^{(s_n)}(t) .$$

If we temporarily fix i and τ, and if we abbreviate $M(\pi(1+\tau^2)b)$ by M, then for n large enough so that

$$s_n > |q_i| + M \log_+(q_i)$$

the ith particle does not strike the wall between time $-\tau$ and τ. Also, any particle such that $F(q_i^{(s_n)}(t) - q_j^{(s_n)}(t)) \neq 0$ must have

$$|q_j| - M \log_+(q_j) \leq |q_i| + M \log_+(q_i) + R$$

(where R is the range of the force), and there are only finitely many (say N) such particles. We may evidently bound

$$|F_i^{(s_n)}(t)| \leq N\|F\|_\infty$$

where

$$\|F\|_\infty = \sup_q |F(q)|.$$

Thus: for each i and each positive τ we have a bound on

$$\left|\frac{dp_i^{(s_n)}(t)}{dt}\right|$$

which is uniform in t on $|t| \leq \tau$ and uniform in n for sufficiently large n. By the Arzelà-Ascoli Theorem, there exists a subsequence of (s_n) along which $p_i^{(s_n)}(t)$ converges uniformly on $\{|t| \leq \tau\}$. Since this is true for each i and each τ, we can use a diagonal procedure to find a subsequence (which we will again denote by (s_n)) along which each $p_i^{(s_n)}(t)$ converges uniformly on each bounded set (of times) to a limit which we will denote by $p_i(t)$. Also

$$q_i^{(s_n)}(t) = q_i + \frac{1}{m}\int_0^t dt_1\, p_i^{(s_n)}(t_1) ,$$

so each $q_i^{(s_n)}(t)$ converges uniformly on bounded sets to a limit which we will denote by $q_i(t)$, and we have

$$q_i(t) = q_i + \frac{1}{m}\int_0^t dt_1\, p_i(t) .$$

For each i and sufficiently large n (large enough so that the ith particle does not strike the wall of Λ_{s_n} between time zero and time t) we have

$$p_i^{(s_n)}(t) = p_i + \int_0^t dt_1 \sum_{j\neq i} F\left(q_i^{(s_n)}(t_1) - q_j^{(s_n)}(t_1)\right).$$

Using the uniform convergence of the $q_j^{(s_n)}(t)$, the continuity of F, and the uniform convergence of the sum over j (for fixed i and a bounded interval of times) we may pass to the limit $n \to \infty$ in this equation to get

$$p_i(t) = p_i + \int_0^t dt_1 \sum_{j \neq i} F\big(q_i(t_1) - q_j(t_1)\big) .$$

We have thus verified that the limits $\big(q_i(t), p_i(t)\big)_{i=1,2,\ldots}$ satisfy the equations of motion in integral form so the proposition is proved.

A few remarks may help to clarify the above argument. Note first of all that there is no measure theory in the proposition; this has all been isolated in the argument that $\bar{B}_\infty < \infty$ almost everywhere. In the proposition, we show the existence of a solution for a particular initial phase-point $\underline{x}$, and the construction is adapted to that phase point. In particular, we have said nothing about any uniformity of the convergence in $\underline{x}$, or even of whether the subsequence (s_n) may be chosen so as to work simultaneously for a set of $\underline{x}$'s of strictly positive measure. We have also not shown that the limiting solution is in any sense unique. These defects in the argument will to some extent be remedied in the next chapter where we discuss, using quite different techniques, the uniqueness of solutions and related questions.

We next summarize the changes needed to adapt the above argument to long-range forces. As before, the potential Φ is assumed to be superstable and continuously differentiable, but instead of assuming that it has finite range we will assume that there exists a bounded nonincreasing function $\chi(|q|)$ such that

$$|\Phi(q)| \leq \chi(|q|) \quad \text{for all } q$$

(2) $$|F(q)| = |-\text{grad}\ \Phi(q)| \leq \chi(|q|) \quad \text{for all } q$$

$$\int^{\infty} r^{\nu-1}\ \chi(r)\ dr < \infty .$$

These conditions say slightly more than that Φ and its gradient are integrable at infinity.

The first obstacle to carrying out the above argument in this more general situation is that, if we fix s and choose an arbitrary phase point $\underline{x}_{\Lambda_s'}$ in $\mathbb{R}^{\nu} \setminus \Lambda_s$, the density of particles at infinity in this phase point may be so large as to make the expression for the force on a particle inside Λ_s diverge. To eliminate this difficulty, we can replace $\mathfrak{X}$ by the subset $\mathfrak{X}_1$ consisting of phase points $\underline{x}$ such that

$$\sup_s \left\{ \frac{n_{\Lambda_s}(\underline{x})}{s^{\nu}} \right\} < \infty .$$

Thus, points in $\mathfrak{X}_1$ have, in a weak sense, bounded densities. One of the technical provisions in the definition of Gibbs state is that every Gibbs state assigns measure one to $\mathfrak{X}_1$.

Once we have cut down $\mathfrak{X}$ to $\mathfrak{X}_1$, we may define the flows $T^t_{(s)}$, and the functions $\bar{B}_{(s)}$, $\bar{B}_{\infty}$ exactly as before, and we have:

Theorem 7. Let Φ be superstable, continuously differentiable, and satisfy (2), and let μ be a Gibbs state for Φ. Then

(a) $\int \bar{B}_{\infty}\, d\mu < \infty$, and in particular $\bar{B}_{\infty}$ is finite μ- almost everywhere.

(b) If $\underline{x} \in \mathfrak{X}_1$, and if $\bar{B}_{\infty}(\underline{x}) < \infty$, there exists a solution $\underline{x}(t) = (q_i(t), p_i(t))$ of the equations of motion with $\underline{x}(0) = \underline{x}$. Moreover, for each positive τ there exists a constant M such that

$$|q_i(t) - q_i| \leq M \log_+(q_i)$$

for all i and all t with $|t| \leq \tau$.

Proof: The proof of (a) is exactly as before and need not be repeated. To prove (b), we begin, as above, by choosing a positive number b and a sequence (s_n) approaching infinity such that

$$\bar{B}_{(s_n)}(\underline{x}) \leq b \quad \text{for all} \quad n.$$

By Lemma 3, we have

$$|q_i^{(s_n)}(t) - q_i| \leq M\left(\pi(1+\tau^2)b\right) \log_+(q_i)$$

for all n, all i, and all t with $|t| \leq \tau$. To apply the argument given above, we have only to show that these estimates imply a bound on $F_i^{(s_n)}(t) = \sum_{j \neq i} F(q_i^{(s_n)}(t) - q_j^{(s_n)}(t))$ for each i which is uniform in n and in t with $|t| \leq \tau$. Fixing i and τ and writing M for $M\left(\pi(1+\tau^2)b\right)$, we choose r_0 so large that

(i) $r_0 \geq |q_i| + M \log_+(q_i)$

(ii) $a > r_k$ implies $a - M \log_+ a > r_{k-1}$, where r_k denotes $2^k r_0$.

To estimate the sum

$$\sum_{j \neq i} F\left(q_i^{(s_n)}(t) - q_j^{(s_n)}(t)\right),$$

we collect the particles into groups labeled by k according to the size of q_j.

$$k = 1: \ |q_j| \leq r_2$$

$$k \geq 2: \ r_k < q_j \leq r_{k+1}.$$

The number of j's in the kth group is equal to the number of particles initially in the spherical shell $r_k < |q| \leq r_{k+1}$, and, since $\underline{x} \in \mathcal{X}_1$, this number is bounded by const $\times\ r_{k+1}^{\nu}$. On the other hand, we have by the choice of r_0 that

$$|q_i^{(s_n)}(t)| \leq r_0$$

and

$$|q_j^{(s_n)}(t)| > r_{k-1} \quad \text{if} \quad |q_j| > r_k.$$

Thus, the contribution of any particle in the kth group to the force on the ith particle is bounded by

$$\chi(r_{k-1} - r_0) = \chi(r_0(2^{k-1} - 1)) \ .$$

Hence, we have

$$|F_i^{(s_n)}(t)| \le \sum_{k=1}^{\infty} \text{const} \quad (r_0 \cdot 2^k)^{\nu} \ \chi(r_0(2^{k-1} - 1))$$

for all n and all t with $|t| \le \tau$. The integrability of $r^{\nu-1}\ \chi(r)$ at infinity implies the convergence of the sum on the right, and this in turn provides the bounds necessary to extend the argument, given above in detail for finite range forces, to the present situation. The estimate

$$|q_i(t) - q_i| \le M \log_+(q_i)$$

for the exact solution of the infinite system equations of motion follows at once from the estimates

$$|q_i^{(s_n)}(t) - q_i| \le M\big(\pi(1 + \tau^2)b\big) \log_+(q_i)$$

which hold uniformly for the approximate solutions.

To complete the chapter, it remains to prove Lemma 2. As in Chapter 2, we let $\Delta(\alpha)$ $(\alpha \in \mathbb{Z}^{\nu})$ denote the unit cube centered at α, and we define

$$\tilde{B}(\underline{x}) = \sup_{\alpha \in \mathbb{Z}^{\nu}} \frac{\tilde{B}_{\alpha}(\underline{x})}{\log_+(\alpha)}$$

where $\tilde{B}_{\alpha}(\underline{x})$ is the maximum of $|p_i/m|$ over particles in $\Delta(\alpha)$. Since B is bounded by a constant times $\tilde{B}$, it is enough to show that $\tilde{B}$ is integrable with respect to any Gibbs state μ.

Now fix α, μ, and estimate the dependence on λ of

$$\mu\{\underline{x}: \tilde{B}_{\alpha}(\underline{x}) \ge \lambda\} = \sum_{n=0}^{\infty} \mu_n \times \text{conditional probability that the largest velocity of any particle in } \Delta(\alpha) \text{ is} \ge \lambda \text{ given that the number of particles in } \Delta_{\alpha}(\alpha) \text{ is } n.$$

Here μ_n denotes the probability that there are exactly n particles in $\Delta(\alpha)$. Because a Gibbs state has a Maxwellian momentum distribution, the conditional probability in the above sum is bounded by

$$c_1 \cdot n \exp[-c_2\lambda^2]$$

where c_1, c_2 are positive constants depending only on the mass and inverse temperature. Thus:

$$\mu\{\underline{x}: \tilde{B}_\alpha(\underline{x}) \geq \lambda\} \leq c_1 \exp[-c_2\lambda^2] \cdot \sum_{n=0}^{\infty} n\mu_n .$$

Now $\sum_n n\mu_n$ is just the mean number of particles in $\Delta(\alpha)$ which, by Ruelle's probability estimates, has a bound independent of α. Absorbing this bound in c_1, we get:

$$\mu\{\underline{x}: \tilde{B}_\alpha(\underline{x}) \geq \lambda\} \leq c_1' \exp[-c_2\lambda^2]$$

$$\mu\{\underline{x}: \tilde{B}(\underline{x}) \geq \lambda\} \leq \sum_\alpha \mu\{\underline{x}: \tilde{B}_\alpha(\underline{x}) \geq \lambda \log_+(\alpha)\}$$
$$\leq c_1' \sum_\alpha e^{-c_2\lambda^2(\log_+(\alpha))^2} .$$

The right-hand side of this inequality goes to zero like $e^{-c_2\lambda^2}$ for large λ, which is much more than is needed to prove $\tilde{B}$ is μ-integrable.

<u>Reference to Chapter 3</u>

The contents of this and the following chapter are taken from unpublished work of the author.

4. Dynamics of Systems of Infinitely Many Particles: Uniqueness

Summary. We first give an example which suggests that an infinite system of particles may be driven at infinity in such a way as to produce nonunique solutions to the equations of motion. Hence, uniqueness can be expected to hold only subject to a regularity condition which may be viewed as a boundary condition at infinity. We impose such a condition and prove uniqueness for finite-range Lipschitz continuous interparticle forces. We also show that the infinite system dynamics is the limit in measure of finite system dynamics.

....

Before we begin on the somewhat involved set of estimates which prove the uniquenss of the solutions constructed in the previous chapter, let us give an example of nonuniqueness which may serve to justify some restrictions we are forced to impose in order to obtain a uniqueness theorem. The example we will give concerns a system of infinitely many hard spheres, interacting only by elastic collisions, rather than systems with smooth inter-particle forces of the sort we have been considering. We do not know how to construct analogous examples for the smooth-forces case, but the example illustrates in any event a phenomenon which can occur for infinite systems and not for finite ones.

To construct the example, we take a sequence of spheres arranged as follows:

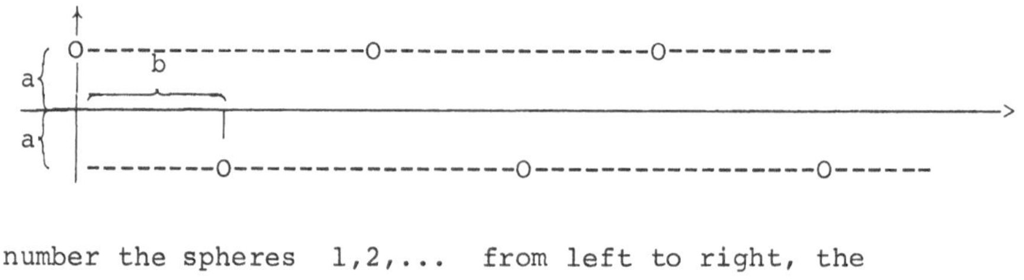

If we number the spheres 1,2,... from left to right, the positions of their centers are (0,a), (b,-a), (2b,a), (3b,-a),... .

At time t_n we start out the nth sphere with a speed v_n and in a direction which leads to collision with the (n-1)st sphere, all other spheres being initially at rest.

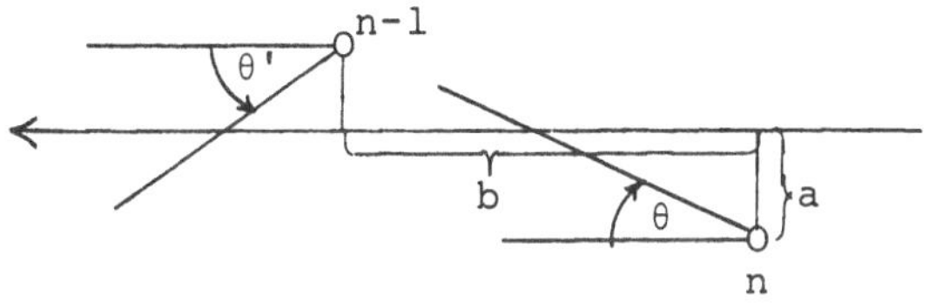

We let θ denote the direction in which the nth particle is started out and θ' the direction in which the n-1 st particle moves after it has been struck by the nth; the sign conventions for θ and θ' are as illustrated. We claim that if b/a is large enough θ may be chosen so that $\theta' = \theta$. This is most easily seen using the intermediate value theorem. If θ is such that the nth sphere collides head-on with the (n-1)st, then $\theta' = -\theta$. As θ increases, so does θ', and when θ is such that the nth particle just grazes the (n-1)st the outgoing velocity of the (n-1)st particle is perpendicular to the incoming velocity of the nth, i.e.

$$\theta' = \frac{\pi}{2} - \theta .$$

If b/a is large enough, the value of θ corresponding to a grazing collision is less than $\pi/4$; hence less than the corresponding value of θ'. Thus, for a head-on collision, $\theta' = -\theta < \theta$, while for a grazing collision $\theta' > \theta$, so somewhere between there must be a value of θ with $\theta' = \theta$.

We choose the initial direction in this way, and we let v_{n-1} be the speed of the (n-1)st particle after collision. By conservation of energy in collisions, and by the fact that the nth particle is not at rest after collision, we have

$$v_{n-1} < v_n .$$

Now the ratio

$$\alpha = \frac{v_{n-1}}{v_n} < 1.$$

and the angle θ, do not depend on the value of v_n (since the outcome of a collision is determined entirely by the geometry immediately before collision and does not depend on the unit of time: in this respect the hard-sphere system differs sharply from the system with smooth forces). Hence, the (n-1)st particle proceeds until it collides with the (n-2)nd, which after collision has speed $v_{n-2} = \alpha^2 v_n$ and direction of motion θ. The process repeats itself until, finally, the first particle undergoes a collision, after which it moves off to infinity with speed $v_1 = \alpha^{n-1} v_n$ and direction θ. We choose

$$v_n = \left(\frac{1}{\alpha}\right)^{n-1}$$

(so that $v_1 = 1$) and t_n so that the first particle undergoes collision at time zero. (If the distance traveled by particles between collisions is denoted by d, then

$$t_n = -\frac{d}{v_2} - \frac{d}{v_3} - \cdots - \frac{d}{v_n} = -d \sum_{k=1}^{n-1} \alpha^k .)$$

To complete the construction, we "let n go to infinity". By this we mean the following: For any $j < n$, the trajectory $(q_j^{(n)}(t),\ p_j^{(n)}(t))$ of the jth particle in the one particle phase space does not depend on n, i.e., on which particle is initially excited. If we denote the common trajectory by $(q_j(t), p_j(t))$, then the sequence of trajectories

$$(q_j(t), p_j(t))_{j=1,2,\ldots}$$

given what has every right to be called a solution of the equations of motion of the system of infinitely many hard spheres. Note that each particle undergoes only two collisions (only one in the case of the first particle); that the collisions are separated from each other in space and time; and that there are no multiple collisions.

For $t \leq t_\infty = - d \sum_{k=1}^{\infty} \alpha^k$, all particles are at rest in their initial positions, so the "normal solution" to the equations of motion is clearly one in which the particles stay where they are at all later times. The solution we have constructed, although agreeing with the normal solution up to time t_∞, deviates sharply from it at all later times; for any $t > t_\infty$, all but finitely many particles are in motion, with a velocity which increases exponentially with initial distance from the origin (but each individual particle remains at rest for some nonzero length of time after t_∞). Note that, although we constructed this anomalous solution as a limit of ones in which a large amount of energy is given to a single particle very far from the origin, all trace of this external excitation has disappeared from the solution itself; instead, the system appears to excite itself very violently at infinity at time t_∞. This can happen because the meaning we have given to a solution to the equations of motion is purely local, i.e, looks at particles only one at a time, and hence cannot detect whether or not the system is being "driven at infinity". We therefore cannot hope to obtain uniqueness without imposing some sort of boundary condition at infinity which rules out such driving. The condition which we will use is the bound

$$|q_i(t) - q_i| \leq M \log_+(q_i) \quad \text{for all} \quad i$$

which is satisfied by the solutions constructed in the preceding chapter.

We now turn to the proof. We will give the argument only for an interparticle force F which is Lipschitz continuous and of finite range; the case of long range forces is much more complicated. We will not need to use the superstability of the potential, or even the fact that F is a gradient, but we will have to impose a restriction on the initial phase point $\underline{x}$: To formulate this condition, we recall that, for any $\alpha \in \mathbb{Z}^\nu$, $\Delta(\alpha)$ denotes the unit cube in $\mathbb{R}^\nu$ centered at α. We are going to assume that our initial phase point $\underline{x}$ satisfies

(*) There exists σ such that

$$\sup_{\alpha \in \mathbb{Z}^\nu} \frac{n_{\Delta(\alpha)}(\underline{x})}{[\log_+(\alpha)]^\sigma} < \infty .$$

This condition places a mild restriction on the density fluctuations in $\underline{x}$. A straightforward argument, using Ruelle's probability estimates, shows that it holds almost everywhere with respect to any Gibbs state, with $\sigma = 1/2$.

We now formulate our uniqueness result.

<u>Theorem 1</u>. Let the interparticle force F be Lipschitz continuous and of finite range R, and let $\underline{x}$ satisfy (*). Then there exists at most one solution $\underline{x}(t)$ of the equations of motion, defined for $0 \leq t \leq \tau$, and with $\underline{x}(0) = \underline{x}$, such that

$$(**) \qquad \sup_{0 \leq t \leq \tau} \sup_i \frac{|q_i(t) - q_i|}{\log_+(q_i)} < \infty$$

<u>Proof</u>: Let $\underline{x}^{(1)}(t)$, $\underline{x}^{(2)}(t)$ be two solutions satisfying (**). Choose M large enough so that

$$|q_i^{(\frac{1}{2})}(t) - q_i| \leq M \log_+(q_i)$$

for all i and all t between 0 and τ. The equations of motion may be rewritten in integral form as

$$q_i^{(\frac{1}{2})}(t) = q_i + \frac{t}{m} p_i + \int_0^t dt_1 \int_0^{t_1} dt_2 \, F_i^{(\frac{1}{2})}(t_2) .$$

where

$$F_i^{(\frac{1}{2})}(t) = \sum_{j \neq i} F(q_i^{(\frac{1}{2})}(t) - q_j^{(\frac{1}{2})}(t)) .$$

Subtracting, we obtain:

$$(1)\qquad q_i^{(1)}(t) - q_i^{(2)}(t) = \int_0^t dt_1 \int_0^{t_1} dt_2\ [F_i^{(1)}(t_2) - F_i^{(2)}(t_2)] ,$$

and we will proceed by estimating the right-hand side of this equation in terms of the $q_j^{(1)}(t_2) - q_j^{(2)}(t_2)$. In doing this, it is convenient to note first that, because of the finite range of the force and the condition (**) on the distance traveled, we can exclude all but finitely many j's from consideration for any given i. More explicitly: suppose $|q_i| \leq r$, and let

$$(2)\qquad r_1(r) = \sup \{s: s - M \log_+(s) \leq r + M \log_+(r) + R\} .$$

Then if $|q_j| \geq r_1(r)$, we have for any t between 0 and τ

$$q_i^{(\frac{1}{2})}(t) \leq r + M \log_+(r)$$

$$|q_j^{(\frac{1}{2})}(t)| \geq |q_j| - M \log_+|q_j| \geq r + M \log_+(r) + R ,$$

and hence

$$|q_i^{(\frac{1}{2})}(t) - q_j^{(\frac{1}{2})}(t)| \geq R \quad \text{so} \quad F(q_i^{(\frac{1}{2})}(t) - q_j^{(\frac{1}{2})}(t)) = 0 .$$

We further reduce the number of relevant particles as follows: For large r, r_1 is not too much larger than r, so

$$\sup_r \frac{\log_+(r_1)}{\log_+(r)} < \infty ,$$

and this implies that there exists M' such that

$$|q_j| \leq r_1 \quad \text{implies} \quad |q_j^{(\frac{1}{2})}(t) - q_j| \leq M' \log_+(r) .$$

Hence

$$F(q_i^{(\frac{1}{2})}(t) - q_j^{(\frac{1}{2})}(t)) = 0 \quad \text{if} \quad |q_i - q_j| \geq 2\,M' \log_+(r) + R .$$

By (*), the number of such j's is bounded by an expression of the form $N(\log_+(r))^{\nu+\sigma}$ (since the q_j's must lie in a sphere of radius $2M' \log_+(r) + R$ about q_i).

We may now estimate

$$|F_i^{(1)}(t)-F_i^{(2)}(t)| = |\sum_{j\neq i} F(q_i^{(1)}(t)-q_j^{(1)}(t))-F(q_i^{(2)}(t)-q_j^{(2)}(t))|$$

$$\leq N(\log_+(r))^{\nu+\sigma} L \sup_{|q_j|\leq r_1} |[q_i^{(1)}(t)-q_j^{(1)}(t)] - [q_i^{(2)}(t)-q_j^{(2)}(t)]|$$

where we have estimated the sum over j by a bound on the number of nonzero terms ($N \log_+(r)^{\nu+\sigma}$) times a bound on the largest term, and where L is a Lipschitz constant for F. We may replace the right-hand side of the above estimate by

$$2NL\ (\log_+(r))^{\nu+\sigma} \sup_{|q_j|\leq r_1} |q_j^{(1)}(t) - q_j^{(2)}(t)| .$$

Recall that this bound depends on i only through the restriction that $|q_i| \leq r$, i.e., we have shown that, if $|q_i| \leq r$, we have

$$|F_i^{(1)}(t) - F_i^{(2)}(t)| \leq 2NL\ (\log_+(r))^{\nu+\sigma} \sup_{|q_j|\leq r_1} |q_j^{(1)}(t) - q_j^{(2)}(t)|.$$

Thus, if we define

$$\delta(r,t) = \sup \{|q_i^{(1)}(t) - q_i^{(2)}(t)|: |q_i| \leq r\} ,$$

we may use the above estimate and (1) to obtain:

$$\delta(r,t) \leq A(\log_+(r))^{\sigma'} \int_0^t dt_1 \int_0^{t_1} dt_2\ \delta(r_1,t_2) \tag{3}$$

where we have written A for $2NL$ and σ' for $\nu+\sigma$. The usual strategy for proving uniqueness would be to show that, for sufficiently small t, the right-hand side of (3) is strictly smaller than the left unless δ is zero. This argument fails because of the presence of the factor $(\log_+(r))^{\sigma'}$ on the right which goes to infinity with r. Nevertheless, the inequality (3), together with the bound

(4) $$\delta(r,t) \le 2M \log_+(r)$$

(which follows from

$$|q_i^{(1)}(t)-q_i^{(2)}(t)| \le |q_i^{(1)}(t)-q_i| + |q_i^{(2)}(t)-q_i| \le 2M \log_+(q_i))$$

does imply that $\delta(r,t) \equiv 0$. To prove this, we iterate (3) n times; then insert (4), to obtain

$$\delta(r,t) \le A^n \prod_{i=0}^{n-1} (\log_+(r_i))^{\sigma'} \int_0^t dt_1 \int_0^{t_1} dt_2 \int_0^{t_2} dt_3 \cdots \int_0^{t_{2n-1}} dt_{2n}\, \delta(r_n,t_{2n})$$

$$\le A^n \prod_{i=0}^{n-1} (\log_+(r_i))^{\sigma'}\, 2M \log_+(r_n)\, \frac{t^{2n}}{(2n)!}$$

where

$$r_0 = r\ ,\quad r_{i+1} = r_1(r_i)\ ,\quad i = 0,1,2,\ldots\ .$$

The important factor here is the $(2n)!$ in the denominator, and the right-hand side of this inequality will go to zero as n goes to infinity provided

$$\lim_{n\to\infty} \frac{(\log_+(r_n))^{\sigma'}}{n^2} = 0\ .$$

Everything depends, then, on estimating the rate of growth of r_n with n. Recall that

$$r_{n+1} = \sup\{s: s - M \log_+ s \le r_n + M \log_+(r_n) + R\}\ .$$

It may be shown by some straightforward but tedious estimates that $r_n \sim 2M\, n \log n$ for large n, but we can make do with a much easier estimate. We claim that for large n we have

(5) $$r_{n+1} \le \left(\sqrt{r_n} + 1\right)^2\ .$$

This estimate will evidently imply $r_n = O(n^2)$, so

$$\frac{(\log_+(r_n))^{\sigma'}}{n^2} \to 0$$

as desired. To prove (5), we have only to check that

$$(\sqrt{r_n}+1)^2 - M \log(\sqrt{r_n}+1)^2 \geq r_n + M \log_+(r_n) + R$$

for all sufficiently large r_n, and this is trivial.

The situation is now as follows: We may let $\hat{\mathfrak{X}}$ denote the set of all points $\underline{x}$ in $\mathfrak{X}$ which satisfy (*) and for which there is a solution $\underline{x}(t)$ of the equations of motion with $\underline{x}(0) = \underline{x}$ satisfying

$$\sup_{|t|\leq\tau} \sup_i \frac{|q_i(t) - q_i|}{\log_+(q)} < \infty \quad \text{for all } \tau \ .$$

The existence theorem tells us that $\mu(\hat{\mathfrak{X}}) = 1$ for every Gibbs state μ and the uniqueness theorem tells us that the solution $\underline{x}(t)$ is uniquely determined. Moreover, very elementary estimates show that $\underline{x}(t)$ is in $\hat{\mathfrak{X}}$ for all t. Thus, the solution mappings for the equations of motion define a flow T^t on $\hat{\mathfrak{X}}$. A number of questions naturally arise

(i) Is every Gibbs state μ invariant under T^t?

(ii) Is the mapping $(t,\underline{x}) \longrightarrow T^t\underline{x}$ jointly measurable?

(iii) Is T^t the limit, in some sense, of $T^t_{(s)}$ as $s \to \infty$?

We will sketch an argument to show that the answer to (iii) is yes, with the sense of convergence being convergence in measure. This implies, by more or less evident approximation arguments, that the answers to (i) and (ii) are also yes.

Proposition 2. Let $\varepsilon > 0$, r, and τ be given, and let μ be a Gibbs state. Then there exists S such that if $s > S$,

$|q_i^{(s)}(t)-q_i(t)| \leq \varepsilon$ for all i with $|q_i| \leq r$ and all t with $|t| \leq \tau$,

for all $\underline{x}$ in the complement of a set of μ-measure ε.

Proof: Recalling that

$$\int \bar{B}_\infty \, d\mu < \int \bar{B}_{(s)} \, d\mu = \int B \, d\mu < \infty \ ,$$

we see that for sufficiently large b,

$$\mu\{\underline{x}: \bar{B}_\infty(\underline{x}) > b\} < \frac{\varepsilon}{3} \quad \text{and} \quad \mu\{\underline{x}: \bar{B}_{(s)}(\underline{x}) > b\} < \frac{\varepsilon}{3} \quad \text{for all } s.$$

Also, for sufficiently large N_1, we have

$$\sup_{\alpha\in\mathbb{Z}^\nu} \frac{n_{\Delta(\alpha)}(\underline{x})}{[\log_+(\alpha)]^{1/2}} \le N_1 \tag{6}$$

on the complement of a set of measure $\varepsilon/3$. The estimates of the preceding chapter show that there exists M such that

$$\sup_{|t|\le\tau} \sup_i \frac{|q_i^{(s)}(t)-q_i|}{\log_+(q_i)} \le M \tag{7}$$

for all $\underline{x}$ such that $\bar{B}_{(s)}(\underline{x}) \le b$, and similarly that

$$\sup_{|t|\le\tau} \sup_i \frac{|q_i(t)-q_i|}{\log_+(q_i)} \le M \tag{8}$$

for all $\underline{x}$ such that $\bar{B}_\infty(\underline{x}) \le b$. The complement of the set where (6), (7), and (8) all hold thus has measure less than ε, so we want to prove the existence of S assuming only (6), (7), and (8).

We next observe that

$$q_i^{(s)}(t) = q_i + \frac{t}{m} p_i + \int_0^t dt_1 \int_0^{t_1} dt_2\, F_i^{(s)}(t_2)$$

$$F_i^{(s)}(t) = \sum_{j\neq i} F(q_i^{(s)}(t) - q_j^{(s)}(t))$$

provided that the ith particle does not strike the wall of Λ_s between time zero and time t. This will be true for all t with $|t| \le \tau$ provided (7) holds and

$$|q_i| + M\log_+(q_i) < s .$$

Writing the similar equation

$$q_i(t) = q_i + \frac{t}{m} p_i + \int_0^t dt_1 \int_0^{t_1} dt_2\, F_i(t_2)$$

and subtracting, we obtain

$$q_i(t) - q_i^{(s)}(t) = \int_0^t dt_1 \int_0^{t_1} dt_2 \; [F_i(t_2) - F_i^{(s)}(t_2)]$$

which we can estimate by using exactly the same techniques used in the proof of uniqueness. That is, if we define

$$\delta^{(s)}(r,t) = \sup \{ |q_i(t) - q_i^{(s)}(t)| : |q_i| \le r \} ,$$

we get

$$\delta^{(s)}(r,t) \le A \log_+(r)^{\nu/2} \int_0^t dt_1 \int_0^{t_1} dt_2 \; \delta^{(s)}(r_1,t_2) \quad \text{for} \quad 0 \le t \le \tau$$

provided

$$r + M \log_+(r) \le s$$

and provided (6), (7), and (8) hold. Here, A depends only on N_1,M. Iterating n times gives

$$\delta^{(s)}(r,t) \le A^n \prod_{i=0}^{n-1} \log_+(r_i)^{\nu/2} \frac{t^{2n}}{(2n)!} \, 2M \log_+(r_n)$$

provided $s \ge r_n$, so if s is large enough, we may take n as large as we like and hence make $\delta^{(s)}(r,t) < \varepsilon$ for all t with $|t| \le \tau$ as desired.

The example given at the beginning of this chapter makes it plausible that some condition like (**) must be imposed in order to guarantee the uniqueness of solutions. It is somewhat unsatisfactory, however, to have the proof depend on a condition as concrete as (**) whose significance is not very clear. It is interesting to note, therefore, that the uniqueness result may be reworked into a form in which such conditions do not appear explicitly:

Proposition 3. Let μ be a Gibbs state for Φ, and let $\bar{T}^t$ be a flow on $\mathfrak{X}$ leaving μ invariant and such that $\bar{T}^t \underline{x}$ is a solution of the equations of motion for μ-almost all $\underline{x}$. Then

$\bar{T}^t = T^t$ μ-almost everywhere.

Proof: If we write $\bar{T}^t\underline{x} = (\bar{q}_i(t), \bar{p}_i(t))$, then Proposition 4 of Chapter 3 implies

$$\sup_{|t|\leq\tau} \sup_i \frac{|\bar{q}_i(t) - q_i|}{\log_+(q_i)} < \infty \quad \text{for all} \quad \tau$$

for μ-almost all $\underline{x}$; Theorem 1 then says that $\bar{T}^t\underline{x} = T^t\underline{x}$ for μ-almost all $\underline{x}$.

Hence, although it may be possible to find numerous anomalous solutions to the equations of motion, it is not possible to piece them together to form a measure-preserving flow.

5. Dynamics of Systems of Infinitely Many Particles: Other Approaches.

Summary. A number of other ideas about the problem of dynamics of systems of infinitely many particles are described briefly, and it is pointed out that replacing the usual kinetic energy by the relativistic expression introduces a cut-off on the velocities which makes the problem of constructing solutions trivial.

....

In the two preceding chapters, we developed in some detail one approach to the problem of the dynamics of the infinite system. In this chapter, we will summarize a number of other ideas about this problem which either apply to other kinds of interactions or which give added information about the solutions.

In [1], existence and uniqueness were proved for one-dimensional systems interacting by smooth finite range two-body forces. Unlike the theorems described above, these results did not require superstability of the interaction, and did not depend on the existence of invariant Gibbs states. Moreover, the set of initial phase points for which solutions can be obtained was described explicitly, with no uncontrollable sets of measure zero. In summary, what was shown was the following: Let $\hat{\mathfrak{X}} \subset \mathfrak{X}$ now denote the sets of phase points such that

(1) $$\sup_i \frac{|p_i|}{\log_+(q_i)} < \infty$$

(2) $$\sup_{k\in\mathbb{Z}} \frac{n_{[k-\log_+(k),k+\log_+(k)]}(\underline{x})}{\log_+(k)} < \infty$$

Condition (1) is strongly reminiscent of estimates appearing above; condition (2) is evidently stronger than

$$\sup_{k\in\mathbb{Z}} \frac{n_{(k-1/2,k+1/2)}(\underline{x})}{\log_+(k)} < \infty$$

which is a special case of (*) of Chapter 4 (with $\sigma = 1$). It is not difficult to show that (1) and (2) hold almost everywhere with respect to a Gibbs state for <u>any</u> superstable interaction (which need not have anything to do with the interparticle force giving the dynamics). The result is now the following: For any $\underline{x} \in \hat{\mathfrak{X}}$, there is a (global) solution $\underline{x}(t)$ of the equations of motion with $\underline{x}(0) = \underline{x}$; this solution satisfies (**) and is hence unique in the class of such solutions. Moreover $\underline{x}(t) \in \hat{\mathfrak{X}}$ for all t.

To show how the argument goes, and also to explain why it fails in more than one dimension, we will derive some rough (and not quite correct) <u>a priori</u> estimates for the solution $\underline{x}(t)$, assuming it exists. We write

$$V_{max}(t) = \sup_i \left\{\frac{p_i(t)/m}{\log_+(q_i(t))}\right\}$$

$$D_{max}(t) = \sup_{k\in\mathbb{Z}} \frac{n_{[k-\log_+(k),k+\log_+(k)]}(\underline{x}(t))}{\log_+(k)} .$$

The force on the ith particle at time t is bounded by a constant times the number of nearby particles, which is no larger than

$$\text{const} \times D_{max}(t) \log_+(q_i(t)) .$$

Thus

$$\left|\frac{p_i(t)}{m}\right| \leq \left|\frac{p_i}{m}\right| + \int_0^t dt_1 \text{ const} \times D_{max}(t_1) \log_+(q_i(t_1))$$

and if we ignore the change in $\log_+(q_i(t))$ due to the particle's motion, we find

$$\left|\frac{p_i(t)}{m}\right| \leq [V_{max}(0) + \text{const} \int_0^t dt_1 D_{max}(t_1)] \log_+(q_i(t)) ,$$

i.e.

$$\text{(1)} \qquad V_{max}(t) \leq V_{max}(0) + \text{const} \int_0^t dt_1 D_{max}(t_1) .$$

Similarly, if a particle is at time t in the interval $[k - \log_+(k), k + \log_+(k)]$ then, approximately, the largest distance it could have traveled is

$$\log_+(k) \int_0^t dt_1 \, V_{max}(t_1)$$

(we again ignore the variation of $\log_+(q_i(t))$), so it must originally have been in the interval

$$\left[k - \log_+(k)\left(1 + \int_0^t dt_1 \, V_{max}(t_1)\right),\ k + \log_+(k)\left(1 + \int_0^t dt_1 \, V_{max}(t_1)\right)\right].$$

The number of such particles is bounded approximately by

$$\log_+(k) \left(1 + \int_0^t dt_1 \, V_{max}(t_1)\right) D_{max}(0) \ ,$$

so we get

$$D_{max}(t) \leq D_{max}(0) \left[1 + \int_0^t dt_1 \, V_{max}(t_1)\right] . \tag{2}$$

The inequalities (1) and (2) imply bounds on $D_{max}(t)$, $V_{max}(t)$ for all t (which depend on $D_{max}(0)$, $V_{max}(0)$). Once such bounds are obtained, either the method of Chapter 3 or a straightforward iterative technique using the estimates of Chapter 4 may be used to construct solutions.

As indicated, the inequalities (1) and (2) are not quite correct, since some approximations were made in deriving them. However, a careful analysis yields more complicated inequalities which are correct and which accomplish the same result.

We can also see that such inequalities fail completely in more than one dimension. We might try, for example, to keep the definition of V_{max}, but to replace the definition of D_{max} by

$$\sup_{q \in \mathbb{R}^\nu} \frac{n_{\Lambda(\log_+(q),q)}(\underline{x})}{[\log_+(q)]^\nu}$$

where $\Lambda(a,q)$ denotes the ball of radius a centered at q. Finiteness of D_{max} implies a bound on forces which grows like $[\log_+(q)]^\nu$ (rather than $\log_+(q)$), and this bound is only good enough to get a bound on velocities at later times which grows like $[\log_+(q)]^\nu$ and which therefore says nothing about V_{max}. If we had instead taken the definition

$$V_{max} = \sup_i \frac{|p_i/m|}{[\log_+(q_i)]^\nu}$$

then a bound on D_{max} would give a bound on V_{max}, but finiteness of V_{max} would imply only a density bound like

$$n_{\Lambda(\log_+(q),q)} \leq \text{const } [\log_+(q)]^{\nu^2}$$

at later times, and this does not control D_{max}. Moreover, the situation is not improved by replacing $\log_+$ systematically by some other increasing function unless that function is bounded, in which case we are reduced to assuming bounded initial velocities and densities.

The next contribution is due to Sinai [2], who again considered one dimensional systems with finite-range interactions. However, he assumed the particles had hard cores in addition to a smooth interparticle potential. This paper introduced two important ideas:

(a) It showed that the time evolution has the property that, over any finite interval of time, $\underline{x}(t)$ breaks into finite clusters of particles which do not interact.

(b) It exploited in a serious way the invariance of Gibbs states. Roughly, the idea is as follows: Choose a large integer n, and consider the evolution $T^t_{(3^n)}$ in which the particles in $[-3^n, 3^n]$ are allowed to move. If n is large enough, then Sinai shows that with probability very close to one

(a) there are intervals I_-, I_+ contained in $[-2\times 3^{n-1}, -3^{n-1}]$ and $[3^{n-1}, 2\times 3^{n-1}]$ respectively and of length $O(n)=O(\log(3^n))$ in which there are initially no particles. The location of these intervals will of course depend on the particular initial phase point.

(b) The maximum velocity attained by any particle between time zero and time one is $O(\sqrt{n})$, and in particular is very small compared with the length of the initially empty intervals I_-, I_+.

Hence, the initially empty intervals cut off from the system a cluster of particles "inside" which do not interact with the particles outside between time zero and time one. The time evolution for the inside cluster, at least from time zero to time one, is therefore the evolution it would have if the outside particles were not present at all. By pushing the arguments a little further, Sinai is able to construct, for almost all initial phase points, a solution to the equations of motion in which each particle is contained in a finite cluster which does not interact with any particles outside itself for times between zero and one. The uniqueness question then reduces to uniqueness for the finite clusters, which is standard. Global existence and uniqueness are obtained by using the invariance of the Gibbs state under the time evolution mappings T^t $(0 \leq t \leq 1)$ and iterating (i.e., constructing $T^{1+t} = T^1 T^t$ $(0 \leq t \leq 1)$, etc.).

We will not discuss how (a) is proved -- since this is simply a matter of calculation with Gibbs states -- but will outline the proof of (b). What needs to be done is to get an estimate on the probability that the maximum velocity between time zero and time one is bigger than λ which goes to zero rapidly as $\lambda/\sqrt{n} \to \infty$. To get this estimate, first note that this probability is no larger than the sum of the probability that the initial velocity at time zero is bigger than λ -- which again can be estimated by elementary means -- plus the probability that some particle's speed increases past λ at some time between zero and one. Moreover, we we can estimate the latter probability by

$$\int_0^1 \Pi_\lambda(t)\, dt ,$$

where $\Pi_\lambda(t)\, dt$ is the probability that some particle's speed increases past λ between time t and $t+dt$. Now because of the invariance of the Gibbs state, $\Pi_\lambda(t)$ does not depend on t, i.e.,

$$\int_0^1 \Pi_\lambda(t)\, dt = \Pi_\lambda(0) ,$$

and $\Pi_\lambda(0)$ is again easy to estimate using the properties of Gibbs states. Note that this argument is quite similar to those used in our proof of existence -- it is shown that some estimates hold on the complement of a set of small measure, but not much more can be said about the set where the estimates fail.

Recently, Sinai [3] has extended his methods to obtain a time evolution theorem in more than one dimension. Again, the technique is to show that the infinite system breaks into finite clusters of particles which do not interact with each other in unit time. The complication in the multidimensional case as compared to the one-dimensional case seems to be that the empty regions bounding large clusters must have large area, and hence a boundary of any particular shape is unlikely. On the other hand, the possible geometries for boundaries are much more varied. The existence of such a decomposition into noninteracting clusters has been proved only for Gibbs states with sufficiently low density.

In another direction, Marchioro, Pellegrinotti, and Presutti [4] have applied a combination of Sinai's technique for estimating maximum velocities and iterative estimates similar to those we have used to prove uniqueness to obtain another construction of the dynamics for multidimensional systems with smooth, finite range superstable interactions. They also show that

$$T^t_{(2^n)}$$

converges almost everywhere to T^t as $n \to \infty$.

Finally, it is worth pointing out that a small modification of the problem of infinite system dynamics makes everything trivial. In relativity theory, the expression for the kinetic energy is

$$\sqrt{p^2c^2 + m^2c^4} \ ,$$

where c is the speed of light. Note that

$$\frac{\partial}{\partial p} \sqrt{p^2c^2 + m^2c^4} = \frac{p}{m} \left(1 + \left(\frac{p}{mc}\right)^2\right)^{-1/2} \approx \frac{p}{m} \ \text{if} \ \frac{p}{m} << c \ ,$$

but that also

$$\frac{p}{m} \left(1 + \left(\frac{p}{mc}\right)^2\right)^{-1/2} \leq c \quad \text{for all} \quad p \ .$$

Thus, if we redefine the Hamiltonian by replacing the usual kinetic energy by the relativistic kinetic energy, i.e., if we change the equations of motion to

$$\frac{dq_i}{dt} = \frac{p_i}{m}\left(1 + \left(\frac{p_i}{mc}\right)^2\right)^{-1/2}$$

$$\frac{dp_i}{dt} = \sum_{j \neq i} F\left(q_i(t) - q_j(t)\right)$$

then c is a uniform bound on the velocity, which gives a uniform bound on the distance particles can travel in any given length of time. This gives even better localization than our probabilistic estimates of Chapter 4 and it is quite a simple matter to obtain an existence and uniqueness theorem for essentially any initial phase-point. It should be emphasized that using the relativistic formula for the kinetic energy does not make the equations of motion "relativistic", since the prescription for determining the forces is not consistent with relativity theory. It is more accurate to think of modifying the kinetic energy simply as a convenient method for introducing a cut-off on the velocities, and other modifications of the kinetic energy would work equally well for this purpose.

References to Chapter 5

[1] Lanford, O.E., III, Classical Mechanics of One-Dimensional Systems of Infinitely Many Particles, Commun. Math. Phys. 9 (1969) 169-181; and 11 (1969) 257-292.

[2] Sinai, Ya. G., Construction of the Dynamics for One-Dimensional Systems of Statistical Mechanics, Teoret. Mat. Fiz. 11 (1972) 248-258.

[3] Sinai, Ya. G., The Construction of Cluster Dynamics for Dynamical Systems of Statistical Mechanics, Vestnik Markov. Univ., Ser. I, Mat. Meh. 29 (1974) 152-159.

[4] Marchioro, C., Pellegrinotti, A., and Presutti, E., Existence of Time Evolution for ν-Dimensional Statistical Mechanics, to appear in Commun. Math. Phys.

6. The Boltzmann Equation: Introduction

Summary. We formulate the problem of constructing a reduced macroscopic description of systems of many classical particles and argue heuristically that the time evolution of the macroscopic description should be given, in an appropriate limit, by the Boltzmann equation. As preparation for a precise proof of the validity of the Boltzmann equation, we discuss the BBGKY hierarchy for the time development of statistical states of many particle systems, and we show that the Boltzmann equation is equivalent to a linear hierarchy of equations closely resembling the BBGKY hierarchy for the hard-sphere system.

....

In the remainder of these notes, we investigate the classical mechanics of large systems of particles from a different point of view. Our objective is to study a particular kind of reduced "macroscopic" description and its time evolution. Schematically, the questions we want to investigate are as follows: Take a system of a large number n of particles contained in a bounded region Λ of $\mathbb{R}^\nu$. Partition the one-particle phase space $\Lambda \times \mathbb{R}^\nu$ into a finite number of cells $\{\Delta_1,\ldots,\Delta_j\}$. These cells should be visualized as small "macroscopically" but large enough so that each contains many particles. A reduced description of the n-particle phase point $\underline{x}$ is provided by the j occupation numbers

$$\left(\frac{n_{\Delta_i}(\underline{x})}{n}\right)_{1\leq i\leq j}$$

We want to know what, if anything, can be predicted about the occupation numbers at time t from their values at time zero.

A little reflection shows that no very informative production can be made which holds for <u>every</u> initial phase point with a given set of occupation numbers. For example, by arranging the particles in pairs which are close together and approaching collision, we can make the velocities change radically, in almost any way consistent with conservation of energy and momentum, almost immediately after starting out. It is nevertheless not unreasonable to hope to be able to make probabilistic predictions holding for "most" phase points with the given occupation numbers. In order to get a sharp prediction, we can expect to have to pass to the limit as the number of particles goes to infinity. We want to do this while holding the region Λ, the partition $\{\Delta_1,\ldots,\Delta_j\}$, and the occupation numbers n_{Δ_i}/n fixed. We must therefore make the particles smaller i.e., decrease the range of the interparticle force, as we make n larger. This gives a procedure for passing to the limit of an infinite system which is not quite the same as the thermodynamic limit of equilibrium statistical mechanics; we will refer to it as the <u>hydrodynamic limit</u>. We are forced to deal with this new kind of limit because we want to consider systems which are spatially inhomogeneous but where the distances over which the system changes significantly are very large compared to the range of the interparticle force.

To make the question more concrete, we might proceed as follows: Start with the partition $\{\Delta_1,\ldots,\Delta_j\}$, with j positive numbers $f_1,\ldots,f_j$ such that $f_1 + \ldots + f_j = 1$, and with a tolerance $\varepsilon > 0$. For each value of n, we consider the set of all n-particle phase points $\underline{x}$ such that

$$\left| \frac{n_{\Delta_i}(\underline{x})}{n} - f_i \right| < \varepsilon \quad \text{for} \quad 1 \leq i \leq j \,.$$

This set has nonzero volume in $(\Lambda \times \mathbb{R}^\nu)^n$, and we can construct a statistical state of the n-particle system by restricting Lebesgue measure to it and normalizing (making some adjustments for the infinite volume of momentum space). We would like to be able to show that, for each $t > 0$, the random variables

$$\frac{n_{\Delta_i}(\underline{x}(t))}{n}$$

converge in probability to constants $f_i(t)$ as $n \to \infty$, and we would also like to have "hydrodynamic equations"* which permit, in principle at least, the computation of the $f_i(t)$ from the f_i. We do not mean to insist on the technical details of the above formulation; for example, it might be desirable to let the partition $\{\Delta_1,\ldots,\Delta_j\}$ become infinitely fine and the tolerance ε go to zero at the same time as n goes to infinity. Also, instead of looking only at statistical states constructed by restricting Lebesgue measure by constraints on occupation numbers and normalizing, it would be desirable to investigate fairly general sequences $(\mu^{(n)})$ of n-particle statistical states having (at least) the property that each n_{Δ_i}/n converges in probability to the constant f_i (at time zero).

The above questions seem both natural and important, but they are also very difficult. Very little is known about them, even on heuristic level. There is however one successful idea in this area, embodied in the Boltzmann equation. We will sketch here an argument fairly close to Boltzmann's original derivation of the equation. It should be noted that we will regard Boltzmann's equation as describing the time evolution of particular phase points (with nice properties) rather than of mean values in well behaved statistical states.

* We are using the term "hydrodynamic equations" loosely here; in the technical terminology of the subject, equations governing the time development of the n_{Δ_i}/n's are more properly referred to as <u>kinetic equations</u>.

To give Boltzmann's argument, we partition $\Lambda \times \mathbb{R}^{\nu}$ into small cells $\Delta\vec{q}\,\Delta\vec{p}$ centered at points $(\vec{q},\vec{p})$. The occupation numbers are given by a positive function f with integral one through

$$f_i = \int_{\Delta_i} f(\vec{q},\vec{p})\, d\vec{q}\, d\vec{p}\ .$$

The cells $\Delta\vec{q}\,\Delta\vec{p}$ are small enough so that f changes little over a number of contiguous cells, i.e., so that sums can be replaced by integrals when convenient.

Now take an initial phase point $\underline{x}$ with

$$\frac{n_{\Delta_i}(\underline{x})}{n} \approx \int_{\Delta_i} f(\vec{q},\vec{p})\, d\vec{q}\, d\vec{p} \quad \text{for all } i\ ,$$

and assume that, at each later time t, we can find another function f_t such that

$$\frac{n_{\Delta_i}(\underline{x}(t))}{n} \approx \int_{\Delta_i} f_t(\vec{q},\vec{p})\, d\vec{q}\, d\vec{p}\ ;$$

we want to derive an equation of motion for f_t, i.e., to compute $\partial f_t/\partial t$. We have already seen that it is not possible to do this without further assumptions on $\underline{x}$, but it will turn out that a more or less plausible set of assumptions leads to a very useful equation of motion. For the purposes of the following argument, we will assume that the particles are hard spheres of diameter d with no additional interparticle forces; extension to other interactions is straightforward.

The change δf in f over a short interval of time δt comes about for two distinct reasons: The particles mostly move in straight lines with uniform velocities, but they occasionally undergo collisions which result in discontinuous changes in their velocities. Thus we can write

$$\delta f = (\delta f)_{flow} + (\delta f)_{collision}$$

A completely straightforward argument gives

$$(\delta f)_{flow} = -\frac{\vec{p}}{m} \cdot \frac{\partial f}{\partial \vec{q}} \delta t .$$

We can further decompose

$$(\delta f)_{collision} = (\delta f)_{in} - (\delta f)_{out}$$

where $(\delta f)_{in}(\vec{q},\vec{p})$ is the contribution of collisions in which one of the outgoing particles has position $\vec{q}$ and momentum $\vec{p}$, i.e., of scatterings into the neighborhood of $\vec{p}$, while $(\delta f)_{out}(\vec{q},\vec{p})$ is the contribution of scatterings out of the neighborhood of $\vec{p}$.

We now must look with some care at the collision process. To describe a collision, we give first of all the incoming momenta $\vec{p}_1,\vec{p}_2$. The outgoing momenta will be denoted by $\vec{p}_1{}',\vec{p}_2{}'$. Conservation of momentum and energy give

$$\vec{p}_1 + \vec{p}_2 = \vec{p}_1{}' + \vec{p}_2{}' \qquad \text{and} \qquad \vec{p}_1^{\,2} + \vec{p}_2^{\,2} = \vec{p}_1{}'^2 + \vec{p}_2{}'^2$$

Except in one dimension, these equations do not suffice to determine $\vec{p}_1{}'$, $\vec{p}_2{}'$ from $\vec{p}_1$, $\vec{p}_2$; to complete the specification of the collision, we give the direction between the centers of the two particles at the instant of collision as a unit vector $\hat{\omega}$ pointing from the center of particle 1 to the center of particle 2. The fact that $\vec{p}_1$, $\vec{p}_2$ are incoming momenta means that the particles are approaching each other, i.e.

$$\hat{\omega}\cdot (\vec{p}_1-\vec{p}_2) \geq 0 . \tag{1}$$

The definition of "elastic collision" can be taken to be the requirement of energy-momentum conservation together with the requirement that the momentum transferred from particle 1 to particle 2 (i.e. $\vec{p}_2{}'-\vec{p}_2$) be in the direction $\hat{\omega}$. (Thus, we may refer to $\hat{\omega}$ as the direction of momentum transfer.) A simple calculation then shows that $\vec{p}_1' - \vec{p}_2'$ is obtained from $\vec{p}_1 - \vec{p}_2$ by reflection through the plane perpendicular to $\hat{\omega}$ and hence that

$$\text{(2)}\qquad \begin{aligned} \vec{p}_1' &= \vec{p}_1 - [\hat{\omega}\cdot(\vec{p}_1-\vec{p}_2)]\cdot\hat{\omega} \\ \vec{p}_2' &= \vec{p}_2 + [\hat{\omega}\cdot(\vec{p}_1-\vec{p}_2)]\cdot\hat{\omega} \end{aligned}$$

We abbreviate this equation as

$$\text{(3)}\qquad (\vec{p}_1',\vec{p}_2') = T_{\hat{\omega}}(\vec{p}_1,\vec{p}_2)$$

where the mapping $T_{\hat{\omega}}$ of $\mathbb{R}^{2\nu}$ to itself is linear (by inspection) and orthogonal (by energy conservation) and hence in particular is volume preserving. Note also: Although $T_{\hat{\omega}}$ is interpreted as giving the outcome of a collision only for $\hat{\omega}\cdot(\vec{p}_1-\vec{p}_2) \geq 0$, $T_{\hat{\omega}}$ is defined for all $\vec{p}_1,\vec{p}_2$ and satisfies

$$T_{\hat{\omega}}^2 = 1 .$$

Hence

$$(\vec{p}_1,\vec{p}_2) = T_{\hat{\omega}}(\vec{p}_1',\vec{p}_2')$$

which shows that, if $\hat{\omega}\cdot(\vec{p}_1-\vec{p}_2) \leq 0$, we may interpret

$$T_{\hat{\omega}}(\vec{p}_1,\vec{p}_2)$$

as the pair of incoming momenta which, after a collision with momentum transfer in the direction $\hat{\omega}$, give outgoing momenta $(\vec{p}_1,\vec{p}_2)$. Finally, reversing the direction of $\hat{\omega}$ and using

$$T_{\hat{\omega}} = T_{-\hat{\omega}} ,$$

we see that even if $\hat{\omega}\cdot(\vec{p}_1-\vec{p}_2) \geq 0$ we may interpret

$$(\vec{p}_1',\vec{p}_2') = T_{\hat{\omega}}(\vec{p}_1,\vec{p}_2)$$

either as

(a) outgoing momenta after a collision with incoming momenta $(\vec{p}_1,\vec{p}_2)$ and momentum transfer in the direction $\hat{\omega}$

(b) incoming momenta which, after a collision with momentum transfer direction $-\hat{\omega}$, give $(\vec{p}_1, \vec{p}_2)$ as outgoing momenta. The multiple interpretations of $T_{\hat{\omega}}$, although confusing, will be needed repeatedly in our discussion of the Boltzmann equation.

Now let $\Delta\vec{q}$ be a small region in position space (containing $\vec{q}$), $\Delta\vec{p}_1$ and $\Delta\vec{p}_2$ two small regions in momentum space (containing $\vec{p}_1$ and $\vec{p}_2$ respectively), and $\Delta\hat{\omega}$ a small region on the unit sphere. To determine the change in f due to collisions, we need to know how many collisions occur in time δt, in the space region $\Delta\vec{q}$, between particles with incoming momenta in $\Delta\vec{p}_1$ and $\Delta\vec{p}_2$ with momentum transfer direction in $\Delta\hat{\omega}$. We can calculate this as:

(number of particles in $\Delta\vec{q}\cdot\Delta\vec{p}_1$) $\times$ (fraction of these particles undergoing a collision of the indicated type)

Now a particular particle with position $\vec{q}^*$ and momentum in $\Delta\vec{p}_1$ will undergo a collision with another particle with momentum in $\Delta\vec{p}_2$ and momentum transfer in $\Delta\hat{\omega}$ if and only if there is a particle with that momentum whose center is in the <u>collision cylinder</u>. This cylinder has base $\vec{q}^* + d\cdot\Delta\hat{\omega}$, edge in the

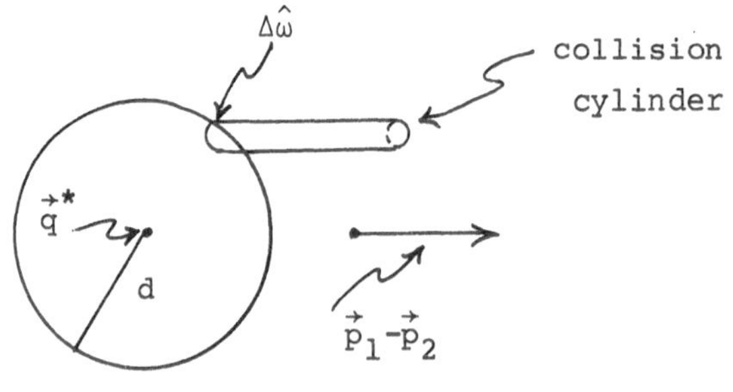

direction $\vec{p}_1-\vec{p}_2$, and slant height $|\vec{p}_1-\vec{p}_2|/m\ \delta t$. Its volume is therefore

$$d^2\cdot|\Delta\hat{\omega}|\ \left(\frac{\vec{p}_1-\vec{p}_2}{m}\right)\cdot\hat{\omega}\cdot\delta t\ .\ ^*$$

* At this point, and for the remainder of these notes, we specialize to $\nu = 3$.

We now make the crucial assumption about our phase point $\underline{x}$: The total number of particles with momentum in $\Delta\vec{p}_2$ occupying such collision cylinders is the total volume of the collision cylinders

$$\left(n\ f(\vec{q},\vec{p}_1)\ \Delta\vec{q}\ \Delta\vec{p}_1\right) \times \left(d^2\ \delta t\ |\Delta\hat{\omega}|\ \left(\frac{\vec{p}_1-\vec{p}_2}{m}\right)\cdot\hat{\omega}\right)$$

multiplied by $n\ f(\vec{q},\vec{p}_2)\ \Delta\vec{p}_2$. This assumption is called the Stosszahlansatz or the Hypothesis of Molecular Chaos; it may be crudely paraphrased to say that, for purposes of computing the number of collisions, one can treat the particles as "uncorrelated". Note that we will need this assumption at all positive times, not just for $t = 0$.

If we make this assumption, we find:

$$n(\delta f)_{out}\ \Delta\vec{q}\ \Delta\vec{p}_1 = n\ f(\vec{q},\vec{p}_1)\ \Delta\vec{q}\ \Delta\vec{p}_1$$

$$\times \sum_{\hat{\omega},p_2} d^2\ \delta t\ |\Delta\hat{\omega}|\ \left(\frac{\vec{p}_1-\vec{p}_2}{m}\right)\cdot\hat{\omega}\ n\ f(\vec{q},\vec{p}_2)\ \Delta\vec{p}_2$$

or, cancelling common terms and replacing sums by integrals,

$$(\delta f)_{out}(\vec{q},\vec{p}_1) = nd^2\ \delta t \int_{\hat{\omega}(\vec{p}_1-\vec{p}_2)\geq 0} d\hat{\omega}\ d\vec{p}_2\ \left(\frac{\vec{p}_1-\vec{p}_2}{m}\right)\cdot\hat{\omega}\ f(\vec{q},\vec{p}_1)\ f(\vec{q},\vec{p}_2)$$

Similarly:

$$n(\delta f)_{in}\ \Delta\vec{q}\ \Delta\vec{p}_2 = \sum_{\hat{\omega},\vec{p}_2} n\ f(\vec{q},\vec{p}_1')\ nf(\vec{q},\vec{p}_2')\ d^2\ \delta t\ |\Delta\hat{\omega}|\ \left(\frac{\vec{p}_1'-\vec{p}_2'}{m}\right)\cdot\hat{\omega}\ \Delta p_1'\ \Delta p_2'\ \Delta\vec{q}$$

Here, the sum runs over all $\vec{p}_2,\hat{\omega}$ with

$$(\vec{p}_1-\vec{p}_2)\cdot\hat{\omega} \leq 0\ ;$$

$(\vec{p}_1',\vec{p}_2') = T_{\hat{\omega}}(\vec{p}_1,\vec{p}_2)$ are the incoming momenta which, in a collision with momentum transfer in the direction $\hat{\omega}$, give outgoing momenta

$\vec{p}_1, \vec{p}_2$; and $\Delta\vec{p}_1{}' \Delta\vec{p}_2{}'$ denotes the volume (in $\mathbb{R}^6$) of the image of $\Delta\vec{p}_1 \Delta\vec{p}_2$ under $T_{\hat{\omega}}$. We now use

$$\Delta\vec{p}_1{}' \Delta\vec{p}_2{}' = \Delta\vec{p}_1 \Delta\vec{p}_2 \quad (T_{\hat{\omega}} \text{ is volume preserving})$$

$$\hat{\omega}\cdot(\vec{p}_1{}' - \vec{p}_2{}') = -\hat{\omega}\cdot(\vec{p}_1 - \vec{p}_2),$$

change summation variable from $\hat{\omega}$ to $-\hat{\omega}$, cancel, and replace sums by integrals to get:

$$(\delta f)_{in}(\vec{q},\vec{p}_1) = nd^2 \, \delta t \int_{\hat{\omega}\cdot(\vec{p}_1-\vec{p}_2)\geq 0} d\hat{\omega} \, d\vec{p}_2 \left(\frac{\vec{p}_1-\vec{p}_2}{m}\right)\cdot\hat{\omega}$$

$$\times \, f(\vec{q},\vec{p}_1{}') \, f(\vec{q},\vec{p}_2{}') \, .$$

Combining the various contributions, we get the <u>Boltzmann equation</u>:

$$\frac{\partial}{\partial t} f_t(\vec{q},\vec{p}_1) = -\frac{\vec{p}_1}{m} \cdot \frac{\partial}{\partial q_1} f_t(\vec{q},\vec{p}_1)$$

(4)

$$+ \, nd^2 \int_{\hat{\omega}\cdot(\vec{p}_1-\vec{p}_2)\geq 0} d\vec{p}_2 \, d\hat{\omega} \left(\frac{\vec{p}_1-\vec{p}_2}{m}\right)\cdot\hat{\omega}\,[f_t(\vec{q},\vec{p}_1{}')f_t(\vec{q},\vec{p}_2{}') - f_t(\vec{q},\vec{p}_1)f_t(\vec{q},\vec{p}_2)]$$

This equation provides an excellent approximation in many physical situations. Analytically, it is not easy to manage, and the question of existence and uniqueness of solutions is not completely settled. It has a number of elementary formal properties. For example:

$$\int d\vec{q} \, d\vec{p} \, f_t(\vec{q},\vec{p}) \quad \text{and} \quad \int d\vec{q} \, d\vec{p} \, f_t(\vec{q},\vec{p}) \, \vec{p}^{\,2}$$

are time independent. The second of these conservation laws reflects conservation of energy and follows easily from the fact that

$$\vec{p}_1^{\,2} + \vec{p}_2^{\,2} = \vec{p}_1{}'^{\,2} + \vec{p}_2{}'^{\,2} \quad (\text{for all } \hat{\omega}) \, .$$

The form of the equation we have written also conserves momentum, but this would no longer be the case if we included (as we should) terms representing the effects of collision with the walls of the container.

Our derivation of the Boltzmann equation makes sense only in a limiting situation where $n \to \infty$ and hence also $d \to 0$. The presence of a factor nd^2 in the equation means that the limit must be taken in such a way that nd^2 remains finite and non-zero. Since d goes to zero, this means in particular that nd^3 goes to zero, i.e., that the fraction of the volume of the container actually occupied by the particles goes to zero. Thus, in the limit we are considering, the molecular size and the density go to zero simultaneously. It might seem more reasonable to hold nd^3 fixed as $d \to 0$, and indeed this is what would have to be done to study nonequilibrium behavior of matter at non-zero density. Nevertheless, holding nd^2 fixed has a very natural meaning, as the following calculation indicates: The <u>mean free path</u> λ in a gas may be estimated by setting the volume of a circular cylinder of height λ and radius d equal to the average volume available per particle:

$$\pi d^2 \lambda \approx \frac{V(\Lambda)}{n} \; ; \qquad \lambda \simeq \frac{V(\Lambda)}{n \pi d^2}$$

Fixing nd^2 corresponds therefore to fixing λ, so the limit in which the Boltzmann equation ought to hold is one in which the molecular size is very small compared to the mean free path. The importance of the limit with nd^2 fixed was emphasized by Grad, and we will refer to it as the <u>Boltzmann-Grad limit</u>.

(It is slightly annoying to have a factor of $V(\Lambda)$ in the expression for λ but not in the Boltzmann equation, as this suggests that something peculiar happens as $V(\Lambda) \to \infty$. The problem actually lies in the normalization

$$\int f\, d\vec{q}\, d\vec{p} = 1$$

chosen for f. If we are interested in the behavior as $V(\Lambda) \to \infty$, it is preferable to normalize by

$$\int f\, d\vec{q}\, d\vec{p} = V(\Lambda) \ ;$$

making this change introduces exactly the desired $V(\Lambda)^{-1}$ in the collision term of the Boltzmann equation. To avoid having to make the change of normalization, it will be convenient to choose units so that $V(\Lambda) = 1$.)

One of the most important facts about the Boltzmann equation is contained in

<u>Boltzmann's H-Theorem</u>. For any positive function $f(\vec{q},\vec{p})$ on $\Lambda \times \mathbb{R}^3$, define

$$H(f) = \int d\vec{q}\, d\vec{p}\, f(\vec{q},\vec{p}) \log f(\vec{q},\vec{p}) \ .$$

Then if $f_t(\vec{q},\vec{p})$ is a solution of the Boltzmann equation,

$$\frac{d}{dt} H(f_t) \le 0$$

and strict inequality holds unless $f = g(\vec{q}) \exp[-\beta(\vec{p}-\vec{p}_0)^2]$ when β, $\vec{p}_0$ may depend on $\vec{q}$ but not on $\vec{p}$.

We give a proof of this result in Appendix 1 to this chapter. To interpret it, let us assume that f does not depend on $\vec{q}$. Then $dH/dt < 0$ unless f is Maxwellian, in which case $\partial f/\partial t = 0$. Moreover, if we assume

$$\int \vec{p}^{\,2} f_0(\vec{p})\, d\vec{p} < \infty$$

then conservation of energy and a simple convexity argument (again see Appendix 1) give a lower bound for $H(t)$ which holds for all time, so H cannot decrease indefinitely. This makes it very plausible that

$$\lim_{t\to\infty} f_t$$

exists and is Maxwellian. Thus, the H Theorem almost implies that any solution of the Boltzmann equation approaches a Maxwellian momentum distribution as time goes to infinity.

The inequality $dH/dt < 0$ shows that the reversibility of the underlying molecular dynamics has been lost in passing to the Boltzmann equation. The irreversibility must have been introduced in the Hypothesis of Molecular Chaos since the rest of the derivation was straightforward mechanics. Indeed, it is not hard to see directly that the Hypothesis of Molecular Chaos is asymmetric in time; it gives a formula for the number of pairs of particles that are about to collide. If we write down an analogous expression for the number of pairs which have just undergone collision and repeat the argument, we obtain the Boltzmann collision term but with the sign reversed. One conclusion which must be drawn is that something more is involved in the Hypothesis of Molecular Chaos than simple statistical independence.

The quantity H is reminiscent of the information-theoretic entropy (except for a sign); it also has a very suggestive interpretation which was discovered by Boltzmann. To obtain this interpretation, we let $\{\Delta_1,\ldots,\Delta_j\}$ be a finite number of non-overlapping cells in $\Lambda \times \mathbb{R}^3$, and we write $|\Delta_i|$ for the volume of Δ_i which is assumed finite. Now choose a large integer n and integers $\{n_1,n_2,\ldots,n_j\}$ with $\sum_i n_i = n$, and ask how much phase-space volume in $(\Lambda \times \mathbb{R}^3)^n$ is available to phase points $\underline{x}$ with $n_{\Delta_i}(\underline{x}) = n_i$ $(1 \le i \le j)$. A straightforward argument shows that this volume is given by

$$V_n = \frac{n!}{\prod_{i=1}^{j} (n_i)!} \prod_{i=1}^{j} |\Delta_i|^{n_i} \approx \left\{ \prod_{i=1}^{j} \left(\frac{n}{n_i} |\Delta_i|\right)^{n_i/n} \right\}^n$$

(where the last approximate equality was obtained by approximating each $n_i!$ by $n_i^{n_i} e^{-n_i}$). Thus, for large n,

$$\frac{1}{n} \log V_n \approx - \sum_{i=1}^{j} \frac{n_i}{n} \log \left(\frac{n_i}{n} \frac{1}{|\Delta_i|} \right)$$

If, now, we take the cells $\{\Delta_1, \ldots, \Delta_j\}$ to be small but numerous (so that they fill up "most" of $\Lambda \times \mathbb{R}^3$), and if we take

$$n_i \approx n \cdot \int_{\Delta_i} f(\vec{q},\vec{p}) \, d\vec{q} \, d\vec{p}$$

the sum on the right becomes approximately $-\int f \log f \, dx = - H(f)$. By refining this argument, it can be shown that the phase space volume available to phase points with occupation numbers given by f is asymptotic (for large n) to $\exp[-n \, H(f)]$. Decrease of H with time thus has a natural physical interpretation; it means that the system moves from regions of lower phase-space volume to "more probable" regions with enormously much larger volume.

Let us now survey where we are. We set out to investigate whether any useful predictions can be made in a reduced description of the system in terms of occupation numbers. We saw that such predictions can at best be statistical in nature, and the derivation of the Boltzmann equation based on the Hypothesis of Molecular Chaos offers a very plausible guess for what the most likely motion of the occupation numbers will be, at least in the Boltzmann-Grad limit. The H-Theorem, however, shows that the reversibility of the molecular dynamics does not hold for the Boltzmann equation and therefore suggests that the derivation of the Boltzmann equation requires close examination. We will give a more controllable derivation in the next chapter; for this purpose, we will need the rudiments of a theory of the time evolution of statistical states of many-particle systems.

We will first discuss particles interacting by smooth forces, rather than hard spheres, since the algebraic manipulations involved are roughly the same in the two cases but the hard sphere system involves a number of technicalities not present for smooth forces. Consider an initial statistical state μ of n particles, i.e., a probability measure on $(\Lambda \times \mathbb{R}^3)^n$. We assume that μ is symmetric under interchange of particles and absolutely continuous with respect to Lebesgue measure and hence can be written

$$\mu(x_1,\ldots,x_n)\; dx_1\ldots dx_n \;.$$

If we let the n-particle system evolve in time, we get a time-dependent statistical state

$$\mu_t(\underline{x}) = \mu(T^{-t}\underline{x})$$

Differentiating with respect to time gives Liouville's equation:

$$\begin{aligned}\frac{\partial}{\partial t}\mu_t(\vec{q}_1,\vec{p}_1,\ldots,\vec{q}_n,\vec{p}_n) &= -\sum_{i=1}^{n}\left\{\frac{\partial\mu_t}{\partial\vec{q}_i}\frac{d\vec{q}_i}{dt}+\frac{\partial\mu_t}{\partial\vec{p}_i}\frac{d\vec{p}_i}{dt}\right\}\\ &= -\sum_{i=1}^{n}\frac{\vec{p}_i}{m}\frac{\partial\mu_t}{\partial\vec{q}_i}+\sum_{i\neq j}\frac{\partial\Phi}{\partial\vec{q}_i}(\vec{q}_i-\vec{q}_j)\frac{\partial\mu_t}{\partial\vec{p}_i}\equiv H_n\,\mu_t\,.\end{aligned}\tag{5}$$

The last equality defines the first order linear differential operator H_n, which we will call the n-particle Liouville opeator.

Because μ_t is a function of a very large number of variables, it contains more information than we can efficiently use. We therefore integrate out most of the variables to obtain the so-called correlation functions. To begin with, we define

$$\rho_1(t;x_1) = n\int_{(\Lambda\times\mathbb{R}^3)^{n-1}} dx_2\ldots dx_n\;\mu_t(x_1,x_2,\ldots,x_n)$$

To interpret this quantity, note that for any subset Δ of $\Lambda\times\mathbb{R}^3$ the probability of finding particle 1 in Δ at time t is

$$\int_\Delta \frac{1}{n} \rho_1(t;x_1)\, dx_1$$

But by symmetry all particles are equally likely to be in Δ, so

$$\int_\Delta \rho_1(t;x)\, dx_1$$

is the mean number of particles in Δ at time t, i.e., the mean value of n_Δ with respect to μ_t. Similarly, if we define

$$\begin{aligned}\rho_j(t;x_1,\ldots,x_j) &= n(n-1)\ldots(n-j+1) \\ &\cdot \int dx_{j+1}\ldots dx_n\, \mu_t(x_1,\ldots,x_j,x_{j+1},\ldots,x_n)\end{aligned} \tag{6}$$

then

$$\int_{\Delta\times\cdots\times\Delta} dx_1\ldots dx_j\, \rho_j(t;x_1,\ldots,x_j)$$

is the mean value of $n_\Delta(n_\Delta-1)\ \ldots\ (n_\Delta-j+1)$ at time t. (Because of the symmetry of $\rho_j(t,x_1,\ldots,x_j)$, it is uniquely determined up to sets of measure zero by its integral over sets of the form $\Delta\times\ldots\times\Delta$.) Hence, if we know both ρ_1 and ρ_2 at all times, we know both the mean values and the mean square fluctuations of all the occupation number observables n_Δ.

We therefore want to determine the time dependence of the ρ_j's. We can obtain an equation of motion by integrating Liouville's equation over some of the variables, viz.,

$$\begin{aligned}\frac{\partial}{\partial t} \rho_j(t;x_1,\ldots,x_j) &= n(n-1)\ldots(n-j+1) \int dx_{j+1}\ldots dx_n \\ &\cdot\left\{-\sum_{i=1}^{n} \frac{\vec{p}_i}{m} \frac{\partial \mu_t}{\partial \vec{q}_i} + \sum_{i\neq j} \frac{\partial \Phi}{\partial \vec{q}_i}(\vec{q}_i-\vec{q}_j) \frac{\partial \mu_t}{\partial \vec{p}_i}\right\} .\end{aligned} \tag{7}$$

The right-hand side can be considerably simplified and expressed in terms of ρ's to give

$$\frac{\partial}{\partial t}\rho_j(t;x_1,\ldots,x_j) = -\sum_{i=1}^{j}\frac{\vec{p}_i}{m}\frac{\partial\rho_j(t)}{\partial\vec{q}_i} + \sum_{i\neq k}\frac{\partial\Phi(\vec{q}_i-\vec{q}_k)}{\partial\vec{q}_i}\frac{\partial\rho_j(t)}{\partial p_i} \tag{8}$$
$$+\sum_{i=1}^{j}\int dx_{j+1}\frac{\partial\Phi(\vec{q}_i-\vec{q}_{j+1})}{\partial\vec{q}_i}\cdot\frac{\partial}{\partial\vec{p}_i}\rho_{j+1}(t)\ .$$

(For the calculation, see Appendix 2.) The first two terms on the right may be recognized as $H_j\rho_j(t)$, where H_j is the j-particle Liouville operator. The second term will be denoted by $C_{j,j+1}\rho_{j+1}(t)$; then we can rewrite the equation of motion (8) as

$$\frac{\partial}{\partial t}\rho_j(t) = H_j\rho_j(t) + C_{j,j+1}\rho_{j+1}(t)\ . \tag{9}$$

Note that n does not occur explicitly in these equations so they are formally well adapted to passing to the thermodynamic limit $n \to \infty$, $V(\Lambda) \to \infty$, or, with appropriate adjustments, to the Boltzmann-Grad limit. On the other hand, the equations are not quite as nice as might have been hoped, as they do not permit the computation of the first few time-dependent correlation functions without determining all the rest -- the equation of motion for ρ_1 involves ρ_2 , that for ρ_2 involves ρ_3 , etc. In fact

$$\rho_n(t;x_1,\ldots,x_n) = n!\ \mu_t(x_1,\ldots,x_n)$$

$$\rho_{n+1}(t;x_1,\ldots,x_{n+1}) \equiv 0$$

so the equation for $j = n$ is exactly Liouville's equation. Nevertheless, we will find this system of equations useful in organizing the behavior in the Boltzmann-Grad limit. The system (8) is called the <u>BBGKY hierarchy</u>, after Bogoliubov, Born, Green, Kirkwood, and Yvon who, among others, discovered it.

We have next to describe the BBGKY hierarchy for hard spheres. From the usual point of view, the hard sphere time evolution is discontinuous, and hence an analysis of the above sort involving differentiation with respect to time would appear to be impossible. Much of the difficulty can, however, be removed by redefining the hard sphere phase space. When two particles collide, their momenta change instantaneously from incoming to outgoing. Thus, two phase points $\underline{x}^{(1)}$ and $\underline{x}^{(2)}$ which differ only by having an incoming collision configuration of some pair of particles replaced by the corresponding outgoing configuration should really be thought of as the <u>same</u> phase point. In other words, we want to form a new phase space by identifying such pairs of phase points in the original "naive" phase space. It is not hard to see that this can be done in such a way as to make the new phase space a smooth manifold and the time evolution a smooth flow, provided that phase points corresponding to triple and higher collisions, and grazing collisions, are deleted. A number of problems remain; although the flow is smooth, not all orbits can be extended indefinitely in time (as some of them lead to the excluded triple collision points). Such orbits are, however, sufficiently infrequent so that the time evolution of statistical states is given by a version of Liouville's equation. If we proceed to integrate out some of the variables, we obtain the BBGKY hierarchy for hard spheres with the same general form as for particles with smooth forces, i.e.

$$(10) \quad \frac{\partial}{\partial t} \rho_j(t) = H_j \rho_j(t) + C_{j,j+1} \rho_{j+1}(t) \ , \qquad j = 1,2,3,\ldots,n \ ,$$

where H_j is the j-particle Liouville operator (including the effects of elastic collisions between these particles) and

$$(11) \quad \begin{aligned} (C_{j,j+1}\rho_{j+1})(x_1,\ldots,x_j) &= d^2 \sum_{i=1}^{j} \int d\vec{p}_{j+1} \, d\hat{\omega} \, \hat{\omega} \cdot \left(\frac{\vec{p}_{j+1} - \vec{p}_i}{m} \right) \\ &\times \rho_{j+1}(x_1,\ldots,x_j, \vec{q}_i + d \cdot \hat{\omega}, \vec{p}_{j+1}) \ . \end{aligned}$$

As before, the $\hat{\omega}$ integral is over the unit sphere. We omit the formal proof of this equation; it is given in Section II of [6].

We note that the above equation for $j = 1$ can be made to look very much like the Boltzmann equation. First of all, the one-particle Liouville operator is simply

$$H_1 = -\frac{\vec{p}_1}{m} \cdot \frac{\partial}{\partial \vec{q}_1}$$

Next, we rewrite the collision term (11) by splitting the integral over $\hat{\omega}$ into the hemisphere $\hat{\omega}\cdot(\vec{p}_2-\vec{p}_1) \geq 0$ and the hemisphere $\hat{\omega}\cdot(\vec{p}_2-\vec{p}_1) \leq 0$. In the first hemisphere, the phase point

$$(\vec{q}_1, \vec{p}_1, \vec{q}_1+d\cdot\hat{\omega}, \vec{p}_2)$$

is an outgoing collision point which can be replaced by the incoming collision point

$$(\vec{q}_1, \vec{p}_1{}', \vec{q}_1+d\cdot\hat{\omega}, \vec{p}_2{}') \quad ; \quad (\vec{p}_1{}', \vec{p}_2{}') = T_{\hat{\omega}}(\vec{p}_1, \vec{p}_2)$$

(since these two are really just different representations of the same phase point). Finally, in this hemisphere we replace $\hat{\omega}$ as integration variable by $-\hat{\omega}$. Thus, we obtain

$$\frac{\partial}{\partial t}\rho_1(t,x_1) = -\frac{\vec{p}_1}{m}\frac{\partial\rho_1}{\partial\vec{q}} + d^2 \int\limits_{\hat{\omega}\cdot(\vec{p}_1-\vec{p}_2)\geq 0} d\vec{p}_2 \, d\hat{\omega} \, \hat{\omega}\cdot\left(\frac{\vec{p}_1-\vec{p}_2}{m}\right) \tag{12}$$

$$\cdot\left\{\rho_2(t;\vec{q}_1,\vec{p}_1{}',\vec{q}_1-d\cdot\hat{\omega},\vec{p}_2{}') - \rho_2(t;\vec{q}_1,\vec{p}_1,\vec{q}_1+d\hat{\omega},\vec{p}_2)\right\}.$$

This goes over to the Boltzmann equation if we

(a) replace $\rho_2(t;x_1,x_2)$ on the right by $\rho_1(t;x_1)\rho_1(t;x_2)$

(b) let d go to zero

(c) rescale by writing $\rho_1(t;x) = n\,f(t;x)$ (which replaces d^2 by nd^2 in the collision term)

Thus it would appear that the Boltzmann equation is obtainable as a limit of the BBGKY hierarchy provided that the factorization assumption (a) can be justified. For a careful discussion of this formal limit, see [1]. It must be pointed out, however, that the factorization assumption, like the Hypothesis of Molecular Chaos to which it is evidently related, is more subtle than it may appear. We obtained (12) by systematically writing collision phase points in their incoming representations. We could equally well have written them systematically in their outgoing representations; if we then assumed factorization we would have obtained the Boltzmann collision term with its sign reversed. It is thus essential, in order to get the Boltzmann equation, to assume

$$\rho_2(x_1,x_2) = \rho_1(x_1)\rho_1(x_2)$$

for incoming collision points (x_1,x_2) and not for outgoing ones.

In the following section, we will find it useful to approach the relation between the Boltzmann equation and the BBGKY hierarchy from another direction by associating a linear hierarchy with the Boltzmann equation. Let $f_1(t,x)$ be a solution of the Boltzmann equation, and define

$$f_j(t;x_1,\ldots,x_j) = \prod_{i=1}^{j} f_1(t;x_i) \,. \tag{13}$$

It is a simple matter to verify that this sequence of functions satisfies the linear hierarchy

$$\frac{\partial}{\partial t} f_j(t) = H_j^{(0)} f_j(t) + C_{j,j+1}^{(0)} f_{j+1}(t) \tag{14}$$

where

$$H_j^{(0)} = -\sum_{i=1}^{j} \frac{\vec{p}_i}{m} \frac{\partial}{\partial \vec{q}_i}$$

and

$$\begin{aligned}(C_{j,j+1} f_{j+1})(x_1,\ldots,x_j) &= nd^2 \sum_{i=1}^{j} \int_{\hat{\omega}\cdot(\vec{p}_i-\vec{p}_{j+1})\geq 0} d\hat{\omega}\, d\vec{p}_{j+1} \\ &\times \hat{\omega}\cdot\left(\frac{\vec{p}_i-\vec{p}_{j+1}}{m}\right) \times \Big\{ f_{j+1}(x_1,\ldots,\vec{q}_i,\vec{p}_i{}',\ldots,\vec{q}_i,\vec{p}_{j+1}{}') \\ &\qquad - f_{j+1}(x_1,\ldots,\vec{q}_i,\vec{p}_i,\ldots,\vec{q}_i,\vec{p}_{j+1}) \Big\} \,;\end{aligned} \tag{15}$$

$$(\vec{p}_i{}',\vec{p}_{j+1}{}') = T_{\hat{\omega}}(\vec{p}_i,\vec{p}_{j+1}) \,.$$

Conversely, if (13) gives a solution of the hierarchy, then $f_1(t,x_1)$ is a solution of the Boltzmann equation. Thus, the nonlinear Boltzmann equation is equivalent to the linear hierarchy (14) plus the factorization condition (13). We will refer to (14) as the Boltzmann hierarchy.

Appendix 1. The H-Theorem.

We have to show that

$$\frac{d}{dt} H(f_t) \leq 0$$

if $f_t(\vec{q},\vec{p})$ is a solution of the Boltzmann equation. Differentiating the formula for $H(f)$, and using

$$\int dx \frac{\partial f_t}{\partial t}(x) = 0,$$

we get

$$\frac{d}{dt} H(f_t) = \int_{\Lambda\times\mathbb{R}^3} dx \frac{\partial f_t}{\partial t} \log (f_t) .$$

We now simply insert the right-hand side of the Boltzmann equation for $\partial f/\partial t$. An integration by parts shows that the term $-\frac{\vec{p}}{m} \cdot \frac{\partial f}{\partial \vec{q}}$ contributes nothing.

(Actually, there is a technical point here about boundary conditions which should be mentioned at least once. We have

$$\int_{\Lambda\times\mathbb{R}^3} dx\, \vec{p} \cdot \frac{\partial f}{\partial \vec{q}} \log f = \int_{\mathbb{R}^3} d\vec{p}\, \vec{p} \cdot \int_{\Lambda} d\vec{q} \frac{\partial}{\partial \vec{q}} \{f \log f - f\}$$

$$= \int_{\mathbb{R}^3} d\vec{p} \int_{\partial\Lambda} d\sigma(\vec{q})\, \vec{p}\cdot\hat{n} \{f \log f - f\} ,$$

where $d\sigma(\vec{q})$ means surface area integration over $\partial\Lambda$ and $\hat{n}$ is the unit external normal to Λ. This integral is supposed to be zero. If Λ is a torus, so $\partial\Lambda$ is empty, there is evidently no problem. If, however, Λ is an ordinary region with smooth boundary, then the equations of motion must be supplemented by a condition of elastic reflection at $\partial\Lambda$. Thus, if $\vec{q} \in \partial\Lambda$, and if $\vec{p}, \vec{p}'$ are vectors which differ only in the sign of their components orthogonal to $\partial\Lambda$ at $\vec{q}$, then $(\vec{q},\vec{p})$ and $(\vec{q},\vec{p}')$ should be viewed as different labellings of the same phase point. Therefore

$$f(\vec{q},\vec{p}) = f(\vec{q},\vec{p}')$$

and

$$\int d\vec{p}\ \hat{n}\cdot\vec{p}\ \{f \log f - f\} = 0$$

since the contribution to the integral from any $\vec{p}$ with $\hat{n}\cdot\vec{p} > 0$ is cancelled by the contribution from the corresponding $\vec{p}'$.)

We thus get

$$\frac{d}{dt} H(f_t) = nd^2 \int_{\hat{\omega}\cdot(\vec{p}_1-\vec{p}_2)\geq 0} d\hat{\omega}\ d\vec{q}\ d\vec{p}_1\ d\vec{p}_2\ \hat{\omega}\cdot\left(\frac{\vec{p}_1-\vec{p}_2}{m}\right) [f_1'f_2'-f_1f_2] \log f_1$$

where

$$f_1 = f_t(\vec{q},\vec{p}_1)\ , \qquad f_2 = f_t(\vec{q},\vec{p}_2)$$

$$f_1' = f_t(\vec{q},\vec{p}_1')\ , \qquad f_2' = f_t(\vec{q},\vec{p}_2')$$

$$(\vec{p}_1',\vec{p}_2') = T_{\hat{\omega}}(\vec{p}_1,\vec{p}_2)\ .$$

Straightforward changes of variables show that the $\log f_1$ on the right may be replaced by $\log f_2$, $-\log f_1'$, or $-\log f_2'$ without changing the integral. For example

$$\int_{\hat{\omega}\cdot(\vec{p}_1-\vec{p}_2)\geq 0} d\hat{\omega}\ d\vec{p}_1\ d\vec{p}_2\ \hat{\omega}\cdot(\vec{p}_1-\vec{p}_2)\ [f_1'f_2'-f_1f_2] \log f_1'$$

$$= - \int_{\hat{\omega}\cdot(\vec{p}_1'-\vec{p}_2')\leq 0} d\hat{\omega}\ d\vec{p}_1'\ d\vec{p}_2'\ \hat{\omega}\cdot(\vec{p}_1'-\vec{p}_2')\ [f_1'f_2'-f_1f_2] \log f_1'$$

(using $\vec{p}_1', \vec{p}_2'$ rather than $\vec{p}_1, \vec{p}_2$ as integration variables and noting $d\vec{p}_1\ d\vec{p}_2 = d\vec{p}_1'\ d\vec{p}_2'$)

$$= \int_{\hat{\omega}\cdot(\vec{p}_1-\vec{p}_2)\geq 0} d\hat{\omega}\, d\vec{p}_1\, d\vec{p}_2\, \hat{\omega}\cdot(\vec{p}_1-\vec{p}_2)\, [f_1 f_2 - f_1' f_2']\, \log f_1$$

(renaming the new integration variables $\vec{p}_1$ and $\vec{p}_2$; replacing $\hat{\omega}$ by $-\hat{\omega}$, and using the fact that $(\vec{p}_1,\vec{p}_2) = T_{\hat{\omega}}(\vec{p}_1{}',\vec{p}_2{}')$.)

Therefore (adding together the four expressions for $\frac{dH(f_t)}{dt}$ and dividing by 4)

$$\frac{d}{dt} H(f_t) = \frac{nd^2}{4} \int_{\hat{\omega}\cdot(\vec{p}_1-\vec{p}_2)\geq 0} d\hat{\omega}\, d\vec{p}_1\, d\vec{p}_2\, d\vec{q}\, \hat{\omega}\cdot\left(\frac{\vec{p}_1-\vec{p}_2}{m}\right)$$
$$\times\ [f_1'f_2'-f_1f_2]\ [\log(f_1f_2)-\log(f_1'f_2')]\ .$$

Since the logarithm is strictly increasing

$$[f_1'f_2'-f_1f_2]\ [\log\,(f_1f_2) - \log\,(f_1'f_2')] < 0 \quad \text{unless} \quad f_1'f_2' = f_1f_2\ .$$

Hence

$$\frac{d}{dt} H(f_t) < 0 \quad \text{unless} \quad f_1'f_2' = f_1f_2 \quad \text{for almost all } \vec{q},\hat{\omega},\vec{p}_1,\vec{p}_2.$$

To complete the proof, we must show that the identity

$$f_1'f_2' = f_1f_2 \quad \text{for almost all} \quad \vec{q},\hat{\omega},\vec{p}_1,\vec{p}_2$$

implies the Maxwellian form for f. To simplify the argument we will assume that $g = \log f$ is smooth. The identity then holds for <u>all</u> $\vec{q}$, $\hat{\omega}$, $\vec{p}_1$, $\vec{p}_2$; since the dependence on $\vec{q}$ will not enter the argument, we will suppress $\vec{q}$ from the notation. Taking logarithms, we have

$$g(\vec{p}_1{}') + g(\vec{p}_2{}') = g(\vec{p}_1) + g(\vec{p}_2)$$

whenever $(\vec{p}_1{}',\vec{p}_2{}') = T_{\hat{\omega}}(\vec{p}_1,\vec{p}_2)$ for some $\hat{\omega}$. It is easy to see that, given $(\vec{p}_1,\vec{p}_2)$ and $(\vec{p}_1{}',\vec{p}_2{}')$, there exists $\hat{\omega}$ such that $T_{\hat{\omega}}(\vec{p}_1,\vec{p}_2) = (\vec{p}_1{}',\vec{p}_2{}')$ if and only if

$$\vec{p}_1 + \vec{p}_2 = \vec{p}_1{}' + \vec{p}_2{}' \ ; \quad \vec{p}_1^{\,2} + \vec{p}_2^{\,2} = \vec{p}_1{}'^{\,2} + \vec{p}_2{}'^{\,2}\ .$$

Thus

$$g(\vec{p}_1+\delta\vec{p}) + g(\vec{p}_2-\delta\vec{p}) = g(\vec{p}_1) + g(\vec{p}_2)$$

provided

$$(\vec{p}_1+\delta\vec{p})^2 + (\vec{p}_2-\delta\vec{p})^2 = \vec{p}_1^{\,2} + \vec{p}_2^{\,2}$$

Taking $\delta\vec{p}$ "infinitesimal",

$$\delta\vec{p}\cdot(\text{grad } g)(\vec{p}_1) = \delta\vec{p}\cdot(\text{grad } g)(\vec{p}_2)$$

if

$$\delta\vec{p}\cdot\vec{p}_1 = \delta\vec{p}\cdot\vec{p}_2 \ .$$

This means that $\delta\vec{p}\cdot\text{grad } g(\vec{p})$ does not change as $\vec{p}$ changes in a direction perpendicular to $\delta\vec{p}$, i.e.,

$$\sum_{i,j=1}^{3} \frac{\partial^2 g}{\partial p_i \partial p_j} \delta p_i \; \delta p_j' = 0 \quad \text{provided} \quad \delta\vec{p}\cdot\delta\vec{p}\,' = 0 \ .$$

Thus

$$\frac{\partial^2 g}{\partial p_i \partial p_j} = \phi(p)\cdot\delta_{ij} \tag{16}$$

and in particular $\partial g/\partial p_i$ depends only on p_i, not on p_j for $j \neq i$. Writing

$$\frac{\partial g}{\partial p_i} = h_i(p_i) \ , \qquad i = 1,2,3,$$

we conclude from (16) that

$$h_i'(p_i) = h_j'(p_j) \quad \text{for} \quad i \neq j \ .$$

Since the two sides are functions of different variables, all h_i' must equal a single constant, which we will denote by 2β. Then

$$\frac{\partial g}{\partial p_i} = 2\beta p_i + \gamma_i \qquad (\gamma_i \text{ constant})$$

and, finally

$$g(\vec{p}) = \beta\vec{p}^{\,2} + \vec{\gamma}\cdot\vec{p} + \alpha$$

as desired.

We must still obtain a lower bound on H.

Let

$$h_0(\vec{q},\vec{p}) = c_1 \exp [-c_2\vec{p}^2]$$

where the constants c_1, c_2 are chosen so that

$$\int d\vec{q}\, d\vec{p}\, h_0(\vec{q},\vec{p}) = 1 \; ; \qquad \int d\vec{q}\, d\vec{p}\, \vec{p}^2\, h_0(\vec{q},\vec{p}) = \int d\vec{q}\, d\vec{p}\, \vec{p}^2\, f_t(\vec{q},\vec{p}) \; .$$

Then

$$H(f_t) = \int f_t \log f_t \, dx = \int f_t \log (f_t/h_0)\, dx + \int f_t \log h_0 \, dx \; .$$

Now

$$\int f_t \log (f_t/h_0)\, dx = \int (f_t/h_0) \log (f_t/h_0)\, h_0 \, dx$$

$$\geq \left[\int (f_t/h_0)\, h_0 \, dx\right] \log \left[\int (f_t/h_0)\, h_0 \, dx\right] = 0 \; ,$$

since the function $y \log y$ is convex.* Also

$$\int f_t \log h_0 \, dx = \log (c_1) \cdot \int f_t \, dx - c_1 \int \vec{p}^2 f_t \, dx$$

$$= \log (c_1) \int h_0 \, dx - c_1 \int \vec{p}^2 h_0 \, dx = \int h_0 \log h_0 dx .$$

Thus

$$H(f_t) \geq H(h_0)$$

for all t.

---

* We are assuming $\int dx\, f_t = 1$.

Appendix 2. The BBGKY Hierarchy.

We start from the equation

$$\frac{\partial}{\partial t}\rho_j(t;x_1,\dots,x_j) = \frac{n!}{(n-j)!}\int_{(\Lambda\times\mathbb{R}^3)^{n-j}} dx_{j+1}\cdots dx_n$$

$$\cdot\left\{-\sum_{i=1}^{n}\frac{\vec{p}_i}{m}\frac{\partial\mu_t}{\partial\vec{q}_i}+\sum_{i\neq k}\frac{\partial\Phi(\vec{q}_i-\vec{q}_k)}{\partial\vec{q}_i}\frac{\partial\mu_t}{\partial\vec{p}_i}\right\}$$

Many of the terms on the right are zero. For example each term $\frac{\vec{p}_i}{m}\cdot\frac{\partial\mu}{\partial\vec{q}_i}$ with $i > j$ contributes nothing since

$$\int_{\mathbb{R}^3} d\vec{p}_i \int_{\Lambda} d\vec{q}_i\ \vec{p}_i\cdot\frac{\partial\mu}{\partial\vec{q}_i} = 0$$

(See the remarks on boundary conditions in Appendix 1.) Similarly, each term in the second sum with $i > j$ contributes nothing. For $1 \le i \le j$,

$$\frac{n!}{(n-j)!}\int dx_{j+1}\cdots dx_n \frac{\vec{p}_i}{m}\frac{\partial}{\partial\vec{q}_i}\mu_t = \frac{\vec{p}_i}{m}\frac{\partial}{\partial\vec{q}_i}\frac{n!}{(n-j)!}\int dx_{j+1}\cdots dx_n\,\mu_t$$

$$= \frac{\vec{p}_i}{m}\frac{\partial}{\partial\vec{q}_i}\rho_j(t;x_1,\dots,x_j)\ ,$$

and for $1 \le i \le j$; $1 \le k \le j$,

$$\frac{n!}{(n-j)!}\int dx_{j+1}\cdots dx_n \frac{\partial\Phi(\vec{q}_i-\vec{q}_k)}{\partial\vec{q}_i}\frac{\partial\mu}{\partial\vec{p}_i} = \frac{\partial\Phi(\vec{q}_i-\vec{q}_k)}{\partial\vec{q}_i}\frac{\partial}{\partial\vec{p}_i}\rho_j(t;x_1,\dots,x_j).$$

Thus,

$$\frac{\partial}{\partial t}\rho(t;x_1,\dots,x_j) = \left\{-\sum_{i=1}^{j}\frac{\vec{p}_i}{m}\frac{\partial}{\partial\vec{q}_i}+\sum_{i\neq k}\frac{\partial\Phi(\vec{q}_i-\vec{q}_k)}{\partial\vec{q}_i}\frac{\partial}{\partial\vec{p}_i}\right\}\rho_j(t;x_1,\dots x_j)$$

$$+\frac{n!}{(n-j)!}\sum_{i=1}^{j}\sum_{k=j+1}^{n}\int dx_{j+1}\cdots dx_n\frac{\partial\Phi(\vec{q}_i-\vec{q}_k)}{\partial\vec{q}_i}\frac{\partial}{\partial\vec{p}_i}\mu_t$$

To rework the last expression, we observe that, because of the symmetry of μ, each term in the sum over k makes the same contribution. Hence, the expression may be rewritten

$$\sum_{i=1}^{j} \int dx_{j+1} \frac{\partial \Phi(\vec{q}_i - \vec{q}_{j+1})}{\partial \vec{q}_i} \frac{\partial}{\partial \vec{p}_i} \left\{ \frac{n!}{(n-j)!} (n-j) \int dx_{j+2} \ldots dx_n \, \mu_t \right\}$$

$$= \sum_{i=1}^{j} \int dx_{j+1} \frac{\partial \Phi(\vec{q}_i - \vec{q}_{j+1})}{\partial \vec{q}_i} \frac{\partial}{\partial \vec{p}_i} \rho_{j+1}(t; x_1, \ldots, x_{j+1})$$

completing the proof of (8).

References to Chapter 6

The literature on the Boltzmann equation is extensive. Among the standard works are:

[1] Grad, H., Principles of the Kinetic Theory of Gases, in Handbuch der Physik, ed. S. Flügge, Vol. 12, Berlin, Göttingen, Heidelberg: Springer Verlag, 1958.

[2] Cercignani, C., Mathematical Methods in Kinetic Theory, New York: Plenum Press, 1969.

For the mathematical theory of the Boltzmann equation, see

[3] Carleman, T., Problèmes Mathematiques dans la Théorie Cinétique des Gaz, Uppsala, 1957.

[4] Povsner, A. Ja., The Boltzmann Equation in the Kinetic Theory of Gases, Mat. Sb. 58 (1962) 63-86, translated in Amer. Math. Soc. Translations, Ser. 2 47 (1965) 193-216.

A proof of approach to equilibrium for a simplified model of the Boltzmann equation which makes precise Boltzmann's argument using the H-Theorem is given in:

[5] McKean, H.P., Jr., Speed of Approach to Equilibrium for Kac's Caricature of a Maxwellian Gas, Arch. Rat. Mech. Anal. 21 (1966) 343-367.

A discussion similar to ours of the formal relation between the BBGKY hierarchy and the Boltzmann equation may be found in:

[6] Cercignani, C., On the Boltzmann Equation for Rigid Spheres, Transport Theory and Statistical Physics 2 (1972) 211-225.

7. The Boltzmann-Grad Limit of Time Dependent Correlation Functions

Summary. We study the limiting behavior of properly normalized time-dependent correlation functions as $n \to \infty$ and $d \to 0$ with nd^2 fixed and show that convergence at time zero implies convergence over a small interval of strictly positive times to a solution of the Boltzmann hierarchy.

....

In the preceding chapter we explored, in a formal way, the relation between the BBGKY hierarchy in the Boltzmann-Grad limit and the Boltzmann equation. In this chapter we will formulate and prove an exact statement about this relation. In outline, what we show is the following: Consider a sequence of particle numbers n converging to infinity and particle diameters d converging to zero; we assume that nd^2 is constant or, at least, approaches a finite nonzero limit. We assume we are given, for each d, a statistical state $\mu^{(d)}$ for a system of n particles of diameter d in a fixed container Λ. (It is possible to allow Λ to grow with n; in this case, we fix $nd^2/V(\Lambda)$ rather than nd^2.) We construct the corresponding correlation functions

$$\rho_1^{(d)}(x_1),\ \rho_2^{(d)}(x_1,x_2),\ \ldots$$

These correlation functions must diverge as $d \to 0$ since

$$\int_{(\Lambda\times\mathbb{R}^3)^j} \rho_j^{(d)}(x_1,\ldots,x_j)\,dx_1\ldots dx_j = n(n-1)\ldots(n-j+1) \to \infty$$

We therefore rescale the ρ's by defining

$$f_j^{(d)}(x_1,\ldots,x_j) = n^{-j}\,\rho_j^{(d)}(x_1,\ldots,x_j)\ ; \tag{1}$$

then

$$\lim_{d\to 0} \int_{(\Lambda\times\mathbb{R}^3)^j} f_j^{(d)}(x_1,\ldots,x_j)\,dx_1\ldots dx_j = 1\ .$$

We further assume that, as $d \to 0$, the f's converge:

$$\lim_{d\to 0} f_j^{(d)}(x_1,\ldots,x_j) = f_j^{(0)}(x_1,\ldots,x_j)$$

(The precise sense in which this convergence is assumed to take place will be spelled out later.)

By solving the equations of motion for the n particle system, we obtain time-dependent rescaled correlation functions

$$f_j^{(d)}(t;x_1,\ldots,x_j)$$

Our main result is that, with all the above hypotheses, there is a strictly positive t_0 (of the order of a fraction of a mean free time) such that:

(a) for $0 < t < t_0$, the $f_j^{(d)}(t;x_1,\ldots,x_j)$ converge almost everywhere as $d \to 0$ to limits $f_j^{(0)}(t;x_1,\ldots,x_j)$

(b) the $f_j^{(0)}(t;x_1,\ldots,x_j)$ are generalized solutions of the Boltzmann hierarchy ((14) of Chapter 6)

(c) if the initial $f^{(0)}$'s <u>factorize</u> (i.e., if

$$f_j^{(0)}(x_1,\ldots,x_j) = \prod_{i=1}^{j} f_1^{(0)}(x_i) \;)$$

then the same is true for the $f_j^{(0)}(t;x_1,\ldots,x_j)$, and $f_1^{(0)}(t;x_1)$ is a (generalized) solution of the Boltzmann equation.

In other words, convergence of initial data for the BBGKY hierarchy in the Boltzmann-Grad limit implies convergence of the solutions for small positive times to solutions of the Boltzmann hierarchy, and factorization of the initial data is preserved at least for small positive times. The factorization condition on the limiting rescaled correlation functions has a simple interpretation; it is equivalent to the fact that all the occupation number observables n_Δ/n $(\Delta \subset \Lambda\times\mathbb{R}^3)$ converge in probability to constants with respect to the statistical states $\mu^{(d)}$ as $d \to 0$.

Reformulated very loosely, the above results say the following: Consider a system of a very large number of very small particles, with nd^2 of moderate size. Construct at time zero a statistical state in which the occupation number observables n_Δ/n for reasonably large Δ's have small fluctuations about mean values given approximately by

$$\int_\Delta f_1(x_1)\, dx_1$$

with f_1 a well behaved function on $\Lambda\times\mathbb{R}^3$. Now follow the statistical state in time. For at least a short while, the fluctuations of the occupation numbers will remain small, and the mean values of the occupation numbers will be given by

$$\int_\Delta f_1(t;x_1)\, dx_1$$

where $f_1(t;x_1)$ is a solution of the Boltzmann equation with $f_1(0;x_1) = f_1(x_1)$. Although these results apply only to small positive times, the times involved are large enough for Boltzmann's H function to decrease a strictly positive amount. Thus our results show unambiguously that there is no contradiction between the reversibility of molecular dynamics and the irreversibility implied by the H-Theorem.

From the physical point of view, the restriction of our results to small positive times is a severe limitation, as the interesting applications of the Boltzmann equation involve the long-time behavior of its solutions. Although our argument fails for large times, the final result could well remain true. In fact, it is probably not difficult to extend the proof to arbitrary times if it is <u>assumed</u> that the time-dependent rescaled correlation functions $f_j^{(d)}(t;x_1,\dots,x_j)$ do not develop singularities. This absence of singularities, while plausible, appears to be hard to prove; one would need, in particular, to obtain considerable control over the behavior of fairly general solutions of the spatially inhomogeneous Boltzmann equation.

The central technical device in our proof is a series expansion for solutions of the BBGKY hierarchy and an analogous expansion for the Boltzmann hierarchy. We have seen that the BBGKY hierarchy has the general form

$$\frac{d}{dt} f_j(t) = H_j f_j(t) + C_{j,j+1} f_{j+1}(t) .$$

Regarding the sequence (f_j) of functions as a single vector $\underline{f}$, we can rewrite the equation more schematically as

$$\frac{d}{dt} \underline{f}(t) = H\, \underline{f}(t) + C\, \underline{f}(t) . \tag{2}$$

If we drop the collision term C, we obtain an equation which is easy to solve (in principle); its solution may be written formally as $S(t)\underline{f}(0)$, where

$$(S(t)\underline{f})_j(x_1,\ldots,x_j) = f_j(T^{-t}(x_1,\ldots,x_j)) . \tag{3}$$

To get a formal solution to the equation including C, we use the time-dependent perturbation series

$$\underline{f}(t) = S(t)\underline{f}(0) + \sum_{m=1}^{\infty} \int_0^t dt_1 \int_0^{t_1} dt_2 \ldots \int_0^{t_{m-1}} dt_m \; S(t-t_1)C\, S(t_1-t_2)\ldots C\, S(t_m)\, \underline{f}(0) . \tag{4}$$

A similar formal solution can be written for the Boltzmann hierarchy:

$$\underline{f}^{(0)}(t) = S^{(0)}(t)\, \underline{f}^{(0)}(0) + \sum_{m=1}^{\infty} \int_0^t dt_1 \int_0^{t_1} dt_2 \ldots \int_0^{t_{m-1}} dt_m \; S^{(0)}(t-t_1)\, C^{(0)} \ldots C^{(0)}\, S^{(0)}(t_m)\underline{f}^{(0)}(0) \tag{5}$$

where $S^{(0)}(t)$ is the free-flow operator:

$$(S^{(0)}(t)\underline{f})_j(x_1,\ldots,x_j) = f_j(\vec{q}_1 - t\,\frac{\vec{p}_1}{m}, \vec{p}_1, \vec{q}_2 - t\,\frac{\vec{p}_2}{m}, \vec{p}_2, \ldots \vec{q}_j - t\,\frac{\vec{p}_j}{m}, \vec{p}_j)$$

and $C^{(0)}$ is the Boltzmann collision operator. We will proceed by making estimates on the convergence of the former series (4) which are uniform in d and then showing that this series approaches the latter series (5) term by term as $d \to 0$. We finally need to know that the latter series converges to give a solution of the Boltzmann hierarchy and that the solution factorizes if the initial $\underline{f}^{(0)}$ does. The limitation of our result to small times arises from the fact that we are able to control the convergence of the series only for small times.

In spite of their formal similarity, the two series (4) and (5) have quite different logical standings. In each case, the mth contribution to f_j depends only on f_{j+m}. For the BBGKY hierarchy $f_j \equiv 0$ for $j > n$, so the series is only a finite sum. Moreover, the left-hand side of (4) has been defined in terms of the n-particle dynamics and the problem of proving that (4) holds reduces basically to an algebraic manipulation, which we will not give. (The issue is complicated somewhat by "measure-zero" difficulties arising from the ambiguity in the outcome of a triple collision, and these same difficulties require attention to some details in the definition of S(t). We will ignore these problems; they can be handled in a straightforward way.) In solving the Boltzmann hierarchy, on the other hand, we do not want to assume that the initial $f_j^{(0)}$ vanish identically for large j (since we want in particular to consider factorizing initial data). The convergence of the series (5) therefore requires investigation. Moreover, we do not know whether solutions to the Boltzmann hierarchy exist for general initial data, so we do not know how to assign a meaning to the left-hand side of (5). If, however, we can prove convergence of the series, it seems plausible that the sum gives at least some sort of generalized solution of the hierarchy, and we can therefore regard (5) as a definition of its left-hand side. This is what we will do. It is in fact not difficult to define a notion of generalized solution to the Boltzmann hierarchy and prove, using the estimates to be given below, that the sum of the series (5) is such a solution, but we will not pursue this point. The problem in showing that we obtain an ordinary solution to the hierarchy lies in the fact that our estimates are not well adapted to showing that the series solution is differentiable with respect to the particle positions if the initial $f_j^{(0)}$ are.

What we need, then, is an estimate on the convergence of the series (4) and (5). To express this estimate efficiently, we first define some normed spaces on which the operators $S(t)$, C, will act. For $\beta > 0$, define

$$\sigma_\beta(\vec{p}) = \left(\frac{\beta}{2\pi m}\right)^{3/2} \exp\left[-\frac{\beta}{2m}\vec{p}^{\,2}\right] \tag{6}$$

and for a function $f(x_1,\ldots,x_j)$ on $(\Lambda\times\mathbb{R}^3)^j$, define

$$_j\|f\|_\beta = \sup_{x_1,\ldots,x_j\in\Lambda\times\mathbb{R}^3} \frac{|f(x_1,\ldots,x_j)|}{\prod_{i=1}^{j}\sigma_\beta(\vec{p}_i)} \tag{7}$$

Let $_jY_\beta$ denote the set of all Borel functions on $(\Lambda\times\mathbb{R}^3)^j$ such that $_j\|f\|_\beta < \infty$. Thus, $f \in {_jY_\beta}$ if and only if

$$|f(\underline{x})| \le \text{const} \times \exp[-\beta H(\underline{x})] \text{ for all } \underline{x}$$

where $H(\underline{x})$ denotes the kinetic energy $\sum_i \vec{p}_i^{\,2}/2m$. By conservation of energy

$$_j\|S_j(t)f\|_\beta = {_j\|f\|_\beta},$$

i.e. $S_j(t)$ is a one-parameter group of isometries on $_jY_\beta$ for any

β. Actually, a little care is required here: By definition,

$$(S_j(t)f)(\underline{x}) = f(T^{-t}\underline{x})$$

and the right-hand side makes no sense unless $\underline{x} = (\vec{q}_1, \vec{p}_1, \ldots, \vec{q}_j, \vec{p}_j)$ is an allowed j-particle phase point, i.e., unless $|\vec{q}_i - \vec{q}_k| \geq d$ for $i \neq k$. Thus, we should regard the f's as defined only where this condition is satisfied. For some purposes, it is convenient to put $f \equiv 0$ when the condition is not satisfied and to build this restriction into the definition of ${}_jY_\beta$. Thus the ${}_jY_\beta$, as well as $S_j(t)$, will depend on d. To avoid overburdening the notation, we will not indicate this dependence explicitly, but it should be kept in mind that many of the objects under consideration change with d. Next, for any positive number z, we define a norm on the space of sequences of functions $\underline{f} = (f_j)_{j=1,2,\ldots}$ by

$$\|\underline{f}\|_{z,\beta} = \sup_j {}_j\|f_j\|_\beta / z^j$$

With this notation, we can formulate our main estimate:

Proposition 1. Let $\beta > \beta' > 0$, and let $z' > (\beta'/\beta)^{3/2} z$. There exists a constant A, depending only on β'/β and z'/z (i.e., not depending on m,β,d,n or z) such that:

$$\|S(t-t_1)CS(t_1-t_2)\ldots CS(t_m)\underline{f}\|_{z',\beta'} \leq m!\,[A\cdot\pi n d^2 z(\tfrac{m\beta}{3})^{-1/2}]^m \|\underline{f}\|_{z,\beta},$$

for all $m = 1,2,\ldots$ and all $\underline{f} \in Y_{z,\beta}$.

To get the sharpest possible estimate (i.e. to make A as small as possible), we should take β' very small and z' very large. To simplify the discussion, however, we will choose z'/z and β'/β once and for all. We will also write

$$(8) \qquad t_0 = [A\,\pi\, n d^2 z(\tfrac{m\beta}{3})^{-1/2}]^{-1}.$$

Corollary 2. Let z, z', β, β' be as above. Then the mth term in the series

$$\sum_{m=1}^{\infty} \int_0^t dt_1 \int_0^{t_1} dt_2 \ldots \int_0^{t_{m-1}} dt_m \; S(t-t_1) C \; S(t_1 - t_2) \ldots C \; S(t_m)$$

is bounded in norm as an operator from $Y_{z,\beta}$ to $Y_{z',\beta'}$ by $(|t|/t_0)^m$. The series therefore converges for $|t| < t_0$, and the convergence is uniform in n, d, z, m, β provided t_0 is held fixed.

Exactly the same estimates hold for the formal series solution to the Boltzmann hierarchy.

The significance of the combination

$$[\pi \, nd^2 \, z \, (\tfrac{m\beta}{3})^{1/2}]^{-1} \tag{9}$$

in the expression for t_0 is as follows: A statistical state has rescaled correlation functions in $Y_{z,\beta}$ if and only if its true correlation functions satisfy

$$\rho_j(x_1, \ldots, x_j) \leq \text{const} \times (nz)^j \prod_{i=1}^{j} \sigma_\beta(\vec{p}_i) \; .$$

The right-hand side, except for the constant, is the correlation function for a noninteracting equilibrium state with density $\rho = nz$ and root mean square velocity $v_{rms} = (3/m\beta)^{1/2}$. Inserting these expressions in the rule of thumb

$$\text{mean free time} \sim \text{mean free path} \; / \; v_{rms} \sim (\pi \rho d^2 v_{rms})^{-1}$$

gives exactly (9), so this expression is a crude estimate of the mean free time. Thus, we should think of t_0 as approximately the pure number A^{-1} times the mean free time, and the uniformity statement in Corollary 2 means that the convergence of the series is determined by the ratio of $|t|$ to the mean free time and is otherwise independent of the physical parameters.

We will not give a detailed proof of Proposition 1 but will make a few comments about the proof. We look first at the properties of $C_{j,j+1}$ acting on ${}_{j+1}Y_\beta$. Recall that

$$(C_{j,j+1}f_{j+1})(x_1,\ldots,x_j) = nd^2 \sum_{i=1}^{j} \int d\vec{p}_{j+1}\; d\hat{\omega}\; \hat{\omega}\cdot\left(\frac{\vec{p}_{j+1}-\vec{p}_i}{m}\right)$$

$$f_{j+1}(x_1,\ldots,x_j,\vec{q}_i+d\cdot\hat{\omega},\vec{p}_{j+1})$$

(This differs from formula (11) of Chapter 6 in that d^2 has been replaced by nd^2 because we are writing the hierarchy for the rescaled f's rather than the ρ's.) Taking absolute values, estimating f_{j+1} by

$$|f_{j+1}(x_1,\ldots,x_j)| \le {}_{j+1}\|f_{j+1}\|_\beta \prod_{i=1}^{j+1} \sigma_\beta(\vec{p}_i) \;,$$

and doing some elementary integrals gives

$$|C_{j,j+1}f_{j+1}(x_1,\ldots,x_j)| \le {}_{j+1}\|f_{j+1}\|_\beta \prod_{i=1}^{j} \sigma_\beta(\vec{p}_i) \left(\frac{2\pi nd^2}{m}\right)$$

$$[\sum_{i=1}^{j} \int d\vec{p}\; \sigma_\beta(\vec{p})\; |\vec{p}-\vec{p}_i|] \;.$$

The term in square brackets is an unbounded function of $p_1,\ldots,p_j$ (so we do not expect $C_{j,j+1}$ to be bounded from ${}_{j+1}Y_\beta$ to ${}_jY_\beta$), but its product with $\prod_{i=1}^{j}\left(\frac{\sigma_\beta(\vec{p}_i)}{\sigma_{\beta'}(\vec{p}_i)}\right)$ is bounded for any $\beta' < \beta$. Thus, $C_{j,j+1}$ is bounded from ${}_{j+1}Y_\beta$ to ${}_jY_{\beta'}$ for any $\beta' < \beta$. For fixed β', the bound obtained grows with j like $(\beta/\beta')^{(3/2)j}j$, and hence C is bounded from $Y_{z,\beta}$ to $Y_{z',\beta'}$ for any $z' > (\beta/\beta')^{3/2}z$. Also, each $S(t)$ is an isometry on each $Y_{z,\beta}$. Straightforward iteration of the above estimate, keeping careful track of the dependence on n,z,d,β,m leads to Proposition 1.

By Corollary 2 applied to the Boltzmann hierarchy, we can use the series (5) to obtain a generalized solution to the Boltzmann hierarchy defined for $|t| < t_0$, provided that the initial $\underline{f}^{(0)}$ belongs to $Y_{z,\beta}$. An easy manipulation shows that, formally, the series solution factorizes if the initial $\underline{f}^{(0)}$ does. For $\underline{f}^{(0)}$ in $Y_{z,\beta}$ and $|t| < t_0$, there is no problem in making this formal calculation into a correct proof, so the solutions we construct in this way preserve factorization.

We are now in a position to formulate precisely the result described at the beginning of this chapter. Consider a sequence of statistical states with rescaled correlation functions $f_j^{(d)}(x_1,\ldots,x_j)$. We will assume:

(C1) For some $z,\beta > 0$, $\|\underline{f}^{(d)}\|_{z,\beta}$ is bounded uniformly in d.

(C2) $f_j^{(0)}$ is continuous on $(\Lambda\times\mathbb{R}^3)^j_{\neq}$ and

$$\lim_{d\to 0} f_j^{(d)}(x_1,\ldots,x_j) = f_j^{(0)}(x_1,\ldots,x_j)$$

uniformly on compact sets in $(\Lambda\times\mathbb{R}^3)^j_{\neq}$, where $(\Lambda\times\mathbb{R}^3)^j_{\neq}$ denotes the set of j-particle phase points with no pair of particles at the same point of physical space.

Theorem 3. Assume (C1) and (C2), and let t_0 be defined by (8) with z,β as in (C1). For $0 < t < t_0$, let $\underline{f}^{(0)}(t)$ denote the generalized solution of the Boltzmann hierarchy with initial condition $\underline{f}^{(0)}$ obtained by summing the series (5). Then for any $j = 1,2,3,\ldots$ and any t between 0 and t_0, we have

$$\lim_{d\to 0} f_j^{(d)}(t;x_1,\ldots,x_j) = f_j^{(0)}(t;x_1,\ldots,x_j) \quad \text{almost everywhere.}$$

This theorem is a precise reformulation of statements (a) and (b) at the beginning of this chapter. We have already discussed statement (c). To prove the theorem, we use the series expansions (4) and (5) for $\underline{f}^{(d)}(t)$ and $\underline{f}^{(0)}(t)$ respectively. By Corollary 2, both series converge in $Y_{z',\beta'}$ and the convergence of the first is uniform in d. It will therefore suffice to prove

$$\lim_{d\to 0} S^{(d)}(t-t_1)C^{(d)}\ S^{(d)}(t_1-t_2)\ldots\ C^{(d)}\ S^{(d)}(t_m)\ f^{(d)}$$

$$= S^{(0)}(t-t_1)C^{(0)}\ S^{(0)}(t_1-t_2)\ldots\ C^{(0)}\ S^{(0)}(t_m)\ f^{(0)} \quad \text{a.e.,}$$

for all m and all $t_0 > t > t_1 > \ldots > t_m > 0$. (In the above expression, we have written $S^{(d)}$, $C^{(d)}$ rather than S, C to emphasize the dependence on d). For simplicity, we will discuss only the two-particle component of the term with m = 1; the extension to general m,j is similar. Let us fix a two particle phase point (x_1,x_2) and examine in detail the definition of

$$(10)\quad (S^{(d)}(t-t_1)C^{(d)}\ S^{(d)}(t_1)\underline{f}^{(d)})_2(x_1,x_2) \quad \text{for} \quad 0 < t_1 < t.$$

To compute this quantity, first follow the two-particle phase point (x_1,x_2) back $t-t_1$ units of time. This gives a new phase point (x_1',x_2'). Then choose a value of $i = 1,2$, a momentum $\vec{p}_3$, and a direction $\hat{\omega}$, and adjoin a particle at the point $\vec{q}_i' + d\cdot\hat{\omega}$ with momentum $\vec{p}_3$. Follow the resulting 3-particle phase point back t_1 units of time, and denote the result by (x_1'',x_2'',x_3''). Note that (x_1'',x_2'',x_3'') depends, in general, on i, $\vec{p}_3$, $\hat{\omega}$, x_1, x_2. Form

$$nd^2\ \hat{\omega}\ \cdot\ \frac{(\vec{p}_3-\vec{p}_i')}{m}\ f_3^{(d)}(x_1'',x_2'',x_3'')\ ,$$

integrate over $\hat{\omega}$, $\vec{p}_3$, and sum over i. The resulting quantity is exactly (10).

Now what happens to this expression as $d \to 0$ with fixed nd^2? Unless (x_1,x_2) belongs to the negligible set of phase points where $\vec{p}_1-\vec{p}_2$ and $\vec{q}_1-\vec{q}_2$ are collinear, there will for small enough d be no collisions in going from (x_1,x_2) to (x_1',x_2'). When the third particle is adjoined, there are two possibilities: If $(\vec{p}_3-\vec{p}_i')\cdot\hat{\omega}>0$, the new particle is created coming out of a collision with the ith particle and hence immediately undergoes a collision with that particle when we start to follow the system backwards in time. This replaces $(\vec{p}_i', \vec{p}_3)$ by the corresponding incoming momenta $T_{\hat{\omega}}(\vec{p}_i', \vec{p}_3)$. If, on the other hand, $(\vec{p}_3 - \vec{p}_i') \cdot \hat{\omega} < 0$, and the third and ith particles move smoothly apart. Continuing backwards in time leads to further collisions only for exceptional choices of $\vec{p}_3, \hat{\omega}, i$, and the contribution of these choices to the integral defining (10) becomes negligible in the limit as $d \to 0$. Thus, for most of $x_1, x_2, i, \hat{\omega}, \vec{p}_3$, and for small d, the only effect of the finite particle size on x_1'', x_2'', x_3'' is a displacement of q_3'' by $d \cdot \hat{\omega}$. Using (C2), it is easy to pass to the limit of (10) as $d \to 0$, and the limit is exactly what would have been obtained by similarly spelling out the calculation of

$$(S^{(0)}(t-t_1)C^{(0)}\; S^{(0)}(t_1)f^{(0)})_2(x_1,x_2)$$

This completes the proof of the results formulated at the beginning of this section, except for the interpretation of the condition

$$f_j^{(0)}(x_1,\ldots,x_j) = \prod_{i=1}^{j} f_1^{(0)}(x_i) \quad . \tag{11}$$

The interpretation we will give assumes (C1) and convergence almost everywhere of the $f_j^{(d)}$ as $d \to \infty$. Assume first the factorization condition (11) for $j = 2$. For any $\Delta \subset \Lambda \times \mathbb{R}^3$,

$$E^{(d)}(n_\Delta) = \int_\Delta \rho_1^{(d)}(x_1)\, dx_1 = n \int_\Delta f_1^{(d)}(x_1)\, dx_1$$

$$E^{(d)}(n_\Delta(n_\Delta - 1)) = \int_{\Delta\times\Delta} \rho_2^{(d)}(x_1,x_2)\, dx_1\, dx_2 = n^2 \int_{\Delta\times\Delta} f_2^{(d)}(x_1,x_2)\, dx_1\, dx_2$$

where $E^{(d)}$ means the expectation with respect to the statistical state $\mu^{(d)}$. Since the $f_j^{(d)}$ converge almost everywhere to $f_j^{(0)}$,

$$\lim_{d\to 0} E^{(d)}\left(\frac{n_\Delta}{n}\right) = \int_\Delta dx_1\, f_1^{(0)}(x_1)$$

$$\lim_{d\to 0} E^{(d)}\left[\left(\frac{n_\Delta}{n}\right)^2\right] = \int_{\Delta\times\Delta} dx_1\, dx_2\, f_2^{(0)}(x_1,x_2) .$$

Therefore, factorization for $j = 2$ implies that

$$\lim_{d\to 0}\left[E^{(d)}\left\{\left(\frac{n_\Delta}{n}\right)^2\right\} - \left\{E^{(d)}\left(\frac{n_\Delta}{n}\right)\right\}^2\right] = 0$$

and hence that n_Δ/n converges in probability to the constant $\int_\Delta f_1^{(0)}(x_1)\, dx_1$ as $d \to 0$.

Conversely, suppose n_Δ/n converges in probability to $\int_\Delta dx_1\, f_1^{(0)}(x_1)$ for each $\Delta \subset \Lambda\times\mathbb{R}^3$. Condition (C1) implies that all moments of n_Δ/n are bounded uniformly in d, and hence that

$$\lim_{d\to 0} E^{(d)}\left[\left(\frac{n_\Delta}{n}\right)^j\right] = \left[\int_\Delta dx_1\, f_1^{(0)}(x_1)\right]^j \quad \text{for all } j.$$

But the left-hand side of this equality is also equal to

$$\int_{\Delta\times\ldots\times\Delta} f_j^{(0)}(x_1,\ldots,x_j)\, dx_1\ldots dx_j ,$$

so we have

$$\int_{\Delta\times\cdots\times\Delta} f_j^{(0)}(x_1,\ldots,x_j)\, dx_1\ldots dx_j = \left[\int_\Delta dx_1\, f_1^{(0)}(x_1)\right]^j$$

for all $\Delta \subset \Lambda\times\mathbb{R}^3$. This identity, together with the symmetry of $f_j^{(0)}(x_1,\ldots,x_j)$, implies (11).

We conclude with a number of remarks:

1. Since the series for $f^{(d)}(t)$ also converges for $-t_0 < t < 0$, we can also study the limiting behavior of $\underline{f}^{(d)}(t)$ as $d \to 0$ for negative times. Exactly the same analysis applies as for positive times except that, in the investigation of the limiting behavior of the series (4), collision occurs immediately if the extra particle is inserted in an incoming collision configuration with the ith particle and not if it is in an outgoing configuration. This has

the effect of changing the sign of the limiting collision term so, for $t < 0$, $f^{(d)}(t)$ converges to a solution of the Boltzmann hierarchy with the sign of the collision term reversed.

2. The Boltzmann hierarchy, like the Boltzmann equation, is not invariant under time-reversal. That is, irreversibility appears in passing to the limit $d \to 0$, not in the assumption that the limiting rescaled correlation functions factorize.

3. In (C2), we assume that the initial $\underline{f}^{(d)}$'s converge uniformly on compact sets to $\underline{f}^{(0)}$, but the conclusion of Theorem 3 is that the $\underline{f}^{(d)}(t)$ converge almost everywhere to $\underline{f}^{(0)}(t)$. That is, we have to assume a stronger kind of convergence at time zero than we can deduce at positive times. The gap between the two kinds of convergence can be narrowed but cannot be closed entirely. This follows from the fact that the BBGKY hierarchy is time-reversal invariant but the Boltzmann hierarchy is not: if the $\underline{f}^{(d)}(t)$ converged to $\underline{f}^{(0)}(t)$ in a sufficiently strong sense so that Theorem 3 could be applied to them as initial data, then, by running the system backwards through t units of time, we would conclude that $\underline{f}^{(d)}(0)$ converges as $d \to 0$ to the value at time zero of a solution of the time-reversed Boltzmann hierarchy equal to $\underline{f}^{(0)}(t)$ at time t. Since this value is generally different from $\underline{f}^{(0)}(0)$, this is impossible.

4. Although (C2) requires a fairly strong kind of convergence of the $\underline{f}^{(d)}$'s, it is not sensitive to the presence or absence of correlations between particles with separations of the order of a few molecular diameters. Thus, fairly strong short-range correlations for finite d are compatible with a limiting $\underline{f}^{(0)}$ which factorizes. Note, however, that (C1) imposes, implicitly, bounds on the strength of such correlations.

5. The estimates of Lemma 1 can be applied to the usual (non-scaled) correlation functions to prove a local existence theorem

for the solutions of the BBGKY hierarchy for an infinite system of hard spheres, and to show that the solutions obtained are thermo-dyanmic limits of finite-volume time-dependent correlation functions.

Reference

The contents of this chapter are taken from unpublished work of the author.

ERGODIC PROPERTIES OF INFINITE SYSTEMS

Sheldon Goldstein*

Institute for Advanced Study, Princeton University, New Jersey

Joel L. Lebowitz† and Michael Aizenman‡†

Belfer Graduate School of Science, Yeshiva University, New York

Abstract

Macroscopic systems are successfully modeled in statistical mechanics, at least in equilibrium, by infinite systems. We discuss the ergodic theoretic structure of such systems and present results on the ergodic properties of some simple model systems. We argue that these properties, suitably refined by the inclusion of space translations and other structure, are important for an understanding of the nonequilibrium properties of macroscopic systems.

1. Introduction

Statistical mechanics attempts to account for the observed behavior of macroscopic physical systems on the basis of the microscopic laws which govern the behavior of their "elementary" constituents (particles or molecules). The central fact of the entire endeavor is that the number of particles is very large ($\sim 10^{26}$), so that the concepts of probability theory play an important role. (For large systems a microscopic statistical description is consistent with deterministic 'laws' and indeed is the only one feasible.) It is an assumption which can be proved in some cases that insofar

* Supported in part by National Science Foundation Grant GP-16147A-1.

† Supported in part by U. S. Air Force Office of Scientific Research 73-2430.

‡ Supported in part by National Science Foundation Grant GP-37069X.
Present address: Courant Institute of Mathematical Sciences, New York University.

as many qualitative aspects of macroscopic behavior are concerned the exact nature of the microscopic laws even whether they are classical or quantum, is of little importance. Since the problems of classical statistical mechanics are mathematically interesting, and since the physical concepts are, in the classical framework, more transparent, classical statistical mechanics continues to be a subject of much interest. [See Lanford,* and Lanford and Lebowitz articles in this book.]

The success of the program, at least from the mathematical or rigorous point of view, has so far been largely limited to the realm of equilibrium statistical mechanics, which deals with the description of matter in thermal equilibrium. It explains in particular how even complex macroscopic systems are susceptible to a complete macroscopic description involving only a small number of parameters (temperature, pressure, energy, entropy, volume,...), which satisfy the simple relationships of thermodynamics. The explanation is obtained by means of the identification, due to Gibbs, of macroscopic states with measures on the phase space of microscopic description and the appropriate utilization of the fact that macroscopic systems are "big".

The situation with respect to nonequilibrium statistical mechanics is far more tentative. Here one is concerned with why and how systems come to equilibrium; i.e., we would like to account for the experimental fact that starting from a nonequilibrium state the system will evolve, under the action of its time evolution, to the appropriate equilibrium state, and furthermore that this approach to equilibrium satisfies the relevant kinetic (e.g.

*Lanford's lectures contain much of the statistical mechanical background for this article.

Boltzmann), hydrodynamic, and transport equations. We shall say that such systems exhibit "good thermodynamic behavior", (gtb). Here, too, it is argued that macroscopic states should be identified with measures on the phase space, but the problems of nonequilibrium statistical mechanics have proven, so far, less tractable than their equilibrium counterparts.

Because of the key role played by probability theory, it is to be expected that only in some appropriate thermodynamic limit (number of particles approaches infinity) is gtb precisely achieved. Hence, if exact mathematical results are desired, infinite systems of particles should be directly investigated; their very large yet finite counterparts, which may for all practical purposes exhibit behavior of the type described by macroscopic laws, will nonetheless exhibit it only approximately, precluding the formulation and proof of the appropriate theorems. (In the same way phase transitions, which are associated with nonanalyticities in thermodynamic functions, do not occur for finite systems, though numerical computations on systems containing just a few hundred particles and experimental observations on macroscopic systems mimic this behavior very closely).

The key ingredients in equilibrium statistical mechanics are the appropriate abstract framework -- to isolate the elements of structure relevant to the phenomenon -- and the right limit: the framework is that of measure theory and the limit is that of infinite volume (and particle number), usually referred to as the thermodynamic limit. A microstate of a finite system (in a finite volume) is completely described by a point x of the phase space Γ; $x = (q_1,\dots,q_n, p_1,\dots,p_n)$ gives the coordinates and momenta of all particles of the system. There exists a natural measure

$d\underline{q}\,d\underline{p}$ (Lebesgue measure) on Γ whose "projection" onto a surface, Γ_E, of constant energy (Hamiltonian function) E, gives a measure μ_E, the microcanonical measure. According to statistical mechanics expectations of observables (functions on Γ_E) are equated to the equilibrium values of thermodynamic quantities of the system at energy E. The exact macroscopic relations are obtained by taking the limit $N \to \infty$, $V \to \infty$, $E \to \infty$, $N/V \to \rho$, $E/V \to \varepsilon$, where N is the particle number, V the volume, E the energy, ρ the particle and ε the energy density of the macroscopic system.

A major difficulty in nonequilibrium statistical mechanics lies in finding the framework and limit appropriate to the task. We shall assume, as has often been asssumed, that the same thermodynamic limit as used in equilibrium statistical mechanics is appropriate, at least for part of the problem. (Different or additional limits may have to be used for obtaining more specific kinetic behavior, see Lanford's lecture for the appropriate limit needed for the derivation of the Boltzmann equation.) The underlying structure is therefore taken to be the phase space of infinite (but locally finite) configurations of particles in $\mathbb{R}^\nu \times \mathbb{R}^\nu$, (where $\nu = 3$ for realistic systems).

Before considering the nonequilibrium behavior of infinite systems, we will review briefly the situation for finite systems. Here, the nonequilibrium behavior is governed by the time evolution T_t determined by the Hamiltonian in terms of which the equilibrium situation is described. Since Liouville's theorem, asserting the invariance of Lebesgue measure under T_t, implies that μ_E is also preserved by T_t, one finds oneself directly within the context of ergodic theory: one has a triple (Γ_E, μ_E, T_t) of just the right kind. Indeed, ergodic theory arose out of the attempt to justify the foundations of statistical mechanics. It has typically provided the framework for the investigation of the nonequilibrium behavior of finite systems [1].

It seems physically plausible to assume that events corresponding to subsets of the phase space which have vanishing Lebesgue measure have vanishing probability of occurring, and should therefore be irrelevant for experimentally observed behavior (which depends on reproducibility). We may then regard as physically reasonable only those measures which are absolutely continuous with respect to the Lebesgue measure (or μ_E).* With this assumption the traditional ergodic properties have the following implications (see Lanford's lectures and [1]):

(a) If (Γ_E, μ_E, T_t) is ergodic, then for systems with energy E, μ_E is the only physically reasonable measure which is stationary (invariant) under the time evolution; furthermore, any physically reasonable state will tend, in the sense of time average, to this uniquely determined equilibrium state.

(b) If the system is in addition mixing, this approach to equilibrium occurs in the strict sense, i.e.,

$$\rho_t(A) = \rho(T_{-t}A) \xrightarrow[t\to\infty]{} \mu_E(A) ,$$

where ρ is an absolutely continuous nonequilibrium measure and A is a bounded observable.

The stronger ergodic properties -- K-system and Bernoulli -- are not so directly susceptible to physical interpretation in terms of good thermodynamic behavior. We mention only that if (Γ_E, μ_E, T_t)

*While μ_E is itself singular with respect to the Lebesgue measure $d\underline{q}\, d\underline{p}$, it may be regarded as the limit of measures concentrated on the 'energy shell' (E, E+ΔE). (The energy being both smooth and universally conserved plays a special role; see Lanford's lectures.)

is a K-system, its behavior will be completely nondeterministic in the following sense: a realistic measurement corresponds to a finite partition of the phase space of the observed system into subsets corresponding to the possible outcomes of that measurement. If we wish to obtain more information about the system we can repeat that measurement at discrete time intervals. If the system is K (and only if) then no such measurements even when carried out infinitely often in the past, determine with certainty the outcome of future measurements. Thus, in a very strong sense, K-systems cannot be "finitely approximated", and since it is not unreasonable to expect of a finite system that it can be approximated by a sufficiently fine coarse graining (into position and momentum cells), the K-property has significant implications for finite systems. (For some interpretations of the Bernoulli property see [1,2]).

Now the KAM theorem and related results sugest that realistic models of finite physical systems may fail to be ergodic. Though this need not prevent large systems from exhibiting, for all practical purposes, good thermodynamic behavior, it renders an exact result rather unlikely. It is therefore natural to investigate directly the nonequilibrium behavior of infinite systems.

The basic structural necessities for such an investigation -- the equilibrium states μ and measure preserving Hamiltonian time evolutions T_t -- can be realized, as Lanford has indicated in his lectures, on the phase space X of infinite (locally finite) configurations. An obvious candidate for the abstract framework for the explication of the good thermodynamic behavior of infinite systems is therefore, as before, the (abstract) triple (X,μ,T_t). We are thus led to the investigation of the ergodic properties of infinite systems. However, some general observations indicate that such ergodic theory by itself is of rather limited potential:*

(a) The previous remarks concerning the implications of ergodicity and mixing can be applied without modification to infinite systems if by physically reasonable we understand absolutely continuous with respect to μ (the equilibrium measure). Because of the quasilocal nature of the σ-algebra of measurable sets employed in the description of these systems, measures absolutely continuous with respect to μ may be roughly interpreted as local perturbations of μ. Therefore, insofar as return to equilibrium is concerned the situation is satisfactory. If we are interested, however, in understanding how a system ever comes to be in equilibrium -- the problem of approach to equilibrium -- then mixing says nothing, since the relevant initial states will typically be singular with respect to μ. Similarly, ergodicity will be of little value in choosing one stationary state over another without some prior basis for selecting one ergodic state from the family of stationary (ergodic) states.

Consider for example the ideal gas, where particles undergo free motion in $\mathbb{R}^\nu \times \mathbb{R}^\nu$. μ is such that the particles are, independently of each other, uniformly scattered in space ($\mathbb{R}^\nu$) and their individual momenta have a Maxwellian (Gaussian) distribution. This state as well as any other state differing from it only by having a different momentum distribution is stationary (since the momenta are constant). Furthermore, as we shall see later, the free time evolution on these states will typically be ergodic, and, indeed, Bernoulli, the point, of course, being that states given by different momentum distributions are mutually singular.

---

* Some additional, hopefully relevant, structure will be discussed later.

(b) For an infinite system to be K need not be a significant constraint upon its behavior, since it does not appear plausible that an infinite system should permit a finite approximation in the sense indicated above.

In view of these remarks it should not be very surprising to find that very simple infinite systems possess very good ergodic properties. We will now turn to the description of the ergodic properties of some such systems, particularly, noninteracting systems.* It will then be shown how the introduction of additional structure (space translations) improves the situation.

2. Ergodic Properties of the Time Evolution of Some Simple Infinite Systems

We shall be discussing two kinds of infinite systems, both of whose time evolutions have very good ergodic properties, but of which one exhibits distinctly better thermodynamic behavior than the other. Correspondingly, there are two different "mechanisms", which are responsible for these ergodic properties. The first of these, which is the only one possessed by the ideal gas, corresponds to the escape of local information to infinity. The second, better mechanism, is a local dissipation of information.

Since we shall deal primarily with systems of noninteracting particles, we first describe a convenient representation of such systems.

Let (X_1, Σ_1, μ_1) be a totally σ-finite nonatomic measure space. We regard X_1 as the phase space for a single particle and wish to construct from (X_1, Σ_1, μ_1) a probability space (X, Σ, μ), representing the independent distribution of identical particles in X_1

* See article by Lanford and Lebowitz for the ergodic properties of harmonic crystals.

according to the measure μ_1. Accordingly we let X be the set of countable subsets of X_1 and define for any set $A \in \Sigma_1$, N_A by

$$N_A(x) = \text{the cardinality of } (A \cap x)$$

for $x \in X$.

We then let $\Sigma = \sigma\{N_A\}$, be the smallest σ-algebra for which all of the N_A are mreasurable, and define μ on Σ as the probability measure representing a Poisson distribution of points in X_1, with density given by μ_1. This is the measure for which the functions N_A have distribution given by

$$\mu(\{x \in X \mid N_A(x) = m\}) = \exp\,[-\mu_1(A)]\ [\mu_1(A)]^m/m!\ ,$$

which, if true for all A, implies that disjoint regions of X_1 are independent [3].

For any automorphism T_1 of (flow T_1^t on) (X_1, Σ_1, μ_1) we define an automorphism T of (flow T^t on) (X, Σ, μ) by

$$Tx = T_1\,x\ , \qquad (T^t x = T_1^t x)\ ,$$

where the right-hand side of the equation is to be understood as the action of the (one-particle) automorphism (flow) on the subset x of X_1. Thus we obtain a time evolution on the infinite system by letting all the particles move independently according to the one-particle time evolution. We call $(X, \mu, T(T^t))$ * the <u>Poisson system</u> built over $(X_1, \mu_1, T_1(T_1^t))$.

The Ideal Gas and the Bernoulli Construction

In its simplest manifestations, such as in the ideal gas, the "escape to infinity" mechanism mentioned earlier has a formal representation directly exhibiting the Bernoulli character of the

---

* We will for the most part drop the reference to the σ-algebra.

flow. We call this representation the <u>Bernoulli construction:</u> For $C \subset X_1$, let $\Sigma_C = \sigma\{N_A | A \subset C\}$ be the local σ-algebra ($\subset \Sigma$) of C. If A^n is a family of disjoint sets ($\in \Sigma_1$) such that $\bigcup_{n=-\infty}^{\infty} A^n = X_1$ and $T_1 A^n = A^{n+1}$ then Σ_{A^0} is an independent generator for T. This means that the sequence $\Sigma_{A^0}, \Sigma_{A^1}, \ldots$ is an independent sequence of σ-algebras and that $\Sigma = \sigma\{\Sigma_{A^n} | n \in \mathbb{Z}\}$. Since a generalized Bernoulli shift may be characterized by the existence of an independent generator, these systems give rise to Bernoulli shifts; we shall say that the Poisson system built over such a system admits of a Bernoulli construction.

(We remind the reader (see [2]) that if T^t is a flow such that T^1 ($\equiv$ T) is a Bernoulli shift then T^τ is Bernoulli for all τ; thus, insofar as the Bernoulli property is concerned it suffices, when investigating a flow T^t, to investigate T^1.)

The simplest example of this type of system is the infinite ideal gas (IG). We can realize IG (X^I, μ^I, T^{It}) as the Poisson system built over $(X_1^I, \mu_1^I, T_1^{It})$ where

$$X_1^I = \mathbb{R}^\nu \times \mathbb{R}^\nu$$

$$d\mu_1^I = \rho\left(\frac{\beta}{2\pi}\right)^{\frac{1}{2}\nu} \exp\left[-\frac{1}{2}\beta p^2\right] dq\, dp \ , \quad \text{and}$$

$$T_1^{It}(q,p) = (q+pt,p).$$

Here $(q,p) \in \mathbb{R}^\nu \times \mathbb{R}^\nu$, ρ is the particle density, β the inverse temperature (of the ideal gas equilibrium state), and we have taken all the masses to be unity.

That IG admits of a Bernoulli construction follows from the fact that the set

$$A^0 = \{(q,p) = ((q_1,\ldots,q_\nu), (p_1,\ldots,p_\nu)) \in X_1^I \mid q_1+p_1 t = 0 \text{ for some } 0 \le t < 1\},$$

has, essentially by construction, the necessary properties.

Thus, we construct a partition of X_1^I on the basis of the time at which the particle has vanishing first coordinate. We could also have partitioned according to the time of nearest approach to the origin. More generally, if our system (X,μ,T) were such that the latter did not determine a unique time, we could try to partition according to the last time of closest approach to the origin. It turns out, by virtue of (2) below, that if this too is essentially inadequate then no Bernoulli construction is possible.

This is related to the observation (suggested by the fact that though the one particle ideal gas has rather poor mixing properties, the Poisson system built over it has such properties of the strongest kind) that insofar as the Bernoulli construction is concerned, the ergodic properties of a Poisson system are roughly inversely proportional to those of the one particle system over which it is built. This is stated more precisely in the following proposition.

<u>Proposition</u>. (1) If (X_1,Σ_1,μ_1,T_1) is ergodic in the sense that if a set $A \in \Sigma_1$ is invariant, either it or its complement has zero measure, no Bernoulli construction is possible.

(2) If there exists a set $A \subset X_1$ of finite measure which is recurrent in the sense that points in a non null Σ_1 subset of A return to A infinitely often under the action of T_1, no Bernoulli construction is possible. If all sets of finite measure are nonrecurrent, one can perform a Bernoulli construction [4].

These possibilities are illustrated by the random walk models (RWν). RWν denotes the Poisson system built over the automorphism representing a particle undergoing a symmetric random walk on $\mathbb{Z}^\nu$. Thus RW1 is built over $(B\times\mathbb{Z}, T_1)$, where B is the space of doubly infinite sequences $\xi_i = \pm 1$ equipped with the measure obtained by assigning equal probability to +1 and -1 and forming the infinite product measure, and T_1 is given by

$$T_1(\xi,n) = (S\xi, \xi_0+n) ,$$

where S is the shift on B and $\xi = \{\xi_i\} \in B$. $B \times \mathbb{Z}$ is to be understood as equipped with the natural product measure. The models RWν, $\nu > 1$, are similarly formed.

Since, with probability one, a particle undergoing a random walk on $\mathbb{Z}^\nu$ will return to the origin (infinitely often) only if $\nu \leq 2$, we have that RWν is recurrent if and only if $\nu \leq 2$, so that a Bernoulli construction (based, say, upon the last time of nearest approach to the origin of $\mathbb{Z}^\nu$, using only the $\mathbb{Z}^\nu$ coordinate to measure distance) is possible only for $\nu \geq 3$.

We observe that RWν has much better "mixing" properties than IG and that this is reflected by the presence in RWν of a local mechanism of mixing, to which we earlier referred. Before pursuing this further, we indicate, in the next section, how the properties of IG are affected if we allow the particles to interact in the simplest possible way. We shall see that while the flow of local information to infinity is hindered by the interactions such a loss of local information still takes place by a mechanism similar to the diffusion in the RW_ν models (without however having a local mechanism for mixing) and gives rise to good ergodic properties (at least K). To make this clearer we digress briefly with a comment about K-systems.

A general characterization of a K-system is by the existence of a measurable partition (or σ-algebra) ξ, called a K-partition, with the following properties:

(1) $$T^t \xi \geq \xi \quad \text{for} \quad t \geq 0$$

(2) $$\bigvee_t T^t \xi = \varepsilon$$

(3) $$\bigwedge_t T^t \xi = \nu \quad .$$

Here $\xi \geq \eta$ means that the elements of the partition η are unions of elements of ξ (ξ is finer than η), $\bigwedge_\alpha \xi_\alpha$ is the finest measurable partition coarser than all the ξ_α, ε is the partition into points (corresponding to the full σ-algebra Σ), ν is the partition whose sole element is the entire space X (corresponding to the σ-algebra of trivial sets, which have measure zero or measure one), and all statements here and to follow are to be understood as up to a set of measure zero.

The connection between this characterization and the definition (for finite entropy) in terms of the existence of a finite generating partition P with trivial tail ($\bigwedge_j \bigvee_{k=j}^{\infty} T^{-k} P = \nu$) is obtained by setting $\xi = \bigvee_{k=1}^{\infty} T^{-k} P$. It should thus not be surprising that ξ ($\bigvee_{k=1}^{\infty} T^{-k} \Sigma_{A0}$ for the one-dimensional IG), the partition according to "origin events" (see below) occurring after t = 1, should be a K-partition. (This is just the zero-one law for tail events.)

Ergodic Properties of the Time Evolution of Some One Dimensional Systems of Hard Rods

The independent generator Σ_{A0} which we obtained for IG may be described for $\nu = 1$ as the σ-algebra determined by events at the origin between times zero and one. The two crucial facts concerning these "origin events" are that

(a) They generate: two phase points having the same origin events must coincide (except perhaps for a set of zero measure), and

(b) origin events in disjoint time intervals are independent.

We wish to know how altering our system by introducing a pair interaction, and making the corresponding change of measure, will affect the ergodic properties of the system, and in particular how it will affect (a) and (b).

We may roughly describe the systems we wish to consider -- one-dimensional infinite systems of hard rods of diameter $d > 0$ -- as having measure spaces which are obtained from that of IG (with general momentum distribution given by $f(p)$) by merely deleting phase points for which some pair of particles is closer than d (we ignore the difficulty that what remains has zero measure), and as having a dynamics differing from free motion by the stipulation that when a pair of particles collide (i.e., approach within a distance d of each other) they exchange momenta (all masses are the same). The dynamics may equivalently be described by requiring particles to exchange positions during collisions, their momenta being unaffected. Since, if we adopt the latter description, the hard rod system (HR) goes, in the limit $d = 0$, into IG even insofar as single particle motion is concerned, this description is the more convenient for the investigation of its ergodic properties.

Sinai [5] has shown that the paritition according to (appropriately defined) future origin events are K-partitions for HR. Thus (a) is preserved by the introduction of the hard core interaction, and though (b) will no longer hold, the K-property implies an approximate independence between sufficiently separated (in time) past and future origin events. If f (which, for simplicity we take to be even) vanishes in a small neighborhood of the origin, it can be shown [6] that enough of (b) remains for HR to be Bernoulli.

More precisely, what is shown is that if one defines "origin event" appropriately, there exist arbitrarily "fine" finite partitions determined by origin events between times zero and one, which satisfy a strong property of approximate independence of past and future called the weak Bernoulli property [7]. Now, if a partition P of X satisfies this property, then $(X, \bigvee_{j=-\infty}^{\infty} T^j P, \mu, T)$ is a Bernoulli shift. Since an increasing union of Bernoulli shifts is Bernoulli [2] (i.e., if (X, Σ_n, μ, T) is Bernoulli for each n and $\Sigma_n \nearrow \Sigma$, then (X, Σ, μ, T) is Bernoulli) these hard rod systems will form Bernoulli flows.

The point is that what is going on in this case in HR is not very different from what is going on in IG. A particle of velocity v moves at this velocity except for moments of collision, when it jumps the distance d in the direction of the other colliding particle. However, the "free distance" between two pulses (obtained by subtracting the total length of rods between them) behaves linearly in time. Thus, if one uses a "reduced description" obtained by removing the volume occupied by the hard cores from the distances between the particles, one obtains the exact IG time evolution. If there were a reflecting wall at the origin, this reduction would be straightforward and one would immediately obtain an isomorphism between HR and IG [8]. In our case, however, the reduction is problematical, having the following defects:

(i) The image of the hard rod measure in the reduced description is Poisson only outisde a neighborhood of the origin.

(ii) The location of the unreduced origin jumps in the reduced description whenever a particle crosses it.

Because of (ii) the same particle (momentum pulse) may appear at the origin more than once, precluding the independence of future origin events from past ones. However, if the momentum distribution is as described, all particles will eventually leave the origin, never to reappear, since the origin undergoes essentially a random walk in the reduced description. The measure in the reduced descrption is then sufficiently close to Poisson to allow the properly defined "origin event" partitions to satisfy the weak Bernoulli property.

For more general velocity distributions, in particular when there are a finite fraction of particles with zero velocity, the loss of local information is no longer fast enough to prove the Bernoulli property. Nevertheless a "diffusive" type of loss to infinity is present giving rise to the K-property without a good local mechanism for loss of information. This is made more precise in the next chapter, where we indicate that a quantity, which may be interpreted as the local rate of dissipation (of, say, information) is zero for both IG and HR. We thus wish to consider systems for which a different mechanism, which is local, is responsible for the "mixing" which occurs, and it is to these that we next devote our attention.

The Lorentz Gas and Systems of Periodic K-Type

Instead of turning on a pair interaction, which, except in the most trivial cases, would make the investigation of ergodic properties exceedingly difficult, we turn on an external field. The particles thus remain independent, though the one particle systems are modified. We thus consider the Lorentz gas (LG), which is like IG except that the particles are excluded from a disconnected region whose components have disjoint convex closures and smooth boundaries from which the particles are elastically reflected.

These convex barriers are distributed throughout $\mathbb{R}^{\nu}$ in a periodic array (with periods $L_1,\ldots,L_{\nu}$). We will primarily consider the case when all particles have unit speed, so that the momentum distribution, instead of Maxwellian, is merely spherically symmetrical. These two systems, Maxwellian and constant speed, which may of course be represented as Poisson systems in an obvious manner, are identical insofar as any of the ergodic properties which we investigate are concerned. Therefore we will denote both systems by LG, though the arguments which we give will directly apply only to the latter system.

An important feature of LG is its periodic structure. A Poisson system will be called <u>periodic</u> if (a) its one-particle measure space can be represented in the form $(X_0,\mu_0)\times\mathbb{Z}^{\nu}$, where μ_0 is a finite measure and $\mathbb{Z}^{\nu}$ is regarded as equipped with counting measure, and (b) the representation of the one-particle dynamics in $X_0\times\mathbb{Z}^{\nu}$ commutes with the translations S_{ℓ}, $\ell = 1,\ldots,\nu$, of $X_0\times\mathbb{Z}^{\nu}$ induced by the natural unit translations of $\mathbb{Z}^{\nu}$. For LG we may take (X_0,μ_0) to be the restriction of the one particle measure to $V_0\times\mathbb{R}^{\nu}$ (momentum space), where V_0 is the ν-dimensional "rectangle" $\times_{i=1}^{\nu}[-\frac{1}{2}L_i, \frac{1}{2}L_i)$ of length L_i centered at the origin. Furthermore, since the dynamics of LG is invariant under translation by L_i in the ith direction, we also have (b).

For periodic systems a set of the form $X_0 \times \Lambda$ will be called a <u>rectangle</u> R if Λ is a (bounded ν-dimensional) rectangle of $\mathbb{Z}^{\nu}$. (Λ must be a product of intervals.) We equip any rectangle R with the natural probability measure ${}_R\mu_1$ induced by the one-particle measure μ_1 and with the automorphism ${}_RT_1$ obtained from the one-particle time evolution T_1 by replacing Z^{ν} by Z^{ν}/Z^{ν}_R, where Z^{ν}_R is the subgroup of Z^{ν} corresponding to the rectangle R. Finally, we let $R_0 = X_0 \times \{0\}$.

For LG any rectangle R (whose projection in $\mathbb{Z}^\nu$ has sides of length $n_1,\ldots,n_\nu$) is represented in $\mathbb{R}^\nu$ by a "rectangle" (with sides of length $n_1L_1,\ldots, n_\nu L_\nu$). The induced dynamics ${}_RT_1^t$ is obtained from T_1^t by identifying the opposite sides of R -- periodic boundary conditions. Thus the systems (R, ${}_R\mu_1$, ${}_RT_1$) are just the finite volume one-particle systems of which the Poisson system built over (X_1,Σ_1,T_1) is the infinite volume (and particle number) limit.

Though all the noninteracting systems considered here have been periodic, they can be distinguished by the degree of mixing exhibited by their finite volume time evolutions ${}_RT_1$. For IG ${}_RT_1$ will be "far from mixing", while for LG, as Sinai has shown (explicitly for $\nu = 2$), the ${}_RT_1$ are K-automorphisms (and are in fact, Bernoulli) [see Gallavotti's lecture and the references cited there]. RWν is similar to LG in this respect: for RW1, for example, ${}_RT_1$ is Bernoulli if R has odd length [4]. The good mixing behavior of these one particle systems should give rise to a (local) mechanism for the production of good mixing properties of the Poisson systems built over them.

LG and RWν thus satisfy:

(a) There exists a sequence R_i of rectangles, whose volumes (in $\mathbb{Z}^\nu$) approach infinity, such that the ${}_{R_i}T_i$ are K-automorphisms.

They also satisfy:

(b) $T_1(R_0)$ is bounded.

(c) ${}_{R_0}T_1$ has finite entropy (see [2] or Section 3 below for definition).

Periodic systems satisfying (a)-(c) will be said to be of <u>periodic-K-type</u>. ((b) and (c) are of purely technical significance.) The main result concerning them is contained in the following

Theorem. Systems of periodic K-type form K-systems.

We give a sketch of the proof. The basic idea is to "lift" the structures responsible for the mixing behavior of the one-particle system to the Poisson system built over it. The essence of the "lifting" is embodied in the following construction: If ξ is a measurable partition of (X_1, μ_1), we obtain a measurable parititon $\bar{\xi}$ of (X, μ) by partitioning X according to the number of particles in the fibers (elements) of ξ.

Lemma. If ξ satisfies

(1') $$T_1 \xi \geq \xi$$

(2') $$\bigvee_n T_1^n \xi = \varepsilon$$

(3') the fibers of $T_1^{-n} \xi$ expand toward infinity as $n \to \infty$, then $\bar{\xi}$ is a K-partition for (X, μ, T).

Here (3') means that for any point $x \in X_1$ and any bounded set $A \subset X_1$, the fraction of the fiber of $T_1^{-n}\xi$ containing x which overlaps A approaches zero.

Proof: (1) and (2) of the definition of K-partition follow easily from (1') and (2'), while (3) holds because (3') guarantees that the only information contained in $\bigwedge_n T^{-n} \bar{\xi}$ must be measurable with respect to the local σ-algebra of the complement of any bounded region of X_1, which, since our system is Poisson, implies (3) [4].

To obtain a partition satisfying (1')-(3') we utilize the K-structure possessed by (X_1, Σ_1, T_1). Let P be a finite generator for ${}_{R_0}T_1$. We form the "product" Q_0P, where Q_0 is the partition of X_1 induced by the partition of $\not{t}$ into points. Q_0P is the partition obtained by refining Q_0 by partitioning each of its elements "according to P". Let $\xi = \bigvee_{j=0}^{\infty} T_1^{-j} Q_0 P$ (the partition according to future Q_0P-names). ξ satisfies (1')-(3'). (1') and (2') are immediate. (3') is satisfied because ξ/R (the restriction of ξ to the rectangle R) is a K-partition for ${}_RT_1$ if ${}_RT_1$ is a K-automorphism. The Doob Martingale Theorem then implies (3') [4].

Finally, we observe that the partition ξ which we have constructed also satisfies

(4') $\quad\quad$ ξ is invariant under translations (S_ℓ),

(5') $\quad\quad$ $\xi \geq Q_0$.

Thus systems of periodic K-type possess translation invariant K-partitions ($\bar{\xi}$) reflecting a local mechanism for the dissipation of disturbances.

3. Space-Time Ergodic Theory of Infinite Systems

We denote by IG1 the system which differs from the one-dimensional IG only in that all particles are constrained to move with unit velocity to the "right", say. In exactly the same manner as for IG, IG1 forms a Bernoulli flow.

We are now in a position to further evaluate the adequacy of the (X,μ,T^t) framework for the account of the emergence of good thermodynamic behavior. Two comments are relevant:

(a) Since IG1, whose thermodynamic behavior must be regarded as rather poor, forms a Bernoulli flow -- the strongest of ergodic properties -- the ergodic properties associated with the (X,μ,T^t) framework cannot possibly bear an adequate relation to gtb.

(b) The situation is even worse. It appears likely that LG forms a Bernoulli flow. (If it does not it affords a rather natural example of a K-system which is not Bernoulli (see [2]). It is also rather plausible that realistic models of physical systems which exhibit gtb should have ergodic properties as strong as those of systems which do not, and hence should be Bernoulli. If this is so, then Ornstein's isomorphism theorem for Bernoulli flows would imply that these systems are completely identical from our abstract point of view, rendering the (X,μ,T^t) framework entirely inadequate for the formalization of the relevant thermodynamic distinctions.

(The relevant theorem asserts that two Bernoulli flows of infinite (K-S) entropy are isomorphic. All infinite systems to which we have thus far referred have infinite entropy. That systems which permit a Bernoulli construction have infinite entropy may be seen by noting that (X,Σ_{A0},μ) is a continuous measure space, and hence given any n we may find a partition P of X having n elements of equal measure which are measurable with respect to Σ_{A0}. Since Σ_{A0} is an independent generator, we have that $h(T) \geq h(P,T) = \log n$.)

Perhaps the simplest way to expand our abstract framework is to replace T by G in the triple (X,μ,T), where G is the abelian group generated by space translations and time evolution. This is sensible because translation invariant equilibrium states have the greatest physical significance, since they represent a homogeneous macroscopic situation. Furthermore, if the temperature and fugacity are such that no phase transition occurs (e.g., a low density gas), the unique equilibrium state must be translation invariant. In addition, since physical interactions are translation invariant, space translations and time evolution will commute. Finally, since the localizing (physical) space $(\mathbb{R}^\nu)$ is a crucial element in the underlying structure of infinite systems [see Lanford's lectures], it is natural that some token of that structure-space translations -- should be embodied in the abstract framework. We also note that the systems which we are considering-- the periodic systems -- fit naturally into the new framework, if by G we understand the group generated by T and the S_ℓ , i.e., the discrete space-time (translation) group.

For simplicity we consider only the discrete space-time group. We shall further limit the detailed description to the case of one spatial dimension, though everything easily extends, at the expense of somewhat more cumbersome notation, to higher dimensions.

Introduction of space translations produces the immediate benefit of alleviating the embarrassment posed by Ornstein's theorem: though the theorem should be generalizable to multi-dimensional abelian groups [9], it should be much more difficult, and, as we shall argue, of much greater significance, for an infinite system to be Bernoulli under the space-time group.

Space-Time Ergodic Properties of Periodic Systems

We will show that the two mechanism (the local and nonlocal) for the production of the "mixing" behavior of our systems, which lead to similar properties within the (X,μ,T) framework, lead to very different sorts of ergodic properties in the space-time framework.

We will denote by (X,Σ,μ,G) both the general dynamical system and specific infinite particle systems, with G the abelian group generated by two automorphisms, S and T, of the probability space (X,Σ,μ). If X is the phase space of a system of particles, S and T will denote respectively unit space translation and unit time evolution. If (X,Σ,μ,T) is a periodic (Poisson) system, (X,Σ,μ,G) will also be called a Poisson system, and we will continue to designate the corresponding one-particle system by means of the subscript "1".

Though the properties which we shall discuss will be formulated in terms of the pair (S,T), they will depend only upon G, unless we indicate otherwise. We shall have to distinguish between the system (X,μ,G) and the system $(X,\mu,(S,T))$ only if S and T are not independent (e.g., IG1) or when considering (S,T) K-systems.

A system (X,μ,G) is <u>ergodic</u> if measurable sets invariant under G have measure zero or measure one. Since this property is weaker than the ergodicity of T alone, we shall have nothing more to say about it.

$(X,\mu,(S,T))$ is said to be mixing if

$$\lim_{(n,m)\to\infty} \mu(S^n T^m A \cap B) = \mu(A)\,\mu(B)$$

for all $A,B \in \Sigma$. By $(n,m) \to \infty$ we mean that $n^2+m^2 \to \infty$. Thus if a system is (S,T) mixing it is not only mixing under both S and T; after a sufficiently long "time" an event A -- and all its translates -- will become approximately independent of a fixed event B. In particular, an observation will tend to become uncorrelated not only with future performances of that observation but, uniformly, with similar observations performed at different locations. Thus systems in which the dissipation of disturbances is purely nonlocal, i.e. in which the disturbances travel unhindered to infinity and are otherwise unaffected, are not (S,T) mixing.

This state of affairs is illustrated by IG1, for which S and T coincide. Here $S^n T^{-n} = I$ (the identity) so that IG1, with purely nonlocal dissipation, is not (S,T)-mixing.

On the other hand, IG is (S,T)-mixing. This is so because

$$\lim_{(n,m)\to\infty} \mu_1(S_1^n\, T_1^m\, A_1 \cap B_1) = 0$$

if A_1 and B_1 are bounded measurable subsets of X_1 $(= \mathbb{R}^2)$. Thus the space-time framework distinguishes two systems identical in the framework of the time evolution. That the mixing which occurs in IG is not purely nonlocal is a consequence of the dispersion due to the continuous momentum distribution.

The mixing which occurs in IG is nonetheless of a very poor type and we would thus like to be able, within the expanded framework, to distinguish IG from systems like LG in which the mixing is of a much "better" sort. To do so we will use the notions of (S,T)- K-system and G-entropy.

The concept of (S,T)-K-system which we will define (see [11]) is a property of an ordered pair (S,T) of automorphisms rather than of the group G which they generate. Insofar as space translations and time evolution play rather different roles in statistical mechanics, this distinction is quite appropriate. The definition may be regarded as obtained by means of a generalization of the natural ordering of $\mathbb{Z}$, on the structure of which the notion of K-system for a single automorphism is implicitly based, to an ordering of $\mathbb{Z}^2$. We will write $(n,m) \leq (p,q)$ if $m < q$ or if $m = q$ and $n \leq p$. $(X,\mu,(S,T))$ is said to be a K-system if there exists a measurable partition ζ which

(1) is increasing: $S^n T^m \zeta \geq \zeta$ if $(n,m) \geq (0,0)$

(2) generates: $\bigvee_{(n,m)} S^n T^m \zeta = \varepsilon$,

(3) has trivial tail: $\bigwedge_{(n,m)} S^n T^m \zeta = \nu$, and

(4) has "continuity" $\bigwedge_n S^n \zeta = T^{-1} \bigvee_n S^n \zeta$.

(1)-(3) imply that there exists an S-invariant K-partition for T. As it does not appear likely that IG should possess such a partition, it is suggested that IG is not a (S,T)-K-system. That this is so we shall later show using G-entropy. Furthermore, if (X,μ,T) is of periodic K-type, it is not difficult to construct from the K-partition $\overline{\xi}$ obtained for this system, using the fact that ξ satisfies (4') and (5') (see end of Section 2), a partition satisfying (1)-(4) above. Thus systems of periodic-K-type, and in particular LG form space-time K-systems, and IG is not isomorphic to LG in the space-time framework.

We mention that we are interested in the concept of (S,T)-K-system primarily because it is a property which is easy to check for systems of periodic-K-type and which implies that the system has completely positive G-entropy, of which we shall soon say more.

The entropy of G is defined in a manner completely analogous to the definition of the entropy of an automorphism T (see [2]). Recall that the entropy H(P) of a partition $P = \{P_i\}$ is given by

$$H(P) = - \sum \mu(P_i) \log \mu(P_i) .$$

h(P,G) is defined by

$$h(P,G) = \lim_{\alpha\to\infty} \frac{1}{|\alpha|} H\Big(\bigvee_{(n,m)\in\alpha} S^n T^m P\Big)$$

where $\alpha \subset \mathbb{Z}^2$ is a parallelogram, $|\alpha|$ is the number of points in α, and $\alpha \to \infty$ means that the distances between parallel sides approach infinity. That this limit exists is a consequence of the subadditivity of the entropy of a partition:

$$H(P_1 \vee P_2) \le H(P_1) + H(P_2) .$$

The entropy $h(G)$ of G is then given by

$$h(G) = \sup_P h(P,G) .$$

If P generates (i.e. if $\bigvee_{(n,m)} S^n T^m P = \varepsilon$) we have that $h(G) = h(P,G)$, just as for a single automorphism. Recall, too, that for a single automorphism we have that

$$h(P,T) = H\Big(P \Big| \bigvee_{j<0} T^j P\Big) .$$

Using the ordering of $\mathbb{Z}^2$ introduced above, one may formulate (see [10]) a similar expression for the G-entropy of a partition P:

$$h(P,G) = H(P \mid \bigvee_{(n,m)<(0,0)} S^n T^m P)$$

This expression implies that $h(G)$ is, rather roughly, a measure of the local (asymptotic) rate of loss of information.

Another, apparently better, measure of the local rate of information dissipation may be obtained as follows: If (X,μ,T) is the time evolution of an equilibrium state of an infinite system of particles (translation invariant) and if Λ is a finite rectangular volume, we may consider the projection $(X_\Lambda,\mu_\Lambda,T_\Lambda)$ of (X,μ,T) into Λ obtained by ignoring all particles not in Λ and by defining T_Λ unambiguously by means of periodic boundary conditions (say) and by allowing the motions of the individual particles to be influenced only by the mutual interaction between particles in Λ and possibly by an external field. If (X,μ,T) is a periodic (Poisson) system, Λ should be taken to be a rectangle (see Section 2).

We define the T-entropy per unit volume to be the

$$\lim_{\Lambda\to\infty} \frac{h(T_\Lambda)}{|\Lambda|},$$

where $|\Lambda|$ denotes the volume of Λ and $\Lambda \to \infty$ in the sense that the smallest side of Λ approaches infinity. This quantity may be regarded as a measure of the loss of information (due to "collisions") occurring in the system per unit volume per unit time (local rate of information loss).*

It is not obvious from the definition that the local rate of information loss is an invariant of the (S,T)-framework, i.e. that two systems which are abstractly isomorphic have the same rate of information loss. If this quantity were given by $h(G)$, this of

---

* We are indebted to O. Penrose for very valuable suggestions and discussions on this point.

course would be so, and, at the same time, we would have a nice interpretation for $h(G)$. In this direction we have the following result:

Proposition. If (X,μ,G) is a Poisson system (satisfying some rather mild additional conditions (see [11])), then $h(G) = h(T_\Lambda)/|\Lambda|$ for any rectangle Λ (no limit necessary).

We conjecture that a similar proposition holds for a general (interacting) infinite system; of course, the limit $\Lambda \to \infty$ would then be essential.

The above proposition implies that IG and LG (say) can also be distinguished on the basis of their space-time entropies: $h(G)$ must be finite and nonvanishing for LG since it is so for the finite volume projections (K-systems have positive entropy) while the finite volume IG has vanishing entropy. We now give a sketch of a direct proof of the vanishing of $h(G)$ for IG. We do this by finding a partition P such that

$$\bigvee_{j=-\infty}^{-1} T^j\Big[\bigvee_n S^n P\Big] = \varepsilon \ .$$

This will suffice because we would have

$$h(G) = h(P,G) = H\big(P|\varepsilon\big) = 0 \ ,$$

since $P \leq \varepsilon$.

We choose for P the partition with elements of the form

$$P_{\{n;m_1,\ldots,m_n;k_1,\ldots,k_n\}} \ .$$

Here n is the number of particles in $[0,1)$ (i.e., in $[0,1) \times \mathbb{R}$, the unit cell at the "right" of the origin in the one-dimensional IG). The particles in any cell $[\ell,\ell+1)$ are indexed according to their distance from the "left" boundary of that cell. m_i labels the cell which at $t = 1$ will contain the particle with index i in $[0,1)$ (the ith particle from the origin) and k_i is the index at $t = 1$

in $[m_i, m_i+1)$ of that particle.

Thus the partition $\bigvee_{j=-\infty}^{\infty} S^j P$ gives not only the number of particles in each "unit cell", but also provides some velocity information by giving the cell membership of each particle at t = 1 along with sufficient index information to keep track of the identity of each particle so that upon successive applications of T we obtain partitions giving information about the trajectories with respect to unit cells of the t = 0 indexed particles. The velocity and the exact position of each particle may then be determined (using, e.g., the Jacobi theorem for the irrational rotation of the circle), proving the desired result. (We remark that though the partition P just defined is not finite, its entroph H(P) is finite [11], and partitions of finite entropy have essentially the same properties as finite partitions [12]). By a similar argument it may be shown that $h(G) = 0$ for the hard rod systems as well, providing some justification for our assertion that the mixing which occurs in the hard rod systems is of essentially the same sort as occurs in IG.

Just as for a single automorphism, (X,μ,G) is said to have completely positive entropy if $h(P,G) > 0$ for all nontrivial partitions P. For a single automorphism it is this concept which is equivalent to that of K-system, upon which the characterization of K-systems in terms of finite approximation and completely nondeterministic behavior given in the introduction is based. It is therefore of interest that (S,T)-K-systems have completely positive G-entropy. (We don't know whether the converse holds.) IG thus does not form a (S,T)-K-system. Systems with completely positive G-entropy are completely nondeterministic in a much stronger sense than would be implied by completely positive T-entropy. These systems do not admit of an approximation, in the sense given in the introduction, by the outcomes of a measurement (coarse graining) which is global and only locally finite, in the sense that it is performed by making a doubly infinite sequence of measurements (corresponding to $\bigvee_n S^n P$) each of which is a translate of some finite measurement (corresponding to a finite partition P). This is evidently a reasonable extension to infinite systems of the concept of "finite approximation" for finite systems.

The final property we wish to mention is that of G-Bernoulli system. This may be defined by the existence of a measurable partition P which is an independent generator for G, i.e. $\{S^n T^m P\}$, $n \in \mathbb{Z}$, $m \in \mathbb{Z}$, forms an independent family of partitions and generats If (X,μ,G) is Bernoulli, then (a) (X,μ,S) is Bernoulli and (b) (X,μ,T) has an S-invariant independent generator (for T, namely, $\bigvee_n S^n P$). G-Bernoulli systems can in fact be characterized by (a) and (b), since factors of Bernoulli shifts are Bernoulli [2]. As the space-time Bernoulli property implies the existence of a (generating) global measurement whose successive (time) iterates are independent, it appears to be an appropriate extension to infinite systems of the time Bernoulli property for finite systems. We have not yet found any particle systems to be G-Bernoulli.

Concluding Remarks

(i) The G-ergodic properties are invariant under (integral) Galilean transformations (this need not be the case for T-ergodic properties -- consider IG1). This is so because from the abstract standpoint we may view the sole effect of a Galilean transformation as the replacement of the pair (S,T) by another pair of generators (S',T') of G (see [11]).

(ii) Space-time ergodic theory, like time ergodic theory,

cannot deal directly with the approach to the equilibrium, as opposed to the return to equilibrium, problem (see Section 1). Thus, if the LG starts out with an initial velocity distribution f(p) which is not spherically symmetric (but is otherwise Poisson) we cannot, on the basis of its good G-ergodic properties say anything about approach to a stationary state, since this initial state is not absolutely continuous with respect to any stationary state. Nevertheless it is easy to see that the same local mixing mechanism which is responsible for LG being a G K-system will also bring the system to a stationary state, with a velocity distribution $\bar{f}(p)$ obtained from f(p) by angular averaging. Similar results can be proven for more general initial states of this kind. Indeed we expect an approach to a stationary state to occur whenever the initial state is 'singular' with respect to the good stationary measure <u>only</u> in a global sense, i.e. its projection on any finite region Λ, would be absolutely continuous with respect to a correspondingly projected stationary measure (with good mixing properties). Of course in LG all Poisson measures with spherical symmetric velocity distributions are stationary and there is therefore no approach to a 'unique' equilibrium state. We may expect however that in a system with mutual particle interactions which also possesses a good local dissipation mechanism, e.g. a hard sphere system, a true approach to equilibrium will take place from initial states 'locally absolutely continuous' with respect to the equilibrium state. We conjecture that good G-ergodic properties (e.g. completely positive G-entropy) will be somehow related, as in LG, to such good (nonergodic theoretic) gtb behavior.

(iii) In attempting to formulate, in the infinite system context, additional properties of physical import similar to that of mixing for finite systems several vague possibilities suggest themselves. We mention a few:

(a) One might consider some sort of topological mixing: Infinite systems come equipped with a natural topology within whose framework measure zero difficulties or consequences thereof may not appear.

(b) The framework of something like Axiom A Attractors might be useful. Perhaps the phase space is largely decomposable into disjoint regions, each with their own "attractor" and a unique limit measure on each "attractor".

(iv) Finally, we note that to completely account for gtb one would almost surely have to use the limit employed for the derivation of the Boltzmann equation [see Lanford's Lectures], in which in particular the particle density becomes infinite while the size of the particle vanishes. We expect however that such a limit should not be necessary for the problem of (nondetailed) approach to equilibrium -- it should be possible to formulate the infinite system "physical" analog of mixing for finite systems.

Acknowledgements

We thank Oscar Lanford III and Oliver Penrose for many very valuable discussions and suggestions.

References

[1] Lebowitz, J. L., and Penrose, O., Physics Today, Vol. 26, No. 2 (1973).

[2] Ornstein, D. S., Ergodic Theory, Randomness, and Dynamical Systems, in New Haven: Yale University Press, 1974; see also his contributions in this volume.

[3] Goldstein, S., Occupation number measures and the uniqueness of the state in classical statistical mechanics (to appear).

[4] Goldstein, S., and Lebowitz, J. L., Commun. Math. Phys. 37, 1 (1974).

[5] Sinai, Y. G., Funkts. Analiz. 6, 1 (1972) 41.

[6] Aizenmann, M., Goldstein, S., and Lebowitz, J. L., Ergodic properties of a one-dimensional system of hard rods with an infinite number of degrees of freedom, Commun. Math. Phys., 39, 289 (1975).

[7] Shields, P., The theory of Bernoulli shifts, University of Chicago Press (1973).

[8] Pazzis, O., de, Commun. Math. Phys. 22 (1971) 121.

[9] Katznelson, Y., and Weiss, B., Israel S. Math. 12 (1972) 161.

[10] Conze, J. P., Z. Wahrscheinlichkeitstheorie verw. Geb. 25 (1972) 11-30.

[11] Goldstein, S., Space-time ergodic properties of systems of infinitely many independent particles, Commun. Math. Phys., 39, 303 (1975).

[12] Parry, W., Entropy and generators in ergodic theory, New York: Benjamin (1969).

TIME EVOLUTION AND ERGODIC PROPERTIES OF HARMONIC SYSTEMS *

Oscar E. Lanford III

Department of Mathematics, University of California, Berkeley, Cal.

Joel L. Lebowitz

Belfer Graduate School of Science, Yeshiva University, New York

Abstract

We prove the existence of a time evolution and of a stationary equilibrium measure for the infinite harmonic crystal. The ergodic properties of the system are shown to be related in a simple way to the spectrum of the force matrix; when the spectrum is absolutely continuous, as in the translation invariant crystal, the flow is Bernoulli. The quantum crystal is also discussed.

1. Introduction

The ergodic properties of infinite systems are of considerable interest. They yield information about the expected (average) time dependent behavior of physical observables when the system is in or near equilibrium. Unfortunately very little is known at the present time about such properties for realistic systems. It seems therefore valuable to study the behavior of model systems. These may shed some light on which properties of the interactions are relevant to ergodic behavior. Previous studies along these lines have dealt with the ideal gas, the one dimensional hard rod system and the general noninteracting system in an external field produced by fixed scatterers (e.g. the Lorentz model). (For a review see the article by Goldstein Lebowitz and Aizenman in this volume.)

* Supported in part by NSF Grant GP 42225 and AFOSR Grant 73-2430B.

In this note we investigate the time evolution and ergodic properties of the infinite harmonic crystal. The harmonic crystal is commonly used by physicists as a model of an ideal solid and is quite successful in describing many features of real solids.* Explicit calculations of the time evolution of specified phase functions have been carried out for a variety of harmonic systems beginning with the work of Newton and Bernoulli on one dimensional lattices with nearest neighbor interactions† (example (a) below). The deay of time dependent correlations in these systems (which is directly related to ergodic properties) has also been studied by various authors.‡ More recently Titular [3] has studied the harmonic crystal from the general point of view of ergodic theory and has arrived at some of the same conclusions that we arrive at by somewhat different methods.

The outline of this paper is as follows: in Section 2 we prove the existence of the time evolution of the infinite harmonic system for a very large class of initial conditions with only very mild restrictions on the force matrix. This time evolution is the natural limit of the finite system time evolution. In Section 3 we prove the existence of an invariant equilibrium (Gaussian) measure for the infinite system which is the limit of the finite system canonical measure. (For this we require some conditions on the force matrix which fail for translation invariant forces in one and two dimension where the measure space has to be modified.)

In Section 4 we investigate the ergodic-theoretic properties of this dynamical system, and reduce the determination of these

---

* For a general reference see Maradudin, Montroll and Weiss [1].

† See Section 4 of Chapter II of [1] for historical background.

‡ See Sections 7 and 8 of Chapter VII of [1] and Section III of [2] for discussion and references to the works of Hemmer, Mazur, Montroll, Rubin and others.

properties to an analysis of the spectrum of the matrix of interparticle force constants. We also show how to modify our general formalism to apply in one and two dimensions. In Section 5, we extend the analysis of Section 4 by investigating the consequences of translation invariance, and we show that the time evolution flow T^t for a translation-invariant crystal with finite-range interactions is a Bernoulli flow except in special pathological cases. If, however, the crystal is not translation-invariant then the flow may not be ergodic, e.g. if there is a particle with a small mass giving rise to a local mode. In Section 6, we sketch the extension of our analysis to quantum harmonic crystals.

2. Time Evolution

We begin by investigating the time evolution of an infinite harmonic crystal. Our discussion will apply to arbitrary lattices and allows for rather general "defects", but at the expense of a slightly complicated notation.

We can describe a general lattice in the ν-dimensional space $\mathbb{R}^\nu$ by specifying the group Γ of translations carrying the lattice onto itself. Γ is a discrete subgroup of the additive group $\mathbb{R}^\nu$; as an abstract group it is isomorphic to $\mathbb{Z}^\nu$. It need not act transitively on the lattice, i.e., given two lattice sites, there need not be a lattice translation carrying one to the other. If we take the quotient space of $\mathbb{R}^\nu$ under the action of Γ, we obtain a ν-dimensional torus which is called the unit cell (for Γ). If we choose representatives for the elements of the unit cell in a straightforward way, we obtain a parallelepiped Δ_0 in $\mathbb{R}^\nu$; the translates of this parallelepiped under Γ are disjoint and cover all of $\mathbb{R}^\nu$. Labelling these parallelepipeds by elements of Γ, (i.e. $\Delta_\alpha = \Delta_0 + \alpha$), we obtain "coordinates" for $\mathbb{R}^\nu$ in which a point q is described by giving an element α of Γ and a point ξ of Δ_0 ; the point q may be expressed in terms of its "coordinates" by $q = \alpha + \xi$. To complete the description of a regular lattice, we need to specify the set of lattice sites in the unit cell Δ_0 , i.e. we must specify a finite subset $X_0=\{\xi_1,\ldots\xi_J\}$ of Δ_0; then the lattice is precisely the set of points of the form $\alpha + \xi_i$, where α is an arbitrary element of Γ and ξ_i an arbitrary element of X_0.

To allow for defects (e.g., for the absence of some of the particles), we loosen the definition a bit by allowing a different finite subset X_α of Δ_0 for each α, so the points of the "lattice" are then all points of the form $\alpha + \xi_i$; $\alpha \in \Gamma$, $\xi_i \in X_\alpha$. (For most applications, one will want to impose some restrictions on the X_α's, e.g., that they should all be subsets of some fixed finite set $\bar{X}$, but we seem to gain nothing by imposing these restrictions explicitly at this point. Some of them will appear implicitly in the assumptions we make later about the forces.)

The points of our (generalized) lattice are supposed to represent the equilibrium positions of the particles making up our crystal. To describe the dynamics, we introduce a position variable $q_{\alpha,i} \in \mathbb{R}^\nu$ for each lattice site giving the displacement of the particle in question from its equilibrium position and a conjugate momentum variable $p_{\alpha,i}$. Thus, the position of the particle with label (α,i) is $\alpha + \xi_i + q_{\alpha,i}$. The equations of motion read:

$$(2.1) \qquad m_{\alpha,i} \frac{dq_{\alpha,i}}{dt} = p_{\alpha,i} , \qquad \frac{dp_{\alpha,i}}{dt} = F_{\alpha,i}$$

where $m_{\alpha,i}$ is the mass of the (α,i) particle and the force $F_{\alpha,i}$ depends on the $q_{\beta,j}$'s. We now assume that the $F_{\alpha,i}$ are linear functions of the q's, and that $q_{\alpha,i} = 0$ for all α,i represents an equilibrium position. (Note that this last assumption places, implicitly, very strong restrictions on the sort of defects we can allow if we are deriving our equations from a realistic model of a crystal.) Thus, we can write

$$F_{\alpha,i} = - \sum_{\beta,j} V_{\alpha,i;\beta,j}\, q_{\beta,j}$$

where each $V_{\alpha,i;\beta,j}$ is a $\nu \times \nu$ (real) matrix.

The above may be formally simplified if we think of the sequence $(q_{\alpha,i})$ as a vector $\underline{q}$ in an infinite-dimensional vector space. Then the equations of motion can be written schematically as

$$\underline{M}\, \frac{d^2\underline{q}}{dt^2} = - \underline{V}\, \underline{q}$$

where $\underline{M}$ is the diagonal matrix with entries $m_{\alpha,i}$ and $\underline{V}$ is the force matrix. Let us look at two simple examples of the above.

(a) The perfect harmonic chain with nearest neighbor interactions: In this case, we take $\nu = 1$, $\Gamma = \mathbb{Z}$, and only one particle in each unit cell. (We may therefore dispense with the subscript i.) The potential energy is then given formally by

$$\frac{\gamma}{2} \sum_{\alpha} (q_{\alpha+1} - q_{\alpha})^2 \, , \quad \gamma \text{ a positive constant.}$$

(The q 's are now just numbers, since $\nu = 1$.) We may rewrite this expression as $\frac{1}{2}(\underline{V}\underline{q},\underline{q})$, where $\underline{V}$ is a tridiagonal matrix whose rows and columns are labelled by $\mathbb{Z}$, $V_{\alpha\alpha} = 2\gamma$, $V_{\alpha\beta} = -\gamma$, for $\beta = \alpha \pm 1$ and is zero otherwise.

From the original formula for $\underline{V}$ it is clearly positive semi-definite but not positive definite in the strict sense (If all the q_α's are equal, $(\underline{V}\underline{q},\underline{q}) = 0$. If, however, $\underline{q} \neq 0$ but only finitely many q_α's are different from zero, then $(\underline{V}\underline{q},\underline{q}) > 0$.). The equations of motion may be written simply as

$$m_\alpha \frac{d^2 q_\alpha}{dt^2} = -\sum_\beta V_{\alpha,\beta}\, q_\beta = \gamma\, (q_{\alpha+1} - 2q_\alpha + q_{\alpha-1})$$

(b) The two dimensional triangular lattice. Here we take $\nu = 2$; we again take only one lattice site in each unit cell, and we take the lattice to look like

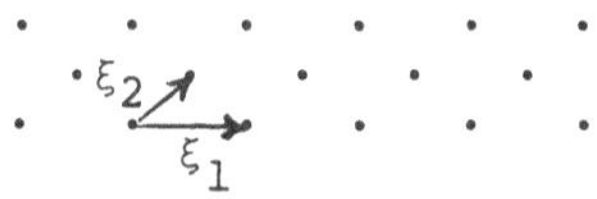

is generated by ξ_1 and ξ_2. If we choose the unit length so that $|\xi_1| = |\xi_2| = 1$, we have $\xi_1 = (1,0)$; $\xi_2 = (\frac{1}{2}, \frac{\sqrt{3}}{2})$, and

$$\Gamma = \{(n_1 + \tfrac{1}{2} n_2, \tfrac{\sqrt{3}}{2} n_2) : n_1, n_2 \in \mathbb{Z}\} .$$

We now associate a particle with each lattice site, and let the particles interact by a pair potential $\phi(r)$ which is repulsive for $r < 1$, has a minimum at $r = 1$ and vanishes for $r > r_0$, $r_0 < \sqrt{3}$, (the distance between second neighbor sites on the lattice). The configuration $q_\alpha = 0$ for all α is an equilibrium configuration, and the potential energy near this equilibrium configuration is approximated by

$$\frac{\gamma}{4} \sum_\alpha \sum_\xi \left[\left((q_{\alpha+\xi} - q_\alpha), \xi\right)\right]^2 + \text{constant} ,$$

where the sum ξ is taken over the six "nearest neighbors of 0",

$$\pm\xi_1, \pm\xi_2, \pm(\xi_1 - \xi_2) , \quad \text{and where } \gamma = \phi''(1) .$$

We now return to our general formalism. To specify the initial conditions, for the solution of (2.1) it is necessary to specify a momentum $p_{\alpha,i}$ as well as a position $q_{\alpha,i}$ for each α, i. If $\underline{x}$ denotes a sequence $(q_{\alpha,i}, p_{\alpha,i})$, we define

$$\|\underline{x}\|_k = \sup_{\alpha,i} \frac{|q_{\alpha,i}| \vee |p_{\alpha,i}|}{(1+|\alpha|^2)^k}$$

for any positive integer k. We let χ_k denote the set of sequences $\underline{x}$ such that

$$\lim_{\alpha\to\infty} \sup_i \frac{|q_{\alpha,i}| \vee |p_{\alpha,i}|}{(1+|\alpha|^2)^k} = 0 \;;$$

χ_k is a Banach space with the norm $\| \ \|_k$. (Note that we have not taken χ_k to be the set of all $\underline{x}$ with $\|\underline{x}\|_k < \infty$. Our choice turns out to be convenient later on; note, however, that if $\|\underline{x}\|_k < \infty$ then $\underline{x} \in \chi_{k+1}$.)

The equation of motion can be written formally as

$$\frac{d}{dt}\,\underline{x}(t) = \underline{A}\;\underline{x}(t) \;,$$

where

$$(\underline{\underline{A}}\underline{x})_{\alpha,i} = \Big(\frac{1}{m_{\alpha,i}}\,p_{\alpha,i}\,,\; -\sum_{\beta,j} V_{\alpha,i;\beta,j}\,q_{\beta,j}\Big) \;.$$

We now assume

(1) $\inf_{\alpha,i} m_{\alpha,i} > 0.$

(2) For each positive integer k, $\sup_{\alpha,i} \sum_{\beta,j} \|V_{\alpha,i;\beta,j}\| \,(1+|\alpha-\beta|^2)^k < \infty.$

The second assumption means that the forces between the unit cells Δ_α and Δ_β drop off rapidly as $|\alpha-\beta| \to \infty$. In the case where there is full translation invariance, so

$$V_{\alpha,i;\beta,j} = V_{i,j}(\alpha-\beta)$$

our assumption amounts to the requirement that $V_{i,j}(\xi)$ goes to zero more rapidly than any inverse power of $|\xi|$ as ξ goes to infinity.

Under these assumptions on the forces and masses, it is trivial to prove

<u>Proposition 1</u>. For each k, $\underline{A}$ is a bounded linear operator on χ_k. Hence if $\underline{x}_0 \in \chi_k$, there is a unique global solution $\underline{x}(t)$ of the equation

$$\frac{d\underline{x}}{dt} = \underline{A}\,\underline{x} \quad \text{on} \quad \chi_k ,$$

with $\underline{x}(0) = \underline{x}_0$; the solution is given by

$$\underline{x}(t) = e^{t\underline{A}}\,\underline{x}_0 \; ; \qquad e^{t\underline{A}} = \sum_{n=0}^{\infty} \frac{t^n}{n!}\,\underline{A}^n .$$

In particular, if $(q_{\alpha,i})$ and $(p_{\alpha,i})$ are arbitrary polynomially bounded sequences, there exist solutions $q_{\alpha,i}(t)$, $p_{\alpha,i}(t)$ of the equations

$$\text{(2.2)} \qquad \frac{dq_{\alpha,i}(t)}{dt} = \frac{p_{\alpha,i}(t)}{m_{\alpha,i}} \; ; \qquad \frac{dp_{\alpha,i}(t)}{dt} = - \sum_{\beta,j} V_{\alpha,i;\beta,j}\, q_{\beta,j}(t)$$

with $q_{\alpha,i}(0) = q_{\alpha,i}$; $p_{\alpha,i}(0) = p_{\alpha,i}$, and the solutions remain polynomially bounded (in α) for each t.

The uniqueness statement in the above proposition is not as strong as it sounds. The point here is that what is asserted is only that there is no other solution of the equations of motion such that

$$\lim_{\delta t \to 0} \left\| \frac{\underline{x}(t+\delta t) - \underline{x}(t)}{\delta t} - \underline{A}\underline{x}(t) \right\|_k = 0$$

i.e., no other solution in which the convergence of difference quotients to derivatives has the right amount of uniformity in α,i. We can improve this situation somewhat:

<u>Proposition 2</u>. Let $\underline{x}_0 \in \chi_k$ for some k, and let $\underline{x}(t)$ be a solution of (2.2) with $\underline{x}(0) = \underline{x}_0$ such that for some k', $\varepsilon > 0$,

$$\|\underline{x}(t)\|_{k'} \text{ is bounded on } 0 \leq t < \varepsilon .$$

Then $\underline{x}(t) \in \chi_k$ and $\underline{x}(t) = e^{\underline{A}t}\,\underline{x}_0$ for $0 \leq t < \varepsilon$.

This follows in an obvious way from the fact that $e^{At}\,\underline{x}_0 \in \chi_k$ if $\underline{x}_0 \in \chi_k \subset \chi_{k'}$ for $k' \geq k$.

We remark here that for finite range interactions the above results extend immediately to the case where the initial values $\underline{x}_0$ are exponentially bounded in α.

Although we have dealt directly with the infinite system, the time evolution we have obtained is the limit of the finite harmonic system in the region Λ with the boundary "tied down", i.e., with $q_{\alpha,i}(t) \equiv 0 \equiv p_{\alpha,i}(t)$ for $\alpha,i \notin \Lambda$. We can see this as follows. Let Λ be a finite subset of the crystal, and let P_Λ be the operator on the space of sequences $(q_{\alpha,i}, p_{\alpha,i})$ which puts all coordinates belonging to lattice sites α,i outside Λ equal to zero and leaves those belonging to lattice sites inside Λ unchanged. The equation of motion for the system in Λ is then

$$\frac{d}{dt} \underline{x}_\Lambda(t) = P_\Lambda \underline{A} P_\Lambda \underline{x}_\Lambda(t) ,$$

i.e.

$$\underline{x}_\Lambda(t) = \exp [t(P_\Lambda \underline{A} P_\Lambda)] P_\Lambda \underline{x}_0$$

But P_Λ converges strongly to 1 on each χ_k, so

$$\lim_{\Lambda\to\infty} \underline{x}_\Lambda(t) = \underline{x}(t)$$

for all t and all $\underline{x}_0 \in \bigcup_k \chi_k$.

To conclude this section, we establish a number of notational conventions to be observed for the remainder of this article.

1. We will assume that there is exactly one particle per unit cell (Bravais lattice), so the lattice may be identified with Γ. We may therefore drop the subscript i. It should be kept in mind, however, that p_α and q_α are elements of $\mathbb{R}^\nu$, not $\mathbb{R}$, and hence that there is another index which we suppress entirely from our notation.

2. We make a canonical transformation

$$q'_\alpha = m_\alpha^{1/2} q_\alpha ; \qquad p'_\alpha = m_\alpha^{-1/2} p_\alpha ,$$

and we introduce

$$S_{\alpha\beta} = (m_\alpha m_\beta)^{-1/2} V_{\alpha\beta}$$

Then the equations of motion read

$$\frac{dq'_\alpha}{dt} = p'_\alpha \; ; \qquad \frac{dp'_\alpha}{dt} = - \sum_\beta S_{\alpha\beta}\, q'_\beta$$

i.e., in the new variables the system looks like a system of particles of unit mass with force matrix $S_{\alpha\beta}$. We will from now on use only the variables q'_α, p'_α and will drop the primes.

3. We will need names for various spaces of sequences $(\xi_\alpha)_{\alpha\in\Gamma}$. Such sequences will usually be assumed to take values in $\mathbb{R}^\nu$. We write

$$\begin{aligned}
\ell^2(\Gamma) &= \{(\xi_\alpha)_{\alpha\in\Gamma} : \sum_\alpha |\xi_\alpha| < \infty\} \\
s(\Gamma) &= \{(\xi_\alpha)_{\alpha\in\Gamma} : \sup_\alpha |\xi_\alpha| \cdot (1 + |\alpha|)^k < \infty \text{ for all } k\} \\
s'(\Gamma) &= \{(\xi_\alpha)_{\alpha\in\Gamma} : \sup_\alpha \frac{|\xi_\alpha|}{(1+|\alpha|)^k} < \infty \text{ for some } k\} \\
d(\Gamma) &= \{(\xi_\alpha)_{\alpha\in\Gamma} : \xi_\alpha = 0 \text{ for all but finitely many } \alpha\}.
\end{aligned}$$

Occasionally, we will want to consider sequences with values in $\mathbb{C}^\nu$ rather than $\mathbb{R}^\nu$; we then use the subscript $\mathbb{C}$, e.g., $\ell^2_{\mathbb{C}}(\Gamma)$.

3. <u>Equilibrium Statistical Mechanics</u>

The existence of the time evolution was proven without any positivity (or even symmetry) assumption on $\underline{V}$. For physical applications the matrix $\underline{V}$ and hence $\underline{S}$ should be, in an appropriate sense, positive (otherwise we would certainly not have a crystal). We shall therefore assume:

(3) The infinite matris $S_{\alpha\beta}$ defines a (bounded) strictly positive operator S on the sequence space $\ell^2(\Gamma)$ i.e. $(\Psi,S\Psi) \geq 0$, for $\Psi \in \ell^2(\Gamma)$ with the equality holding only if $\Psi = 0$.

It follows from (3) that $H_\Lambda(\underline{q},\underline{p}) > 0$ unless $\underline{q} = \underline{p} = 0$. *

---

* $H_\Lambda(\underline{q},\underline{p}) = \frac{1}{2} \sum p_\alpha{}^2 + \frac{1}{2} \sum S_{\alpha\beta} q_\alpha q_\beta$, $\alpha,\beta \in \Lambda$ is the Hamiltonian of the (finite) crystal in the domain Λ with "tied down" boundary conditions.

The equilibrium canonical ensemble* for the finite system (with temperature set equal to unity in appropriate units) may now be described economically as the Gaussian measure on the space of finite sequences $(q_\alpha, p_\alpha)_{\alpha\in\Lambda}$ with mean zero and covariance

$$E_\Lambda\{p_\alpha\ p_\beta\} = \delta_{\alpha\beta}\ ; \quad E_\Lambda\{q_\alpha\ q_\beta\} = (S_\Lambda)^{-1}_{\alpha\beta}\ ; \quad E_\Lambda\{q_\alpha\ p_\beta\} = 0.$$

The matrix S_Λ^{-1} is the matrix obtained by restricting $S_{\alpha,\beta}$ to α,β in Λ and inverting the resulting finite matrix. Thus $(S_\Lambda)^{-1}_{\alpha\beta}$ is not at all the same as $(S^{-1})_{\alpha\beta}$ for $\alpha,\beta \in \Lambda$. By assumption (3) zero is not an eigenvalue of the (positive) operator S_Λ and hence S_Λ^{-1} exists.

Assumption (3) also implies that S^{-1} makes sense as a (usually unbounded) densely defined self-adjoint operator on $\ell^2(\Gamma)$. We denote the positive sequre root of this operator by $S^{-1/2}$. We can now pass (formally at least) to the infinite system if we assume

(4) The vector $e^{(\alpha)} \in \ell^2(\Gamma)$ which has 1 in the αth place and zero elsewhere is in the domain of $S^{-1/2}$ for all α, and the sequence of numbers

$$\| S^{-1/2}\ e^{(\alpha)} \|$$

is polynomially bounded.

It seems to be hard to verify (or disprove) assumption (4) in realistic situations, except that it can normally be expected to fail in one and two dimensions and hold in higher dimensions. We will return to this point later, when we investigate the consequences of assuming that the force matrix $\underline{V}$ is translation invariant. It will be seen that (4) can be relaxed.

* See Lanford's lectures.

It seems natural to define the canonical ensemble for the infinite harmonic crystal to be the Gaussian probability measure P on the space of pairs (q_α, p_α) of polynomially bounded sequences with mean zero and covariance given by

$$E\{p_\alpha\, p_\beta\} = \delta_{\alpha,\beta}\ , \quad E\{q_\alpha\, q_\beta\} = (S^{-1/2}\, e^{(\alpha)},\, S^{-1/2}\, e^{(\beta)}),\ E\{p_\alpha\, q_\beta\}=0.$$

We will shortly justify this definition by showing that this measure is indeed the limit of the finite system canonical ensemble as Λ becomes infinitely large. Two questions now suggest themselves.

(a) Is the measure P invariant under the time evolution defined by solving the equations of motion

$$\frac{dq_\alpha}{dt} = p_\alpha\ , \qquad \frac{dp_\alpha}{dt} = -\sum_\beta S_{\alpha\beta}\, q_\beta \quad ?$$

(b) If so, what are the ergodic properties of the dynamical system thus defined?

We will give an affirmative answer to (a), and show how (b) can be reduced in a certain sense to an analysis of the spectral properties of S acting on $\ell^2(\Gamma)$.

Invariance of the Measure

If $(q_\alpha(t),\, p_\alpha(t))$ denotes the solution of the equations of motion with initial data (q_α, p_α), it follows easily from the proof of Proposition 1 that the solutions can be written in the form

$$(3.1)\qquad \begin{aligned} q_\alpha(t) &= \sum_\beta K^{(1)}_{\alpha,\beta}(t)\, q_\beta + \sum_\beta K^{(2)}_{\alpha,\beta}(t)\, p_\beta \\ p_\alpha(t) &= \sum_\beta K^{(3)}_{\alpha,\beta}(t)\, q_\beta + \sum_\beta K^{(4)}_{\alpha,\beta}(t)\, p_\beta \end{aligned}$$

where the kernels $K^{(i)}_{\alpha,\beta}(t)$ are rapidly decreasing in β for each fixed α. Because the $\{q_\alpha, p_\alpha\}$ are Gaussian, the random variables

$q_\alpha(t)$ and $p_\alpha(t)$ are also Gaussian. Thus, the measure P_t obtained by evolving the initial measure P to time t is Gaussian with mean zero and covariance

$$E\{q_\alpha(t)\ q_\beta(t)\}\ ,\quad \text{etc.}$$

Since a Gaussian measure is uniquely determined by its mean and covariance, in order to prove that $P_t = P$ (i.e. that P is invariant), it suffices to prove that

$$E\{q_\alpha(t)\ q_\beta(t)\} = E\{q_\alpha\ q_\beta\}\ ,\quad \text{etc.}$$

From the fact that the equations of motion are linear and autonomous, we see that the quantities

$$\frac{d}{dt} E\{q_\alpha(t)\ q_\beta(t)\}\ ,\quad \frac{d}{dt} E\{p_\alpha(t)\ p_\beta(t)\}\ ,\quad \frac{d}{dt} E\{q_\alpha(t)\ p_\beta(t)\}\ ,$$

can be expressed as linear combinations of the same derivatives at $t = 0$. Thus to prove that the derivatives are equal to zero at all times, it is enough to prove it at $t = 0$. Now, at $t = 0$

$$\frac{d}{dt} E\{q_\alpha(t)\ q_\beta(t)\}\Big|_{t=0} = E\{p_\alpha\ q_\beta\} + E\{q_\alpha\ p_\beta\} = 0$$

$$\frac{d}{dt} E\{p_\alpha(t)\ p_\beta(t)\}\Big|_{t=0} = E\{-\sum_\gamma S_{\alpha\gamma}\ q_\gamma\ p_\beta\} + E\{-\sum_\gamma S_{\beta\gamma} p_\alpha q_\gamma\} = 0$$

$$\frac{d}{dt} E\{q_\alpha(t)\ p_\beta(t)\}\Big|_{t=0} = E\{p_\alpha\ p_\beta\} - \sum_\gamma S_{\beta\gamma}\ E\{q_\alpha\ q_\gamma\}$$

$$= \delta_{\alpha\beta} - \sum_\gamma S_{\beta\gamma}\left(S^{-1/2}\ e^{(\alpha)}, S^{-1/2}\ e^{(\gamma)}\right).$$

Now, because of the rapid decrease of $S_{\beta\gamma}$ in γ, $\sum_\gamma S_{\beta\gamma}\ S^{-1/2} e^{(\gamma)}$ converges in $\ell^2(\Gamma)$, as does $\sum_\gamma S_{\beta\gamma}\ e^{(\gamma)} = S\ e^{(\beta)}$. Thus, since $S^{-1/2}$ is a closed operator,

$$\sum_\gamma S_{\beta\gamma}\left(S^{-1/2} e^{(\alpha)}, S^{-1/2} e^{(\gamma)}\right) = \left(S^{-1/2} e^{(\alpha)}, S^{-1/2} S\ e^{(\beta)}\right)$$

$$= \left(S^{-1/2}\ e^{(\alpha)},\ S^{1/2}\ e^{(\beta)}\right) = \delta_{\alpha\beta}\ ,$$

so $\frac{d}{dt} E\{q_\alpha(t)\ p_\beta(t)\}_{t=0} = \delta_{\alpha\beta} - \delta_{\alpha\beta} = 0$, which completes the proof that P is invariant under the time-evolution.

The Thermodynamic Limit

We investigate now the sense in which the Gaussian probability measure P on the space of polynomially bounded sequences $\{q_\alpha, p_\alpha\}$ is the limit of the finite-system canonical ensembles.

Proposition 3.1. Assume (1) - (4). Then the joint distribution of any finite set of p's and q's, $p_{\alpha_1},\ldots,p_{\alpha_n},\ q_{\beta_1},\ldots,q_{\beta_m}$ with respect to the canonical ensemble in Λ converges as $\Lambda \uparrow \infty$ to the joint distribution of the same set of p's and q's with respect to the infinite-system equilibrium measure P.

Proof: Because of the Gaussian character of all the measures involved, it suffices to prove that the covariances converge. Since

$$E_\Lambda(p_\alpha\ p_\beta) = \delta_{\alpha\beta}\ , \qquad E_\Lambda(p_\alpha\ q_\beta) = 0$$

provided $\alpha,\beta \in \Lambda$, and since the same formulas hold for the infinite system equilibrium measure, it is only necessary to prove that

$$\lim_{\Lambda\to\infty} (S_\Lambda)^{-1}_{\alpha\beta} = \left(S^{-1/2}\, e^{(\alpha)},\ S^{-1/2}\, e^{(\beta)}\right) \tag{3.2}$$

for all α,β. By polarization, it suffices in fact to prove that

$$\lim_{\Lambda\to\infty} \sum_{\alpha,\beta} (S_\Lambda)^{-1}_{\alpha\beta}\ \xi_\alpha\ \xi_\beta = \left(S^{-1/2}\ \xi,\ S^{-1/2}\ \xi\right)$$

for all $\xi \in d(\Gamma)$ (the set of finite sequences).

We now fix $\xi \in d(\Gamma)$, and consider only Λ's such that $\xi_\alpha = 0$ for all $\alpha \notin \Lambda$. We prove (3.2) by proving the following four statements:*

---
* The argument is similar to that used in ferromagnetic spin systems, cf. Lebowitz and Martin-Löf [4].

(a) If $\Lambda \subset \Lambda_1$, then $(S_\Lambda^{-1}\xi,\xi) \leq (S_{\Lambda_1}^{-1}\xi,\xi)$

(b) If $\kappa > 0$, then

$((S_\Lambda+\kappa 1_\Lambda)^{-1}\xi,\xi)$ increases to $((S+\kappa 1)^{-1}\xi,\xi)$ as Λ increases to ∞.

(c) $((S+\kappa 1)^{-1}\xi,\xi)$ increases to $(S^{-1/2}\xi, S^{-1/2}\xi)$ as κ decreases to zero.

(d) $((S_\Lambda+ \kappa 1_\Lambda)^{-1}\xi,\xi)$ increases to $(S_\Lambda^{-1}\xi,\xi)$ as κ decreases to zero.

Here 1_Λ is the unit matrix in Λ, $1_\Lambda \to 1$ as Λ increases to infinity.

Let us first assume these four statements and verify (3.2). Combining (b) and (c) shows that

$$((S_\Lambda + \kappa 1_\Lambda)^{-1} \xi,\xi) \leq (S^{-1/2} \xi, S^{-1/2} \xi) \quad \text{for all } \kappa,\Lambda.$$

Hence, by (d)

$$(S_\Lambda^{-1} \xi,\xi) \leq (S^{-1/2} \xi, S^{-1/2} \xi) \quad \text{for all } \Lambda.$$

From this bound, and (a), we conclude that

$\lim_{\Lambda\to\infty} (S_\Lambda^{-1} \xi,\xi)$ exists and is no larger than $(S^{-1/2} \xi, S^{-1/2} \xi)$.

On the other hand, let $\varepsilon > 0$. By (c) choose κ so that

$$((S + \kappa 1)^{-1} \xi,\xi) \geq (S^{-1/2} \xi, S^{-1/2} \xi) - \frac{\varepsilon}{2};$$

then (b) implies that, for sufficiently large Λ,

$$((S_\Lambda + \kappa 1_\Lambda)^{-1} \xi,\xi) \geq (S^{-1/2} \xi, S^{-1/2} \xi) - \varepsilon.$$

But

$$((S_\Lambda + \kappa 1_\Lambda)^{-1} \xi,\xi) \leq (S_\Lambda^{-1} \xi,\xi) \qquad \text{by (d)},$$

so

$$(S_\Lambda^{-1} \xi,\xi) \geq (S^{-1/2} \xi, S^{-1/2}\xi) - \varepsilon$$

for all sufficiently large Λ, i.e.

$$\lim_{\Lambda\to\infty} (S_\Lambda^{-1} \xi,\xi) = (S^{-1/2} \xi,S^{-1/2} \xi).$$

Thus we have only to prove (a), (b), (c), (d). Statements (c) and (d) are straightforward consequences of the spectral theorem. To prove (a) we consider

$$\left(S_{\Lambda_1} + \kappa\, 1_{\Lambda_1/\Lambda}\right)^{-1}$$

where $1_{\Lambda_1/\Lambda}$ is the projection of $\ell^2(\Lambda_1)$ onto $\ell^2(\Lambda_1/\Lambda)$. For $\kappa = 0$, this is simply $(S_{\Lambda_1})^{-1}$. Moreover, as κ increases

$$\left(\left(S_{\Lambda_1} + \kappa\, 1_{\Lambda_1/\Lambda}\right)^{-1} \xi, \xi\right)$$

decreases (since $A \to A^{-1}$ is operator monotone-decreasing). Finally,

$$\lim_{\kappa\to\infty} \left(\left(S_{\Lambda_1} + \kappa\, 1_{\Lambda_1/\Lambda}\right)^{-1} \xi, \xi\right) = \left(S_{\Lambda}^{-1}\, \xi, \xi\right) . \tag{3.1}$$

This is true since

$$\left(\left(S_{\Lambda_1} + \kappa\, 1_{\Lambda_1/\Lambda}\right)^{-1} \xi, \xi\right)$$

$$= \frac{\int d\underline{q}\, \exp\left(-\frac{1}{2} \sum_{\alpha,\beta\in\Lambda_1} S_{\alpha\beta} q_\alpha q_\beta - \frac{1}{2}\kappa \sum_{\alpha\in\Lambda_1/\Lambda} q_\alpha^2\right) \left(\sum_\alpha \xi_\alpha q_\alpha\right)^2}{\int dq\, \exp\left(-\frac{1}{2} \sum_{\alpha,\beta\in\Lambda_1} S_{\alpha\beta} q_\alpha q_\beta - \frac{1}{2}\kappa \sum_{\alpha\in\Lambda_1/\Lambda} q_\alpha^2\right)}$$

and letting $\kappa \to \infty$ simply has the effect of putting $q = 0$ for $\alpha \in \Lambda_1/\Lambda$. This is also intuitively reasonable, since $S_{\Lambda_1} + 1_{\Lambda_1/\Lambda}$ is a force matrix with an interaction differing from the original one by the addition of a restoring force $-\kappa\, q_\alpha$ at each $\alpha \in \Lambda_1/\Lambda$; as $\kappa \to \infty$, the oscillators at these lattice sites become more and more firmly tied to their equilibrium positions, so in the limit they are rigidly fixed. Hence, as κ increases from 0 to ∞,

$$\left(\left(S_{\Lambda_1} + \kappa\, 1_{\Lambda_1/\Lambda}\right)^{-1} \xi, \xi\right)$$

decreases from $(S_{\Lambda_1}^{-1}\xi, \xi)$ to $(S_{\Lambda}^{-1}\, \xi, \xi)$; in particular,

$$(S_{\Lambda}^{-1}\, \xi, \xi) \leq (S_{\Lambda_1}^{-1}\, \xi, \xi) .$$

To prove (b) we note that (a) implies that

$$\left((S_\Lambda + \kappa\, 1_\Lambda)^{-1}\, \xi, \xi\right)$$

increases with Λ, so we want to show

$$\lim_{\Lambda\to\infty} \left((S_\Lambda + \kappa\, 1_\Lambda)^{-1}\, \xi, \xi\right) = \left((S + \kappa\, 1)^{-1}\, \xi, \xi\right) .$$

As a function of (complex) κ,

$$g_\Lambda(\kappa) = \left((S_\Lambda + \kappa\, 1_\Lambda)^{-1}\, \xi, \xi\right)$$

is analytic in the complement of the spectrum of $-S_\Lambda$. Since the spectrum of S_Λ is contained in $[0, \|S\|]$, $g_\Lambda(\kappa)$ is bounded uniformly in Λ on any closed set which does not intersect the interval $[-\|S\|, 0]$ on the real axis. Also, expanding $(S_\Lambda + \kappa\, 1_\Lambda)^{-1}$ in its Neumann series for large κ shows that

$$\lim_{\Lambda\to\infty} g_\Lambda(\kappa) = \left((S + \kappa\, 1)^{-1}\, \xi, \xi\right)$$

for sufficiently large $|\kappa|$. Vitali's theorem then implies

$$\lim_{\Lambda\to\infty} g_\Lambda(\kappa) = \left((S + \kappa\, 1)^{-1}\, \xi, \xi\right)$$

for all strictly positive κ.

This, then, completes the proof of Proposition 3. The argument actually proves more than is included in the statement of the proposition: If we drop Assumption (4) but continue to assume that zero is not an eigenvalue of S, we can still prove that, for any $\xi \in d(\Gamma)$,

$$\left(S_\Lambda^{-1}\, \xi, \xi\right) = E_\Lambda\left(\left(\sum_\alpha \xi_\alpha\, q_\alpha\right)^2\right)$$

increases to a limit as Λ increases to ∞, but that the limit is infinite if $\xi \notin \mathcal{D}(S^{-1/2})$. The interpretation in this case is simply that the square fluctuation of $\sum_\alpha \xi_\alpha\, q_\alpha$ becomes infinite with Λ.

4. Ergodic Properties

We have seen that the solution mappings to the equations of motion define a flow T^t on the space $s'(\Gamma) \oplus s'(\Gamma)$ of polynomially bounded sequences of displacements and momenta, and that, subject to assumptions (3), (4) of Section 3, the thermodynamic limit of finite-volume ensembles exists and defines a centered Gaussian probability measure on $s'(\Gamma) \oplus s'(\Gamma)$ invariant under the solution flow. We want next to analyze the ergodic properties of the dynamical system so constructed. The essential step in doing this will be the observation that this problem may be recast into a question about one-parameter groups of orthogonal mappings on Hilbert space.

As in the preceding section, we let P denote the infinite-volume equilibrium state as a probability measure on $s'(\Gamma) \oplus s'(\Gamma)$. We let $h_1 \subset L^2(P)$ denote the subspace of <u>linear</u> real valued random variables, i.e., the closure in $L^2(P)$ of the set of finite linear combinations of p_α's and q_α's. We have seen that the action of T^t on $L^2(P)$ carries h_1 into (hence, onto) itself. Let $U_1(t)$ denote the one-parameter group of orthogonal transformations on h_1 given by the action of T^t. We will proceed by showing first how to determine most of the ergodic-theoretic properties of T^t from Hilbert space properties of the orthogonal group $U_1(t)$; we will then introduce coordinates in h_1 which enable us to reduce the study of $U_1(t)$ to the analysis of the spectral properties of S acting on $\ell^2(\Gamma)$.

To begin, we fit the situation under consideration into a more general context. Let $(X, \mathfrak{A}, P)$ denote a probability space. By a <u>generating Gaussian subspace</u> of $(X, \mathfrak{A}, P)$ we mean a closed subspace h_1 of $L^2(P)$ such that

(1) Every Ψ in h_1 is a centered Gaussian random variable.
(2) h_1 generates the σ-algebra $\mathfrak{A}$, i.e., $\mathfrak{A}$ is the smallest σ-algebra with respect to which every Ψ in h_1 is measurable. It follows easily that any finite set of Ψ's in h_1 is jointly Gaussian.

Now let $(X, \mathfrak{A}, P)$ and $(\hat{X}, \hat{\mathfrak{A}}, \hat{P})$ be two probability spaces equipped with generating Gaussian subspaces h_1 and $\hat{h}_1$ respectively, and let U be an orthogonal mapping of h_1 onto $\hat{h}_1$ (i.e., a Hilbert-space isomorphism of h_1 onto $\hat{h}_1$). We claim that there is a unique measure algebra isomorphism T of $(\hat{X}, \hat{\mathfrak{A}}, \hat{P})$ onto $(X, \mathfrak{A}, P)$ such that

$$U \Psi = \Psi \circ T$$

for all $\Psi \in h_1$. (If the two measure spaces in question are Lebesgue spaces [5], then T gives rise to an essentially unique measure-preserving transformation). The claim is easy to prove if h_1 and $\hat{h}_1$ are finite-dimensional. In the general case, we may use the finite-dimensional result to construct an isomorphism between the subalgebra of $\hat{\mathfrak{A}}$ generated by $U\mathcal{K}$ and the subalgebra of $\mathfrak{A}$ denerated by $\mathcal{K}$, for any finite-dimensional subspace $\mathcal{K}$ of h_1. By uniqueness, these partial isomorphisms fit together to define an isomorphism of all of $\hat{\mathfrak{A}}$ onto all of $\mathfrak{A}$.

Thus, in particular, any orthogonal transformation U of h_1 onto itself induces a corresponding automorphism T of $(X, \mathfrak{A}, P)$, and any group of orthogonal transformations induces an anti-isomorphic group of automorphisms. (Not every automorphism T of $(X, \mathfrak{A}, P)$ arises in this way; it is necessary that $\Psi \circ T \in h_1$ if $\Psi \in h_1$.) If U, U' are two orthogonal transformations which are orthogonally equivalent, i.e., if there exists an orthogonal transformation V such that

$$U' = V^{-1} U V ,$$

then the induced automorphisms T, T' are isomorphic. A part of the problem of determining the ergodic properties of the time-development of the infinite harmonic crystal is contained in the more general problem of determining the ergodic properties of an automorphism T induced by an orthogonal transformation U in terms of the spectral properties of U. The following results give a fairly complete answer to this general problem.

Proposition 4.1. Let $(X, \mathfrak{A}, P)$ be a probability space with a generating Gaussian subspace h_1. Then there is an isomorphism between $L^2(P)$ and $\bigoplus_{n=0}^{\infty} (h_1)^{\otimes n}_{\text{symm}}$ such that, if U is any orthogonal transformation on h_1 and if T is the induced automorphism of $(X, \mathfrak{A}, P)$, the action of T on $L^2(P)$ corresponds to

$$\bigoplus_{n=0}^{\infty} \underbrace{U \otimes_s \cdots \otimes_s U}_{n \text{ times}} \quad \text{on} \quad \bigoplus_{n=0}^{\infty} (h_1)^{\otimes n}_{\text{symm}} .$$

Proposition 4.2. Let the notation be as in the preceding proposition. Then

(1) T is ergodic if and only if U, acting on the complexification $(h_1)_{\mathbb{C}}$ of h_1, has no point spectrum.

(2) T is mixing if and only if

$$\text{weak} \lim_{|n| \to \infty} U^n = 0 .$$

(3) T has Lebesgue spectrum if and only if U, acting on $(h_1)_{\mathbb{C}}$ has Lebesgue spectrum.

Proposition 4.3. T is a Bernoulli automorphism if and only if U has Lebesgue spectrum.

Remarks on terminology:

1. As we have defined them, generating Gaussian subspaces are real Hilbert spaces, not complex Hilbert spaces. To obtain a simple spectral theory for orthogonal transformations, we pass to

a complex Hilbert space, called the complexification of h_1 obtained formally as $h_1 \oplus h_1$. An orthogonal transformation of h_1 extends by complex linearity to a unitary operator on $(h_1)_{\mathbb{C}}$.

2. If U is a unitary operator on a complex Hilbert space, the spectral theorem implies that there is a unique representation

$$U = \int_0^{2\pi} e^{i\theta} E(d\theta)$$

where $E(d\theta)$ is a projection-valued measure. We say that U has Lebesgue spectrum if $E(d\theta)$ is absolutely continuous with respect to Lebesgue measure.

The proof of the first half of Proposition 4.1 is a standard argument about Gaussian random variables; we will only sketch it. Since the Ψ's in h_1 are Gaussian, any polynomial in them is square-integrable, and the set of all polynomials is dense in $L^2(P)$. Let $h^{\leq n}$ denote the closed subspace generated by the polynomials of degree no greater than n, and let $h_n = h^{\leq n} \ominus h^{\leq (n-1)}$. Then

$$L^2(P) = \bigoplus_{n=0}^{\infty} h_n ,$$

and our two uses of the symbol h_1 are consistent. We will construct for each n an orthogonal mapping of $(h_1)^{\otimes n}_{symm}$ onto h_n as follows: Let π_n be the projection onto h_n. For $\Psi_1,\ldots,\Psi_n$ in h_1, map

$$\Psi_1 \otimes_s \cdots \otimes_s \Psi_n \longmapsto \Psi_1 \cdot \Psi_2 \cdot \ldots \cdot \Psi_n \longmapsto \pi_n (\Psi_1 \cdot \ldots \cdot \Psi_n)$$

and extend by linearity. A straightforward calculation shows that the resulting mapping, divided by $\sqrt{n!}$ is an isometry. (See, for example, Section 6.3 of [6].) Since π_n applied to the set of all nth degree monomials is a total set in h_n, we have constructed the desired orthogonal mapping of $(h_1)^{\otimes n}_{symm}$ onto h_n. The statement that $U^{\otimes n}_{symm}$ goes over to the action of T on h_n is immediate since

$$(U\Psi_1) \otimes_s (U\Psi_2) \otimes_s \cdots \otimes_s (U\Psi_n) \longmapsto (\Psi_1 \cdot \ldots \cdot \Psi_n) \circ T \longmapsto [\pi_n(\Psi_1 \cdot \ldots \cdot \Psi_n)] \circ T$$

where the last identity follows from the fact that composition with T carries over $h^{\leq n}$ onto itself and hence commutes with π_n.

The proofs of statements (2) and (3) of Proposition 4.2 are straightforward from Proposition 4.1 and will be omitted. To prove (1), first assume that U has an eigenvector Ψ in $(h_1)_{\mathbb{C}}$. Since $(h_1)_{\mathbb{C}}$ may be regarded as the space of complex-valued functions on (X, ,P) which are complex linear combinations of elements of h_1, we may regard Ψ as an element of $L^2_{\mathbb{C}}(P)$. Then $|\Psi|$ (the pointwise absolute value) is invariant under T, so, if it is not constant, T is not ergodic. But since the real and imaginary parts of Ψ are Gaussian random variables (and therefore unbounded), $|\Psi|$ cannot be constant.

Conversely, suppose U has no point spectrum. It is then easy to see (using the spectral theorem) that $\underbrace{U \otimes \ldots \otimes U}_{n \text{ times}}$ has no point spectrum for any $n > 0$. Since $U \otimes_s \ldots \otimes_s U$ is the restriction of $U \otimes \ldots \otimes U$ to the symmetric subspace, the symmetric tensor product also has no point spectrum so it follows from Proposition 4.1 that T has no eigenfunctions except the constants. Thus, if U has no point spectrum, T is weakly mixing and hence ergodic. Note that we have shown that an automorphism which arises from an orthogonal transformation on a generating Gaussian subspace is weakly mixing if it is ergodic.

To prove Proposition 4.3, we note first that, since a Bernoulli automorphism has Lebesgue spectrum, statement (3) of Proposition 4.2 implies that T cannot be a Bernoulli automorphism unless U has Lebesgue spectrum. To prove the converse, we consider first the special case in which U has Lebesgue spectrum of uniform multiplicity on the unit circle. There then exists a direct sum decomposition

$$h_1 = \bigoplus_{m=-\infty}^{\infty} g_m ,$$

such that $U g_m = g_{m+1}$ for all m (i.e., U is equivalent to a shift operator). Let F_m be the smallest σ-algebra with respect to which all Ψ in g_m are measurable. Then

(a) the orthogonality of the subspaces g_m implies that the F_m are independent (orthogonal centered Gaussian random variables are independent);

(b) the fact that the g_m span h_1, together with the fact that h_1 generates $\mathfrak{A}$, implies

$$\bigvee_m F_m = \mathfrak{A} ;$$

(c) the equation $U g_m = g_{m+1}$ implies

$$T F_m = F_{m+1}$$

Hence, F_0 is an independent generator for T, so T is a Bernoulli automorphism. Since the σ-algebra F_0 is non-atomic, the entropy of T must be infinite.

To finish the proof we must eliminate the assumption of uniform multiplicity of the spectrum of U. Thus, let U be any orthogonal transformation with Lebesgue spectrum on a generating Gaussian subspace h_1. It follows easily from the spectral theorem that there exists $\hat{U}$ on h_1 with "complementary spectrum" such that $U \oplus \hat{U}$ on $h_1 \oplus h_1$ has Lebesgue spectrum of uniform multiplicity. We may identify $h_1 \oplus h_1$ with the generating Gaussian subspace

$$h_1 \otimes_{id} \oplus_{id} \otimes h_1$$

on $(X \times X, \mathfrak{A} \otimes \mathfrak{A}, P \otimes P)$. The automorphism of the product space induced by $U \oplus \hat{U}$ is $T \times \hat{T}$, which is a Bernoulli automorphism since $U \oplus \hat{U}$ has uniform Lebesgue spectrum. Thus, T is a factor of a Bernoulli automorphism. But Ornstein has proved that any factor of a Bernoulli automorphism is again a Bernoulli automorphism, so T must be a Bernoulli automorphism, as desired. (For a statement and proof of Ornstein's result, see [7].)

The above Propositions 4.1 - 4.3 give a nearly complete description of the ergodic-theoretic properties of the time-development flow of the infinite harmonic crystal provided we can determine the spectral properties of the one-parameter unitary group $U_1(t)$ on h_1. While we cannot do this explicitly in general, we can by introducing appropriate coordinates reduce it to determining the spectral properties of the operator S on $\ell^2(\Gamma)$. We equip $\mathcal{D}(S^{-1/2}) \subset \ell^2(\Gamma)$ with the norm

$$\|\Psi\|_{-1/2} = \|S^{-1/2}\Psi\| ;$$

this makes $\mathcal{D}(S^{-1/2})$ into a Hilbert space. Now we map $d(\Gamma) \oplus d(\Gamma)$ onto a dense subspace of h_1 by sending $\xi \oplus \eta$ to $\sum_\alpha (\xi_\alpha q_\alpha + \eta_\alpha p_\alpha)$. By a straightforward calculation,

$$\| \sum_\alpha (\xi_\alpha q_\alpha + \eta_\alpha p_\alpha)\|^2 = \|\xi\|^2_{-1/2} + \|\eta\|^2$$

(where the norm on the left means the norm on $L^2(P)$).

Taking into account the following lemma, we may extend this mapping to an orthogonal mapping of $\mathcal{D}(S^{-1/2}) \oplus \ell^2(\Gamma)$ onto h_1.

Lemma 4.4. $d(\Gamma)$ is dense in $\mathcal{D}(S^{-1/2})$.

Proof: Assume not. Then there exists a non-zero vector Ψ in $\mathcal{D}(S^{-1/2})$ such that $(S^{-1/2}\Psi, S^{-1/2}\xi) = 0$ for all ξ in $d(\Gamma)$. By taking limits, using the fact that $S^{-1/2}$ is a closed operator and the assumed polynomial boundedness of $\|S^{-1/2} e^{(\alpha)}\|$, we see that this remains true for all ξ in $s(\Gamma)$. But S maps $d(\Gamma)$ into $s(\Gamma)$, so replacing ξ by $S\xi'$ we get

$$0 = (S^{-1/2}\Psi, S^{-1/2} S\xi') = (S^{-1/2}\Psi, S^{1/2}\xi') = (\Psi,\xi') ,$$

for all $\xi' \in d(\Gamma)$. This evidently contradicts the assumption that $\Psi \neq 0$. We are thus able to identify h_1 with $\mathcal{D}(S^{-1/2}) \oplus \ell^2(\Gamma)$, and can carry over the orthogonal group $U_1(t)$ under this identification to obtain a one-parameter orthogonal group on $\mathcal{D}(S^{-1/2}) \oplus \ell^2(\Gamma)$. To identify the group so obtained, we compute its infinitesimal generator:

$$\frac{d}{dt} \sum_{\alpha} \left(\xi_\alpha q_\alpha(t) + \eta_\alpha p_\alpha(t)\right)\Big|_{t=0} = \sum_{\alpha} \xi_\alpha p_\alpha - \sum_{\alpha,\beta} \eta_\alpha S_{\alpha\beta} q_\beta \ , \ (\xi,\eta) \in d(\Gamma))$$

and hence the infinitesimal generator on $d(\Gamma) \oplus d(\Gamma)$ is given by:

$$\begin{pmatrix} 0 & -S \\ 1 & 0 \end{pmatrix}$$

This operator is easily checked to be skew-adjoint and bounded on $\mathcal{D}(S^{-1/2}) \oplus \ell^2(\Gamma)$ so the transformed orthogonal group must be

$$\exp\left[t\begin{pmatrix} 0 & -S \\ 1 & 0 \end{pmatrix}\right]$$

We can further simplify matters by identifying $\mathcal{D}(S^{-1/2})$ with $\ell^2(\Gamma)$ through the mapping $\xi \longmapsto S^{-1/2}\xi$; this sets up an isomorphism between h_1 and $\ell^2(\Gamma) \oplus \ell^2(\Gamma)$ which carries $U_1(t)$ to the group generated by $\begin{pmatrix} 0 & -S^{1/2} \\ S^{1/2} & 0 \end{pmatrix}$.

All these rearrangements were carried out by treating the Hilbert spaces involved as real. Propositions 4.2 and 4.3, however, express the ergodic-theoretic properties of T^t in terms of the spectral properties of the unitary group obtained by complexifying the underlying Hilbert space. If we allow complex changes of coordinates, then the generator $\begin{pmatrix} 0 & -S^{1/2} \\ S^{1/2} & 0 \end{pmatrix}$ may be converted into $\begin{pmatrix} iS^{1/2} & 0 \\ 0 & -iS^{1/2} \end{pmatrix}$. Combining these remarks with Propositions 4.2 and 4.3, we get

Proposition 4.5. The time-evolution flow T^t for the infinite harmonic crystal is

(1) ergodic if and only if S, acting on $\ell^2_{\mathbb{C}}(\Gamma)$, has no point spectrum

(2) a Bernoulli flow if and only if S has Lebesgue spectrum.

Let us now see what happens when assumption (3) of Section 3 holds, but (4) does not. Abstracting from the above results, we may describe the infinite system equilibrium state as an orthogonal mapping from $\mathcal{D}(S^{-1/2}) \oplus \ell^2(\Gamma)$ onto a generating Gaussian subspace of a probability space, with the interpretation that the random variable corresponding to $\xi \oplus \eta$ is $\sum_\alpha (\xi_\alpha q_\alpha + \eta_\alpha p_\alpha)$. The time evolution flow is induced by the one-parameter orthogonal group on $\mathcal{D}(S^{-1/2}) \oplus \ell^2(\Gamma)$ with infinitesimal generator $\begin{pmatrix} 0 & -S \\ 1 & 0 \end{pmatrix}$. If (ξ_α) is a sequence which does not belong to $\mathcal{D}(S^{-1/2})$, the series $\sum_\alpha \xi_\alpha q_\alpha$ does not make sense as a random variable.

This description of the equilibrium state and time evolution flow works perfectly well whether or not the coordinate vectors $e^{(\alpha)}$ are in the domain of $S^{-1/2}$. However, if $e^{(\alpha)} \notin \mathcal{D}(S^{-1/2})$, then the position of the αth particle does not make sense as a random variable. Indeed, as we have seen, if $e^{(\alpha)} \notin \mathcal{D}(S^{-1/2})$, then

$$\lim_{\Lambda\to\infty} (S_\Lambda^{-1})_{\alpha\alpha} = \infty ,$$

i.e., the variance of the position of the αth particle goes to infinity with Λ. A typical situation, exemplified by the one-dimensional harmonic chain with nearest neighbor interaction, is that $\xi \in d(\Gamma)$ is in $\mathcal{D}(S^{-1/2})$ if and only if $\sum_\alpha \xi_\alpha = 0$, i.e., if and only if $\sum_\alpha \xi_\alpha q_\alpha$ may be written as a linear combination of difference variables $q_\alpha - q_\beta$. In this case, the variances of difference variables approach finite limits as $\Lambda \to \infty$ even though the variances of individual positions do not. Moreover, the joint distribution of the difference variables for the finite system canonical ensemble converges to the corresponding distributions for the infinite system.

5. Translation Invariance

The preceding section shows how to determine the behavior of the harmonic crystal in the thermodynamic limit provided we know enough about the operator S. Specifically, we need to know

(1) is 0 an eigenvalue of S ;

(2) are the coordinate vectors $e^{(\alpha)}$ in $\mathcal{D}(S^{-1/2})$;

(3) what is the spectral type of S ?

These equations are hard to answer in general. However, if the matrix S is translation-invariant, Fourier transformation can be used to simplify matters somewhat.

We let $\hat{\Gamma}$ denote the "first Brillouin zone", i.e., the compact dual group of the discrete additive group Γ, and let V denote the volume of $\hat{\Gamma}$. Fourier transformation gives a unitary mapping of $\ell^2_{\mathbb{C}}(\Gamma)$ onto $L^2_{\mathbb{C}}(\hat{\Gamma})$ sending ξ to $\tilde{\xi}$, where ξ and $\tilde{\xi}$ are related by [1]

$$\xi_\alpha = \frac{1}{\sqrt{V}} \int_{\hat{\Gamma}} e^{ip\alpha}\, \tilde{\xi}(p)\, dp \, .$$

Translation invariance means simply that $S_{\alpha\beta}$ depends only on $\alpha-\beta$; we write $S(\alpha-\beta)$ for $S_{\alpha\beta}$ in this case. The operator S becomes a convolution operator:

$$(S\xi)_\alpha = \sum_\beta S(\alpha-\beta)\, \xi_\beta$$

which under Fourier transformation becomes a multiplication operator:

$$(S\xi)\,(p) = \tilde{S}(p)\, \tilde{\xi}(p)$$

with

$$\tilde{S}(p) = \sum_\alpha e^{ip\alpha}\, S(\alpha) \, .$$

Recall, however, that $S(\alpha-\beta)$ is a $\nu \times \nu$ matrix, so the same is true for $\tilde{S}(p)$ and we have not yet completely diagonalized S; to do this, we would have to diagonalize each $\tilde{S}(p)$. This can only be done on a case-by-case basis, but there are a number of remarks which clarify the situation without requiring intricate computations:

(i) The symmetry and reality of S imply that $\tilde{S}(p)$ is self-adjoint for each p and that $\tilde{S}(-p) = \tilde{S}(p)$. Since $S(\alpha)$ is assumed to be rapidly decreasing in α, $\tilde{S}(p)$ is an infinitely differentiable function of p.

(ii) In order that S be positive semidefinite it is necessary and sufficient that $\tilde{S}(p)$ be positive semidefinite for all p; similarly, S does not have zero as an eigenvalue if and only if $\tilde{S}(p)$ is strictly positive definite for almost all p.

(iii) Assumption 4 of Section 3 holds if and only if $\int dp \, \|\tilde{S}(p)^{-1}\| < \infty$.

(iv) In most interesting cases $\tilde{S}(0) = 0$. This reflects the fact that a uniform displacement of all the atoms in the crystal costs no energy. Because $\tilde{S}(-p) = \tilde{S}(p)$, this implies that each eigenvalue of $\tilde{S}(p)$ vanishes at least as fast as $|p|^2$ as p approaches zero. Thus, we must expect that

$$\int dp \, \|\tilde{S}(p)^{-1}\| = \infty$$

in one and two dimensions. This argument does not apply in three or more dimensions, and it is not hard to check that the above integral converges at zero for many reasonable interactions. The problem in showing that the overall integral is finite is that some eigenvalues may have "accidental" zeros at non-zero values of p.

The preceding analysis may be applied if the force matrix V is translation invariant even if strict translation invariance is destroyed by variation of the masses. From the definition of S it follows readily that $\| S^{-1/2} e^{(\alpha)} \|^2 = m_\alpha \| V^{-1/2} e^{(\alpha)} \|^2$. Hence, as long as the masses are bounded, and bounded away from zero, assumptions (3) and (4) hold if and only if they hold with S replaced by V.

The spectrum of S is almost arbitrary under our assumptions; we may simply choose any periodic infinitely differentiable matrix-valued function $\tilde{S}(p)$ which is strictly positive definite everywhere and satisfies the qualitative conditions of (i) and construct $S(\alpha)$ as its inverse Fourier transform. If we take $\tilde{S}(p)$ constant on some non-empty open set, then S has an eigenvalue with infinite multiplicity. On the other hand, if $S(\alpha) = 0$ for all sufficiently large α (or, more generally, if $S(\alpha)$ decreases exponentially with α), then $\tilde{S}(p)$ is a real-analytic function of p. We claim that, in this case, either $\tilde{S}(p)$ has at least one p-independent eigenvalue or else S has Lebesgue spectrum. To see this, we argue as follows: If $\tilde{S}(p)$ is an analytic function of p then, on the complement of a closed nowhere dense set of Lebesgue measure zero, the eigenvalues of $\tilde{S}(p)$ are analytic functions of p. Then if no eigenvalue is constant, the set where some eigenvalue has zero derivative is again a closed set of Lebesgue measure zero. By excluding these closed sets of measure zero, we see that S is unitarily equivalent to multiplication by a smooth function with nowhere vanishing derivative on an open set in Euclidean space; such an operator has Lebesgue spectrum. In concrete cases it is frequently easy (e.g. by looking at behavior near $p = 0$) to show that there is no constant eigenvalue and hence that T^t is a Bernoulli flow.

This is indeed the expected behavior of physically reasonable models of harmonic crystals with strict translation invariance [1,8]. It is true in particular for those systems for which the time dependence of expectations of the type $E(p_\alpha(t)\ p_\beta(0))$ has been calculated explicitly [1], [2] (see footnote ‡ in section 1). Since these systems are Bernoulli (and hence mixing) thexe correlations will go to zero as $|t| \to \infty$. The rate of decay cannot of course be obtained from the general theory; the calculations show it behaves generally as $t^{-(1/2)\nu}$ [1,2,9]. The way this inifnite volume behavior is approached as $\Lambda \to \infty$ (for finite Λ the system is not even ergodic) is also discussed in these studies.

The situation may be quite different however when the system does not have strict translation invariance. In this case the crystal may have localized modes in the neighborhood of the defect (deviation from regularity) giving rise to a point spectrum for S. An example of this is the case where one of the masses, say m_γ, is (sufficiently) smaller than the other masses, $m_\gamma < m_\alpha = m$ for all $\alpha \neq \gamma$, while V is translation invariant. It follows then from Rayleigh's theorems (Chapter V of [1]) that there will be an isolated frequency, or discrete eigenvalue of S. (This remains true if there are a finite number of such mass defects [1], [2].) Hence such a system will not be ergodic (as found also by Cukier and Mazur [10] from explicit calculations in the one dimensional case).

6. Quantum Mechanics of the Harmonic Crystal.

Because the equations of motion are linear the quantum-mechanical time evolution of the infinite harmonic crystal is easily deduced from the classical time evolution. Equation (3.1) gives, in the classical case,

$$q_\alpha(t) = \sum_\beta K^{(1)}_{\alpha\beta}\ q_\beta + K^{(2)}_{\alpha\beta}\ p_\beta\ ,$$

and similarly for $p_\alpha(t)$. To get the quantum theory (in the Heisenberg picture) it is only necessary to substitute a set of operators Q_α , P_α satisfying the canonical commutation relations for the numerical initial data q_α , p_α. This gives the time evolution, formally at least, as a one-parameter group of automorphisms of an algebra of unbounded operators. It is technically preferable to pass to a one-parameter group of automorphisms of an C^* algebra "of bounded operators" associated with representation of the commutation relations. This can be done without difficulty -- the time evolution automorphisms become Bogoliubov transformations -- but we will not give the details [11].

The equilibrium statistical mechanics of the quantum crystal is also simple to obtain. For a finite crystal Λ, a straightforward calculation shows that the canonical ensemble at inverse temperature β is a quasi-free state ρ_Λ (i.e., the truncated expectation values of degree greater than two vanish) with two-point function given by [1]

$$\rho_\Lambda(Q_\alpha\ Q_\beta) = \left[\frac{S_\Lambda^{-1/2}}{2} \coth\left(\frac{\beta S_\Lambda^{1/2}}{2}\right)\right]_{\alpha\beta}$$

$$\rho_\Lambda(P_\alpha\ P_\beta) = \left[\frac{S_\Lambda^{+1/2}}{2} \coth\left(\frac{\beta S_\Lambda^{1/2}}{2}\right)\right]_{\alpha\beta}$$

$$\rho_\Lambda(P_\alpha\ Q_\beta + Q_\beta\ P_\alpha) = 0 .$$

(Note that $\rho(P_\alpha Q_\beta)$ can be computed from the third equation and the commutation relations.) We have put $\hbar = 1$ in these expressions. To obtain the equilibrium state for the infinite crystal, we must let $\Lambda \to \infty$. Let us look first at the expression for $\rho_\Lambda(P_\alpha P_\beta)$. The function

$$z \longmapsto \frac{z^{1/2}}{2} \coth\left(\frac{\beta z^{1/2}}{2}\right)$$

is analytic in a neighborhood of the real line (including $z = 0$). We may regard S_Λ as a positive operator on $\ell^2(\Gamma)$ which is zero on the orthogonal complement of $\ell^2(\Gamma)$. Then S_Λ converges strongly to S as $\Lambda \to \infty$, and hence

$$\frac{S_\Lambda^{1/2}}{2} \coth \left(\frac{\beta S_\Lambda^{1/2}}{2}\right)$$

converges strongly to

$$\frac{S^{1/2}}{2} \coth \left(\frac{S^{1/2}}{2}\right)$$

Therefore

$$\lim_{\Lambda\to\infty} \rho_\Lambda(P_\alpha P_\beta) = \left[\frac{S^{1/2}}{2} \coth \left(\frac{S^{1/2}}{2}\right)\right]_{\alpha\beta}$$

The limit for $\rho_\Lambda(Q_\alpha Q_\beta)$ involves one more step. The function

$$z \mapsto \frac{z^{-1/2}}{2} \coth \left(\frac{\beta z^{1/2}}{2}\right)$$

is not analytic on the real axis; it has a pole at the origin. However, the difference between this function and $\beta^{-1}z^{-1}$ is analytic. Hence, by the above argument,

$$\lim_{\Lambda\to\infty} \{\rho_\Lambda(Q_\alpha Q_\beta) - \beta^{-1}(S_\Lambda^{-1})_{\alpha\beta}\}$$

exists. If, in addition, Assumption (4) of Section 3 holds, the results of that section imply that

$$\lim_{\Lambda\to\infty} (S_\Lambda^{-1})_{\alpha\beta} = (S^{-1/2} e^{(\alpha)}, S^{-1/2} e^{(\beta)}),$$

so we can conclude that

(1) each $e^{(\alpha)}$ is in the domain of $\left[\frac{S^{-1/2}}{2} \coth \left(\frac{\beta S^{1/2}}{2}\right)\right]^{1/2}$

(2) $\lim_{\Lambda\to\infty} \rho_\Lambda(Q_\alpha Q_\beta) = \left(\left[\frac{S^{-1/2}}{2} \coth \left(\frac{\beta S^{1/2}}{2}\right)\right]^{1/2} e^{(\alpha)},\right.$

(*) $$\left.\left[\frac{S^{-1/2}}{2} \coth \left(\frac{\beta S^{1/2}}{2}\right)\right]^{1/2} e^{(\beta)}\right).$$

Thus: the infinite-system equilibrium state is the quasi-free state of the canonical commutation relations with two-point function given by

$$\rho(Q_\alpha Q_\beta) = \left[\frac{S^{-1/2}}{2}\coth\left(\frac{\beta S^{1/2}}{2}\right)\right]_{\alpha\beta}$$ (defined by the right-hand side of (*))

$$\rho(P_\alpha P_\beta) = \left[\frac{S^{1/2}}{2}\coth\left(\frac{\beta S^{1/2}}{2}\right)\right]_{\alpha\beta}$$

$$\rho(P_\alpha Q_\beta + Q_\beta P_\alpha) = 0 .$$

Calculations similar to those used to prove the invariance of the classical equilibrium state show that this state of the quantum system is invariant under the time-development automorphisms.

If Assumption (4) of Section 3 fails, a part of the theory of equilibrium states for the infinite crystal can be salvaged as indicated in Section 4 for the classical system. In the resulting theory, positions of the individual oscillators are not observables, but appropriate linear combinations of the positions are.

References

[1] A. A. Maradudin, E. W. Montroll and G. H. Weiss (with I. P. Ipatova), Theory of Lattice Dynamics in the Harmonic Approximation, Academic Press (1963,(1971)).

[2] J. L. Lebowitz in Statistical Mechanics: New Concepts, New Problems, New Applications, S. A. Rice, K. F. Freed and J. C. Light, editors, p. 41, University of Chicago Press (1972).

[3] U. M. Titulaer, Physica 70, 257, 70, 276, 70, 456 (1973).

[4] J. L. Lebowitz and A. Martin-Löf, Commun. Math. Phys. 25, 272 (1973).

[5] V. A. Rohlin, Amer. Math. Soc. Transl. 10, 1 (1962); 49, 171 (1966).

[6] T. Hida, Stationary Stochastic Processes, Princeton University Press, 1970.

[7] D. Ornstein, Ergodic Theory, Randomness, and Dynamical Systems, Yale University Press, 1974.

[8] C.f. A. J. O'Connor and J. L. Lebowitz, J. Math. Phys. 15, 692 (1974), Section 8.

[9] P. Mazur and E. W. Montroll, J. Math. Phys. 1, 70 (1960).

[10] R. I. Cukier and P. Mazur, Physica 53 (1971) 157

[11] C.f. D. Ruelle, Statistical Mechanics, Benjamin (1969).

THE LASER: A REVERSIBLE QUANTUM DYNAMICAL SYSTEM WITH IRREVERSIBLE CLASSICAL MACROSCOPIC MOTION

Klaus Hepp

Department of Physics, E. T. H., CH-8049 Zürich, Switzerland

Elliott H. Lieb

Departments of Mathematics & Physics, M.I.T., Cambridge, Mass. 02139

We shall describe a mathematical machine which behaves like a laser but which, admittedly, is much simpler than any laser found in the laboratory. Our machine will consist of two main parts: the first one, which is the laser proper, consists of atoms and photons which are the quantized proper vibrations of a resonant cavity. It is essential that the number, N, of atoms be allowed to become very large, but the number of different photon modes can be infinite or finite. This number can in fact be one in order for the machine to operate, but in real life the number of modes is infinite. We can handle the infinite mode case provided that essentially only a finite number of modes is macroscopically excited. As the one-mode case is simpler, we shall mainly discuss it.

The second part consists of reservoirs. They are dynamical systems which in some sense are infinitely large compared to N, and they are so constructed that they can both pump energy into and drain energy out of the laser. The main point, however, is that the energy flux from reservoir to laser is thoroughly chaotic. The interesting property of the laser is that it can take this energy and structure it. In other words, the laser, when suitably coupled to the reservoir, will go into a cooperative oscillatory state in which all the N atoms and the photons vibrate coherently and periodically. In this regard the laser behaves like a clock or a steam engine or a scientist's brain.

To describe the operation of our machine we first have to show how each component operates separately. The analysis has several parts: one is the quantum mechanical level; another is the $N \to \infty$ limit in which the quantum equations go over into ordinary classical differential equations which, being nonlinear, have interesting bifurcation properties; a third level is the analysis of the quantum fluctuations around the classical orbit.

1. The Atoms

What we basically want to have for an atom is a very anharmonic oscillator. While one could possibly construct a laser with classical oscillators as "atoms," there exist quantum mechanical systems which quite naturally fulfill our requirements, namely two-level atoms. Each atom has two levels, each of which can either be occupied or not by one electron. Thus, there are four possible states for an atom. This is described by a four dimensional (complex) vector space, $\mathcal{V} = \mathcal{V}_+ \otimes \mathcal{V}_-$, where $\mathcal{V}_+$ (resp. $\mathcal{V}_-$) is a two-dimensional space for the upper (resp. lower) level. As an orthogonal basis in $\mathcal{V}_+$ (resp. $\mathcal{V}_-$) we take e_+^0 and e_+^1 (resp. e_-^0 and e_-^1) where 0 stands for unoccupied and 1 for occupied.

As generators for the operator algebra on $\mathcal{V}_+$, we take b_+, b_+^* and I, where

$$b_+ e_+^1 = e_+^0 \, , \quad b_+ e_+^0 = 0 \, . \tag{1}$$

Note: the superscript * will always mean adjoint when referring to operators or complex conjugate when referring to numbers. b_+ is called the annihilation operator and b_+^* the creation operator. Likewise, we have $b_-^\# = b_-$, b_-^* for $\mathcal{V}_-$. A particularly important subspace $\mathcal{W}$ of $\mathcal{V}$ is the one spanned by $e_+^1 \otimes e_-^0$ and $e_+^0 \otimes e_-^1$. In $\mathcal{W}$, each atom has exactly one electron. On $\mathcal{W}$, the operators

$$S^+ = b_+^* b_- = (S^-)^* \tag{2}$$
$$S^3 = \frac{1}{2}(b_+^* b_+ - b_-^* b_-) = \frac{1}{2}[S^+, S^-]$$

form an irreducible representation of SU(2) with spin 1/2. The Hilbert space $\mathcal{H}_{(N)}^A$ for N atoms is the N-fold tensor product of $\mathcal{V}$. Let $\varepsilon > 0$. The Hamiltonian

$$H_{(N)}^A = \varepsilon S_{(N)}^3 \equiv \varepsilon \sum_{n=1}^{N} S_n^3 \tag{3}$$

measures the total energy of the electrons when the values $\pm \varepsilon/2$ are given to the upper and lower level. Note that the subscript n labels the atoms. Under the unitary transformation $A \to A(t) = \exp(i H_{(N)}^A t) A \exp(-i H_{(N)}^A t)$ the $S_n^i(t)$ satisfy the Heisenberg equations of motion

$$\dot{S}_n^3(t) \equiv \frac{d}{dt} S_n^3(t) = 0 , \tag{4}$$

$$\dot{S}_n^+(t) = i \varepsilon S_n^+(t) . \tag{5}$$

We conclude our introduction to quantum mechanics by mentioning the concept of a <u>state</u>. For N finite, a state of the atomic system is described by a <u>density matrix</u>, i.e. a positive operator, $\omega_{(N)}$ on $\mathcal{H}_{(N)}^A$ with trace one. The <u>expectation value</u> of an <u>observable</u>, i.e. of an operator A on $\mathcal{H}_{(N)}^A$, in the state $\omega_{(N)}$ is given by

$$\omega_{(N)}(A) = \mathrm{Tr}\, A\omega_{(N)} \tag{6}$$

A <u>pure state</u> is a $\omega_{(N)}$ which is a one-dimensional projection onto a normalized vector $\Psi \in \mathcal{H}_{(N)}^A$, i.e.

$$\omega_{(N)}(A) = (\psi, A\psi) \tag{7}$$

for pure states.

2. The Radiation Field (Photons)

The atoms are in a three-dimensional cavity which is a box of length L. The free classical electromagnetic field with periodic boundary condition has normal modes of vibration, called photons, of the form $\exp i(\underset{\sim}{k}\cdot\underset{\sim}{x} - |\underset{\sim}{k}|t)$, where the vector $\underset{\sim}{k} \in \Gamma(L) = \left\{\frac{2\pi}{L}\underset{\sim}{n} \mid \underset{\sim}{n} \in \mathbb{Z}^3\right\}$. We neglect the fact that the photons have a polarization. If $a(\underset{\sim}{k})$ is the amplitude of the $\underset{\sim}{k}$-th mode, the total energy (in proper units, such that $\hbar = c = 1$) of the field is

$$(8) \qquad H^P_{(L)} = \sum_{\underset{\sim}{k}\in\Gamma(L)} |\underset{\sim}{k}|\; a^*(\underset{\sim}{k})\; a(\underset{\sim}{k})$$

In a fully quantum theory, the electromagnetic field should be quantized. This means that $a(\underset{\sim}{k})$, instead of being a complex number, is an operator on a Hilbert space $\mathcal{H}^P_{(L)}$ and satisfies the canonical commutation relations

$$(9) \qquad [a(\underset{\sim}{k}), a^*(\underset{\sim}{q})] = \delta_{\underset{\sim}{k},\underset{\sim}{q}}, \qquad [a(\underset{\sim}{k}), a(\underset{\sim}{q})] = 0,$$

where δ is the Kronecker delta. $\mathcal{H}^P_{(L)}$, the Fock space, is the span of all vectors $a^*(\underset{\sim}{k}_1) \ldots a^*(\underset{\sim}{k}_n)\,\Omega^P$, where Ω^P, the photon vacuum state, satisfies

$$(10) \qquad a(\underset{\sim}{k})\,\Omega^P = 0 \quad \text{for all} \quad \underset{\sim}{k} \in \Gamma(L).$$

In $\mathcal{H}^P_{(L)}$, $H^P_{(L)}$ is an unbounded selfadjoint operator. As a consequence of (9) the $a_{(L)}(\underset{\sim}{k},t) \equiv \exp(iH^P_{(L)}t)\, a(\underset{\sim}{k})\, \exp(-iH^P_{(L)}t)$ satisfy Maxwell's equations

$$(11) \qquad \dot{a}_{(L)}(\underset{\sim}{k},t) = -i|\underset{\sim}{k}|\, a_{(L)}(\underset{\sim}{k},t),$$

and they are the unique solutions of (11) with $a_{(L)}(\underset{\sim}{k},0) = a(\underset{\sim}{k})$ on the domain of $H^P_{(L)}$ [1].

3. The Atom-Field Interaction

To begin with, we shall henceforth assume that $L^3 = N \in \mathbb{Z}_+$ and use N instead of L(N) as a label for the cavity. The atoms are regarded as being distributed inside the cavity with small density fluctuations for $N \to \infty$ and $\underset{\sim}{x}_n \in \mathbb{R}^3$ is the coordinate of the n-th atom.

The Hilbert space $\mathcal{H}^S_{(N)}$ for the laser proper, i.e. for N atoms in a cavity of length L(N), is $\mathcal{H}^A_{(N)} \otimes \mathcal{H}^P_{(N)}$. The Hamiltonian has the form

$$H_{(N)} = H^A_{(N)} + H^P_{(N)} + H^{AP}_{(N)} \tag{12}$$

where $H^A_{(N)}$ and $H^P_{(N)}$ are as before [(3) and (8)] and

$$\begin{aligned} H^{AP}_{(N)} = N^{-1/2} \sum_{k\in\Gamma(N)} \sum_{n=1}^{N} \Big\{ & e^{i\underset{\sim}{k}\cdot\underset{\sim}{x}_n} a(\underset{\sim}{k}) [\lambda(\underset{\sim}{k}) S^+_n + \mu(\underset{\sim}{k}) S^-_n] \\ & + e^{-i\underset{\sim}{k}\cdot\underset{\sim}{x}_n} a^*(\underset{\sim}{k}) [\lambda^*(\underset{\sim}{k}) S^-_n + \mu^*(\underset{\sim}{k}) S^+_n] \Big\} , \end{aligned} \tag{13}$$

where $\lambda(\underset{\sim}{k})$ and $\mu(\underset{\sim}{k})$ are continuous functions on $\mathbb{R}^3$ with rapid decrease at infinity. This interaction Hamiltonian is a well known [2] and reasonable approximation to quantum electrodynamics and we shall not try to justify it here. Physically, one usually has that $|\lambda(\underset{\sim}{k})| = |\mu(\underset{\sim}{k})|$, but we shall not assume this. In fact, it is often mathematically convenient and physically justified to suppose that $\mu(\underset{\sim}{k}) = 0$. This is called the <u>rotating wave approximation to quantum electrodynamics</u>. It leads to the conservation law that the operator

$$C_{(N)} = \sum_{\underset{\sim}{k}\in\Gamma(N)} a^*(\underset{\sim}{k}) a(\underset{\sim}{k}) + S^3_{(N)} \tag{14}$$

commutes with $H_{(N)}$.

Using $H_{(N)}$, one easily computes the new Heisenberg equations of motion to be

$$
\begin{aligned}
\dot{a}_{(N)}(\underset{\sim}{k},t) &= -i|\underset{\sim}{k}|\,a_{(N)}(\underset{\sim}{k},t) - iN^{-1/2}\Big\{\lambda^*(\underset{\sim}{k})S^-_{(N)}(-\underset{\sim}{k},t) \\
&\qquad + \mu^*(\underset{\sim}{k})S^+_{(N)}(-\underset{\sim}{k},t)\Big\} \\
\dot{S}^-_{(N)}(\underset{\sim}{k},t) &= -i\varepsilon S^-_{(N)}(\underset{\sim}{k},t) + 2iN^{-1/2}\sum_{\underset{\sim}{q}\in\Gamma(N)}\Big\{\lambda(\underset{\sim}{q})a_{(N)}(\underset{\sim}{q},t) \\
&\qquad \cdot S^3_{(N)}(\underset{\sim}{k}+\underset{\sim}{q},t) \\
&\qquad + \mu^*(\underset{\sim}{q})a^*_{(N)}(\underset{\sim}{q},t)S^3_{(N)}(\underset{\sim}{k}-\underset{\sim}{q},t)\Big\} \\
\dot{S}^3_{(N)}(\underset{\sim}{k},t) &= iN^{-1/2}\sum_{\underset{\sim}{q}\in\Gamma(N)}\Big\{\mu(\underset{\sim}{q})a_{(N)}(\underset{\sim}{q},t)S^-_{(N)}(\underset{\sim}{k}+\underset{\sim}{q},t) \\
&\qquad + \lambda^*(\underset{\sim}{q})a^*_{(N)}(\underset{\sim}{q},t)S^-_{(N)}(\underset{\sim}{k}-\underset{\sim}{q},t) - \mu^*(\underset{\sim}{q})a^*_{(N)}(\underset{\sim}{q},t) \\
&\qquad \cdot S^+_{(N)}(\underset{\sim}{k}-\underset{\sim}{q},t) - \lambda(\underset{\sim}{q})a_{(N)}(\underset{\sim}{q},t)S^+_{(N)}(\underset{\sim}{k}+\underset{\sim}{q},t)\Big\}\,.
\end{aligned} \tag{15}
$$

Here we have introduced for $i\in\{+,-,3\}$ the notation

$$
S^i_{(N)}(\underset{\sim}{k},t) = \sum_{n=1}^{N} e^{i\underset{\sim}{k}\cdot\underset{\sim}{x}_n}\, S^i_n(t)\,. \tag{16}
$$

4. Properties of the Equations of Motion

The equations (15) are nonlinear operator differential equations which have unique global solutions on the domain of $H_{(N)}$, which is the domain of $H^P_{(N)}$. To understand their physical content it is helpful to keep track of the powers of N in (15). The intensive observables

$$
\begin{aligned}
\sigma^i_{(N)}(\underset{\sim}{k},t) &= N^{-1}\, S^i_{(N)}(\underset{\sim}{k},t) \\
\alpha_{(N)}(\underset{\sim}{k},t) &= N^{-1/2} a_{(N)}(\underset{\sim}{k},t)
\end{aligned} \tag{17}
$$

satisfy the equations of motion

$$\dot{\alpha}(\underset{\sim}{k},t) = -i|\underset{\sim}{k}|\ \alpha(\underset{\sim}{k},t) - i\lambda^*(\underset{\sim}{k})\ \sigma^-(-\underset{\sim}{k},t) - i\mu^*(\underset{\sim}{k})\ \sigma^+(-\underset{\sim}{k},t)$$

$$\dot{\sigma}^-(\underset{\sim}{k},t) = -i\varepsilon\sigma^-(\underset{\sim}{k},t) + 2i\sum\Big\{\lambda(\underset{\sim}{q})\alpha(\underset{\sim}{q},t)\ \sigma^3(\underset{\sim}{k}+\underset{\sim}{q},t) + \mu^*(\underset{\sim}{q})\alpha^*(\underset{\sim}{q},t)\ \sigma^3(\underset{\sim}{k}-\underset{\sim}{q},t)\Big\}$$

(18)
$$\dot{\sigma}^3(\underset{\sim}{k},t) = i\sum\Big\{\mu(\underset{\sim}{q})\alpha(\underset{\sim}{q},t)\ \sigma^-(\underset{\sim}{k}+\underset{\sim}{q},t) + \lambda^*(\underset{\sim}{q})\alpha^*(\underset{\sim}{q},t)\sigma^-(\underset{\sim}{k}-\underset{\sim}{q},t) - \mu^*(\underset{\sim}{q})\alpha^*(\underset{\sim}{q},t)\sigma^+(\underset{\sim}{k}-\underset{\sim}{q},t) - \lambda(\underset{\sim}{q})\alpha(\underset{\sim}{q},t)\sigma^+(\underset{\sim}{k}+\underset{\sim}{q},t)\Big\}.$$

We have suppressed the index N, since we shall see immediately that in a suitable topology the operators $\sigma^i_{(N)}(\underset{\sim}{k},t)$ and $\alpha_{(N)}(\underset{\sim}{k},t)$ (which have equal time commutators bounded by 1/N) will converge towards complex number valued functions $\sigma^i(\underset{\sim}{k},t)$ and $\alpha(\underset{\sim}{k},t)$, which are solutions of the equations of motion (18) considered as classical dynamical equations.

For this limit we must, however, restrict the modes coupled in (18) to belong to a <u>fixed</u> cavity M and sum over $\underset{\sim}{q} \in \Gamma(M)$. We assume that $\lambda(\underset{\sim}{0}) = \mu(\underset{\sim}{0}) = 0$ and that $\sum_{\underset{\sim}{k}\in\Gamma(M)} \{|\lambda(\underset{\sim}{k})|+|\mu(\underset{\sim}{k})|\} < \infty$. Let $\mathcal{G}^M$ be the Hilbert space of all sequences

(19)
$$\rho = \Big\{\rho^i\Big\}_{i=1}^{\infty} \equiv \{\alpha(\underset{\sim}{k}) \in ¢,\ \sigma^+(\underset{\sim}{k}) = \sigma^-(-\underset{\sim}{k})^*,\ \sigma^3(\underset{\sim}{k}) = \sigma^3(-\underset{\sim}{k})^* \in ¢ \mid \underset{\sim}{k} \in \Gamma(M)\Big\},$$

with

$$\|\rho\|^2 = \sum_{\underset{\sim}{k}\in\Gamma(M)} \{|\alpha(\underset{\sim}{k})|^2 + |\sigma^+(\underset{\sim}{k})|^2 + |\sigma^3(\underset{\sim}{k})|^2\} < \infty \tag{20}$$

$\mathcal{G}^M$ is the classical phase space for our system.
For every initial condition $\rho \in \mathcal{G}^M$, the classical equations (18) have global unique solutions. For small t, this follows by the contraction mapping theorem. Quantities invariant under the flow

are

$$h = \sum_{\underset{\sim}{k}\in\Gamma(M)} |\underset{\sim}{k}|\ |\alpha(\underset{\sim}{k})|^2 + \varepsilon\,\sigma^3(\underset{\sim}{0})$$

$$+ \sum_{\underset{\sim}{k}\in\Gamma(M)} \{\alpha(\underset{\sim}{k})\ \big(\lambda(\underset{\sim}{k})\sigma^+(\underset{\sim}{k}) + \mu(\underset{\sim}{k})\ \sigma^-(\underset{\sim}{k}) + \text{c.c.}\},$$

(21)

$$s = \sum_{\underset{\sim}{k}\in\Gamma(M)} \{|\sigma^+(\underset{\sim}{k})|^2 + |\sigma^3(\underset{\sim}{k})|^2\}.$$

Since $\lambda(\underset{\sim}{0}) = \mu(\underset{\sim}{0}) = 0$, no solution in $\mathcal{G}^M$ can escape to infinity. It is to be noted that the equations (18) can be derived from h as a Hamiltonian, if one introduces the Poisson brackets

$$\{\alpha(\underset{\sim}{k}),\ \alpha^*(\underset{\sim}{q})\} = i\ \delta_{\underset{\sim}{k},\underset{\sim}{q}}\ ,\qquad \{\alpha(\underset{\sim}{k}),\ \alpha(\underset{\sim}{q})\} = 0\ ,$$

(22) $$\{\alpha(\underset{\sim}{k}),\ \sigma^i(\underset{\sim}{q})\} = 0\qquad ,\quad \{\sigma^+(\underset{\sim}{k}),\sigma^-(\underset{\sim}{q})\} = 2i\ \sigma^3(\underset{\sim}{k}+\underset{\sim}{q})\ ,$$

$$\{\sigma^3(\underset{\sim}{k}),\ \sigma^\pm(\underset{\sim}{q})\} = \pm\ i\ \sigma^\pm(\underset{\sim}{k} + \underset{\sim}{q})\ .$$

We remark that if N is a cavity with $L(N) = 2^n\ L(M)$ for some $n \in \mathbb{Z}_+$, then $\Gamma(N) \supset \Gamma(M)$ and the modes in M form a subset of those in N. Similarly, $\mathcal{G}^M$ is a subspace of $\mathcal{G}^N$, which is invariant under the time evolution (18).

Let us fix M and consider a sequence $N \to \infty$ with $\Gamma(N) \supset \Gamma(M)$. Let $H^M_{(N)}$ be the Hamiltonian (12), where one retains in (13) only the coupling between modes in $\Gamma(M)$. Consider the observables

(23) $$\rho_{(N)} = \{\rho^i_{(N)}\} = \{\alpha^\#_{(N)}(\underset{\sim}{k}),\ \sigma^i_{(N)}(\underset{\sim}{k})\ \mid\ \underset{\sim}{k} \in \Gamma(M)\}\ .$$

We shall see that for all t, the operators

$$\rho^i_{(N)}(t) = \exp\ (i\ H^M_{(N)}\ t)\ \rho^i_{(N)}\ \exp\ (-i\ H^M_{(N)}\ t)$$

converge in a very weak sense towards the classical solutions $\rho^i(t)$ of (18).

A sequence $\omega_{(N)}$ of density matrices on $\mathcal{H}^A_{(N)} \otimes \mathcal{H}^P_{(N)}$ (i.e. of states of the atoms and of the radiation field in an increasing family of cavities) is called <u>2-classical</u> (see [3], [4]) with respect to the $\rho_{(N)}$ of the form (23) with value at $\rho \in \mathcal{G}^M$, if

$$(24) \qquad \lim_{N\to\infty} \sup_i \omega_{(N)}\left((\rho^i_{(N)} - \rho^i)^* (\rho^i_{(N)} - \rho^i)\right) = 0$$

By the Schwarz inequality

$$(25) \qquad \begin{aligned} &\lim_{N\to\infty} \omega_{(N)}(\rho^i_{(N)}) = \rho^i , \\ &\lim_{N\to\infty} \omega_{(N)}(\rho^i_{(N)}\, \rho^j_{(N)}) = \rho^i \rho^j , \end{aligned}$$

and hence the observables $\rho_{(N)}$ assume the classical value ρ at least in mean and covariance.

<u>Theorem 1</u>. If $\omega_{(N)}$ is 2-classical for $\rho_{(N)}$ at $\rho \in \mathcal{G}^M$, then $\omega_{(N)}$ is 2-classical for all $\rho_{(N)}(t)$ at $\rho(t)$, the unique solution of (18) with $\rho(0) = \rho$.

<u>Proof</u>: The difference $\rho_{(N)}(t) - \rho(t)$ satisfies a linear differential equation

$$(26) \quad \left(i\omega_k + \frac{d}{dt}\right) \left(\rho^k_{(N)}(t) - \rho^k(t)\right) = \sum_{j=1}^{\infty} A^{kj}_{(N)}(t) \left(\rho^j_{(N)}(t) - \rho^j(t)\right)$$

where the $\omega_k \in \mathbb{R}$ and the $A^{kj}_{(N)}(t)$ are bounded linear operators on $\mathcal{H}_{(N)}$ with

$$(27) \qquad \sup_{k,N,t} \sum_j \| A^{kj}_{(N)}(t) \| = A < \infty .$$

This follows immediately from the fact that $\lambda(\underset{\sim}{k})$ and $\mu(\underset{\sim}{k})$ are summable and from writing, e.g.,

$$(28) \qquad \left(i\varepsilon + \frac{d}{dt}\right) \left(\sigma^-_{(N)}(\underset{\sim}{k},t) - \sigma^-(\underset{\sim}{k},t)\right) =$$

$$= 2i \sum_{\underset{\sim}{q}\in\Gamma(M)} \lambda(\underset{\sim}{q})\ \sigma^3_{(N)}(\underset{\sim}{k}+\underset{\sim}{q},t)\ (\alpha_{(N)}(\underset{\sim}{q},t) - \alpha(\underset{\sim}{q},t))$$

$$+ 2i \sum_{\underset{\sim}{q}\in\Gamma(M)} \lambda(\underset{\sim}{q})\alpha(\underset{\sim}{q},t)\ \left(\sigma^3_{(N)}(\underset{\sim}{k}+\underset{\sim}{q},t) - \sigma^3(\underset{\sim}{k}+\underset{\sim}{q},t)\right)$$

$$+ \text{similar terms},$$

and by observing that $\|\sigma^i_{(N)}(\underset{\sim}{q},t)\| = 1$. By iteration one obtains the estimate

$$(29)\quad \sup_i \omega_{(N)}\left((\rho^i_{(N)}(t) - \rho^i(t))^* (\rho^i_{(N)}(t) - \rho^i(t))\right)$$

$$\leq e^{2A|t|} \sup_i \omega_{(N)}\left((\rho^i_{(N)}-\rho^i)^* (\rho^i_{(N)}-\rho^i)\right).$$

QED

Examples of classical states are "coherent states" (see reference [3]). Let e.g. $\sum_{\underset{\sim}{k}\in\Gamma(M)} |\alpha(\underset{\sim}{k})|^2 < \infty$ and consider the pure state

$$(30)\quad \Omega_{(N)} = \exp\left(\sqrt{N} \sum_{\underset{\sim}{k}\in\Gamma(M)} \{\alpha(\underset{\sim}{k})a^*(\underset{\sim}{k}) - \alpha^*(\underset{\sim}{k})a(\underset{\sim}{k})\}\right)$$

$$\times\ \Omega^P \otimes \bigotimes_{n=1}^{N} \left(e^0_{+n} \otimes e^0_{-n}\right).$$

By (1), (2), (9) and (10) one can verify that $\omega_{(N)} = (\Omega_{(N)}, \cdot\,\Omega_{(N)})$ is classical at $\{\alpha(\underset{\sim}{k}),\ \sigma^+(\underset{\sim}{k}) = 0,\ \sigma^3(\underset{\sim}{k}) = -\frac{1}{2}\,\delta_{\underset{\sim}{k},\underset{\sim}{0}}\}$.

Later we shall formulate a much stronger result about the approach to the classical limit.

The classical equations (18) simplify considerably in the rotating wave approximation, if at $t = 0$ only one mode $\alpha(\underset{\sim}{k}) = \alpha$, $\sigma^+(\underset{\sim}{k}) = \sigma^+ = (\sigma^-)^* = \sigma^-(-\underset{\sim}{k})^*$ and $\sigma^3(\underset{\sim}{0}) = \sigma^3$ is excited. It is easily seen that this state of affairs will persist for all times, and we obtain the classical one-mode laser equations (in the rotating wave approximation and without reservoirs):

$$\begin{aligned} \dot{\alpha} &= -\,i\nu\alpha - i\lambda\sigma^- \\ \dot{\sigma}^- &= -\,i\varepsilon\sigma^- + 2i\lambda\sigma^3\alpha \\ \dot{\sigma} &= i\lambda(\sigma^-\alpha^* - \sigma^+\alpha) \end{aligned} \tag{31}$$

Besides h and s, (31) has the integral $c = |\alpha|^2 + \sigma^3$ (see (14)) and $\{h,s\} = \{h,c\} = \{s,c\} = 0$. It can be shown that (31) is an integrable system with invariant tori. The allowed physical values of s are $0 \leq s \leq 1/4$.

Simple special solutions of (31) are

(a) Trivial stationary solutions

$$-\frac{1}{2} \leq \sigma^3 = \text{const.} \leq \frac{1}{2}, \quad \sigma^+ = \alpha = 0 . \tag{32}$$

(b) Nontrivial stationary solutions

$$\sigma^3 = -\frac{\varepsilon\nu}{2\lambda^2}, \quad \alpha = -\lambda\sigma^-/\nu = \text{const.} , \tag{33}$$

which are physically admissible only for large coupling constants, for $\varepsilon\nu \leq \lambda^2$.

(c) Harmonic solutions $\sigma^-(t) = e^{-i\omega t}\sigma^-$, $\alpha(t) = e^{-i\omega t}\alpha$,

$$\sigma^- = \frac{\omega-\nu}{\lambda}\alpha = \text{const.} , \quad \sigma^3 = -(\varepsilon-\omega)(\nu-\omega)/2\lambda^2 . \tag{34}$$

The stationary solutions (32) and (33) are particularly important for equilibrium statistical mechanics: the Gibbs state at the reciprocal temperature β is of the form (33) averaged over arg α, if β is large enough and if $\varepsilon\nu < \lambda^2$. If β is small, then the Gibbs state is of the form (32). Thus, if $\varepsilon\nu < \lambda^2$, a phase transition occurs from a state in which the expected photon number is of O(N), to a state in which it is O(1).

The critical value of β for the equilibrium phase transition is

$$\beta_c = \frac{2}{\varepsilon}\tanh^{-1}\left(\frac{\nu\varepsilon}{\lambda^2}\right) . \tag{35}$$

We shall not discuss here the thermodynamics of the laser, which can be found in [5], [6], [7].

5. The Atomic Reservoirs

The physical picture which we have in mind of the atomic dissipation and pumping is a coupling of the localized electron states of each atom to the surrounding electrons which has the property of draining or filling the two atomic energy levels. We shall do this in such a way that the mean electron number of each atom is one, and will remain so for all time. While there are fluctuations in the individual atomic electron numbers of order one, the total number fluctuation will be only of order $N^{1/2}$. This fact will be demonstrated later.

Consider the nth atom, $1 \leq n \leq N$, with creation and annihilation operators $b^{\#}_{+n}$ and $b^{\#}_{-n}$. We shall drop the label n until it becomes necessary. The reservoir for this atom consists of four fermion fields labelled by the continuous parameter $w \in \mathbb{R}$ with creation and annihilation operators $B^{\#}_{\pm}(w)$, $C^{\#}_{\pm}(w)$. The reservoir Hamiltonian is $H^{AR} = H^{AR}_0 + H^{AR}_1$,

$$H^{AR}_0 = \sum_{\pm} \int_{-\infty}^{+\infty} dw\; w\; [B^*_{\pm}(w)\; B_{\pm}(w) + C^*_{\pm}(w)\; C_{\pm}(w)]$$

(36)

$$H^{AR}_1 = \sum_{\pm} \int_{-\infty}^{+\infty} dw\; [g^B_{\pm}(w)\; B^*_{\pm}(w) b_{\pm} + g^C_{\pm}(w)\; C^*_{\pm}(w) b^*_{\pm} + \text{h.c.}] ,$$

where $g^B_{\pm}$, $g^C_{\pm}$ are real functions to be specified later.

These fermion fields anticommute with the $b_{\pm}$ operators of the same index $\pm$ and commute with all operators of a different index. For the same index, the anticommuators, $\{A,B\} = AB + BA$, are

$$\{B^*(w), B(w')\} = \delta(w-w') = \{C^*(w), C(w')\} ,$$

(37) .

$$\{b^*, b\} = 1 ,$$

and all other anticommutators between such $B^\#$, $C^\#$, $b^\#$ are zero. It is important to note that H_0^{AR} is selfadjoint in the Fock representation, where

$$B_\pm(w)\ \Omega^{AR} = C_\pm(w)\ \Omega^{AR} = 0 \tag{38}$$

for all w. If $g_\pm^B$, $g_\pm^C \in L^2$, then H_1^{AR} is bounded.

The Heisenberg equations of motion of the $B^\#$ and $C^\#$ operators are

$$\begin{aligned} \dot{B}_\pm(w,t) &= -iw\ B_\pm(w,t) - ig_\pm^B(w)\ b(t) \\ \dot{C}^*_\pm(w,t) &= iw\ C^*_\pm(w,t) + ig_\pm^C(w)\ b(t) \end{aligned} \tag{39}$$

Thus

$$B_\pm(w,t) = e^{-iwt}\ B_\pm(w) - ig_\pm^B(w) \int_0^t ds\ e^{-iw(t-s)}\ b_\pm(s) \tag{40}$$

and similarly for $C_\pm(w,t)$.

The Heisenberg equations of motion for the $b_\pm$-operators are

$$\begin{aligned} \dot{b}_\pm(t) &= i[H^A + H^{AP} + H^{AR}, b_\pm(t)] \\ &= i[H^A + H^{AP}, b_\pm(t)] \\ &\quad - i \int_{-\infty}^{+\infty} dw\ [g_\pm^B(w)\ B\ (w,t) - g_\pm^C(w)\ C^*_\pm(w,t)] . \end{aligned} \tag{41}$$

The expressions for $B_\pm$ and $C_\pm$ obtained above can be inserted into (41). In this section we primarily wish to study the effects of the reservoir on the $b_\pm$-operators, and to that end shall temporarily drop the terms involving H^A and H^{AP}. Equation (41) then becomes

$$(42)\quad \dot{b}_{\pm}(t) = -\int_0^t ds\, b_{\pm}(s) \int_{-\infty}^{+\infty} dw\, \left(g_{\pm}^{B}(w)^2\, e^{-iw(t-s)} + g_{\pm}^{C}(w)^2 e^{iw(t-s)}\right)$$

$$- i \int_{-\infty}^{+\infty} dw\, \left(g_{\pm}^{B}(w)\, e^{-iwt}\, B_{\pm}(w) - g_{\pm}^{C}(w)\, e^{iwt}\, C_{\pm}^{*}(w)\right) .$$

The second term is a definite time-dependent operator, $f_{\pm}(t)$, which depends only on the observables of the reservoir at time zero, and which we shall call a "fluctuation force." To make further progress we shall assume that

$$(43)\qquad g_{\pm}^{B,C}(w) = g_{\pm}^{B,C}\, \tau(w^2 + \tau^2)^{-1/2}$$

with $g_{\pm}^{B,C} \geq 0$, $\tau > 0$. Then with $\gamma_{\pm} = \pi((g_{\pm}^{B})^2 + (g_{\pm}^{C})^2)$ equation (42) becomes

$$(44)\qquad \dot{b}_{\pm}(t) = -\gamma_{\pm}\tau \int_0^t ds\, e^{-\tau(t-s)}\, b_{\pm}(s) + f_{\pm}(t) .$$

The first term is a damping term with memory. For our Lorentzian cutoff, (44) is a second order linear inhomogeneous differential equation

$$(45)\qquad \ddot{b}_{\pm}(t) = -\gamma_{\pm}\tau\, b_{\pm}(t) - \tau\, \dot{b}_{\pm}(t) + \dot{f}_{\pm}(t) + \tau f_{\pm}(t)$$

with initial conditions

$$(46)\qquad \begin{aligned} b_{\pm}(0) &= b_{\pm} , \\ \dot{b}_{\pm}(0) &= f_{\pm}(0) . \end{aligned}$$

For large τ, the solutions to the homogeneous equation decay as $e^{-\tau t}$ and $e^{-\gamma_{\pm} t}$.

In the spirit of dynamical systems, we have to specify the initial state of the fermion and its reservoir in order to determine the expectation values of the observables at all later times. Mathematically, the assumptions that at time $t = 0$ one puts together the system and reservoir in an uncorrelated way means that the initial state ω is a tensor product, $\omega = \omega^S \otimes \omega^R$, of a state of the system, ω^S, and of the reservoir, ω^R. A state ω^R of a quantum system of infinitely many degrees of freedom with an algebra of observables, $\mathfrak{A}^R$, generated by all polynomials in $B_\pm^\#(w)$, $C_\pm^\#(w)$ (smeared out over L^2 test functions in w) is a positive normalized linear functional on $\mathfrak{A}^R$ [8]. We may assume that at $t = 0$ the reservoir is in thermal equilibrium, and hence in the Gibbs state for the H_0^{AR}-time evolution of $\mathfrak{A}^R$. However, in order to avoid technical complications we shall restrict ourselves to the Fock state (38).

The atom, when coupled to this reservoir, shows an exponential approach to equilibrium. If ω^S is any density matrix on $\mathcal{V}$, then as $t \to +\infty$,

$$\omega^S \otimes \omega^R \, (b_\pm^\#(t)) \to 0 \tag{47}$$

$$\omega^S \otimes \omega^R \, (b_\pm^\#(t) \; b_\mp^\#(t)) \to 0$$

and

$$\omega^S \otimes \omega^R \, (b_\pm^*(t) \; b_\pm(t)) \to I_\pm \; . \tag{48}$$

The algebraic computations simplify greatly if one takes the limit $\tau \to \infty$, $g_\pm^{B,C}(w) \to g_\pm^{B,C}$, in which case the time evolution continues to make sense as a 1-parameter group of automorphisms of $\mathfrak{A} = \mathfrak{A}^S \otimes \mathfrak{A}^R$, where $\mathfrak{A}^S$ are the bounded operators on $\mathcal{H}^S$:

$$\dot{b}_\pm(t) = - \gamma_\pm b_\pm(t) + f_\pm(t) \tag{49}$$

$$b_\pm(t) = e^{-\gamma_\pm t} \, b_\pm + \int_0^t ds \; e^{-\gamma_\pm (t-s)} f_\pm(s) \; ,$$

and in this case

$$I_\pm = (g_\pm^C)^2 \left((g_\pm^B)^2 + (g_\pm^C)^2 \right)^{-1} \; . \tag{50}$$

As one sees from (48) the quantities $I_\pm$ are the $t \to \infty$ limit of the occupations in the upper and lower atomic levels. These quantities are determined by the parameters of the reservoir, namely $(g_\pm^B)^2$ and $(g_\pm^C)^2$. We wish to adjust them so that three conditions are satisfied:

(i) The asymptotic electron number is unity. This requires that

(51) $$I_+ + I_- = 1$$

(ii) The asymptotic "atomic inversion," namely the average difference between the number of electrons in the upper level and in the lower level, is some specified number η. This requires that

(52) $$\frac{1}{2}(I_+ - I_-) = \eta$$

(iii) The third condition is one whose significance will only become apparent later when we couple the atoms to the radiation field. As a consequence of (38) and (44), the equations of motion for $N_\pm \equiv \omega^S \otimes \omega^R\left(b_\pm^*(t)\, b_\pm(t)\right)$ are

(53) $$\dot{N}_\pm(t) = -2\gamma_\pm(N_\pm(t) - I_\pm)$$

The equations of quantum electrodynamics (15) only involve the operators $S^i(\underset{\sim}{k})$. By definition, the expectation value of S^3 is $\frac{1}{2}(N_+ - N_-)$. Owing to the interaction with the radiation field, $N_+ - N_-$ may not approach the value $I_+ - I_-$ for $t \to +\infty$, which as we demonstrated happens for an isolated atom. By (53) this would force $N_+ + N_-$ to approach a value other than 1, unless we set

(54) $$\gamma_+ = \gamma_- = \gamma/2 \ .$$

This choice also has the important advantage that for the <u>isolated</u> atom $N_+ + N_- = 1$ for all times if $N_+ + N_- = 1$ at $t = 0$.

An elementary computation shows that one has to choose

$$(g_+^B)^2 = (g_-^C)^2 , \quad (g_+^C)^2 = (g_-^B)^2 . \tag{55}$$

Theré remain two free parameters:

$$\begin{aligned} \gamma &= 2\pi\left((g_+^B)^2 + (g_-^B)^2\right) , \\ \eta &= \left((g_-^B)^2-(g_+^B)^2\right)/2\left((g_+^B)^2+(g_-^B)^2\right) , \end{aligned} \tag{56}$$

and the whole region $-1/2 \leq \eta \leq 1/2$ can be reached.

The results of this section can be summarized as follows: Each electron state of each atom is driven by its reservoir in such a way that the mean electron number per atom approaches unity, if it was not at that value initially, the mean value of S^3 is pumped to the value η, and the mean value of $S^{\pm}$ goes to zero.

6. The Photon Reservoirs

Similarly to the atom-reservoir system, we introduce a photon reservoir which will have the effect of damping the photon modes. This photon damping is necessary; otherwise the equations of motion we shall subsequently develop will have no bounded stationary states for $\eta > 0$. Damping occurs, of course, in a real laser in the form of losses at the cavity walls and mirrors.

As a mathematical model of a photon reservoir we take a separate reservoir for each photon mode designated by the wave vector $\underset{\sim}{k}$. This reservoir has boson creation and annihilation operators $A^{\#}(\underset{\sim}{k},w)$ with commutation relations

$$[A(\underset{\sim}{k},w) , A^*(\underset{\sim}{k}',w')] = \delta_{\underset{\sim}{k},\underset{\sim}{k}'} \delta(w-w') \tag{57}$$

and all other commutators vanish. In the above, the first δ is a Kronecker delta and the second δ is a Dirac delta function.

The Hamiltonian for these reservoirs is

$$H^{PR} = \sum_{\underset{\sim}{k}} \int_{-\infty}^{+\infty} dw \Big\{ w\, A^*(\underset{\sim}{k},w)\, A(\underset{\sim}{k},w) \tag{58}$$
$$+ \tau\, h(\underset{\sim}{k})\, (w^2+\tau^2)^{-1/2} [A^*(\underset{\sim}{k},w)\, a(\underset{\sim}{k}) + a^*(\underset{\sim}{k})\, A(\underset{\sim}{k},w)] \Big\}$$

In this section we shall suppress the index $\underset{\sim}{k}$. The equations of motion for a photon coupled to its reservoir can be derived as in Section 5. In the absence of a coupling to the atoms, the solution is the integro-differential equation

$$\dot{a}(t) = -\tau\kappa \int_0^t ds\, e^{-\tau(t-s)}\, a(s) \tag{59}$$
$$- ih \int_{-\infty}^{+\infty} \frac{\tau\, dw}{(w^2+\tau^2)^{1/2}}\, A(w)\, e^{-iwt}$$

with $\kappa = \pi h^2$. The solution of (59) is similar to the one of (45). In the limit $\tau \to \infty$ we obtain

$$a(t) = e^{-\kappa t} a + \int_0^t ds\, e^{-\kappa(t-s)} f(s) \tag{60}$$

where $f(s) = -ih \int dw\, A(w)\, e^{-iwt}$ is again a fluctuation force.

As the state of the reservoir we take a pure state, the Fock vacuum Ω^{PR} which satisfies for all w

$$A(w)\, \Omega^{PR} = 0 . \tag{61}$$

Then, for an arbitrary state ω^S of the photon, for which $\omega^S(a^* a) < \infty$ at $t = 0$, one has

$$\lim_{t\to\infty} \omega^S \otimes \omega^R \big(a^*(t)\, a(t)\big) = 0 . \tag{62}$$

In [4] we took another model for the photons and their reservoirs. There we built the photons out of infinitely many fermions, each of which had a fermion field as reservoir. Then the construction of Section 5 is applicable, including finite temperature Gibbs states for the reservoirs.

7. The Fully Coupled Laser

In order to gain an understanding of the cooperative properties of the full laser-reservoir system, we shall first describe the one mode laser in the rotating wave approximation. Now the Hamiltonian is

$$H_{(N)} = H^{P}_{(N)} + H^{A}_{(N)} + H^{AP}_{(N)} + H^{AR}_{(N)} + H^{PR}_{(N)} . \tag{63}$$

The first three terms together are the Hamiltonian of the laser proper (with restriction to the modes of a fixed large cavity M). The last two terms are the reservoir Hamiltonians; these are the Hamiltonians described in equations (36) and (58) summed over all the atoms. Again, to simplify matters, we shall first take the singular reservoirs, by passing to the limits $\tau \to \infty$ in all reservoirs.

As before we are interested in the intensive observables $\sigma^{i}_{(N)} = N^{-1} S^{i}_{(N)}$ and $\alpha_{(N)} = N^{-1/2} a$, $i \in \{+,-,3\}$. The Heisenberg equations of motion are

$$\begin{aligned}
\dot{\alpha}_{(N)}(t) &= -(\kappa+i\nu)\,\alpha_{(N)}(t) - i\lambda\,\sigma^{-}_{(N)}(t) + \Phi_{(N)}(t) \\
\dot{\sigma}^{-}_{(N)}(t) &= -(\gamma+i\varepsilon)\,\sigma^{-}_{(N)}(t) + 2i\lambda\,\sigma^{3}_{(N)}(t)\,\alpha_{(N)}(t) + \chi^{-}_{(N)}(t) \\
\dot{\sigma}^{3}_{(N)}(t) &= -\gamma(\sigma^{3}_{(N)}(t)-\eta) + i\lambda[\sigma^{-}_{(N)}(t)\,\alpha^{*}_{(N)}(t) - \sigma^{+}_{(N)}(t)\alpha_{(N)}(t)] \\
&\qquad + \chi^{3}_{(N)}(t) .
\end{aligned} \tag{64}$$

The influence of the photon reservoir shows up in two ways: in a systematic damping term, $-\kappa\,\alpha_{(N)}(t)$, which depends only on the state of the small system and which is hopefully $O(1)$ as $N \to \infty$, and in a fluctuation force

(65) $$\Phi_{(N)}(t) = -ih \int_{-\infty}^{+\infty} dw\, e^{-iwt} A(w) \,/\, \sqrt{N}$$

which is manifestly $O(N^{-1/2})$ and which only depends on the initial state of the reservoir. In the equations for the atomic observables the damping terms, $-\gamma\, \sigma^{i}_{(N)}(t)$, arise from contributions from all electrons according to (49). The fluctuation forces $\chi^{i}_{(N)}(t)$ come from the second term on the right side of (49), and they are very complicated for finite N:

(66) $$\chi^{-}_{(N)}(t) = N^{-1} \sum_{n=1}^{N} \left\{ b^{*}_{-n}(t)\; f_{+n}(t) + f^{*}_{-n}(t)\; b_{+n}(t) \right\} ,$$

where, according to (42),

(67) $$f_{\pm n}(t) = -i \int_{-\infty}^{+\infty} dw\; [g^{B}_{\pm} B_{\pm n}(w) e^{-iwt} - g^{C}_{\pm}\, C^{*}_{\pm n}(w)\; e^{iwt}]$$

and

(68) $$\chi^{3}_{(N)}(t) = (2N)^{-1} \sum_{n=1}^{N} \Big\{ b^{*}_{+n}(t) f_{+n}(t) + f^{*}_{+n}(t) b_{+n}(t) - b^{*}_{-n}(t) f_{-n}(t) - f^{*}_{-n}(t) b_{-n}(t) \Big\} - \gamma\eta$$

The constant $\gamma\eta$ has been subtracted on the right side of (68) in order that for $i \in \{+,-,3\}$ and all t

(69) $$\omega^{R}(\chi^{i}_{(N)}(t)) = 0 \;.$$

We have previously computed η to be (56) for noninteracting atoms. It can be shown from the anticommutation relations of the $b^{\#}$'s, which are preserved in time, and from (38) and (49) that (69) holds with the same value of η for all times.

If the atoms were uncoupled, then the $\chi^{i}_{(N)}$ would each be a sum of identically distributed independent random variables with mean zero. By the law of large numbers the $\chi^{i}_{(N)}$ would be $O(N^{-1/2})$ for $N \to \infty$. In our model, the $b^{\#}_{\pm n}(t)$ are weakly dependent on each other, through the radiation field, but it is a theorem (see References [4], [5] and Theorem 2 below) that for a suitable choice of classical states the $\chi^i_{(N)}$ can be neglected in the equations (64) as $N \to \infty$. Furthermore, the $\alpha_{(N)}(t)$ and $\sigma^i_{(N)}(t)$ converge, as they did in Theorem 1, to complex solutions of the classical laser equations

$$\begin{aligned} \dot{\alpha}(t) &= -(\kappa+i\nu)\alpha(t) - i\lambda\,\sigma^-(t) \\ (70) \quad \dot{\sigma}^-(t) &= -(\gamma+i\varepsilon)\,\sigma^-(t) + 2i\lambda\,\sigma^3(t)\,\alpha(t) \\ \dot{\sigma}^3(t) &= -\gamma(\sigma^3(t)-\eta) + i\lambda(\sigma^-(t)\alpha^*(t) - \sigma^+(t)\alpha(t)) \end{aligned}$$

with initial value $\alpha(0) = \alpha$, $\sigma^i(0) = \sigma^i$ determined by the state ω^S.

There is one additional intensive observable of the atoms to which we should pay attention, the mean number of electrons

$$(71) \quad \nu_{(N)}(t) = N^{-1} \sum_{n=1}^{N} \left\{ b^*_{+n}(t)b_{+n}(t) + b^*_{-n}(t)b_{-n}(t) \right\} .$$

In the fully coupled laser (63), $\nu_{(N)}(t)$ satisfies

$$(72) \qquad \dot{\nu}_{(N)}(t) = -\gamma(\nu_{(N)}(t)-1) + \Psi_{(N)}(t) .$$

The term $(-\gamma)(-1)$ is a result of (55). The $\Psi_{(N)}(t)$ is a fluctuation force of the type given before. If one believes the previous argument that fluctuation forces can be negelcted on the macroscopic level as $N \to \infty$, one sees that the dissipation maintains the mean electron number at unity.

Next we shall describe the properties of these classical equations. There is a Liapunov function

$$(73) \qquad L(t) = |\alpha(t)|^2 + \sigma^3(t) + |\sigma^-(t)|^2 + |\sigma^3(t)|^2 ,$$

which is bounded from below and satisfies $\dot{L}(t) < 0$ for

$|\alpha(t)|^2 + |\sigma^+(t)|^2 + |\sigma^3(t)|^2$ outside of some sphere in the 5-dimensional phase space. Hence all solutions of (70) stay bounded for all times.

The only stationary solution is

$$\sigma^3 = \eta \ , \quad \sigma^- = \alpha = 0 \ . \tag{74}$$

This solution is asymptotically stable for

$$\eta < \eta^C = \frac{\kappa\gamma}{2\lambda^2}\left(1 + \frac{(\varepsilon-\nu)^2}{(\kappa+\gamma)^2}\right) \tag{75}$$

If η passes the threshold η^C, a Hopf bifurcation occurs and a one-parameter family of periodic solutions becomes attracting:

$$\begin{aligned} &\alpha(t) = \alpha\, e^{-i\omega t} \ , && \sigma^-(t) = \sigma^-\, e^{-i\omega t} \ , \\ &\sigma^3(t) = \eta^C \ , && \omega = \frac{\kappa\varepsilon + \gamma\nu}{\kappa + \gamma} \ , \\ &(\kappa + i(\nu-\omega))\alpha = -i\lambda\sigma^- \ , && |\alpha|^2 = \frac{\gamma}{2\kappa}(\eta-\eta^C) \ . \end{aligned} \tag{76}$$

It is possible that there are further bifurcations and, indeed, for the multimode laser one can exhibit such phenomena.

The physical interpretation of η^C is that it is the critical pumping constant for a reservoir-driven phase transition from a non-radiating to a coherently radiating oscillatory state. In the non-vanishing stationary state the atoms and photons behave as though they are uncoupled and influenced only by their reservoirs. The oscillatory state is truly cooperative and has a frequency ω which is a convex combination of the cavity and atomic frequencies.

There is an obvious generalization of (70) to the case of infinitely many modes $\underset{\sim}{k} \in \Gamma(M)$ associated with a cavity of fixed length $L(M)$ and without making the rotating wave approximation. It is clear that there again exists a transition from a stationary to a radiating state. However, the detailed description of the bifurcations becomes truly complicated.

Another level of complication arises if one takes regular reservoirs for the atoms. We have derived only the equations for the Lorentzian cut-off $g(w) = g\tau(w^2+\tau^2)^{-1/2}$. By (45) the photon and atomic observables satisfy second order differential equations in this case. Hence more intensive observables are necessary to describe the macroscopic behavior of the atoms. In addition to the $\sigma^i_{(N)}$ and $\alpha_{(N)}$ they are

$$\sigma^{+'}_{(N)}(t) = N^{-1}\sum_{n=1}^{N} \dot{b}^*_{+n}(t)\dot{b}_{-n}(t) = \sigma^{-'}_{(N)}(t)^*$$

$$\sigma^{3'}_{(N)}(t) = (2N)^{-1}\sum_{n=1}^{N}\left(b^*_{+n}(t)\dot{b}_{+n}(t) - \dot{b}^*_{-n}(t)b_{-n}(t)\right) \tag{77}$$

$$= \sigma^{3''}_{(N)}(t)^*$$

$$\sigma^{3'''}_{(N)}(t) = (2N)^{-1}\sum_{n=1}^{N}\left(\dot{b}^*_{+n}(t)\dot{b}_{+n}(t) - \dot{b}^*_{-n}(t)\dot{b}_{-n}(t)\right)$$

In the rotating wave approximation and for maximal pumping, $\eta = \frac{1}{2}$, the classical laser equations become (for $N \to \infty$):

$$\begin{aligned}
\ddot{\sigma}^+ &= -\tau(\gamma-i\varepsilon)\sigma^+ + \left(\tfrac{i\varepsilon}{2} - \tau\right)\dot{\sigma}^+ + 2\sigma^{+'} \\
&\quad - 2i\lambda\tau\alpha^*\sigma^3 - 2i\lambda\dot{\alpha}^*\sigma^3 - 2i\lambda\alpha^*\sigma^{3'} \\
\dot{\sigma}^{+'} &= \tfrac{\tau}{2}(i\varepsilon-\gamma)\dot{\sigma}^+ + (i\varepsilon-2\tau)\sigma^{+'} - 2i\lambda(\dot{\alpha}^*+\tau\alpha^*)\sigma^{3''} - 2i\lambda\alpha^*\sigma^{3'''} \\
\dot{\sigma}^3 &= \sigma^{3'} + \sigma^{3''} \\
\dot{\sigma}^{3'} &= -\tfrac{\tau}{2}(i\varepsilon+\gamma)\sigma^3 - \left(\tau+\tfrac{i\varepsilon}{2}\right)\sigma^{3'} + \sigma^{3'''} \\
&\quad - i\lambda\dot{\alpha}\sigma^+ - \tfrac{i\lambda}{2}\alpha\dot{\sigma}^+ - i\lambda\tau\alpha\sigma^+ \\
\dot{\sigma}^{3'''} &= -\tfrac{\tau}{2}(\gamma-i\varepsilon)\sigma^{3'} - \tfrac{\tau}{2}(\gamma+i\varepsilon)\sigma^{3''} - 2\tau\sigma^{3'''} \\
&\quad - i\lambda\alpha\sigma^{+'} + i\lambda\alpha^*\sigma^{-'} - \tfrac{i\lambda}{2}\dot{\alpha}\dot{\sigma}^+ + \tfrac{i\lambda}{2}\dot{\alpha}^*\dot{\sigma}^- + \tfrac{\gamma\tau^2}{2} \\
\ddot{\alpha} &= -(\tau+i\nu)\dot{\alpha} - \tau(i\nu+\kappa)\alpha - i\lambda\dot{\sigma}^- - i\lambda\tau\sigma^- .
\end{aligned} \tag{78}$$

One sees by inspection that these equations have a stationary solution, where the only nonvanishing amplitudes are

$$\sigma^3 = \frac{1}{2}\,, \quad \sigma^{3'} = -\frac{i\varepsilon}{4}\,, \quad \sigma^{3'''} = \frac{\gamma\tau}{4} + \frac{\varepsilon^2}{8}. \tag{79}$$

This solution is stable for small coupling $|\lambda|$. For $\varepsilon = \nu = 0$, $\eta = 1/2$ and τ sufficiently large, (79) becomes unstable for $\lambda^2 = \lambda_C^2 = \kappa\,\gamma$, just as in (75). For large τ and $\varepsilon > 0$ and $\nu > 0$ the constant term in the characteristic polynomial for (78) linearized around (79) does not vanish. Hence, the unstable attractor bifurcates into a stable periodic solution, and the qualitative behavior of the one-mode laser remains unchanged for these more complicated but physically more realistic reservoirs.

What we have said so far, without proof, is that the quantities $\lim_{N\to\infty} \omega_{(N)}\left(\sigma^i_{(N)}(t)\right)$ satisfy the classical equations of motion given above when the sequence of states $\omega_{(N)}$ is classical. We now wish to investigate the next lower order in $N^{1/2}$, namely expectations of the algebra generated by the <u>fluctuation observables</u> $a_{(N)}(t) = N^{1/2}[\alpha_{(N)}(t)-\alpha(t)]$ and $s^i_{(N)}(t) = N^{1/2}[\sigma^i_{(N)}(t)-\sigma^i(t)]$. Clearly the fluctuation forces which have heretofore been neglected will now play a decisive role. The equal time commutators between fluctuation observables are intensive observables, e.g.

$$\begin{aligned} [a_{(N)}(t),\, a^*_{(N)}(t)] &= 1 \\ [s^+_{(N)}(t),\, s^-_{(N)}(t)] &= 2\sigma^3_{(N)}(t)\,. \end{aligned} \tag{80}$$

Hence in the limit $N \to \infty$ they should become numbers.

One would hope that equations of motion for the $a_{(N)}(t)$ and $s^i_{(N)}(t)$ could be obtained as follows: Take the full equations of motion, including the fluctuation forces, and linearize them around the classical orbits which are regarded as having been determined given the initial expectation value for the intensive observables in $\omega_{(N)}$. In this way one obtains linear equations for the operators $a_{(N)}(t)$ and $s^i_{(N)}(t)$ which have time dependent coefficients given by the numerical functions $\alpha(t)$ and $\sigma^i(t)$ and operator valued inhomogeneous terms $F_{(N)}(t) = \sqrt{N}\,\Phi_{(N)}(t)$, and $G^i_{(N)}(t) = \sqrt{N}\,\chi^i_{(N)}(t)$. Then, in some weak sense for large N, and for the one-mode laser in the rotating wave approximation,

$$\dot{a}_{(N)}(t) = -(\kappa+i\nu)a_{(N)}(t) - i\lambda\, s^-_{(N)}(t) + F_{(N)}(t) ,$$

$$\dot{s}^-_{(N)}(t) = -(\gamma+i\varepsilon)s^-_{(N)}(t) - 2i\lambda\, \alpha(t)\, s^3_{(N)}(t)$$

$$(81) \qquad + 2i\lambda\, \sigma^3(t)\, a_{(N)}(t) + G^-_{(N)}(t) ,$$

$$\dot{s}^3_{(N)}(t) = -\gamma\, s^3_{(N)}(t) + i\lambda\, [\alpha^*(t)\, s^-_{(N)}(t) + \sigma^-(t)\, a^*_{(N)}(t)$$

$$- \alpha(t)\, s^+_{(N)}(t) - \sigma^+(t)\, a_{(N)}(t)] + G^3_{(N)}(t)$$

We have taken the one-mode laser with rotating wave approximation as an illustrative example, but similar equations could be derived in the other cases.

The meaning of (81) is clear. One solves these linear equations in terms of the initial conditions for the $a_{(N)}(t)$ and $s^i_{(N)}(t)$ and using the $F_{(N)}(t)$, $G^i_{(N)}(t)$ for all t. Assuming that the states are such that the $N \to \infty$ limits of $a_{(N)}(0)$, $s^i_{(N)}(0)$, $F_{(N)}(t)$ and $G^i_{(N)}(t)$ for all t exist, we could calculate the fluctuations and their moments for all t. Note that the same reasoning would apply to the mean electron number and ensures that the fluctuations in the mean electron number are $O(N^{-1/2})$.

If the foregoing reasoning can be justified -- and we can in fact do so -- the expectation values of the fluctuations of the $a_{(N)}(t)$ and $s^i_{(N)}(t)$ can be computed in terms of the expectation values of the fluctuations $F_{(N)}(t)$ and $G^i_{(N)}(t)$. These in turn require knowledge of the $b_{\pm n}(t)$. It turns out again that the obvious simplification for $N \to \infty$ can be shown to be correct. That is that one returns to the elementary Heisenberg equations of motion for the $b_{\pm n}(t)$ and decouples them by replacing the operator valued function $\alpha_{(N)}(t)$ by the classical function $\alpha(t)$. The terms coming from the time zero reservoir operators are retained, of course. Call this approximate solution $\hat{b}_{\pm n}(t)$ and use it to compute approximate fluctuation forces $\hat{G}^i_{(N)}(t)$ in (66) and (68). The expectation values of these fluctuation forces can easily be calculated (see [3], Chapter 3) and shown to be Gaussian and δ-correlated in time as $N \to \infty$. The relevant fact is that as $N \to \infty$ this simple calculation becomes exact, for $\lim_{N\to\infty} (G^i_{(N)}(t) - \hat{G}^i_{(N)}(t)) = 0$ in a suitable topology described below.

In the above heuristic discussion we have outlined in principle a method of calculating the fluctuations around the classical orbit. The theorem that justifies this will now be given without proof. This latter can be found in references [3] and [4].

Let $\rho_{(N)} = \{\rho^i_{(N)}\}$ be the sequence of intensive observables of the laser model under consideration, namely (23) for the ∞-mode case and $\rho_{(N)} = \{\alpha^\#_{(N)}, \sigma^i_{(N)}, \nu_{(N)}\}$ for the 1-mode laser. Let $\mathcal{G}$ be the phase space of the model. $\mathcal{G}$ is the Hilbert space $\mathcal{G}^M \times \mathbb{R}^1$, if one takes all modes in $\Gamma(M)$ and $\mathcal{G} = \mathbb{R}^6$ for the 1-mode laser. A sequence of density matrices $\omega^S_{(N)}$ on $\mathcal{H}^S_{(N)} = \mathcal{H}^A_{(N)} \otimes \mathcal{H}^P_{(N)}$ is called <u>classical</u> at $\rho \in \mathcal{G}$, if for all monomials of elements in $\rho_{(N)}$

(82) $$\lim_{N\to\infty} \omega^S_{(N)} \left(\rho^{i_1}_{(N)} \dots \rho^{i_n}_{(N)}\right) = \rho^{i_1} \dots \rho^{i_n}.$$

In reference [4] such a sequence of states was called pure classical and convex combinations of pure classical states were called classical.

A sequence $\omega^S_{(N)}$ has <u>normal fluctuations</u> around ρ, if for all monomials

$$\lim_{N\to\infty} \omega^S_{(N)} \left(\sqrt{N}\,(\rho^{i_1}_{(N)} - \rho^{i_1}) \ldots \sqrt{N}\,(\rho^{i_n}_{(N)} - \rho^{i_n})\right) \tag{83}$$

exists. Clearly, a state with normal fluctuations around ρ is classical at ρ, by the Schwarz inequality.

Let $\omega^R_{(N)}$ be a state of the reservoir of the type mentioned in Sections 5 and 6.

<u>Theorem 2</u>. Let $\omega^S_{(N)}$ be classical at $\rho \in \mathcal{G}$. Let $\rho_{(N)}(t)$ be the solution of the Heisenberg equations for the fully coupled laser and let $\rho(t)$ be the solutions of the corresponding classical equations with $\rho(0) = \rho$. Then for all monomials and all times $t_1,\ldots,t_n$

$$\lim_{N\to\infty} \omega^S_{(N)} \otimes \omega^R_{(N)} \left(\rho^{i_1}_{(N)}(t_1) \ldots \rho^{i_n}_{(N)}(t_n)\right) = \rho^{i_1}(t_1) \ldots \rho^{i_n}(t_n) \tag{84}$$

uniformly over bounded time intervals.

<u>Theorem 3</u>. Let $\omega^S_{(N)}$ have normal fluctuations around ρ. Let $r^i_{(N)}(t) = \sqrt{N}\,\left(\rho^i_{(N)}(t) - \rho^i(t)\right)$. Let $\hat{r}_{(N)}(t)$ be the solution of the linearized Heisenberg equations of motion as described above with the fluctuation forces for the photons given explicitly, and with the fluctuation forces for the atomic observables calculated by the decoupling scheme mentioned above for the $\hat{b}_{\pm n}(t)$. Then for all times

$$\lim_{N\to\infty} \omega^S_{(N)} \otimes \omega^R_{(N)} \left(r^{i_1}_{(N)}(t_1) \ldots r^{i_n}_{(N)}(t_n)\right) \tag{85}$$

exists and is equal to

$$\lim_{N\to\infty} \omega^S_{(N)} \otimes \omega^R_{(N)} \left(\hat{r}^{i_1}_{(N)}(t_1) \ldots \hat{r}^{i_n}_{(N)}(t_n)\right). \tag{86}$$

This latter limit exists by explicit computation. The limit is uniform on bounded time intervals.

In reference [4] these theorems were proved for a modified dynamics in which the photon operators were replaced by truncated operators having norms $\leq \zeta N$ and behaving as real photons for photon numbers $\ll \zeta N$. In the limit $N \to \infty$ followed by $\zeta \to \infty$ we obtained results exactly as described in Theorems 2 and 3.

An alternative to the cutoff is to treat the photon as being classical, with a classical reservoir, and to study the equations of semiclassical laser theory. Then the equations of motion for the atoms remain formally unchanged, while e.g.

$$(87)\quad \dot{\alpha}_{(N)}(t) = -(\kappa+i\nu)\,\alpha_{(N)}(t) - i\lambda\,\omega_A^S \otimes \omega_A^R\,(\sigma^-_{(N)}(t)) + \Phi_{(N)}(t)\ .$$

In reference [3] we proved both theorems without cutoffs for special sequences $\omega^S_{(N)}$ with normal fluctuations, namely coherent states of the type (30). In this case the interchange of the limit $\zeta \to \infty$ and $N \to \infty$ can be justified. We believe that Theorems 2 and 3 hold for a large class of states $\omega^S_{(N)}$ provided some uniformity properties are imposed as in (24). Theorem 1 shows that this hope is justified for the laser without reservoirs. The technical problem in generalizing Theorem 1 is to show that the covariance of the fluctuation forces $\chi^i_{(N)}(t)$ vanishes for $N \to \infty$.

8. Conclusion

From time reversible quantum dynamics we have constructed a machine which is irreversible and classical, yet displays cooperative behavior and does not necessarily tend to a state of maximum chaos. We could just as well have derived equations of motion for negative time and would have found that all the damping constants change signs. There is thus a time reversal symmetry around $t = 0$, but a break of time translation invariance. Several factors cause our system to be dissipative and self-oscillatory instead of being conservative:

(a) Coarse graining in the sense that we concentrate on very few observables, these being averages over many microscopic degrees of freedom.

(b) The limit $N \to \infty$ in a dynamical system with a mean field type of interaction.

(c) A special choice of time zero states in which the reservoirs and the system under observation are decoupled and the reservoir is in equilibrium.

The equations we have derived and their physical interpretation are not original. There is an extensive literature on the subject, with the first contribution going back to Einstein [9]. Several modern encyclopedic references are [2] and [10]. What we have demonstrated is that the irreversible and classical equations of this complex dynamical system can be derived from Hamiltonian dynamics with complete rigor and without recourse to ad hoc statistical assumptions about the nature of the influence of the reservoir on the laser proper.

References

[1] T. Kato, "Perturbation Theory for Linear Operators," Springer, Berlin, 1966.

[2] H. Haken, "Handbuch der Physik," Vol. XXV, 2c, Springer, Berlin, 1970.

[3] K. Hepp and E. H. Lieb, in: "Constructive Quantum Field Theory," Erice Lectures 1973, G. Velo and A. S. Wightman Editors, Springer, Berlin, 1973.

[4] K. Hepp and E. H. Lieb, Helv. Physica Acta 46, 573, 1973.

[5] K. Hepp and E. H. Lieb, Annals of Physics 76, 360, 1973.

[6] K. Hepp and E. H. Lieb, Phys. Rev. A 8, 2517, 1973.

[7] E. H. Lieb, Physica, 73, 226, 1974.

[8] D. Ruelle, "Statistical Mechanics," Benjamin, New York, 1969.

[9] A. Einstein, Physik. Zeitschr. 18, 121, 1917.

[10] F. T. Arecchi, E. O. Schulz-Dubois, "Laser Handbook," North-Holland Publ. Co., Amsterdam (1972).

II. ERGODIC THEORY

WHAT DOES IT MEAN FOR A MECHANICAL SYSTEM TO BE ISOMORPHIC TO THE BERNOULLI FLOW?

Donald S. Ornstein

Mathematics Department, Stanford University, California 94305

It has been shown recently that the geodesic flow on a surface of negative curvature, and the flows arising from a hard sphere gas in a box or from the motion of a billiard ball on a square table with a finite number of convex obstacles are isomorphic to the Bernoulli flow. The proof rests on an analysis of the above systems and also on some abstract ergodic theory. The analysis of the above systems will be described by Weiss and Gallavotti. I would like to describe some of the abstract ergodic theory involved and to discuss the meaning of the above results.

1. Abstract Dynamical Systems

A standard mathematical approach to a problem is to abstract the essential features of a problem and then study the resulting abstract mathematical object. The attempt to study the statistical properties of mechanical systems has led to the study of measure preserving transformations or flows.

Take the case of a billiard ball moving on a square table with a finite number of convex obstacles. If we assume that the ball always moves with unit speed then we can describe the configuration of the system by the position of the ball and the direction in which it is moving (that is, by a point in a three-dimensional manifold called the phase space of the system). The time evolution of the system is represented by the movement of this point in phase space. (If x is the point in the phase space representing our system at time 0, then the point representing our configuration at time t will be denoted by $S_t(x)$. It is clear that $S_{t_1}(S_{t_2}(x)) = S_{t_1+t_2}(x)$.) It turns out that Lebesgue measure on the phase space is preserved by the flow (that is, if E is a measurable set in phase space then the measure of E and $S_t(E)$ are the same).

The measure of a set E in phase space is supposed to model the probability that the configuration of the system lies in E. The justification for this is the following: We first assume that if the Lebesgue measure of E is 0, then the configuration will lie in E with probability 0. This seems reasonable and we take it on faith. Now because of the ergodic theorem and because of Sinai's theorem on the ergodicity of the above billiard system we get that except for a set of points x in the phase space, of Lebesgue measure 0, the fraction of time that $S_t(x)$ spends in E tends to the Lebesgue measure of E as t tends to ∞.

A similar analysis holds for the geodesic flow and hard sphere gas.

It is felt that a measure preserving flow (sometimes called an abstract dynamical system) is a good model for the statistical properties of the time evolution of a mechanical system and that if two systems give rise to the same abstract flow then they are, in some sense, statistically the same.

The idea of isomorphism is a precise way of formuating the above, (or of telling us what to ignore). We say that S_t acting on X is isomorphic to (or statistically the same as) $\bar{S}_t$ acting on $\bar{X}$, if X and $\bar{X}$ are each the union of two disjoint invariant sets X_1, X_2 and $\bar{X}_1$, $\bar{X}_2$ where X_2 and $\bar{X}_2$ have measure 0, and if there is a 1-1 invertible measure preserving map ϕ from X_1 onto $\bar{X}_1$ such that $\bar{S}_t(\phi(x)) = \phi(S_t(x))$ for all x in X_1 and all t.

A class of transformations, playing a central role in ergodic theory are the Bernoulli shifts. For each finite set of positive numbers $p_1 \dots p_k$ such that $\sum P_i = 1$ we define the Bernoulli shift $T_{p_1 \dots p_k}$ as follows: Let Y be a measure space with k points having measures $p_1 \dots p_k$. Let Y_i, $-\infty < i < \infty$, be copies of Y and let $X = \prod_{-\infty}^{\infty} Y_i$ with the product measure. (Points of X are doubly infinite sequences of points in Y.) $T_{p_1 \dots p_k}$ is the transformation that shifts each of the above sequences one to the right. (X could be thought of as the measure space formed by doubly infinite consecutive outcomes of spins of a roulette wheel. This is the standard probability model for spinning a roulette wheel. T will be the time shift.)

It is not hard to see that $T_{p_1 \dots p_k}$ has the following geometric representation: Divide the unit square into horizontal strips of widths $p_1 \dots p_k$. Deform the P_i strip in a linear way so that the height is now P_i and the width 1. Now reassemble as shown below:

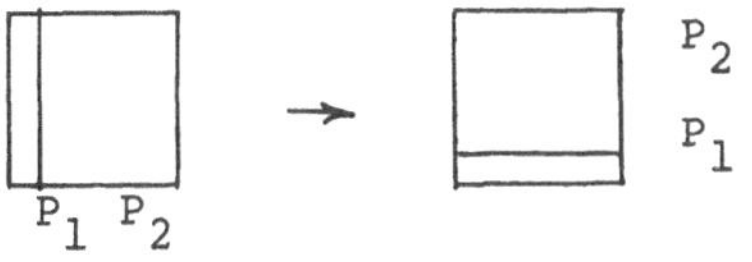

We say that a flow S_t is Bernoulli if for some t_0, S_{t_0} is isomorphic to a Bernoulli shift.

The following theorem was proved recently, see [1,2,3].

<u>Theorem</u>. $T_{p_1 \dots p_k}$ is isomorphic to $T_{g_1 \dots g_\ell}$ if and only if $\sum_1^h p_i \log p_i = \sum_1^\ell g_i \log g_i$ (the only if part is older and was proved by Kolmogorov about fifteen years ago).

The method of proof of the isomorphism theorem yields criteria for showing various transformations to be Bernoulli. In particular the flow $S_t^{(c)}$ which we will define below is isomorphic to a Bernoulli shift for each fixed time: Let Y be the space on which $T_{1/2,1/2}$ acts and let E_0 and E_1 be the points of Y whose first coordinate is respectively 0 or 1. Let f be a function on Y that takes the value α_1 on E_1 and α_2 on E_2 and assume that α_1/α_2 is irrational, and $(\alpha_1+\alpha_2)/2 = 1$. Let X be the part of Y × R that lies below the graph of f. Pick a number c. Define $S_t^{(c)}$ as follows: each point r,t moves up at speed c until it hits the graph of f. (tc = f(r)). Then it goes to $T_{1/2,1/2}(y)$, 0 and continues to move up at speed c.

Theorem. $S_t^{(c)}$ (as defined above) is isomorphic to a Bernoulli shift for each fixed t.

A further extension of the isomorphism theorem yields

Theorem. If S_t is a Bernoulli flow then S_t is isomorphic to $S_t^{(c)}$ for some c.

The results which we started out to explain are the following.

Theorem. Geodesic flow on a surface of negative curvature and the flow arising from a billiard ball moving on a square with a finite number of convex obstacles are isomorphic to Bernoulli flows.

The above systems are singled out because they are good examples and because they will be dealt with by Weiss and Gallavotti. Many other systems have been shown to be Bernoulli such as the hard sphere gas. For example, see [8, 9].

The above mentioned results are proved by applying the criterion arising from the isomorphism theorem. To check this criterion requries a deep analysis of the specific systems, which will be described by Weiss and Gallavotti.

2. Measurements on a System

We will explore the implications of proving that a mechanical system is isomorphic to the Bernoulli flow by examining the character of measurements made on such a system. We will look only at measurements that have only a finite number of possible outcomes (and from now on "measurement" will mean such a measurement). It will be convenient to consider what happens when we make the measurement at fixed time intervals, nt_0, $-\infty < n < \infty$, so that we can restrict our attention to the shift S_{t_0} instead of the whole flow.

A measurement corresponds to a finite ordered partition P of the phase space according to the outcome of the measurement. If x is the configuration of the system at time 0, then the measurement at time nt_0 amounts to observing which atoms of P, $(S_{t_0})^n(x)$ lies in.

A measurement on a system made at times nt_0 can be thought of as a stationary process (the mathematical model for a stationary process being a transformation together with an ordered partition). A stationary process is completely described by the probabilities of all finite strings of consecutive outputs. In our case this is the measure of the set of x such that $S_{t_0}^n(x)$ lies in specified atoms of P for specified n. (It turns out that any stationary process has such a model. The points of the measure space are all possible doubly infinite sequences of outcomes of the process. The measure is determined by the probability of finite strings and the transformation is the shift corresponding to the passage of time. In these terms the Bernoulli shifts are exactly those transformations arising from independent spins of a roulette wheel with probabilities $p_1 \cdots p_k$.)

If the mechanical system is ergodic (which is the case for all the systems we are considering), then except for a set of x of measure 0, the frequency of any fixed string of consecutive outcomes of the measurement will equal its probability. (That is, with probability one, all the statistical properties of the measurement can be read off from each doubly infinite sequence of observations).

There is a natural notion of distance between processes (which we will call $\bar{d}$ distance) that we will need in our description of measurements. (Roughly speaking, if we change a process very infrequently, then the new process should be close to the old one.) We say that two processes have distance less than ε if and only if there exists a stationary process that prints out two letters at each unit of time; and if we look only at the first letter we get the first process, and if we look only at the second letter we get the second process. Furthermore, the probability that the two letters differ is less than ε.

In terms of the model for a process consisting of a transformation T and an ordered partition P we have that $\bar{d}((T,P),(\bar{T},\bar{P})) < \varepsilon$ if and only if there is an ergodic transformation T_1 and partitions P_1 and $\bar{P}_1$ such that the measure of the set of points labeled differently by P_1 and $\bar{P}_1$ is less than ε, and P_1, T_1 and $\bar{P}_1$, T_1 give the same process as P, T and $\bar{P}$, $\bar{T}$ respectively. (That is, for each n corresponding atoms in $\bigvee_1^n T^i P$ and $\bigvee_1^n T_1^i P_1$ have the same measure. Since our partitions are ordered, it makes sense to talk about corresponding atoms.)

There is another definition in terms of typical sequences (where a typical sequence for a process is a sequence such that each finite string occurs with a frequency equal to its probability). If two processes differ by less than ε, then for almost every (typical) sequence of the first process we can find a typical sequence of the second that differs from it on a set of frequency less than ε. Furthermore, if we can find a typical string from each process, such that the two strings differ with frequency less than ε then the $\bar{d}$ distance is less than ε.

Before describing the processes arising from the Bernoulli flow, we will try to give some overall view of the kind of randomness and determinacy that measurements on a mechanical system can have.

A process is said to be deterministic if by knowing the printout for all past times we can predict the next printout with probability one. A transformation T is said to be deterministic if for all P the process P, T is deterministic. (For those who know about entropy this means that T has entropy 0.) A flow S_t is deterministic if S_t is deterministic for each fixed t.

On the other hand, there are transformations (called K-transformations or K-automorphisms) that are characterized by the property that no measurement on them is deterministic. S_t is said to be K if it is K for each fixed t.

A process is said to satisfy the 0-1 law if the only events that can be predicted from the arbitrarily distant past have probability 0 or 1. (That is, the sigma algebra, $\bigcap_{n=1}^{\infty} (\bigvee_{i=n}^{\infty} T^i P)$ only has sets of measure 0 or 1. A result of Sinai and Rochlin [6] says: The K-transformations as defined below are those transformations such that every measurement made on them satisfies the 0-1 law.

It was once thought that K-transformations were the most random possible and that every K-transformation was a Bernoulli shift. (It is easy to see that Bernoulli shifts are K.) This turns out to be false, see [3]. It was also thought that every transformation was either K, or deterministic or else a direct product of these. This also turns out to be false, see [4].

We will now try to characterize the measurements on a

Bernoulli shift.

An independent process can be thought of as a process without memory. A multistep Markov process can be thought of as a process with finite memory. (A k-step Markov process is a process with the property that given its last k printouts the next printout is independent of any printout occurring before time -k.) A k-step Markov process is said to be mixing if for all n large enough and any 2 k-strings that can be printed out with nonzero probability there is a nonzero probability of printing out the first starting at time 0 and the second starting at time n.

We now have: The processes arising from measurements on a Bernoulli shift are exactly those processes that can be approximated arbitrarily well (in the $\bar{d}$ sense) by multistep mixing Markov processes. (This result follows from an extension of the isomorphism theorem.)

We can say a little more. Every stationary process has a canonical sequence of multistep Markov processes approaching it in some sense. The nth approximate being the unique n step Markov process giving the same measure to n strings as the original process. (We simply don't allow the process to remember more than n steps.) The processes arising from Bernoulli shifts are exactly those whose canonical sequence of Markov approximates approach in the $\bar{d}$ sense.

Returning to the geodesic flow, billiard, or hard sphere gas we see that even though the laws governing the time evolution of these systems are completely deterministic, as soon as we make a measurement with only a finite number of possible outcomes (watch it on a TV screen for example) we get a process that is essentially indistinguishable from a process with finite memory. We are thus getting random behavior from a deterministic model whereas in most of statistical mechanics the randomness is built into the mathematical model.

3. Two More Characterizations of Measurements on Bernoulli Systems

In this section we will give a more detailed and technical discussion of various properties of stationary processes and in particular discuss the criterion that allows us to prove that the geodesic flow and billiard system are Bernoulli.

It will be convenient to get some definitions out of the way before we begin.

We would like to extend the definition of $\bar{d}$ distance so that it applies to two measures on sequences of length n for fixed n. Let us assume that each term in our sequences is a number between 1 and k. Then a measure on sequences of length n can be though of as n ordered partitions $P_1 \ldots P_n$ of a measure space, each partition having k atoms. (The sequences themselves form a measure space and the P_i is the partition according to the ith coordinate. On the other hand, if we start with the P_i then they assign to each point in the measure space a unique sequence of length n.)

Let P_i and Q_i, $1 \le i \le n$ be ordered partitions of X and Y respectively, each partition having k atoms (assume X and Y non-atomic). We define $\bar{d}(\{P_i\}_1^n \{Q_i\}_1^n)$ as follows: Let ϕ and ψ be 1-1 measure preserving maps of X and Y respectively onto Z. Let $\bar{d}_{\phi,\psi}(\{P_i\}_1^n \{Q_i\}_1^n) = \frac{1}{n} \sum |\phi(P_i) - \psi(Q_i)|$. (Here $|\phi(P) - \psi(Q))|$ denotes the measure of the set of points labeled differently by $\phi(P)$ and $\psi(Q)$. Note that $\phi(P)$ and $\psi(Q)$ are ordered partitions of the same measure space). We define $\bar{d}(\{P_i\}_1^n\{Q_i\}_1^n)$ as $\inf_{\phi\ \psi} \bar{d}_{\phi,\psi}(\{P_i\}_1^n\{Q_i\}_1^n)$. (It is not hard to show that $\bar{d}((P,T),(\bar{P},\bar{T})) = \lim_{n\to\infty} \bar{d}(\{T^iP\}_1^n \{\bar{T}^i\bar{P}\}_1^n))$.

The $\bar{d}$ distance can be used to get a measure of how independent the partitions $\{P_i\}_1^n$ are from a fixed partition Q. We say that $\{P_i\}_1^n$ are independent of Q in the $\bar{d}$ sense, or $(\{P_i\}_1^n \perp^{\varepsilon} Q)$, $(\bar{d})$, if except for a collection of atoms A of Q, the measure of whose union is less than ε, we have $\bar{d}(\{P_i\}_1^n, \{P_{i/A}\}_1^n) < \varepsilon$. (By $P_{i/A}$ we mean the partition P_i restricted to A where A is normalized to have measure 1.)

If T is a transformation and P a partition we say that P generates (under T) if every measurable set E can be approximated arbitrarily well by sets in $\bigvee_{-K}^{K} T^i P$ (for larger and larger K). (That is: if we take the σ-algebra generated by all the $T^i P$ and add all sets of measure 0 we get the entire σ-algebra.) If the original σ-algebra separates points and if P generates them we can throw out a set of measure 0 and for any two points that are left there will be some $T^i P$ that distinguishes them. Thus, a generating measurement is a measurement such that with probability one the doubly infinite sequence of outcomes of the measurement determine the configuration of the system exactly.

With a little routine measure theory it can be seen that the <u>Bernoulli shifts are exactly those transformations having a generator P such that the $T^i P$ are independent</u>. (We assume here that the measure space is "reasonable", i.e., Lebesgue, see Appendix A of [3].)

<u>The K-transformations turn out to be exactly those transformations that have a generator satisfying the 0-1 law</u>. (This is not routine but follows from a result of Sinai and Rochlin, see [6].)

We are now ready to discuss two other characterizations of the measurements on a Bernoulli shift.

We say that a process P,T is finitely determined or F.D. if given ε there is a δ and an n such that if $\bar{P},\bar{T}$ satisfies (a) P and $\bar{P}$ have the same number of atoms, (b) the entropies of P,T and $\bar{P},\bar{T}$ differ by less than δ, (c) $\bar{d}(\{T^iP\}_1^n \{\bar{T}^i\bar{P}\}_1^n) < \delta$, then $\bar{d}((P,T),\bar{P},\bar{T})) < \varepsilon$. (This says if another process has entropy close to that of P,T and looks like P,T for n times then it looks like P,T forever.)

Independent processes are F.D. and this is the only property of independence used in the isomorphism theorem. Thus, if <u>two transformations of the same entropy have F.D. generators then they are isomorphic (and are Bernoulli shifts)</u>. It is then shown that <u>if T is a Bernoulli shift and P any partition then</u> P,T <u>is F.D. See</u> [3].

P,T is said to be very weak Bernoulli or V.W.B. if given ε there is an n such that $\{T^iP\}_1^n \perp^{\varepsilon} \bigvee_{-\ell}^{0} T^iP(\bar{d})$ for all $\ell \geq 0$. (That is $\{T^iP\}_1^n$ is ε independent, in the $\bar{d}$ sense, of the past. We will write this as $\{T^iP\}_1^n \perp^{\varepsilon} \bigvee_{-\infty}^{0} T^iP(\bar{d})$.)

The relevant theorem is <u>P,T is V.W.B. if and only if P,T is F.D.</u> (See [7] for the if part and [3] for the only if.)

It is the V.W.B. condition that one is able to verify in the case of geodesic flow or billiard. The way this is done is described in [5]. A very rough idea, however, is the following: The phase spaces of these mechanical systems are foliated by contracting curves (and by nonexpanding curves, given by the flow trajectories). We fix a partition P such that the atoms of P have smooth boundaries. We then look at a typical atom $A \in \bigvee_{-m}^{0} T^iP$, normalized to have measure 1, and the map A by ϕ onto the whole space so that ϕ is measure preserving and we can go from x to $\phi(x)$ by moving a small distance along a contracting curve and then a small distance along a nonexpanding one. This means that T^ix and $T^i\phi(x)$ are close for all $i > 0$. Thus $T^i(x)$ and $T^i\phi(x)$ lie in the same atoms of P for most $i > 0$. Therefore, the P process conditioned on A is close to the unconditioned P process in the $\bar{d}$

sense and this is the V.W.B. condition.

4. Comparisons of Various Properties of Measurements

We would now like to compare the V.W.B. characterization of measurements on a Bernoulli shift with the 0-1 law characterization of measurements on a K system. We will write these conditions in such a way that V.W.B. is an obvious strengthening of the 0-1 law.

P,T satisfies the 0-1 law if and only if it satisfies the following: given $\varepsilon>0$ and $K>0$ there is an $N > 0$ such that $\{T^i P\}_N^{N+K} \perp^{\varepsilon} \bigvee_{-\infty}^{0} T^i P(\bar{d})$ (the only if part is easy and the if part follows easily from the Martingale convergence theorem).

Since $\bar{d}$ involves an average an equivalent deinition of V.W.B. is the following: Given $\varepsilon > 0$ there is a $K > 0$ and an $N > 0$ where $N < \varepsilon K$ and such that $\{T^i P\}_N^{N+K} \perp^{\varepsilon} \bigvee_{-\infty}^{0} T^i P(\bar{d})$.

Both of these conditions are flow conditions in the following sense: Suppose we have a flow S_t and a partition P and a time t_0 such that P generates under S_{t_0} and P, S_{t_0} is V.W.B. (or satisfies the 0-1 law), then all Q, S_t are V.W.B. (or satisfy the 0,1 law).

V.W.B. is a different kind of condition from the 0-1 law in the sense that if all measurements are V.W.B., then the flow itself is determined exactly whereas in the case of the 0-1 law there is still a very wide variety of possible flows.

There is another strengthening of the 0-1 law that I would like to mention. For this purpose phrase the 0-1 law as follows: P,T satisfies the 0-1 law if the σ-algebra $\bigcap_{n=1}^{\infty} (\bigvee_{i=n}^{\infty} T^i P)$ contains only sets of measure 0 or 1. We say that P,T has trivial double tail if $\bigcap_{n=1}^{\infty} (\bigvee_{|i|>n} T^i P)$ contains only sets of measure 0 or 1. One might think that any partition of a K-transformation or at least of a Bernoulli shift had trivial double tail. This is false, see [8], and in fact, any ergodic transformation (and in particular a

Bernoulli shift) has a partition P such that $\bigcap_{n=1}^{\infty} \bigvee_{|i|>n} T^iP$ is the entire σ-algebra. This means that with probability one all of the outputs of the process can be recovered from the outputs in the arbitrarily distant past and future.

Because of the above, we have a V.W.B. process which by definition satisfies $\{T^iP\}_{-n}^{n} \perp^{\varepsilon} \bigvee_{-\infty}^{-n-1} T^iP(\bar{d})$, but the $\{T^iP\}_{-n}^{n}$ are in $(\bigvee_{-\infty}^{-n-1} T^iP) \vee (\bigvee_{n+1}^{\infty} T^iP)$.

In this context we should mention that the Sinai-Rochlin characterization of K-transformations implies the following: If P,T satisfies the 0-1 law, then so does P, T^{-1}; (i.e., if $\bigcap_{n=1}^{\infty} \bigvee_{n}^{\infty} T^iP$ is trivial then so is $\bigcap_{n=1}^{\infty} \bigvee_{-\infty}^{-n} T^iP$).

There is a natural variant of the V.W.B. definition which is easily shown to be equivalent. That is, given ε there is an N such that $\{T^iP\}_1^n \perp^{\varepsilon} \bigvee_{-\ell}^{0} T^iP(\bar{d})$ for all $n > N$ and all $\ell > 0$. This is easily seen to be equivalent to the following: Given ε there is an N such that $\{T^iP\}_N^{N+\ell} \perp^{\varepsilon} \bigvee_{-\ell}^{0} T^iP(\bar{d})$ for all $\ell > 0$.

There is one more condition (called Weak Bernoulli or W.B.) which is a strengthening of the last definition of V.W.B. To define W.B. we will introduce another measure of independence. We will say that $Q \,{}^{\varepsilon}\!\perp \bar{Q}$ if $\sum_i \sum_j |m(Q_i \cap \bar{Q}_j) - m(Q_i)m(\bar{Q}_j)| < \varepsilon$ (here Q_i, $\bar{Q}_i$ are atoms of Q and $\bar{Q}$ respectively and m denotes measure). To make this sound more like the $\bar{d}$ measure of independence we note that $Q \,{}^{\varepsilon}\!\perp \bar{Q}$ for small ε if and only if the distribution of Q on most atoms of $\bar{Q}$ is close to the distribution of Q) P,T is said to be W.B. if given ε there is an N such that $\bigvee_{N}^{N+\ell} T^iP \,{}^{\varepsilon}\!\perp \bigvee_{-\ell}^{0} T^iP$ for all . (This is the same as the last definition of V.W.B. except for the way of measuring independence. Note that this measure of independence depends only on the partition $\bigvee_{N}^{N+\ell} T^iP$ while the $\bar{d}$ measure depended on writing $\bigvee_{N}^{N+K} T^iP$ as the span of the T^iP.)

W.B. ⇒ V.W.B. but not every partition of a Bernoulli shift is

W.B. (The partition whose double tail generated is an example.) Thus W.B. does not share the property of V.W.B. and 0-1 says that if a generator satisfies that property then all partitions do.

The W.B. property can be compared convenient to the 0-1 law by writing the latter condition in the form: given ε and K, there is an N such that $\bigvee_{N}^{N+K} T^{i}P \overset{\varepsilon}{\perp} \bigvee_{-\infty}^{0} T^{i}P$. (This is equivalent to the definition where the $\bar{d}$ sense of independence was used be ause ε can be taken small compared to K.)

5. Epilogue

Because mechanical systems that are very different turn out to be abstractly the same, we should take a closer look at what we have abstracted away in setting up our mathematical model. What we have lost is the "meaning" of a particular measurement. There is a difference between being told whether or not there are more spheres in the right half of the box than the left and being told that some measurement is giving us a positive or negative reading. We somehow lose the meaning of a measurement when we regard it as a partition of the phase space. Ergodic theory gives us information about all possible measurements on the system (or the existence of some measurement) and this has some philosophical interest, but from the point of view of physics if we want to use ergodic theory to get more specific results about specific systems it would have to be used in conjunction with specific properties of specific measurements.

From a mathematical viewpoint one of the nicer features of ergodic theory is its wide applicability (to a large variety of mechanical system, automorphism of compact groups, number theoretic transformation, the Ising model etc.) On the other hand, this wide applicability puts limits on how much information it will give us in each specific case.

References

[1] Smorodinsky, M., Ergodic Theory, Entropy, Berlin, New York: Springer Verlag, 1971.

[2] Shields, P., The Theory of Bernoulli Shifts, University of Chicago Press, 1973.

[3] Ornstein, D., Ergodic Theory, Randomness, and Dynamical Systems, Yale University Press, 1973.

[4] Ornstein, D., A Mixing Transformation for which Pinsker's Conjecture Fails, 10, No. 1 (February 1973), Advances in Mathematics, 103-123.

[5] Weiss, B., and Ornstein, D., Finitely Determined Implies Very Weak Bernoulli, Israel J. Math., 17, No. 1 (1974) 94-104.

[6] Parry, W., Entropy and Generators in Ergodic Theory, Benjamin, New YOrk, 1969.

[7] Weiss, B., and Ornstein, D., Geodesic Flows are Bernoullian, Israel J. Math., 14, No. 3 (1973) 184-198.

[8] Ratner, M., Anosov K-flows are Bernoulli, Israel J. Math, 16 (1973).

[9] Bowen, R., Lectures on Equilibrium Statistical Mechanics, Minnesota, Spring 1974.

THE GEODESIC FLOW ON SURFACES OF NEGATIVE CURVATURE

Benjamin Weiss

Hebrew University, Jerusalem, Israel

One of the basic examples that has accompanied ergodic theory throughout its development within the last fifty years has been the geodesic flow on surfaces of negative curvature. An excellent account of the early work on surfaces of constsant negative curvature may be found in E. Hopf's book Ergodentheorie (Berlin, 1937), while the more recent developments of the Russian school is described in D. V. Anosov's Geodesic Flows on Closed Riemann Manifolds with Negative Curvature, Proc. Steklor Inst. 90 (1967). In these lectures I shall try to give an introduction to the most recent developments, in particular the discovery that the geodesic flow is a Bernoulli flow. After a brief description of the geodesic flow I shall digress briefly to state some facts needed for the later discussions about K-automorphisms and the turn to a simpler model of the phenomena to be analyzed in the form of automorphisms of the torus. Finally, returning to the geodesic flow we'll see how one improves on Hopf's arguments that proved ergodicity to get that the flow is successively a K-flow and even a Bernoulli flow.

1. Let M be a compact 2-dimensional manifold with a Riemannian metric of class C^2 such that the curvature is everywhere negative. Such manifolds cannot be realized in 3-space but can be though of as surfaces that resemble a saddle point everywhere. If the curvature is constant then locally such a surface looks like the non-euclidean hyperbolic plane, say the Poincaré model with the open unit disk. In fact any such surface is obtained from the hyperbolic plane by factoring out a discrete group of isometrices --

much like the torus is obtained from the plane by identifying points modulo the integer lattice. Let X denote the space of line elements of M, that is X consists of pairs (x,θ) where $x \in M$ and θ is a direction in the tangent plane to M at x, in modern parlance X is the unit tangent bundle of M. It is a consequence of the geometry of complete surfaces that each $(x,\theta) \in X$ determines a unique geodesic, path of shortest length, that emanates from x in the direction θ. Distances along a geodesic will always be measured in terms of arc length. The geodesic flow $\{\phi_t\}$ is a flow on X defined by $\phi_t(x,\theta) = (x',\theta')$ where (x',θ') is obtained by passing through (x,θ) the geodesic determined by (x,θ) and moving along it for a distance t to a point x', and direction θ' which is tangent to this geodesic at x'. In other words the trajectories of ϕ_t are simply the geodesics with their tangent vectors in X, and the motion is with unit speed along the geodesic. The natural volume λ on X is preserved by this flow and it's this system (X,ϕ_t,λ) that we wish to study. In spite of the deterministic nature of this flow it turns out to exhibit very random behavior, and the key to this behavior lies in the existence of asymptotic geodesics.

In Euclidean geometry, geodesics are straight lines and they either never meet -- and then are parallel and remain a fixed distance apart -- or meet at one point and then diverge. In the hyperbolic plane one encounters a different phenomenon. The geodesics there, in the unit disk model, are arcs of circles that are orthogonal to the unit circle. Now if two distinct geodesics intersect on the unit circle it may be seen that although they never intersect they approach each other at an exponential rate. It was J. Madamand apparently who first discovered that such asymptotically approaching geodesics occur whenever the curvature is negative, and Anosov abstracted this property to define what we now call Anosov flows. Without going into the general theory let's try to describe what these asymptotic geodesics lead to.

Through every point p of X one can construct a one-dimensional manifold that consists of all points p' such that

$$\lim_{t\to\infty} d(\phi_t p, \phi_t p') = 0 ,$$

where d is the metric on X. These one-dimensional manifolds fiber X into a foliation called the contracting foliation. Reversing time one obtains the expanding foliation, whose fibers consist of equivalence classes of points under the equivalence relation that sets $p \sim p'$ if

$$\lim_{t\to-\infty} d(\phi_t p, \phi_t p') = 0$$

It turns out that these fibers are transversal at every point and also transversal to the flow direction. Another useful fact is that one has an exponential estimate for the decay of $d(\phi_t p, \phi_t p')$ for two points in the same contracting or expanding fiber. In general, for Anosov flows, the fibers themselves are quite smooth but the foliation may be only continuous -- and need not be continuously differentiable. It has, however, some kind of regularity that we shall discuss soon. We can now describe Hopf's proof that the geodesic flow is ergodic.

Theorem 1. The geodesic flow on a surface of negative curvature is ergodic.

Proof (Hopf): To show that there are no nonconstant invariant functions under the flow, it suffices to show that the projection P onto the space of invariant functions sends every function to a constant function. Since P is bounded it suffices to check this on a dense class of functions and we do this then for any smooth functionf on X. Now the ergodic theorem gives us several representations for Pf, namely:

$$Pf = \lim_{t\to+\infty} \frac{1}{t} \int_0^t f(\phi_s x)\, ds = \lim_{t\to-\infty} \frac{1}{t} \int_0^t f(\phi_s x)\, ds$$

where the equality holds of course only a.e. with respect to λ. Now from the first representation of Pf which we denote by f^+ and the smoothness of f it follows that f^+ is always constant on fibers of the contracting foliation. Similarly f^- is constant on the fibers of the expanding foliation. Naturally both are invariant under ϕ_t, i.e. constant along the orbit foliation. Now if the equality $f^+ = f^-$ were to hold everywhere we would immediately conclude that Pf had to be a constant since locally any two points could be joined by a 3-part path consisting of expanding, contracting fibers and a piece of the orbit. However, the equality $f^+= f^-$ holds only a.e. so that the following picture is conceivable.

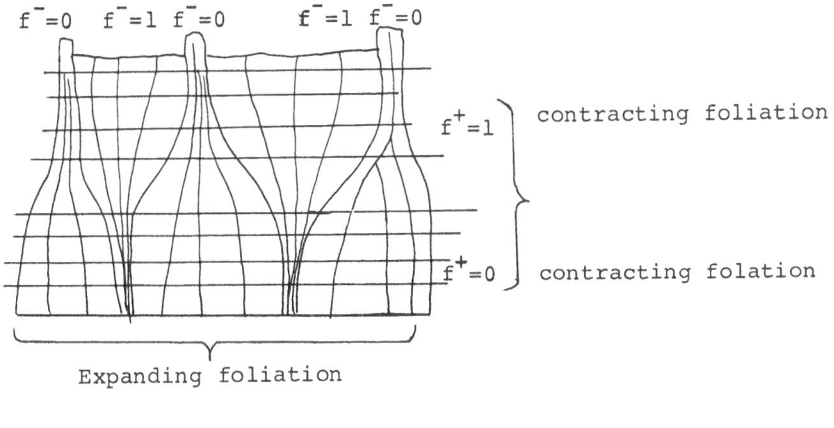

Figure 1

The pinched necks of the bottles of the expanding foliation are to be thought of as having zero measure, neither f^+ nor f^- are constant a.e. but they are equal a.e. To rule out this kind of pathology one invokes the <u>absolute continuity</u> of the folations which says roughly that you cannot transform a set of positive measure on the contracting fibers to a set of zero measure on nearby contracting fibers by moving along the fibers of the expanding foliation. (For fuller details see the lectures by G. Gallavotti on Billiards in these Proceedings.)

For manifolds of dimension 2, Hopf actually showed that the foliations in question are C^1, and thus in that case it's easy to conclude from $f^+ = f^-$ a.e. that Pf must be a constant a.e. □

Before studying the geodesic flow in greater depth we make two digressions, one to general ergodic theory and the other to a discrete time model for the geodesic flow.

2. Let $(X,\mathcal{B},\mu,\phi)$ be a measure preserving transformation with a finite generator $\alpha = \{A_1,\ldots,A_a\}$. Recall that ϕ is said to be a K-automorphism if

$$\bigcap_n \bigvee_{-\infty}^{-n} \phi^{-i}\alpha = \text{trivial } \sigma\text{-algebra.}$$

The basic theorem characterizing such transformations gives a characterization of this remote past $\bigcap_n \bigvee_{-\infty}^{-n} \phi^{-i}\alpha$, namely:

<u>Theorem 2</u> (Pinsker-Rohlin-Sinai). A finite partition $\beta \subseteq \bigcap_n \bigvee_{-\infty}^{-n} \phi^{-i}\alpha$ if and only if $h(\phi,\beta) = 0$. In particular ϕ is a K-automorphism if and only if it has completely positive entropy, that is no nontrivial partition β satisfies $h(\phi,\beta) = 0$.

The ϕ-invariant σ-algebra generated by all partitions β such that $h(\phi,\beta) = 0$ is referred to as the Pinsker algebra. A useful corollary of the P-R-S theorem is that up to sets of measure zero the Pinsker algebra is both the remote past of a generator as well as the remote future.

Corollary 3. For any α, the Pinsker subalgebra of $\bigvee_{-\infty}^{\infty} \phi^{-n}\alpha$ is both $\bigcap_n \bigvee_{-\infty}^{-n} \phi^{-i}\alpha$ and $\bigcap_n \bigvee_{n}^{\infty} \phi^{-i}\alpha$.

This follows since $h(\phi,\beta) = 0$ if and only if $h(\phi^{-1},\beta) = 0$. There is another useful consequence of the triviality of the remote past, that is in fact equivalent to the definition of K-aut. A transformation is mixing if for every A,B, $\lim_{n\to\infty} \mu(A \cap \phi^{-n}B) = \mu(A)\mu(B)$. The consequence I'm referring to is a uniform version of this mixing:

Proposition 4. For any α, $\bigcap_n \bigvee_{-\infty}^{-n} \phi^{-i}\alpha$ is trivial if and only if one of the following equivalent conditions holds:

(1) for all $B \subset \bigvee_{-\infty}^{\infty}\phi^{-i}\alpha$, and any $\varepsilon > 0$ there is an n_0 such that for all $m > n \geq n_0$, but for a set of atoms of $\bigvee_{-m}^{-n} \phi^{-i}\alpha$ of total measure at most $\varepsilon > 0$, each atom A of $\bigvee_{-m}^{-n} \phi^{-i}\alpha$ satisfies

$$|\mu(B|A) - \mu(B)| < \varepsilon .$$

(2) same as (1) for any finite number of sets.

(3) same as (1) for any set in $\bigcup_m \bigvee_{-m}^{m} \phi^{-i}\alpha$.

(4) for any $B \subset \bigvee_{-\infty}^{\infty} \phi^{-i}\alpha$ we have

$$\lim_{n\to\infty} \sup_{A \subset \bigvee_{-\infty}^{-0}\phi^{-i}\alpha} |\mu(A \cap \phi^{-n}B) - \mu(A)\mu(B)| = 0 .$$

One easily proves this proposition by using the martingale convergence theorem, the equivalences between (1)-(4) are standard. We formulate yet one more equivalent way of defining a K-automorphism:

Proposition 5. A transformation ϕ is a K-aut. if and only if for any partition β, the process (ϕ^n,β) tends to independence, in the sense that

$$\lim_{n\to\infty} H(\beta|\phi^{-n}\beta \vee \phi^{-2n} \vee \ldots) = H(\beta) .$$

<u>Proof</u>: Clearly if the property holds then ϕ has completely positive entropy which implies that it is a K-aut. Conversely if ϕ is a K-aut. then the property follows easily from the preceing proposition. □

The various characterizations enable one to prove easily the following useful proposition:

<u>Proposition 6</u>. Suppose that ϕ has finite entropy and that A_n is an increasing sequence of ϕ-invariant σ-algebras such that (X, A_n, μ, ϕ) is a K-aut. for each n and $\vee A_n = \mathcal{B}$, then $(X, \mathcal{B}, \mu, \phi)$ is a K-aut.

<u>Proof</u>: It suffices to show that if $H(\beta) > 0$ then $h(\phi,\beta) > 0$. Since $\vee A_n = \mathcal{B}$, there is a sequence $\{\alpha_n \subseteq A_n\}$ such that

(1) $$\lim_{n\to\infty} \rho(\alpha_n,\beta) \equiv H(\alpha_n|\beta) + H(\beta|\alpha_n) = 0 .$$

Since each α_n is in a K-aut. we have by Proposition 6 an integer k_n such that

(2) $$\lim_{n\to\infty} |h(\phi^{kn},\alpha_n) - H(\alpha_n)| = 0 .$$

However, if $h(\phi,\beta)$ were to vanish we would have also $h(\phi^{kn},\beta) = 0$ for all n. Since

$$|h(\phi^{kn},\beta) - h(\phi^{kn},\alpha_n)| \leq \rho(\alpha_n,\beta)$$

(1) and (2) would contradict $H(\beta) > 0$. □

This last proposition means that one needn't worry too much about looking for generators when proving that a given transformation is a K-aut. We now return to the geometric examples, and will first examine automorphisms of the torus.

3. We shall now give a geometric, rather than algebraic, analytic proof of the fact that certain ergodic autohorphisms of the torus are K-aut. In fact it is known that all ergodic automorphisms are K but the proof we will give here will apply specifically to the hyperbolic automorphisms, i.e. those with no eigenvalues on the unit circle. Suppose then that ϕ is such a hyperbolic automorphism. It preserves the Lebesgue measure and has both expanding and contracting foliations which are transversal. Indeed let C be the subspace of Euclidean space corresponding to the eigenvalues of ϕ that are less than one, and E the subspace corresponding to the eigenvalues greater than one. By our assumption C and E togerher span the covering space of the torus, and the foliations $\mathcal{C}$, $\mathcal{E}$ of the torus by planes parallel to C and E respectively are clearly respectively contracting and expanding.

<u>Theorem 7</u>. A hyperbolic automorphisms of the n-torus is a K-aut.

<u>Proof</u>: By Proposition 6 it suffices to show that any smooth partition α of the torus has a trivial Pinsker algebra. We should really check that the entropy is finite but this is not difficult -- in fact any diffeomorphism of a compact manifold has finite entropy. The main observation is that any function measurable with respect to the remote past of $\bigcap_n \bigvee_{-}^{-n} \phi^{-n}\alpha$ is a.e. equal to a function constant along fibers of the foliation $\mathcal{E}$. In fact if p,p' lie on the same fiber of $\mathcal{E}$ and are separated by $\bigvee^{-n} \phi^{-n}\alpha$ for all n, then for arbitrarily large n we have $\phi^{-n}p$ and ϕ^{-n} p' lying in distinct sets of α. Moreover since $d(\phi^{-n}p,\phi^{-n}p')$ is decaying exponentially and the boundaries of sets in α are smooth, the measure of the set of points that satisfy this condition is zero by the Borel Cantelli lemma.

On the other hand, any function measurable with respect to the distant future $\bigcap_n \bigvee_n^{\infty} \phi^{-n}\alpha$ is a.e. equal to a function that is constant along fibers of the foliation C. By Corollary 3 then, any function f measurable with respect to the Pinsker algebra of $\bigvee_{-\infty}^{\infty} \phi^{-i}\alpha$ is constant a.e. along C and E simultaneously. Now here the foliations are into parallel planes so that it follows at once that f is constant a.e., i.e. the Pinsker algebra is trivial and hence ϕ is a K-aut. □

Such hyperbolic automorphisms are the standard model for Anosov diffeomorphisms which have the same kind of expanding and contracting foliations. The proof we just gave applies to show that Anosov diffeomorphisms that preserve a smooth measure, and such that the foliations are "absolutely continuous" -- the property that enables the last step in the above argument to work -- are K-automorphisms. It was first used by Sinai in a slightly more general context of measurable foliations in his paper: Dynamical Systems with Countable Multiplicity Lebesgue Spectrum II, Izv. Akad. Nauk (1966) 15-68, in English: AMS Transl. (2) 68 (1968) 34-88 We can now return to the geodesic flow and apply the ideas just discussed to the flow situation. First we should say what we mean by a K-flow. There is a definition due to Sinai which goes like this: $(X,\mathcal{B},\phi_t,\mu)$ is a K-flow if there is a σ-algebra $A \subset \mathcal{B}$ such that

(i) for all $t > 0$, $\phi_t A \subset A$

(ii) $\bigvee_t \phi_t A = \mathcal{B}$

(iii) $\bigcap_{t>0} \phi_t A =$ trivial σ-algebra.

A consequence of this definition is that for each fixed t_0, ϕ_{t_0} is a K-automorphism. I do not know if this latter property implies that ϕ_t is a K-flow in Sinai's sense, but will work with this weaker property and take it to define a K-flow. The reason for this choice is the analogy with Bernoulli flows. There one can't make a sensible definition involving all t, for example in a separable measure space there never is a partition α such that $\{\phi_t\alpha\}$ are independent for <u>all</u> t. Thus a Bernoulli flow has been defined to be a flow ϕ_t such that for each t_0, ϕ_{t_0} is a Bernoulli transformation. One of the results of Ornstein's theory is that if for some t_0, ϕ_{t_0} is Bernoulli then for every t, ϕ_t is Bernoulli. The corresponding fact also holds for the property of being a K-aut., namely if for some t_0, ϕ_{t_0} is a K-aut. then for every t, ϕ_t is a K-aut. This follow readily from Abramov's formula $h(\phi_t) = |t|h(\phi_1)$.

4. Having established in Theorem 1 that a geodesic flow ϕ_t is ergodic we turn now to the question of stronger degrees of mixing. We fix a t_0, and then to show that $\phi_{t_0} = \phi$ is a K-aut. it suffices by Proposition 6 to show that for every smooth partition α the Pinsker subalgebra A of $\bigvee_{-\infty}^{\infty} \phi^{-i}\alpha$ is trivial. Let then f be measurable with respect to A, then from the representation $A = \bigcap_n \bigvee_{-\infty}^{-n} \phi^{-i}\alpha$ it follows that f is a.e. equal to a function constant on fibers of the expanding foliation, and in a similar way f is a.e. equal to a function constant on fibers of the contracting foliation. Now, however, since we don't know that f is constant along trajectories we don't have enough to conclude that f is constant a.e., and indeed from the information on hand so far that conclusion cannot be drawn.

To proceed further we need to make a general remark concerning ergodic flows. It is not necessarily true for an ergodic flow ϕ_t, that ϕ_{t_0} is ergodic for each t_0. The easiest way to see this is by a <u>suspension</u> of an ergodic transformation. That is to say, let

$(X,\mathcal{B},\psi,\mu)$ be an ergodic transformation and on $X \times [0,1]$ with $\mu \times$ Lebesgue measure. Define a flow ϕ_t by moving along (x,t) at unit speed in the t direction, identifying $(x,1)$ with $(\phi x,0)$. Clearly ϕ_t is an ergodic flow but ϕ_1 has lots of invariant sets -- any nontrivial set $E \subset [0,1]$ yields $X \times E$ an invariant set for ϕ_1. One can see that essentially this is the only way ϕ_{t_0} can fail to be ergodic for a fixed time, and we call such flows <u>periodic</u>.

Returning now to our partition α and its Pinsker algebra $\mathcal{A}$, let's look at the flow ϕ_t on $(X,\mathcal{A},\mu)$. Clearly $\mathcal{A}$ is invariant under ϕ_t so that this makes sense. We now know that (up to a.e. considerations which are handled by the absolute continuity) together with any $x \in A \subset \mathcal{A}$, all points of the expanding and contracting fibers through x also lie in A. Fixing attention now on a small set P fibered nicely by the flow and the expanding and contracting foliations, we see that there is a time t_0, such that any set A of positive measure that is measurable with respect to $\mathcal{A}$, $\bigcup_{|t|<t_0} \phi_t A$ fills up P up to a set of measure zero. Indeed one takes t_0 to be larger than the longest piece of a trajectory that passes through P. Now if ϕ_t on $\mathcal{A}$ were not periodic with trivial base this couldn't happen by the continuous analogue of the Kabutani-Rohlin tower theorem. Thus we have arrived at the following alternative:

<u>Either for some fixed</u> t, ϕ_t <u>is not ergodic or</u> ϕ_t <u>is a K-flow</u>.

In general we cannot say more about Anosov flows. Indeed by taking a suspension of a hyperbolic torus automorphism one can obtain precisely an Anosov flow that is <u>not</u> a K-flow. To rule out this possibility for geodesic flows requires a more careful analysis of the geometry. One path is that followed by Hopf who showed that the flow along the contracting foliation -- the so-called horocycle flow is ergodic. Naturally with this one doesn't need to use the P-R-S theorem and one sees directly that the remote future is trivial. Another path is described at length by Anosov (op. cit.) and Arnold and amounts to showing that the vector fields associated with the two foliations don't commute, or form a "nonintegrable pair". In any even these considerations are beyond the scope of this lecture.

For the next stage in the development, namely to show that the geodesic flow is Bernoulli I refer to my joint paper with D. Ornstein, Geodesic Flows are Bernoulli, Israel Journal of Mathematics 14 (1973), pp. 184-198, which is also reprinted in the Yale Lecture Notes by D. Ornstein, Randomness, Ergodic Theory and Dynamical Systems. For further developments along these lines one can consult:

1. Marina Ratner, Anosov K-flows are Bernoulli, Israel J. Math. 16 (1973).
2. R. Bowen, Lectures on Equilibrium Statistical Mechanics, Minnesota, Spring 1974.

LECTURES ON THE BILLIARD

Giovanni Gallavotti
Istituto di Fisica Teorica, Universita di Napoli, Italia

Abstract

These lectures together with the appendices contain all the essential steps for a proof of the ergodicity of the dynamical system described below as (M,T,v) or "periodic billiard's section" when the obstacles are so situated that no collisionless paths are possible.

1. Introduction

A point mass moves, without friction, on a table; the particle is either elastically reflected by the boundary of the table (reflecting table) or, upon hitting the boundary, disappears to reappear on the opposite side with the same velocity (toric table).

The surface of the table contains some obstacles: upon collision with the obstacles the particle's velocity changes its direction according to the laws of elastic reflection (equal angle of impact and departure). The motion of the particle is clearly very complicated and we shall deal with the problem of discussing some of its qualitative features.

This problem hardly needs motivation: many authors have already noticed its interest and relevance for many applications.

Here we shall content ourselves with mentioning its relation to the Lorentz model.

The model consists of a particle that moves on a plane covered by randomly thrown obstacles. The quantity of interest is the value of the square of the distance $r(t)$, from a starting point, of a particle moving with unit speed on the plane and elastically colliding with the obstacles

We denote the quantity by $r(t)^2$ and define $D(t)$ as

$$t\,D(t) = E(r(t)^2)$$

where the average has to be taken with respect to the distribution of the random obstacles.

All the attempts to obtain rigorous estimates for $D(t)$ have, so far, failed even for the simplest obstacles' distribution. There is, however, considerable work done on this problem by theoretical physicists and nonrigorous results as well as computer experiments are available.

For instance, assuming that the obstacles are distributed according to the Gibbs' distribution for a hard discs gas, it is believed that

$$\lim_{t\to\infty} D(t) = D_\rho < +\infty$$

where D_ρ is a value which depends on the density ρ of the distribution of the hard discs and $D_\rho > 0$ if ρ is small enough.

It is also conjectured that $\rho\, D_\rho$ is not analytic near $\rho = 0$. It can be seen, at least formally, that the value $D(t)$ is given by

$$D(t) = t^{-1} \int_0^1 d\tau \int_0^\tau E(\underline{v}(0)\cdot\underline{v}(t'))\, dt' = t^{-1} \int_0^t \tilde{D}(\tau)\, d\tau$$

where $\underline{v}(0)$ denotes the initial velocity of the test particle and $\underline{v}(t)$ its velocity at time t; the average is over the obstacles' distribution.

The above formula should make clear that the real problem one is facing is to control, in some sense, the average behavior of a trajectory in the long time region: for fixed time it is clear that the integrand in the above formula is analytic in the density ρ of the scatterers near $\rho = 0$ (at least if the scatterers are Poisson distributed with density ρ).

A similar but simpler problem can be obtained by assuming that the obstacles are periodically and not randomly distributed: so we have a nonrandom lattice structure for the centers of the obstacles and we look for $E(r(t)^2)$ or $E(\underline{v}(0) \cdot \underline{v}(t))$, where the average is, now, over the possible initial unit velocities and position in the unit cell of the obstacles' lattice. It is assumed that the initial positions are equally distributed in the "free" region of a basic cell of the lattice, while the velocity is homogeneously distributed over the angles.

The above problem is clearly a problem concerning a toric billiard and the statement that the toric billiard is ergodic implies, among other things that

$$t^{-1}\int_0^t E(\underline{v}(0)\cdot v(t'))\,dt' \xrightarrow[t\to\infty]{} 0$$

which means that $D(t)$ does not grow as fast as t in this simplified version of the Lorentz model.

On the other hand there are no more general results on $D(t)$ even in this simple case and, amazingly, the problem is still completely open.

We shall be mainly concerned in an exposition of the main steps of the long proof of the basic ergodic properties of toric billiards.

2. Geometry.

Z^2 is a square lattice in R^2 and its basic unit cell contains s inequivalent points: therefore Z^2 is divided into s sublattices of equivalent points. Each point of Z^2 is covered by an obstacle with a shape and orientation and relative position which is the same for all the obstacles covering equivalent points. The obstacles are assumed to be open strictly convex sets with a C^3-boundary furthermore their closures are supposed to be mutually disjoint.

We can regard, in a natural way, the space R^2 covered by the set O of the obstacles as a torus Q with obstacles. A particle, moving in R^2/O freely except when it collides elastically on the surface ∂O, can be thought of as moving on the torus Q.

We call "phase space" the manifold V (with boundary) consisting in the points (q,θ), $q \in Q/O$, $\theta \in [0,2\pi]$: q will represent the position of the "billiard ball" and θ the angle between its unit velocity and a fixed axis.

So V is a 3-dimensional closed manifold with boundary $\partial V = \{(q,\theta) \mid q \in \partial O\}$. V consists of "arrows".

If $x = (q,\theta) \in \partial V$ we call x "colliding" if the angle $\phi(x)$ which x forms with the outer normal to ∂O in q is between $\pi/2$ and $3\pi/2$ (counted counterclockwise). The points of the set $M = \{x \mid x \in \partial V_0 , x \text{ collides}\}$ can be parametrized by three numbers: if $x = (q,\theta) \in \partial V$ then $x = (i,r,\phi)$ where

(1) i is an index, $i = 1,2,\ldots,s$, denoting the obstacle O_i of O to which q belongs.

(2) r is a coordinate measuring the curvilinear abscissa on ∂O_i of $q \in \partial O_i$ counted, clockwise, from a fixed point of ∂O_i. If L_i = length of ∂O_i then two points on ∂O_i whose abscissae differ by L_i are identical.

(3) ϕ is an angle, $\pi/2 \leq \phi \leq 3\pi/2$, measuring the angle between the arrow x and the outer normal to ∂O_i in q, measured counterclockwise.

Therefore M splits naturally into a disjoint union of s cylinders $M_1,M_2,\ldots,M_s$. The billiards flow $S_t\colon V \to V$ is defined in the obvious way and conserves the measure

$$\mu(dq\, d\theta) = \frac{dq\, d\theta}{\int_V dq\, d\theta}$$

This completes the description of the billiards' dynamical system (V,S_t,μ). The system (V,S_t,μ) can also be realigned as a "special flow" in a natural way which, in some sense (see below), isolates the collisions as the main object of interest.

Let $M = \{x \mid x \in \partial V, x \text{ collides}\}$ be parametrized as above and let $T: M \to M$ be the transformation which maps $x \in M$ into the collision $Tx \in M$ which precedes it and let $\tau(x) < 0$ be the time interval between x and Tx: our assumptions imply that there is a lower bound $\tau_0 > 0$ to the flight time $-\tau(x)$.

If $y \in V$ we can clearly use as one of the coordinates of y the time $\tau(y) \leq 0$ which elapses between the instant the particle is in y and the last former collision $x(y) \in M$ on the past trajectory of y. Then y is uniquely determined, outside a set of μ-measure zero, by $(x(y),\tau(y))$ i.e. by a set (i,r,ϕ,t), $i = 1,2,\ldots,s$, $0 \leq r \leq L_i$, $\pi/2 \leq \phi \leq 3\pi/2$, $0 \geq t \geq \tau(T^{-1}x(y))$.

It is easy to check that for $t \geq 0$:

$$S_{-t}y \equiv S_{-t}(x(y),\tau(y)) = (x(y),\tau(y)+t)$$

if $\tau(y)+t \leq 0$ and

$$S_{-t}y = (Tx(y),\ t+\tau(y)+\tau(x(y)))$$

if $\tau(y) + t > 0$ but $\tau(y) + t + \tau(x(y)) < 0$.

Furthermore simple trigonometric considerations show that

$$\left\| \frac{\partial(q,\theta)}{\partial(r,\phi,t)} \right\| = -\cos\phi$$

so that

$$\mu(dr\ d\phi\ dt) = \frac{-\cos\phi\ dr\ d\phi\ dt}{\text{normalization}}$$

The S_t invariance of μ implies then the T-invariance of the measure ν on M defined by

$$\nu(dz\ d\phi) = \frac{-\cos\phi\ dz\ d\phi}{\text{normalization}} = \frac{-\cos\phi\ dz\ d\phi}{2\sum_i L_i}$$

We are therefore led to the dynamical system (M,T,ν) which, above the function $\tau(x)$ generates (V,S_t,μ) as a special flow [1], [2].

We shall discuss the proof that the section (M,T,ν) is

ergodic [1], [2]: it can actually be proved further. Namely it can be proved that (M,T,ν) and (V,S_t,μ) are, respectively, a Bernoulli automorphism and a Bernoulli flow: the proof of such properties will not be considered here [12, [2], [3].

To proceed let us study the geometry of the collisions. The reader is advised to spend some time in checking the geometrical meaning of the statements below in the case of a billiard with only one circular obstacle and with the help of the geometric relations derived in the next lecture.

First we observe that T is not smooth and the space M is divided by the singularity lines into several disconnected parts on each of which T acts (very) smoothly (i.e. where T is as many times differentiable as the boundaries ∂O), see Figure 1.

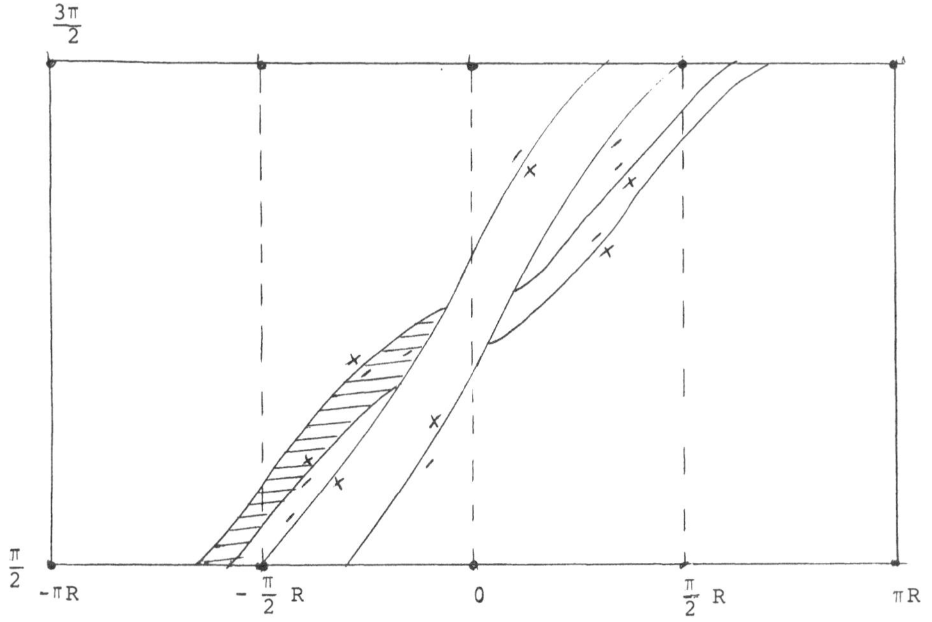

Figure 1

It is easy to see that the singularity set of T (which consist in the set $\partial M \cup T^{-1} \partial M$) which is outside ∂M consists in several (finitely many) connected disjoint sets. Each connected part can be thought of as a union of smooth "singularity lines" (at most in countable number); more precisely it consists in at least one line Σ_0, smooth and whose equation $\phi = \phi(r)$ monotonically increases from $\frac{\pi}{2}$ to $\frac{3\pi}{2}$, and in several other monotonic smooth lines merging into each other. A closer investigation (cf. next section) shows that the slope $d\phi/dr$ of the equation of a singularity line is bounded away from zero and ∞ and, furthermore, if x is a point in which two singularity lines merge ("confluence" point) the angle α between the tangents of the two lines at x is strictly positive. When a connected family of discontinuity lines is denumerable then the singularity lines accumulate at the intersection of Σ_0 with $\phi = \pi/2$ or $3\pi/2$ and nowhere else.

The singularity lines have "two sides": the side on which T is continuous is denoted +, the other -.

The accumulation points discussed above are points $x \in \partial M$ such that $\tau(y)$ is unbounded in the neighborhood of x.

As an example of the above structure we draw a few of the singularity lines of the periodic billiards in which there is only one obstacle of circular shape with radius R large enough (compared to the side of the torus). In this case there are 8 denumerable families of discontinuity lines. The curves of each family accumulate around one of the 8 points which are marked in the picture.

The 8 points marked in the picture correspond to the eight possible ways a particle can "escape" to ∞ without colliding. We shall restrict our attention to the case of billiards in which there is no free straight path to ∞: such billiards have only finitely many discontinuity lines and finitely confluence points in which they intersect.

The reason for this restriction is twofold: these billiards contain a large part of the general problem and mainly only leave out technical difficulties which tend to obscure the ideas of the proofs, on the other hand there seems to be no detailed treatment of the additional difficulties, present in the more general billiards, in the published literature.

To get an idea of the transformation T it is useful to see the image under T of the region shaded in Figure 1. It is easy to see that such an image has a shape which is a mirror image, around $\phi = \pi$, of the initial region. However the images of the short sides are the long sides and vice versa: i.e., the transformation T stretches in one direction the region under investigation and compresses it in another direction. This happens for all the regions in which the singularity lines partition M.

Let $\mathcal{P}$ be this partition: if we remember that the slopes of the "long sides" of the atoms of $\mathcal{P}$ have a slope away from zero and infinity, we can remark that the atoms of $T\mathcal{P}$ have a slope transversal to that of the atoms of $\mathcal{P}$. This means that the atoms of $T\mathcal{P}$ will be cut by singularity lines of T, and therefore, split upon action of T and subjected to a new stretching and contracting. Thus the transformation T acts on $\mathcal{P}$ in a way which vaguely resembles the baker's transformation, at least locally.

To investigate in more detail the action of T is is natural to study its action on smooth curves γ which are cut by singularity lines.

Let $\phi = \phi(r)$, $\bar{r} \leq r \leq \bar{\bar{r}}$, define a smooth (C^1) curve in M. And assume that γ is not cut by the singularity lines of T.

So the image $T\gamma$ will be a smooth curve, since T is a diffeomorphism in the complement of the singularity lines.

From the picture it is easy to obtain a differential relation between the functions $\phi = \phi(r)$ and $\phi' = \phi'(r')$ representing the curves γ and $T\gamma$ respectively.

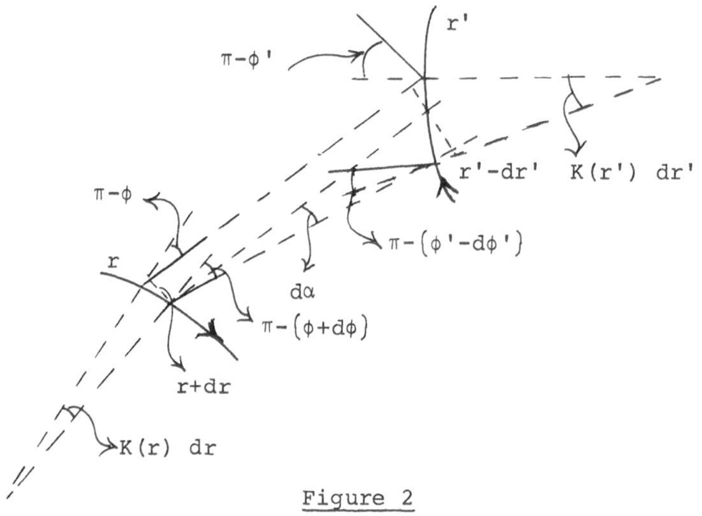

Figure 2

It is simple to deduce from the picture in Figure 2 the following relations:

$$-\cos\phi\, dr + d\alpha(-\tau(r,\phi)) = -\cos\phi'\, dr'$$

$$d\alpha = K(r)\, dr - d\phi$$

$$d\alpha = -K(r')\, dr' - d\phi'$$

By eliminating $d\alpha$ we can deduce:

$$\frac{-\cos\phi'\, dr'}{-\cos\phi\ \ dr} = \left(1 + (K(z) - \frac{d\phi}{dr})\frac{\tau(r,\phi)}{\cos\phi}\right) = \left(1 + (K(r') + \frac{d\phi'}{dr'})\frac{\tau(r,\phi)}{\cos\phi'}\right)^{-1}$$

$$\frac{1}{-\cos\phi'}\frac{d\phi'}{dr'} = \frac{K(r')}{\cos\phi'} + \cfrac{1}{\tau(r,\phi) + \cfrac{1}{\frac{1}{-\cos\phi}\left(\frac{d\phi}{dr} - K(r)\right)}}$$

$$\frac{1}{\cos\phi}\ \frac{d\phi}{dr} = \frac{K(r)}{\cos\phi} + \cfrac{1}{\tau(r,\phi) + \cfrac{1}{\frac{1}{\cos\phi'}\left(\frac{d\phi'}{dr'} + K(r')\right)}}$$

Remarks

(1) If γ is a decreasing curve with equation $\phi = \phi(r)$, $\frac{d\phi}{dr} \leq 0$, then $\gamma' = T\gamma$ is also decreasing and

$$R_+^{-1} \leq -\frac{d\phi'}{dr'} \leq R_-^{-1} + \tau_0^{-1}$$

where $R_+^{-1} = \min K(r)$, $R_-^{-1} = \max K(r)$, $\tau_0 = \min |\tau(x)|$.

(2) Similarly if γ is an increasing curve then the curve $T^{-1}\gamma = \gamma'$ is also increasing and

$$R_+^{-1} \leq \frac{d\phi'}{dr'} \leq R_-^{-1} + \tau_0^{-1}$$

(3) If γ and $T\gamma$ are both decreasing

$$\frac{\cos\phi \;\; dr}{-\cos\phi' \;\; dr'} \leq \left(1 + \frac{\tau_0}{R_+}\right)^{-1} \equiv \lambda_0^{-1}$$

where λ_0 is defined by the last equation

(4) If $\gamma' = T^{-1}\gamma$ and γ are both increasing

$$\frac{-\cos\phi' \;\; dr'}{\cos\phi \;\; dr} \geq \lambda_0$$

It is therefore natural to introduce the notion of p-length of a curve γ,

$$p(\gamma) = -\int_\gamma \cos\phi(r)\, dr$$

as well as the notion of r-length

$$r(\gamma) = \int_\gamma dr$$

(3') If γ, $T^{-1}\gamma$, ..., $T^{-k}\gamma$ is a sequence of smooth curves such that $T^{-k}\gamma$ is decreasing (so that $T^{-i}\gamma$ is decreasing for all $0 \leq i \leq k$) then

$$p(T^{-i}\gamma) \leq \lambda_0^{-1}\, p(T^{-i+1}\gamma) \leq \ldots \leq \lambda_0^{-i}\, p(\gamma)$$

$$r(T^{-i}\gamma) \leq \frac{\lambda_0^{-(i-1)}}{\lambda_0^{-1}}\, r(\gamma) \;,$$

$i = 0,1,\ldots,k$.

(4') If $\gamma, T\gamma, \ldots, T^k\gamma$ is a sequence of smooth curves such that $T^k\gamma$ is increasing (so that $T^i\gamma$ increases for $0 \le i \le k$) then

$$p(T^i\gamma) \le \lambda_0^{-1}\, p(T^{i-1}\gamma) \le \ldots \le \lambda_0^{-i}\, p(\gamma),$$

$$r(T^i\gamma) \le \frac{\lambda_0^{-(i-1)}}{\lambda_0 - 1}\, r(\gamma) \quad ,$$

$i = 0,1,\ldots,k$.

Finally we remark that if $\phi = \phi(r)$ is the equation of $T^{-k}\gamma$ in (3') or of $T^k\gamma$ in (4') then $T^{-(k-1)}\gamma,\ldots,\gamma$ or $T^{k-1}\gamma,\ldots,\gamma$ verify the remarkable inequality

$$R_+^{-1} \le \left|\frac{d\phi}{dr}\right| \le R_-^{-1} + \tau_0^{-1} ,$$

so that, on such curves, the r-length is equivalent to the Euclidean length.

The above considerations show how natural it is to try to define a "contracting fiber" as a curve γ_c such that $T^{-1}\gamma_c, T^{-2}\gamma_c$, ..., are a sequence of decreasing curves; similarly one should naturally look for an expanding curve γ_e as an increasing curve such that $T\gamma_e, T^2\gamma_e$, ..., are all increasing.

It is also worth noting that if such contracting curves exist they must obey the following differential equations:

$$\frac{1}{-\cos\phi}\,\frac{d\phi}{dr} = f^{(\infty)}(a_1,b_1,a_2,b_2,\ldots)$$

where $f^{(k)}(\ldots)$ is the continued fraction

$$f^{(k)}(a_1,b_1,a_2,b_2,\ldots,b_{k-1},a_k) = \frac{a_1}{2} + \cfrac{1}{b_1 + \cfrac{1}{a_2 + \cfrac{1}{b_2 + \cdots \cfrac{1}{b_{k-1} + \cfrac{1}{a_k}}}}}$$

and

$$a_i = \frac{2K(T^{-(i-1)}x)}{\cos\phi(T^{-(i-1)}x)} \le -R_+^{-1}; \qquad b_i = \tau(T^{-i}x) \le -\tau_0 .$$

Similarly an expanding curve should verify some analogous equation. It is, however, clear that the r.h.s. of the above differential equations is not continuous in $x = (r,\phi)$ but, instead, it is expected to have a dense set of singular (discontinuity) points.

3. The Fibers

Let x be a point in M and define

$d_T(x)$ = distance of x from the singularity lines of T

$d_{T^{-1}}(x)$ = distance of x from the singularity lines of T^{-1}

$d(x) = \inf (d_T(x), d_{T^{-1}}(x))$.

1 <u>Definition</u>. Let $q > 1$ and let k be a nonnegative integer. Put

$$M_q^{(K)+} = \{x \mid x \in M; -\cos\phi(T^{-h}x) \text{ and } d(T^{h}x) \geq q^{-1}(1+h^2)^{-1}, \forall h = 0,\dots,k\}$$

$$M_q^{(k)-} = \{x \mid x \in M, -\cos\phi(T^{-h}x) \text{ and } d(T^{-h}x) \geq q^{-1}(1+h^2)^{-1}, \forall h = 0,\dots,k\}$$

$$M_q^{+} = \bigcap_{k=0}^{\infty} M_q^{(k)+}, \qquad M_q^{-} = \bigcap_{k=0}^{\infty} M_q^{(k)-},$$

$$M_q^{(k)} = M_q^{(k)+} \cap M_q^{(k)-}, \qquad M_q = M_q^{+} \cap M_q^{-}.$$

The following lemma is a consequence of the Borel-Cantelli lemma.

2 <u>Lemma</u>. $\nu(\bigcup_q M_q) \equiv \lim_{q\to\infty} \nu(M_q) = 1.$

The following definition is fundamental.

3 <u>Definition</u>. If $x \in M$, $\cos\phi(x) > 0$, we define the 0-order contracting fiber $\hat{\gamma}_c(x)$ and the 0-order expanding fiber $\hat{\gamma}_e(x)$ as the curves with respective equations:

$$\frac{d\phi}{dr} = K(r)\ , \qquad \phi(r(x)) = \phi(x)\ , \qquad \text{for}\ \ \hat{\gamma}_e(x)\ ,$$

$$\frac{d\phi}{dr} = -K(r)\ , \qquad \phi(r(x)) = \phi(x)\ , \qquad \text{for}\ \ \hat{\gamma}_c(x)\ ,$$

and the above differential equations are meant to define $\phi(r)$ in an interval $[r',r'']$ around $r(x)$ which is maximal in the sense that the values of $\phi(r')$ and $\phi(r'')$ are either $\frac{\pi}{2}$ or $\frac{3\pi}{2}$.

It should be noted that this definition has a very simple geometrical meaning: $\hat{\gamma}_c(x)$ is the equation of a set of colliding arrows which, after collision, are parallel, while $\hat{\gamma}_e(x)$ is a set of colliding arrows which are parallel (i.e. which, before collision, are parallel).

4 Proposition. Given $x \in M_q^{(k)^+}$ there exists a $\Delta > 0$ such that the connected part of $T^k\hat{\gamma}_c(T^{-k}x)$ which contains x is a smooth curve lying, at least, above the interval $[r(x)-\Delta, r(x)+\Delta]$.

The following theorem is the first remarkable result.

5 Theorem. Given $q > 1$, there is $\Delta_q > 0$ such that if $x \in M_q^{(k)^+}$ the connected part of the set $T^k\hat{\gamma}_c(T^{-k}x)$ which contains x is a smooth curve above the interval $[r(x)-\Delta_q, r(x)+\Delta_q] \equiv J_x$ at least. We call $\gamma_c^{(k)}(x)$ this curve above J_x. Its equation $\phi = \phi_c^{(k)}(r)$ has a negative derivative bounded away from zero and infinity ($R_+^{-1} \leq -\,d\phi_c^{(k)}/dr \leq R_-^{-1} + \tau_0^{-1}$). Furthermore Δ_q can be so chosen that if $y \in \gamma_c^{(k)}(x)$ then $y \in M_{2q}^{(k)^+}$.

The above theorem allows us to define the germ of a contracting fiber:

6 Definition. If $x \in M_q^{(k)^+}$ the curve $\gamma_c^{(k)}(x)$ will be called a "germ" of a contracting fiber of order k.

Notice the difference between $\gamma_c^{(0)}(x)$ and $\hat{\gamma}_c(x)$.

Similar theorems and definitions hold for the expanding fibers: we shall not write them but we shall refer to them as statements (5'), (6').

Notice also that the geometrical meaning of $\gamma_e^{(k)}(x)$ and $\gamma_c^{(k)}(x)$: they represent a set of arrows which, if followed k steps steps in the past or in the future, respectively, become parallel. They can also be defined by the differential equation

$$\frac{1}{-\cos\phi}\frac{d\phi}{dr} = f^{(k)}\left(\frac{2K(x)}{\cos\phi(x)}, \tau(T^{-1}x), \frac{2K(T^{-1}x)}{\cos\phi(T^{-1}x)}, \ldots, \tau(T^{-k+1}x), \frac{2K(T^{-k}x)}{\cos\phi(T^{-k}x)}\right)$$

for $\gamma_c^{(k)}(x)$ and, similarly for $\gamma_e^{(k)}(x)$.

The above theorem tells us that, if $x \in M_q^{(k)+}$, a solution to the differential equation just written can be found locally and extended at least to the interval $[r(x)-\Delta_q, r(x)+\Delta_q]$.

Let us give an idea of the proof of the above theorem: The points $T^{-\ell}x$ approach "very slowly" the singularity lines of T as ℓ varies between 0 and k, because $x \in M_q^{(k)+}$.

Suppose we consider the smooth part γ_k' of $T^k\hat{\gamma}_c(T^{-k}x)$ defined on an interval of size 2Δ around $r(x)$. Then

$$r(T^{-i}\gamma_k') \le \frac{\lambda_0^{-i}}{\lambda_0 - 1} 2\Delta, \qquad i = 0,1,\ldots,k,$$

but we also know that $d(T^{-i}x) \ge q^{-1}/(1+i^2)$ and $-\cos\phi(T^i x) > q^{-1}/(1+i^2)$ so that if Δ is chosen small enough (depending on q but not on k) we can obtain that, if $y \in T^{-i}\gamma_k'$, the values $-\cos\phi(y)$, $d(y)$ stay larger than $(2q)^{-1}/(1+i^2)$ for $i = 0,1,\ldots,k$.

This is so because the curves $T^{-i}\gamma_k'$ contract exponentially rapidly while their centers approach the singularity with polynomial speed. So the "difficulties" (i.e. having a piece of

$T^{-i}\gamma_k'$ cut by a singularity line of T) can arise only for small values of i and will not arise at all if Δ is chosen small enough at the beginning (roughly so that $q^{-1}(1+i^2)^{-1} - \lambda_0^{-i}(\lambda_0-1)^{-1}\Delta > (2q)^{-1}(1+i^2)^{-1}$, $\forall i = 0,\dots,\infty$, i.e. $\Delta \le \Delta_q \sim q^{-1} \cdot$ const.).

It is very easy to make the above ideas into a rigorous proof by using an inductive procedure, see Appendix A.

4. Smoothness

The smoothness properties of the approximate fibers are of two types. The first type concerns the smoothness of functions defined on the fibers as, for instance, the slope of $\gamma_c^{(k)}(x)$ at y as a function of $y \in \gamma_c^{(k)}(x)$ or $\cos\phi(y)$ or $\tau(y)$ or $K(y)$ etc. The second type of smoothness concerns the x-dependence of $\gamma_c^{(k)}(x)$. Since the sets $M_q^{(k)\pm}$ are open and T is differentiable on them it is clear that $\gamma_c^{(k)}(x)$ has very good smoothness properties of the above types (for instance, if the boundary of the obstacles is C^∞, $\gamma_c^{(k)}(x)$ is C^∞ in any reasonable sense).

Therefore the above problems have to be interpreted as questions concerning the uniformity in k of the smoothness. While the problem of the first type of smoothness is rather well defined, the second is susceptible of several interesting interpretations.

The guide to a more precise formulation has to come from the use we wish to make of the smoothness properties. Consider a point $x \in M_q$ and draw through x the fibers $\gamma_c^{(k)}(x)$ and $\gamma_e^{(k)}(x)$, see Figure 3. Then consider a point $y \in M_q^{(k)} \cap U$ where U is a small neighborhood of x, so small that the fibers $\gamma_e^{(k)}(y)$ and $\gamma_c^{(k)}(y)$ intersect $\gamma_c^{(k)}(x)$ and $\gamma_e^{(k)}(x)$, respectively, in points b and a. Observe also that the restriction on the diameter of U that this implies depends solely on q. Call α, β the r-lengths of the arcs ax and bx respectively.

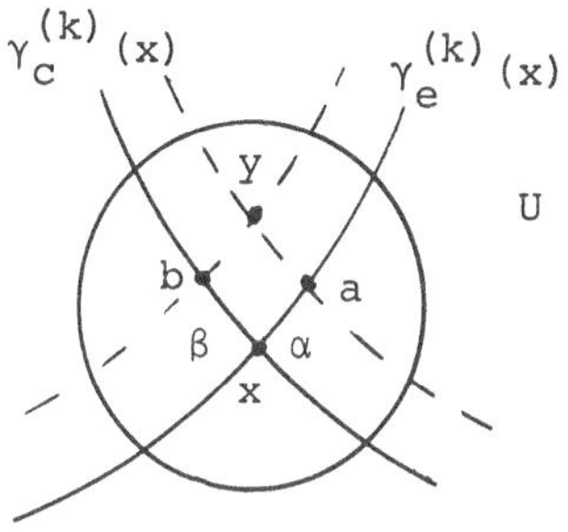

Figure 3

Our purpose is to use the coordinates (α,β) to describe the points of $U \cap M_q^{(k)}$ or of a large subset of it, in the sense that we wish to write, for all $E \subset M_q^{(k)} \cap U$ and with an obvious abuse of notation:

$$\nu(E) = \int_E \rho^{(k)}(\alpha,\beta)\, d\alpha\, d\beta .$$

Our smoothness property will be the statement that $\rho^{(k)}(\alpha,\beta)$ can be uniformly bounded, above and below, on a subset of $M_q^{(k)} \cap U$ which is a "sizable" fraction of it and has a simple geometric structure; i.e., it can be thought of, in the (α,β) coordinates, as the intersection of U with a rectangle. Furthermore the ratio between the upper and lower bounds of $\rho^{(k)}(\alpha,\beta)$ on the good subset of $U \cap M_q^{(k)}$ tends to 1, as the diameter of U tends to zero, uniformly in k and $x \in M_q$.

Let us make the above description more concrete: let $\tilde{\gamma}_c^{(k)}(x) = \gamma_c^{(k)}(x) \cap U \cap M_{q/2}^{(k)^-}$ and $\tilde{\gamma}_e^{(k)}(x) = \gamma_e^{(k)}(x) \cap U \cap M_{q/2}^{(k)^+}$. Let (α,β) be, respectively, the natural coordinates (r-arc lengths) on $\tilde{\gamma}_c^{(k)}(x)$ or $\tilde{\gamma}_e^{(k)}(x)$.

Pick a point $y_e \in \tilde{\gamma}_e^{(k)}(x)$ and $y_c \in \tilde{\gamma}_c^{(k)}(x)$; then define $y = \gamma_c^{(k)}(y_e) \cap \gamma_e^{(k)}(y_c)$, which certainly makes sense if y_e and y_c are in a suitably small neighborhood U of x (so that the fibers of the points $y \in M_{q/2}^{(k)} \cap U$ are longer than diameter (U)).

It is clear that the set $\hat{U}$ of y's so defined is, in a natural sense, the intersection of a "rectangle" with U, and, furthermore,

$$U \cap M_{q/4}^{(k)} \subseteq \hat{U} \subseteq M_q^{(k)} \cap U \subseteq U$$

We wish to show that $\rho^{(k)}(\alpha,\beta)/\rho^{(k)}(\alpha',\beta')$, (α,β), $(\alpha',\beta') \in \hat{U}$ differs from 1 by a quantity which approaches 1 as the diameter of $U \to 0$, uniformly in k and in (α,β), $(\alpha',\beta') \in U$.

Of course sometimes the set $U \cap M_{q/4}^{(k)}$ may be empty or very small, but this does not affect the proof: if q is large $M_{q/4}^{(k)}$ has large measure and so for many U's the content of the result will be nontrivial. There are also other natural systems of coordinates on suitable subsets of U: let $x \in U \cap M_q$, draw $\gamma_c^{(k)}(x)$; let $\tilde{\gamma}_c^{(k)}(x) = \gamma_c^{(k)}(x) \cap M_{q/2}^{(k)-}$; let $\hat{U}' = \bigcup_{y \in \tilde{\gamma}_c^{(k)}(x)} \{\gamma_e^{(k)}(y)\}$; then

$$U \cap M_{q/4}^{(k)-} \subset \hat{U}' \subset M_{q/2}^{(k)-} \cap U \subset U$$

and if $z \in \hat{U}'$ we can attach to z the coordinates (α,β) where

α = r-length along $\gamma_c^{(k)}(x)$ of the segment between x and $\gamma_c^{(k)}(x) \cap \gamma_e^{(k)}(z)$;

β = r-length along $\gamma_e^{(k)}(z)$ of the segment between z and $\gamma_c^{(k)}(x) \cap \gamma_e^{(k)}(z)$.

Similarly we can define a set $\hat{U}''$ and introduce on it coordinates in such a way that the roles of the expanding and contracting fibers are interchanged.

We shall be interested in the Radon-Mykodim derivatives of the measures $d\alpha\, d\beta$ on $\hat{U}'$, $\hat{U}''$ with respect to ν in the same sense discussed in the preceding case of $\hat{U}$. This derivative will be denoted with the same symbol $\rho^{(k)}(\alpha,\beta)$: when we shall talk of natural coordinates and of the associated $\rho^{(k)}(\alpha,\beta)$ it will be clear from the context which of the three types is meant.

To understand which are the properties that have to be proven let us calculate $\rho^{(k)}(\alpha,\beta)$: it is of great help to note that $\hat{U}$ is open, so that we can easily calculate $\rho^{(k)}(\alpha,\beta)$ by some local geometric constructions. Let $y \in \hat{U}$ and let δ_1, δ_2 be two "very small"

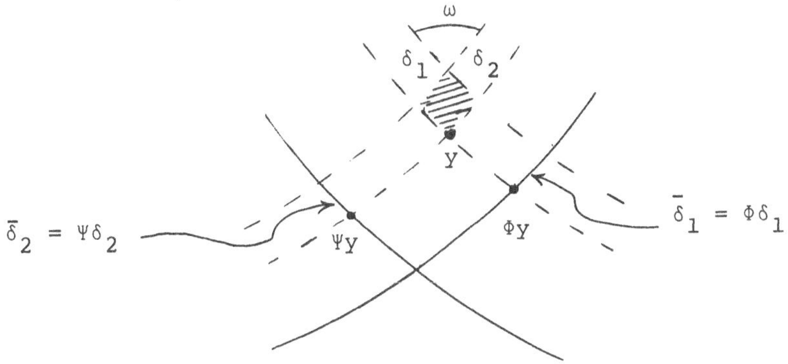

Figure 4

segments of $\gamma_e^{(k)}(y)$, $\gamma_c^{(k)}(y)$ respectively. Let $\bar{\delta}_1$, $\bar{\delta}_2$ be their images on $\tilde{\gamma}_e(x)$, $\tilde{\gamma}_c(x)$ under the mappings

$$\Phi: \hat{U} \cap \gamma_e^{(k)}(y) \to \tilde{\gamma}_e^{(k)}(x) \quad ; \quad \Psi: \hat{U} \cap \gamma_c^{(k)}(y) \to \tilde{\gamma}_c^{(k)}(x)$$

which associate to a point $z \in \gamma_e^{(k)}(y) \cap \hat{U}$ the point $\Phi z = \gamma_e^{(k)}(x) \cap \gamma_c^{(k)}(z)$ and to $z \in \hat{U} \cap \gamma_c^{(k)}(y)$ the point $\Psi z = \gamma_c^{(k)}(x) \cap \gamma_e^{(k)}(z)$.

It is easy to see that Φ and Ψ are, in the natural sense, diffeomorphisms in the neighborhood of y.

Let E be the shaded region: some elementary geometrical considerations on Figure 4 allow us to check that:

$$\nu(E) = \frac{-\cos\phi(r)\ \text{area } E}{\text{normalization}} = -\frac{(\text{length } \delta_1)\cdot(\text{length } \delta_2)\,|\sin\omega|\cos\phi(y)}{\text{normalization}}$$

$$= \frac{\cos\phi(\Phi y)\ \cos\phi(\Psi y)}{-\cos\phi(y)} \left(\frac{d\phi_e}{dr} - \frac{d\phi_c}{dr}\right)_y \frac{f_c^{(k)}(y)\ f_e^{(k)}(y)}{\text{normalization}}\ r(\bar{\delta}_1)\cdot r(\bar{\delta}_2)$$

where $\phi_e(r)$, $\phi_c(r)$ are, respectively, the equations of $\gamma_e^{(k)}(y)$ and $\gamma_c^{(k)}(y)$ and $f_c^{(k)}(y)$, $f_e^{(k)}(y)$ are the Radon-Nykodim derivatives

$$\frac{p(\delta_1)}{p(\bar{\delta}_1)} \quad \text{and} \quad \frac{p(\delta_2)}{p(\bar{\delta}_2)}$$

Our problem is, thus, "reduced" to the search for a bound for $f_e^{(k)}(y)$ and $f_c^{(k)}(y)$ in a sufficiently small neighborhood of points $x \in M_q^{(k)}$ and to an estimate of $\frac{d\phi_e}{dr}$ and $\frac{d\phi_c}{dr}$.

The problem of studying $p(\delta_1)/p(\bar{\delta}_1)$ (say) can be naturally attacked by observing that, if we apply the transformation T^{-k} to $\gamma_c^{(k)}(z)$, $z \in \delta_1$, we obtain a family of curves which form a very smooth foliation, i.e. the foliation generated by the smooth differential equation $d\phi/dr = -K(r)$. This means that $p(T^{-k}\delta_1)/p(T^{-k}\bar{\delta}_1)$ is very easy to compute and is given, in the simple case $K(r) = R^{-1}$ = const., by

$$\frac{p(T^{-k}\ \delta_1)}{p(T^{-k}\ \Phi\delta_1)} = \frac{\cos\phi(T^{-k}\ \Phi y)}{\cos\phi(T^{-k}\ y)}\ \frac{R^{-1} + \frac{d\phi_e^{(k)}}{dr}(T^{-k}\Phi y)}{R^{-1} + \frac{d\phi_c^{(k)}}{dr}(T^{-k}\ y)} \equiv F_1^{(k)}(y)\ ,$$

where $\phi_e^{(k)}$ denoted the equations of $T^{-k}\ \gamma_e^{(k)}(y)$ or $T^{-k}\ \gamma_e^{(k)}(\Phi y)$.

Then the ratio $p(\delta_1)/p(\Phi\delta_1)$ can be written as

$$\frac{p(\delta_1)}{p(\Phi\delta_1)} = F_1^{(k)}(y) \prod_{i=0}^{k-1} \frac{\dfrac{p(T^{-i}\delta_1)}{p(T^{-i-1}\ \delta_1)}}{\dfrac{p(T^{-i}\ \Phi\delta_1)}{p(T^{-i-1}\Phi\delta_1)}} \equiv F_1^{(k)}(y)\ F_2^{(k)}(y)\ .$$

The expressions for the $\rho^{(k)}(\alpha,\beta)$ for the other natural systems of coordinates on $\hat{U}'$ or $\hat{U}''$ can be studied in a similar way: e.g. in the case of $\hat{U}'$ the $\rho^{(k)}(\alpha,\beta)$ has an expression which can be conveniently derived along the same lines as in the case of U by introducing an auxiliary smooth foliation which contains $\gamma_c^{(k)}(x)$ and which consists of decreasing curves with slope verifying the usual bounds (i.e. between $-(R_-^{-1}+\tau_0)^{-1}$ and $-R_+^{-1}$). We shall discuss the details of a slightly more complicated discussion in the proof of Lemma 13 and therefore we omit the details here.

As we have seen in the geometrical considerations the expansion coefficient at a point w of the p-length of a curve γ depends only on quantities like $\cos\phi(w)$, $\cos\phi(T^{-1}w)$, $k(w)$, $K(t^{-1}w)$, $\tau(w)$, $d\phi(w)/dr$: therefore it is through the study of the variation of these quantities as w varies along the curves $T^{-i}\gamma_c^{(k)}(y)$ and $T^{-i}\gamma_e^{(k)}(y)$ that the desired bounds can be obtained.

The key theorem is the following:

7 Lemma. Given $q > 1$, $k > 0$, $x \in M_q^{(k)}$ there exist constants $c_q > 0$, $\lambda > 1$ such that if $y,y' \in \gamma_c^{(k)}(x)$ and $\gamma_{yy'}$ denotes the arc between y,y' on $\gamma_c^{(k)}(x)$, then:

(i) $p(T^{-\ell}\gamma_{yy'}) \leq \lambda^{-\ell}p(\gamma_{yy'})$; $r(T^{-\ell}\gamma_{yy'}) \leq \dfrac{\lambda^{-\ell+1}}{\lambda-1}r(\gamma_{yy'})$

for $\ell = 0,1,\ldots,k$.

(ii) if $\phi = \phi^{\ell}(r)$ is the equation of $T^{-\ell}\gamma_c^{(k)}(x)$ then

$$R_+^{-1} \leq -\frac{d\phi^{\ell}}{dr} \leq R_-^{-1}+\tau_0^{-1} \quad \text{for } \ell = 0,1,\ldots,k.$$

(iii) the quantities

$$\frac{\cos\phi(T^{-\ell}y)}{\cos\phi(T^{-\ell}y')}, \quad \frac{\tau(T^{-\ell}y)}{\tau(T^{-\ell}y')}, \quad \frac{K(T^{-\ell}y)}{K(T^{-\ell}y')}$$

are between $\exp \pm c_q\lambda^{-\ell}r(\gamma_{yy'})$.

(iv) the quantities

$$\frac{d\phi^{\ell}}{dr}(y) \Big/ \frac{d\phi^{\ell}}{dr}(y')$$

are between $\exp \pm c_q\lambda^{-\ell}r(\gamma_{yy'})$.

(v) $\sup_{|r-r(x)|\leq\Delta_q} |\phi^{(\ell)}(r) - \phi^{(\ell+1)}(r)| \leq c_q\lambda^{-\ell}$, $\ell = 0,1,\ldots,k.$

Proof: See Appendix A.

8 Corollary. Let $x \in M_q^+$ and let $\phi_c^{(k)}(r)$ denote the equation of $\gamma_c^{(k)}(x)$ defined on $[r(x)-\Delta_q, r(x)+\Delta_q]$. Then $\lim_{k\to\infty} \phi_c^{(k)}(r) = \phi_c(r)$ exists in the C^1-sense and the properties (i), (ii), (iii) and (iv) in the above lemma hold for the curve $\gamma_c(r)$ defined by $\phi_c(r)$ for all $\ell = 0,1,\ldots$. The limit is reached exponentially rapidly in the C^0-sense.

The above lemma and corollary tell us that each single k-contracting germ of a fiber is very smooth and does not behave wildly as $k \to \infty$.

The following theorem provides the estimates needed for the bounds on $\rho^{(k)}(\alpha,\beta)$:

9 Theorem. Let $k > 0$, $q > 1$. Let U be a neighborhood with diameter $D(U)$ so small that a k-fiber's germ (contracting or expanding) through any point of $M_q^{(k)} \cap U$ (if such a set is not empty), has a length larger than $D(U)$. Let $x \in M_q^{(k)} \cap U$ and let $\hat{U}$ $(\hat{U}',\hat{U}")$ be the set defined above and introduce on it the natural coordinates (α,β). Then there is a θ_q: $(0,+\infty) \to (0,+\infty)$, infinitesimal near zero, and a $C_q > 0$ such that

$$\frac{\rho^{(k)}(\alpha,\beta)}{\rho^{(k)}(\alpha',\beta')}, \qquad (\alpha,\beta),\ (\alpha',\beta') \in \hat{U},\ (\hat{U}',\hat{U}")$$

is between $\exp \pm \theta_q(D(G))$ and $\rho^{(k)}(\alpha,\beta) \leq C_q$, $\forall(\alpha,\beta) \in \hat{U},\ (\hat{U}',\hat{U}")$.

Proof: See Appendix A.

The above theorem and corollary are enough to deduce that the system (M,T,ν) is a B-system (i.e. isomorphic to a Bernoulli shift) once it is known that it is a K-system [3].

The reader may notice that the statements of the last theorem would make sense also if $K = +\infty$: It would be interesting to have an explicit proof of such a result (see concluding remarks).

In the remaining lectures we sketch the proof of the ergodicity of (M,T,ν).

5. Ergodicity

As we saw in the preceding sections the billiards possess approximate foliations of contraction and expansion which are absolutely continuous in the sense of the last theorem.

It is therefore tempting to try the classical Hopf method to prove ergodicity.

A first step is described by the following theorem and corollary which express rather weak ergodic properties.

Given a continuous function $f \in C(M)$, put

$$W^f_{k,\rho} = \{x \mid x \in M, \; |h^{-1} \sum_{i=0}^{h-1} (f(T^i x) - \bar{f}^{\pm}(x))| < \rho, \; \forall h \geq k\},$$

where $\bar{f}^+$, $\bar{f}^-$ denote the future and the past averages of f. Birchoff-Von Neumann's theorem tells us that

$$\lim_{k\to\infty} \nu(W^f_{k,\rho}) = 1 \;, \qquad \forall \rho > 0 \;.$$

10 Lemma. Let $x \in M_q$ and $U \ni x$ be a neighborhood of x of sufficiently small diameter. Let $V = M_{q/4} \cap U$ and let $f \in C(M)$.

Given $\theta > 0$, $\rho > 0$, $\exists K_{\rho,\theta}$ such that if $K > K_{\rho,\theta}$ there is $\bar{W} \subset W^f_{k,\rho} \cap V$, and

(1) $\nu(\bar{W}) > (1-\theta)\nu(V)$;

(2) if $x,y \in \bar{W}$ there is a sequence $\gamma_1,\gamma_2,\gamma_3,\gamma_4$ of alternatingly ℓ-expanding and ℓ-contracting fibers such that $x \in \gamma_1$, $y \in \gamma_4$, $\gamma_i \cap \gamma_{i+1} \in W^f_{k,\rho}$, $i = 1,2,3$, where ℓ is an arbitrary integer.

This lemma easily implies that $\bar{f}^{\pm}$ has an "essential oscillation" $\leq 4\rho$ on V i.e., that $\bar{f}^{\pm}$ is a constant on V. Hence

11 Corollary. The transformation T is ergodic on $\bigcup_{n=-\infty}^{+\infty} T^n V$.

By an application of Vitali's covering theorem the following can be deduced.

12 Proposition. The dynamical system (M,T,V) break in, at most, denumerably many ergodic components.

The proof of the above lemma shall not be explicitly given since Lemma 10, Corollary 11, Proposition 12 will not be used in the proof of global-ergodicity. It is not hard and involves only a few technical difficulties: it can be deduced from the information contained in statements 8 and 9 above without "further" use of the structure of the billiards: it is also the strongest result that we can obtain without further investigation of the special properties of our system. In fact, in the theory of the Anosov diffeomorphisms, the ergodicity is obtained from a local ergodic property through a covering argument. In our case the family of the sets of type V, of the above theorem and corollary, is a priori too small to allow such a covering argument (notice that the sets M_q are "badly" disconnected). It may just not be possible to construct, for all couples, x,y chosen outside a set of zero measure, a sequence $V_1, V_2, \ldots, V_s$ of finitely many sets of a form to which Proposition 11 applies and such that $x \in V_1$, $y \in V_s$ and $\nu(V_i \cap V_{i+1}) \neq 0$, $i = 1,2,\ldots,s-1$.

The following Theorem 13 will allow us to make a construction (see Lemma 17) of a chain of alternatingly ℓ-expanding and ℓ-contracting fibers connecting two points x,x' chosen outside a subset of measure zero of a sufficiently small neighborhood of a point x_0

such that $d(T^i x_0) \neq 0$ for all but one integer i.

The number $K(x,x')$ of elements of such a chain will now depend upon the two points x,x' but, if x,x' are outside a suitable set of measure zero, will not depend on ℓ.

Furthermore the fibers can be chosen to have the property that their successive intersections always take place in points in which, an a priori given continuous function, attains its past and future averages within $\rho K(x,x')^{-1}$ in $\bar{K}$-steps, provided $\bar{K}$ is large enough. This allows us to prove that the average of every continuous function is constant in U.

Now we can deduce global ergodicity by means of a simple covering argument. The complement of the set of the x 's whose trajectory is at most once tangent is denumerable: therefore the intersections of this set with the connected components of M are linearly connected. This means that each connected component of M is entirely contained in an ergodic component of T. This, of course, implies that T is ergodic. We shall indicate some details of the derivation of the just described results from the main Theorem 13 (below) after describing its proof in the next lecture. The complete proof is however in Appendix C.

The main lemma for the proof of global ergodicity is the following.

13 <u>Theorem</u>. Let x_0 be a point whose trajectory $x_0, Tx_0, T^2x_0, \ldots$ is never "tangent". Given $\varepsilon > 0$, $C > 0$ there is a neighborhood $U(\varepsilon,C)$ of x_0 such that if $\gamma \subset U$ is a decreasing curve with $r(\gamma) = \delta_0$ and with slope between $-(R_-^{-1} + \tau_0^{-1})$ and R_+^{-1} there is a quadrilateral G whose boundary consists of $\gamma_\ell \equiv \gamma, \gamma_r, \gamma_\mu, \gamma_d$ where $\gamma_r, \gamma_\ell \equiv \gamma$ are two disjoint decreasing curves while γ_d, γ_μ are two disjoint increasing curves with slope between R_+^{-1}

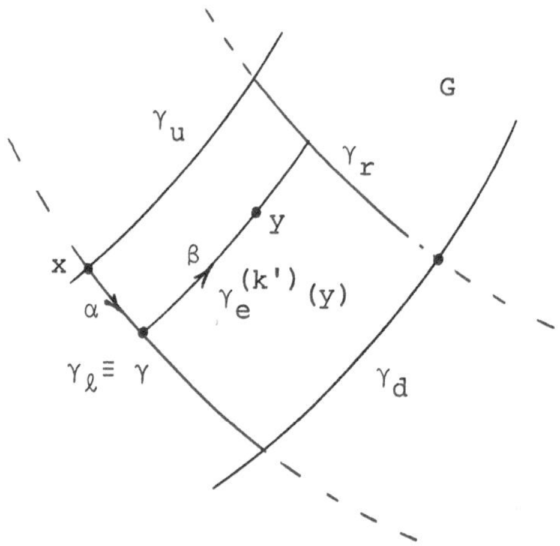

Figure 5

and $(R_-^{-1}+\tau_0^{-1})$. Furthermore $U(\varepsilon,C) \supset U(\varepsilon,C')$ if $C < C'$ and, if k is an arbitrary integer:

(i) there is an open subset $\hat{G}_k \subset G$ which admits a foliation with k'-expanding curves ($k' \geq k$) which join the opposite sides γ_ℓ, γ_r without intersecting γ_u, γ_d and have an r-length $\geq C\delta_0$.

(ii) $\nu(\hat{G}_k) > (1-\varepsilon)\nu(G)$

(iii) Let $y \in \hat{G}_k$ be parametrized by the r-coordinates (α,β) of $\bar{y} = \gamma_e^{(k')}(y) \cap \gamma_\ell$ along γ_ℓ and of y along $\gamma_e^{(k')}(y)$ from $\bar{y}$ (see Figure 5).
Then the Radon-Nykodim derivative $-\cos\phi\ dr\ d\phi/d\alpha\ d\beta$ is a function $\rho^{(k)}(\alpha,\beta)$ with the property that for all (α,β), $(\alpha',\beta') \in \hat{G}_k$ the ratio $\rho^{(k)}(\alpha,\beta)/\rho^{(k)}(\alpha',\beta')$ is between $\exp \pm \eta(\delta_0)$ where $\eta(\delta_0)$ is a suitable real function.

(iv) The side γ_r can be chosen to be a curve contained in a parallel translate, along γ_u of the curve γ_ℓ(extended, if too short (see Figure 5), beyond its lowest extreme in an arbitrary way which keeps it smooth and with slope between $-(R_-^{-1}+\tau_0^{-1})$ and $-R_+^{-1})$.

A similar result holds if $d(T^{-i}x_0) > 0$, $i.= 0,1,\ldots,$ and the role of the expanding and contracting fibers is inter changed. Similarly it is possible to prove an almost identical lemma in which $\gamma = \gamma_r$ instead of $\gamma = \gamma_\ell$. When we shall refer to Theorem 13 we shall also mean these last results.

Strictly speaking the properties of absolute continuity of the approximate fibers are not needed in the proof of this theorem sketched in the next lecture. However a crucial step of the proof relies upon a lemma whose proof is, essentially, a repetition of the arguments needed to prove the theorem 8 on the absolute continuity (see next lecture).

6. About the Main Lemma

The proof of this theorem (statement 13 above) proceeds in a very natural way. It needs a little extra information about the structure of the singularity lines of the transformations $T, T^2, \ldots T^L$. This information is contained in the next lemma.

14 Lemma. Consider the set, $S_L = \partial M \cup T^{-1}\,\partial M \cup \ldots \cup T^{-L}\,\partial M$ (which is the set of singularity lines of $\tilde{T} = T^L$). Suppose the billiard is "without horizon" (i.e. there is no straight path to ∞). Then the set S_L consists of finitely many smooth curves. The set X_L of points of S_L belonging to different curves (confluence points) is finite and the lines meet at $x \in X_L \setminus \partial M$ with a nonzero angle; if $x \in X_L \cap \partial M$ they have a contact of second order. So the ν-measure of the area trapped in a circle with radius $\sqrt{\delta}$ centered at $x \in X_L \cap \partial M$ by the two merging lines is of the order δ^2 (while its Lebesgue measure would be $\sim\delta^{3/2}$).

The proof is based on the differential expression of the singularity lines γ which is of the form

$$\frac{d\phi}{dr} = -K(r) - \frac{\cos\phi}{\tau_s(r,\phi)}$$

and $\tau_s(r,\phi) = \tau(r,\phi)$ if $\gamma \subset T^{-1}\,\partial M$ or, more generally, $\tau_s(\cdot)$ is given by a finite continued fraction (see formula in Lecture 2). The main remark is that the fact that the two curves merging in $x \in X_L$ are different implies that the limit of $\tau_s(r,\phi)$ as $(r,\phi) \to x$ is different along the two lines. This easily implies the Lemma 14.

We now come to the proof of Theorem 13. It will be convenient to put

$$Z = R_+(R_-^{-1}+\tau_0^{-1}) > 1$$

Consider a power $\tilde{T} = T^L$ so large that

$$\lambda^{Lm}(\lambda-1) > \lambda_1^m > 2\ , \qquad\qquad m = 1,2,\ldots,$$

(see Lemma 7 for the definition of λ).

One starts with a neighborhood U of a point x_0, such that $-\cos\phi(T^i x_0) > 0$, $i = 0,1,\ldots$, so small that it is possible to draw through every point of U a k_0L-expanding fiber with r-length exceeding $4ZCD(U)$. The number k_0 will be at least such that $4CZ\,\lambda_1^{-k_0} < \lambda_2^{-k_0}$ where λ_2 is arbitrarily chosen so that $\lambda_1 > \lambda_2 > 2$; however the final choice of k_0 will be made at the end of the proof (to give its value now would be too complicated since it involves a large number of different geometrical constants which would have to be defined before their interest really appears).

Let $\gamma_0 \subset U$ be a decreasing curve and $r(\gamma_0) = \delta_0$. Draw through the points of γ_0 the k_0L-expanding fibers with r-length exceeding $4ZCD(U)$. Pick the point situated on the expanding fiber through the upper vertex of γ_0 and at an r-distance $2ZCD(U)$ from it (see Figure 6).

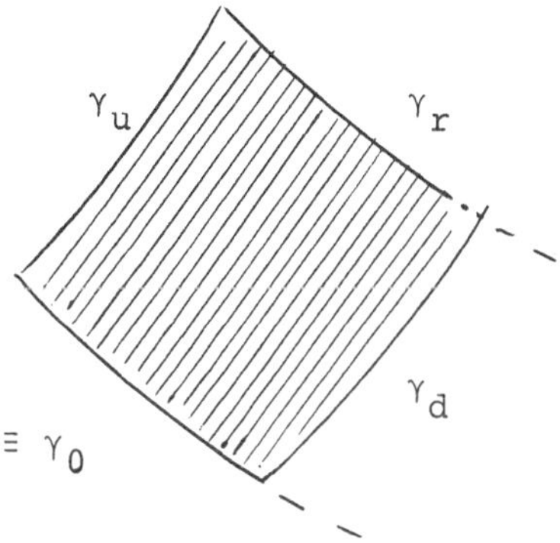

Figure 6

If the above construction makes sense the lines $\gamma_u, \gamma_d, \gamma_r, \gamma_\ell$ determine a quadrilateral G: in order that the above construction makes sense we may be forced to further reduce D(U), ahd hence U, and possibly extend γ_0 slightly beyond the lower vertex (in an arbitrary way which keeps the slope bounded between $-(R_-^{-1}+\tau_0^{-1})$ and and $-R_+^{-1}$) so that γ_r is long enough to cross γ_d.

Choosing U even smaller, if necessary, we can and shall assume that for $b > 0$ to be fixed later:

$$-\cos\phi\big|_{T^k G} > \delta_0\left(\frac{K}{L}\right)^{-4}, \qquad k = 1,\dots,k_0 L$$

$$d_{\tilde{T}}(x)\big|_{\tilde{T}^k G} > \delta_0\, K^{-16b}, \qquad k = 1,2,\dots k_0-1$$

the exact value of b is irrelevant provided it is large enough (from the estimates below b = 9 will be large enough); $d_{\tilde{T}}$ is the distance from the singularities of $\tilde{T}$.

If $k \leq k_0$ the set $\hat{G}_k$ will coincide with G

If $k > k_0$ we shall obtain $\hat{G}_k$ by subtracting from G points in such a way that $\hat{G}_k$ is the union $\hat{G}_k = \bigcup_{s=1} \hat{G}_{k,s}$ of quadrilaterals which are unions of intersections of G with KL-expanding fibers (i.e. are fibered by kL-expanding fibers).

Furthermore,

$$\cos \phi|_{T^i \hat{G}_{k,s}} > \delta_0 (\frac{i}{L})^{-4}, \qquad i = 1,\dots,kL ;$$

$$d_{\tilde{T}}(x)|_{\tilde{T}^i G_{k,s}} > \delta_0 \, i^{-16b}, \qquad i = 1,\dots,k-1.$$

Suppose $\hat{G}_{k-1,m}$ have been constructed. Put $\tilde{G}_{k-1,m} = \tilde{T}^{k-1} \hat{G}_{k-1,m}$. The set $\tilde{G}_{k-1,m}$ will, in general, be cut into "components" $O_{k-1,m,\ell}$ by the singularity lines of $\tilde{T}$. The sets $\tilde{T}^{-1}\tilde{G}_{k,s}$ will be subsets of some of these components which will be called the "good" components.

We immediately declare "bad" components all those which are contained in the triangular regions defined by the lines merging into a point $x \in X_L$ and a circle around x with radius $\delta_0(k-1)^{-b/2}$, if $x \in X_L \setminus \partial M$, and with radius $(\delta_0(k-1)^{-b/2})^{1/2}$, if $x \in X_L \cap \partial M$.

Lemma 14 above expresses that the ν-measure of the set of points so far discarded does not exceed

$$\text{const.} \frac{\delta_0^2}{(k-1)^b}$$

Divide the remaining components into regular and irregular. Regular will be those which have both bases on singularity lines.

It follows from the definition that the regular components 0 are such that

$$r(\gamma_r(0)) \geq \text{const.} \; \delta_0 \; (k-1)^{-b/2}$$

(provided the D(U) had been chosen small compared to the minimal distance between disjoint singularity lines of $\tilde{T}$).

Let us call "big" a component 0 such that

$$r(\gamma_r(0)) \geq \delta_0 (k-1)^{-4b} .$$

The regular components are all big if the k_0 had been chosen large enough.

If a component is big it is possible to draw the L-expanding fibers which cut $\gamma_r(0)$ in points at a distance $\delta_0(k-1)^{-8b}$ from the bases. This is easy to see if account is taken of the fact that the r-length of an increasing line joining $\gamma_r(0)$ to $\gamma_\ell(0)$ and verifying the usual bound on the slope is bounded by $\lambda_2^{-k}\delta_0$. We call $0'$ the set obtained by cutting, off 0, the two sets between the bases of 0 and the just drawn fibers. We discard 0 as bad if $T\,0' \subset V_{k-1+1/L,4} = \{x \mid -\cos\phi(x) \le \delta_0(k-1+\frac{1}{L})^{-4}\}$ or if $T0' \cap V_{k-1+1/L,4} \neq \emptyset$ but $T0' \cap V_{k,3/2} = \{x \mid -\cos\phi(x) \le \delta_0 k^{-3/2}\}$.

If $T\,0' \cap V_{k-1+1/L,4} \neq \emptyset$ and $T0' \not\subset V_{k,3/2}$ we draw through $\gamma_r(T0') \cap \partial V_{k-1+1/L,4}$ and through $\gamma_\ell(T0') \cap \partial V_{k-1+1/L,4}$ an $(L-1)$-expanding fiber and cut out of $T0'$ the points between the bases of $T0'$ and the just drawn fibers. We repeat this process $(L-1)$ more times using for the ith stage the set $V_{k-1+i/L,4}$ instead of $V_{k-1+1/L,4}$. Let $\tilde{G}_{k,m}$ be the sets obtained in this way from the big 0's and let $\bar{0} = \tilde{T}^{-1}\tilde{G}_{k,m}$: $\bar{0} \subset 0$.

Lemma 15 and Corollary 16 below allow, after simple considerations, the deduction that

$$\frac{\nu(0 \setminus \bar{0})}{\nu(0)} \le \text{const.}\ (k-1)^{4b} ,$$

in fact the Lemma 15 allows us to estimate the above ratio in terms of $r(\gamma_r(0 \setminus \bar{0}))/r(\gamma_r(0))$.

Consider now the irregular small components. They will sometimes be kept and sometimes be discarded.

To decide whether to keep them or not we define $\kappa(0)$ to be the number of times $\tilde{T}^{-1}$ has to be applied to an irregular small component 0 before seeing it contained as a part of a big component $0_{k-1-\kappa(0),m_1,\ell_1}$ cut out of some $\tilde{G}_{k-1-\kappa(0),m_1}$.

Define also n(k) so that $(\frac{\lambda_2}{2})^{n(k)} = k^3$. Then if $\kappa(0) \geq n(k)$ we declare 0 bad. If $\kappa(0) < n(k)$ we declare it bad if also

$$\frac{r(\gamma_r(\tilde{T}^{-\kappa(0)}\ 0_{k-1,m,\ell}))}{r(\gamma_r(0_{k-\kappa(0)-1,m_1,\ell_1}))} < k^{-3}$$

Notice that

(1) If $0 = 0_{k-1,m,\ell}$ is irregular and if $\tilde{T}^{-\kappa(0)}\ 0 \subset 0_{k-\kappa(0)-1,m_1,\ell_1}$, then all the sets $\tilde{T}^{-i}\ 0$, $0 \leq i < \kappa(0)$, are contained in irregular components and, therefore, the component $0_{k-\kappa(0)-1,m_1,\ell_1}$ can contain $\tilde{T}^{-\kappa(0)}\ 0$ only for $2^{\kappa(0)}$ different 0's with given value of $\kappa(0)$.

(2) If $0_{k-1,m,} = 0$ is irregular and small it follows that

$$r(\tilde{T}^{-\kappa(0)}\ \gamma_r(0)) \leq \lambda_2^{-\kappa(0)} r(\gamma_r(0)) \leq \lambda_2^{-\kappa(0)}\ \delta_0(k-1)^{-4b}$$
$$\leq \lambda_2^{-\kappa(0)}\ r(\gamma_r(0_{k-\kappa(0)-1,m_1,\ell_1}));$$

hence, if $\hat{0} = 0_{k-\kappa(0)-1,m_1,\ell_1}$, the Lemma 15 below shows

$$\frac{\nu(0)}{\nu(\hat{0})} \leq \lambda_2^{-\kappa(0)}\ \text{const.}$$

(3) If $0_{k-1,m,\ell}$ is irregular but not yet bad it follows that

$$r(\gamma_r(0_{k-1,m,\ell})) \geq \lambda_2^{\kappa(0)}\ r(\tilde{T}^{-\kappa(0)}\ \gamma_r(0_{k-1,m,\ell}))$$
$$\geq \frac{\lambda_2^{\kappa(0)}}{k^3}\ r(\gamma_r(0_{k-1-\kappa(0),m_1,\ell_1})) \geq \delta_0(k-1-\kappa(0))^{-4b}\ k^{-3}$$

so that

$$r(\gamma_r(0_{k-1,m,\ell})) \geq \delta_0(k-1)^{-5b}$$

(if k_0, b had been chosen large enough).

We continue our construction by looking at those irregular small components which are not yet bad according to the criteria so far exposed. Let 0 be such a component: We apply to it the same treatment applied to the big components and construct out of 0 $0'$ and then $\bar{0}$ (if $T0'$ is not such that 0 itself has to be declared bad, see case of the big components). If 0 is good the same arguments used for the case of big 0's allows us to say, via simple considerations based on Lemma 15 and Corollary 16 below, that

$$\frac{\nu(0 \setminus \bar{0})}{\nu(0)} \leq \text{const.}\ (k-1)^{-3b}.$$

The sets $\tilde{T}\bar{0} = \tilde{G}_{k,s}$ are added to the similar sets obtained from the big components and the union of this family of sets is $\tilde{G}_k = \tilde{T}^k \hat{G}_k$.

Remarking that $\nu(\mathcal{V}_{k,3/2}) \leq \text{const.}\ \delta_0^2/k^3$ and collecting all the above estimates we realize that the amount of points in $G \setminus \hat{G}_k$ has a measure not exceeding

$$\text{const.}\ \delta_0^2 \sum_{k \geq k_0} K^{-3}.$$

Thus Lemma 13 is proved, at least as far as the first two items are concerned. The proof of the third item is briefly commented upon after Lemma 15 below.

The measures of the little pieces of G discarded in the inductive construction have been estimated above by the following lemma which, essentially, tells us that, in the natural coordinates, a connected subset $\bar{G} \subset \tilde{G}_{k,m}$ consisting of expanding fibers of kL-order and embedded with the measure ν can be treated as a rectangle with ordinary Lebesgue measure in the natural coordinates (α,β), provided its side is not too long: so that the ratio of the ν-measures of two strips 0 and $\tilde{0}$ in $\bar{G}$ is close to the ratio of the r-lengths of their sides. More precisely,

15 Lemma. Given $C > 0$, $\delta_0 > 0$ let G_0 be a quadrilateral such that all the increasing lines which join its two decreasing sides $\gamma_\ell(G_0)$, $\gamma_r(G_0)$ have r-length $\geq C\,\delta_0$ but $\leq 4CZ\delta_0$. Assume

also that the slope of $\gamma_\ell(G_0)$ is between $-(R_-^{-1}+\tau_0^{-1})$ and $-R_+^{-1}$.

Let k_0 be such that $\lambda_1^{-k_0}$ 4 C Z $\leq \lambda_2^{-k_0}$, where λ_1,λ_2 are the same defined at the beginning of this lecture.

Suppose that for some $k \geq k_0$ the sets $\tilde{T}G_0,\ldots,\tilde{T}^kG_0$ are quadrilaterals and $\tilde{T}^kG_0$ consists of 0-expanding fibers (i.e. the 0-expanding fiber through every point of $\tilde{T}^kG_0$ joins the opposite decreasing sides of $\tilde{T}^kG_0$ without crossing the other two sides) and:

$$r(\gamma_r(\tilde{T}^hG_0)) < \delta_0(1+h)^{-b}, \qquad h = 0,1,\ldots,k$$

$$-\cos\phi|_{T^hG_0} \geq \delta_0(1+\tfrac{h}{L})^{-4}, \qquad h = 0,1,\ldots,kL$$

Then we introduce on $\tilde{T}^hG_0$ the natural coordinates (α,β) the Radon-Nykodim derivative of the Lebesgue measure $d\alpha\, d\beta$ on $\tilde{T}^hG_0$ with respect to the ν-measure is constant within $\exp \pm \theta(c)/(1+h)^{6-4}$, $\forall h \leq k$, where $\theta(\cdot)$ is a suitable function.

Proof: See Appendix B.

16 Corollary. In particular if $0_1 \supseteq 0_2$ are two connected subsets of $\tilde{T}^hG_0$, $h = 0,\ldots,k-1$, which are unions of $(k-h)L$-expanding fibers the ratio $\nu(0_1)/\nu(0_2)$ is bounded by const. $r(\gamma_r(0_1))/r(\gamma_r(0_2))$ if $r(\gamma_r(0_2)) \geq \delta_0(1+h)^{-5b}$.

The proof of the above lemma is an easy repetition of the absolute continuity proof sketched in Section 3): one has to follow the same chain of inequalities taking some care in not being too generous in the estimates and making full use of the strong convergence properties of the continued fraction which appears in the game. This proof yields also, implicitly, the third part of the main lemma.

The proof of Lemma 15 also yields, as a byproduct, item (iii) of the main lemma (actually this part can be proved directly and is much simpler).

In the Appendix B, Lemma 15 is derived in detail.

7. More About Ergodicity

Let $f \in C(M)$. Let $\bar{f}^{+}, \bar{f}^{-}$ be the pointwise future and past averages of f. Let

$$M_f = \{x \mid \bar{f}^{+}(x) = \bar{f}^{-}(x)\}$$

$$W^f_{k,\rho} = \{x \mid x \in M,\ \ h^{-1} \sum_{i=0}^{h-1} f(T^{\pm i}x) - \bar{f}^{\pm}(x)| < \rho,\ \forall h \geq k\}.$$

Theorem 13 allows us to prove the following lemma.

17 Lemma. Let the trajectory $T^i x_0$, $i = 0, \pm 1, \ldots$ be never tangent or once tangent.

There is a neighborhood U of x_0 such that if $f \in C(M)$ $\exists$ a subset $V_f \subset U$ with full measure with the property that if $x, y \in V_f$ and $\rho > 0$ then one can find $k(x,y)$ and $s(x,y,f,\rho)$ such such that for all $s > s(x,y,f,\rho)$ there is a sequence of alternatingly ℓ-expanding and ℓ-contracting fibers $\gamma_1, \ldots, \gamma_k$ with $k \leq k(x,y)$ and $x \in \gamma_1$, $g \in \gamma_k$, $\gamma_i \cap \gamma_{i+1} \in W^f_{s,\rho}$ where ℓ can be arbitrarily chosen.

This lemma implies the

18 Corollary. The transformation T is ergodic on $\bigcup_{n=-\infty}^{+\infty} T^n U$, hence it is ergodic. In fact the lemma implies that the function $\bar{f}^{+}(x) = \bar{f}^{-}(x) =$ constant on V_f i.e. almost everywhere in U. This implies global ergodicity through the already mentioned covering argument.

The proof of Lemma 17 is quite simple in principle but very messy in its details. The main reason for the origin of complications is the fact that our fibers are only approximate. Of course they can be as approximated as one wishes and their absolute continuity is in some sense "uniform" and this allows us to overcome the problems that arise.

To give an idea of the proof we assume that the natural version for $K = +\infty$, i.e. for exact fibers, of Lemma 13 is true, and also that for almost all points $x \in M_q$ $(q = 1,2,\ldots)$, the fibers $\gamma_{c,e}(x)$ intersect in a set of full arc-length a fixed set of full measure . Then we show how the analogous version of Lemma 17 would arise.

This argument is interesting because it shows the path to follow in our more complicated case in which only approximate fibers are available.

Let $f \in C(M)$ and $U = U(\varepsilon, C)$ where the value of ε, C will appear later and let x,y be two points in $M_f \cap \left(\bigsqcup_{q=1}^{\infty} (M_q \setminus M_{q-1})\right) \cap U$ such that $\gamma_c(x)$, $\gamma_e(y)$ have full intersection, with respect to their arc lengths, with M_f. Let $x \in M_{q_x} \setminus M_{q_x-1}$ and $y \in M_{q_y} \setminus M_{q_y-1}$.

Draw through x the lower segment γ_1 of $\gamma_c(x)$ (which extends

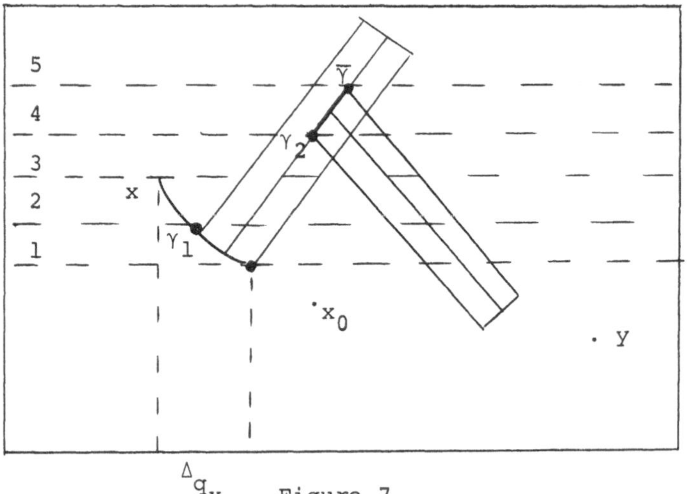

Figure 7

on the r-axis by an r-length Δ_{q_x}, if $x \in M_{q_x} \setminus M_{q_x-1}$). Let γ_1' be its intersection with the lowest strip of the four strips with equal width drawn in Figure 7 (i.e. with region 1).

We apply to it the main lemma and find an expanding line $\bar{\gamma}$ which extends beyond region 4 and has full intersection with M_f and intersects γ_1 in a point of M_f (remember that the slopes of the fibers are uniformly bounded away from zero and ∞ and C can be suitably chosen (say, C = 10 Z would be large enough)). Let $\bar{\gamma}_2$ = intersection of $\bar{\gamma}$ with the region (1 ∪ 2 ∪ 3 ∪ 4) in Figure 7. We call $\gamma_2' = \gamma_2 \cap$ (region 4) and apply again the main lemma to γ_2', etc.

A similar construction is made starting from y. Clearly two sequences of alternatingly expanding and contracting curves with successive intersections in M_f and containing at most const. $(\Delta_{q_x}^{-1} + \Delta_{q_y}^{-1})$ lines can be constructed using the above procedure. Let γ_ℓ and $\tilde{\gamma}_m$ be the first two fibers of the strings starting in x and y, respectively, such that $\gamma_\ell \cap \tilde{\gamma}_m = z \neq \emptyset$. Let z be a point (if it is an arc the argument is much simpler).

We consider two little segments δ_1, δ_2 of equal r-arc length having z as common vertex and contained in γ_ℓ, $\tilde{\gamma}_m$ respectively. We construct two quadrilaterals G and G' by applying the main lemma to δ_1 and δ_2 (with $Z^{-1} = {}_0$ (say) replacing C and the same ε) .

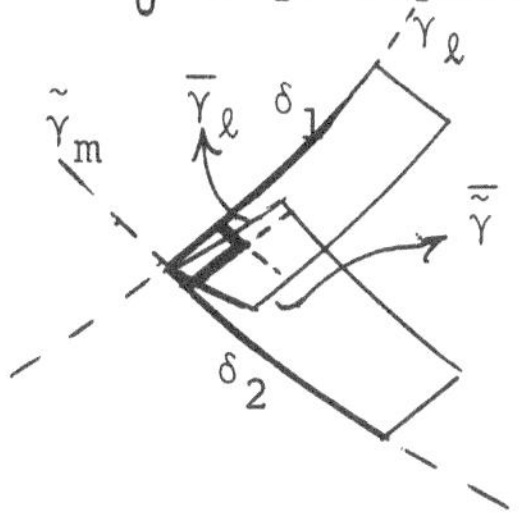

Then, if ε had been conveniently chosen, close to zero, the sets $\hat{G}$, $\hat{G}'$ have an intersection $\hat{G} \cap \hat{G}'$ such that $\nu(\hat{G} \cap \hat{G}') > 0$: we can assume that $\hat{G}$ and $\hat{G}'$ are constituted by fibers with full intersection with M_f and intersecting δ_1 and δ_2 (respectively) in M_f: in this way we can construct two expanding contracting segments $\bar{\gamma}_\ell$ and $\bar{\tilde{\gamma}}_m$ such that $\bar{\gamma}_\ell \cap \gamma_\ell$, $\bar{\gamma}_\ell \cap \bar{\tilde{\gamma}}_m$, $\bar{\tilde{\gamma}}_m \cap \tilde{\gamma}_m$ are in M_f. This completes the construction if x_0 is never tangent.

The case when x_0 is once tangent can be treated in a similar way: the reason the proof does not fail is linked to the fact that the main lemma requires only that $T^i x_0$ is never tangent or $T^{-i}x_0$ is never tangent if $i \geq 0$: so one has to be careful in constructing the string of curves in such a way that when the string has to cross the singularity line through x_0 the segment that does the crossing is the one whose existence can be guaranteed by the lemma.

The proof of Lemma 17 which is of interest to us, just follows the above ideas: it is of course complicated by the fact that we have to deal with approximate fibers and therefore the role of M_f has to be played by $W^f_{k,\rho}$ which does not have full measure. See Appendix C for a complete proof.

8. Concluding Remarks

To go beyond ergodicity one can proceed along the general lines discussed in Weiss' lectures. The easiest way would be to prove the validity of the $k = \infty$ version of Theorem 9. Such a result would provide us with a set V of positive measure which is foliated by germs of contracting and expanding fibers which on V are absolutely continuous. Then a smooth partition of P, thought of as a parition of $\hat{M} = \bigcup_{i=1}^{\infty} T^i V$ would have a tail of σ-algebra whose sets are trivial and, therefore, T restricted to $\hat{M}$ is a K-system. Ergodicity implies that $\nu(\hat{M}) = 1$, so (M,T,ν) is a K-system. This, in turn, implies that (M,T,ν) is isomorphic to a Bernoulli scheme.

The version for $K = \infty$ of Theorem 9 (and of Lemma 13) does not involve too much work after the proof for $K < \infty$. The details of this proof are, however, missing in the literature. This is the reason why we have not relied on such a result in these lectures.

The mixing character of (M,T,ν) insures that (V,S_t,μ) is an ergodic flow (in the sense that there is no S_t-invariant $(\forall t \in (-\infty,+\infty))$ nontrivial set in V).

To prove that (V,S_t,μ) is a Bernoulli (or a K-) flow one has to construct foliations of expansion and contraction in (V,S_t,μ) and prove their absolute continuity and their "nonintegrability" in the sense discussed in [4] or in Weiss' lectures.

The existence and absolute continuity is easily obtained without much extra work while the nonintegrability does not seem to have been derived in detail in the published literature.

References

[1] Sinai, J., Russian Math. Surv. 25 (1970) 132.

[2] Bunimovitch, L., J. Sinai, Mat. Schor. 90, 416, 197.

[3] Gallavotti, G., and Ornstein, D., Comm. Math. Phys. 38 (1974) 83.

[4] Sinai, J., Isvestia Math. Nauk 30 (1966) 15.

Acknowledgements

I wish to express my gratitude to D. Ornstein for his substantial help in the understanding of many points of the billiards theory: without his advice and encouragement, I could not have given these lectures.

I have also taken advantage of discussions with L. Bunimowitch and J. Sinai whom I wish to warmly thank.

Appendix A. Proof of Theorems 5, and 9, and Lemma 7.

(i) If $\bar{\gamma}$ is smooth and has equation $\phi = \bar{\phi}(r)$ and $d\bar{\phi}/dr \leq 0$ and $T, T^2, \ldots, T^k$ are smooth on $\bar{\gamma}$, then a repeated application of the formulas expressing the equation of the curve in terms of its T-image yield

$$\frac{1}{-\cos\bar{\phi}(r)} \frac{d\bar{\phi}}{dr}(x) = f^{(k)}\left(\frac{2\,K(x)}{\cos\phi(x)}, \tau(T^{-1}x), \ldots, \tau(T^{-k}x), \left(-\frac{d\bar{\phi}}{dr}\bigg|_{T^{-k}x} + K(T^{-k}x)\quad\frac{1}{\cos\phi(T^{-k}\,x)}\right)\right)$$

where $d\bar{\phi}/dr|_{T^{-k}x}$ abbreviates the r-derivative of the equation $T^{-k}\gamma$ computed at $r(T^{-k}x)$ and

$$f^{(k)}(a_1, b_1, \ldots, b_{h-1}, a_h) = \frac{a_1}{2} + \cfrac{1}{b_1 + \cfrac{1}{a_2 + \ddots \cfrac{1}{b_{h-1} + \cfrac{1}{a_h}}}}$$

(i) The only property of $f^{(h)}$ that we shall really need is that if h is large, then $f^{(h)}$ depends very little on the variables with large index provided the entries of the continued fraction are not too small; more precisely, we shall need the following statement: if $0 < \sigma < \min\limits_{1\leq i\leq k-1} |b_i\ a_{i+1}|$ and $0 < \sigma < \min\limits_{1\leq i\leq k} |a_i\ b_i|$ and if the entries a_i, b_i all have the same sign there is a constant Q such that

$$\left|\frac{\partial \log f^{(k)}}{\partial a_i}\right| \leq \frac{Q}{(1+\sigma)^{2i}} \frac{1}{a_i}, \qquad i = 1,2,\ldots,k$$

$$\left|\frac{\partial \log f^{(k)}}{\partial b_i}\right| \leq \frac{Q}{(1+\sigma)^{2i}} \frac{1}{b_i}, \qquad i = 1,2,\ldots k-1$$

(Q could actually be chosen as $2(1+\sigma^2)$). In the case a_i, b_i have the values in (i) above the parameter σ could be $\tau_0 R_+^{-1}$.

(iii) From simple trigonometric considerations it is possible to see that $\exists\, c > 0$ such that

$$- \cos \phi(x) \leq C\, d(x)$$

$$- \cos \phi(T^{\pm 1} x) \geq C\, d(x)$$

$$|\tau(x)| \leq \frac{C}{d(x)}$$

(iv) The proof of Theorem 5 proceeds as follows. Let $x \in M_q^{(k)+}$ and put $d_i = c_i = q^{-1}(1+i^2)^{-1}$. Let $\tilde{\gamma}_c^{(k)}(x) = T^k \tilde{\gamma}_c (T^{-k} x)$. The set $\tilde{\gamma}_c^{(k)}(x)$ consists of a union of smooth curves. Let $\approx\!\gamma^{(k)}$ be the smooth curve which contains x. Let $\Delta > 0$ be so small that the curve $\approx\!\gamma^{(k)}(x)$ is at least above all the points of the interval $I = [r(x)-\Delta,\ r(x)+\Delta]$. Call $\tilde{c}_h = \min_{y \in I} |\cos \phi(T^{-h} y)|$ and $\tilde{d}_h = \min_{y \in I} d(T^{-h} y)$. The formula for the contraction coefficient $p(T\, d\gamma) / p(d\gamma)$ in Section 2 implies (see the remarks in section 2):

$$p(T^{-h} \approx\!\gamma^{(k)}) \leq \lambda^{-h} \quad (\approx\!\gamma^{(k)}) \ , \qquad h = 0,\ldots,k,$$

where $\lambda > (1+\tau_0/R_+) > 1$ (note that we are using that, by construction, the curves $T^{-h} \approx\!\gamma^{(k)}$ are decreasing $\forall h = 0,\ldots,k$).

The a priori bound on the slope of $T^{-h} \approx\!\gamma^{(k)}$ imply that the Euclidean length $|T^{-h} \approx\!\gamma^{(k)}|$ is related to $p(T^{-h} \approx\!\gamma^{(k)})$ by (if $\phi(r)$ is the equation of $T^{-h} \approx\!\gamma^{(k)}(x)$):

$$|T^{-h} \approx\!\gamma^{(k)}| = \int_{T^{-h} \approx\gamma^{(k)}} dr \sqrt{1+\left(\frac{d\phi}{dr}\right)^2} \leq C' \int_{T^{-h'} \approx\gamma^{(k)}} dr \leq \frac{c'}{c_h}\, p(T^{-h} \approx\!\gamma^{(k)}) \ ,$$

for some suitable C'.

Hence

$$\tilde{d}_{h+1} \geq d_{h+1} - |T^{-h-1} \approx\!\gamma^{(k)}| \geq d_{h+1} - \frac{2\, C'}{\tilde{c}_{h+1}} \lambda^{-(h+1)} \Delta \ ,$$

and (iii) above implies for suitable " > 0:

$$\tilde{d}_{h+1} \geq d_{h+1} - \frac{c''}{c_h} \lambda^{-h-1} \Delta.$$

Similarly for some c" > 0,

$$\tilde{c}_{h+1} \geq c_{h+1} - \int_{T^{h-1}\tilde{\tilde{\gamma}}(k)} |\sin \phi| |\frac{d\phi}{dr}| dr \geq c_{h+1} - \frac{c''}{\tilde{c}_h} \lambda^{-h-1} \Delta;$$

from the last two formulas it is easy to infer by induction that there is $\Delta_q > 0$ such that if $\Delta < \Delta_q$ then $\tilde{c}_h > c_h/2$, $\tilde{d}_h > d_h/2$. Clearly this implies Theorem 5.

(v) Suppose that $k' > k \geq 1$ and consider the two curves $\gamma_c^{(k)}(x)$ and $\gamma_c^{(k')}(x)$; they are two decreasing curves stemming from x and defined "above" the interval $[r(x)-\Delta_q, r(x)+\Delta_q] = J_x$.

Let $\bar{r} \in J_x$ and define the vertical segment $\gamma_{\bar{r}} = \{r, \phi \mid r = \bar{r},\ \phi$ is between $\phi_c^{(k)}(\bar{r})$ and $\phi_c^{(k')}(\bar{r})\}$: this segment can be regarded as an increasing curve; so that $T^{-k}\gamma_{\bar{r}}$ is increasing and with slope between R_+^{-1} and $R_-^{-1} + \tau_0^{-1}$. Furthermore $T^{-k}\gamma_{\bar{r}}$ is connected: otherwise, if $\bar{h} < k$ is the maximum h for which $T^{-h}\gamma_{\bar{r}}$ is connected, there would exist a singularity line for T^{-1} which cuts $T^{-\bar{h}}\gamma_{\bar{r}}$. This would imply that there is a singularity line which crosses either $T^{-\bar{h}}\gamma_c^{(k)}$ or $T^{-\bar{h}}\gamma_c^{(k')}$ which is impossible impossible because $\tilde{d}_h \geq (2q)^{-1}(1+h^2)$, $\forall h = 0,\dots,k$ on such curves.

The r-length of $T^k\gamma_{\bar{r}}$ has an upper bound $B = \max_{i \leq i \leq s} L_i$. (Where L_i = length $\partial 0_i$); therefore

$$r(T\gamma_{\bar{r}}) \leq \frac{\lambda_0^{-(k-2)}}{\lambda_0 - 1} B$$

which easily implies that there is a geometric constant B' such that

$$|\gamma_{\bar{r}}| = |\phi^{(k)}(\bar{r}) - \phi^{(k')}(\bar{r})| \leq \lambda_0^{-k} B' \qquad \forall k' \geq k.$$

This proves (v) of Lemma 7.

(vi) The statements (i), (ii) and (iii) of Lemma 7 are either already implicitly proved or are a simple consequence of the exponential contraction of the length of $T^{-h}\gamma_c^{(k)}(x)$ as a function of h and the remark that on $T^{-h}\gamma_c^{(k)}(x)$ the quantities c_h, d_h are bounded below by $\frac{1}{2}c_h$ and $\frac{1}{2}d_h$ respectively. Statement (iv) of Lemma 7 is also a simple consequence of (i), (ii) and (iii) in Lemma 7 and of the continued fraction formula for the slope $d\phi^{(h)}/dr$ of $T^{-h}\gamma_c^{(k)}(x)$ and of the convergence property of the continued function discussed in (i), (ii) of this appendix.

(vii) Theorem 9 for the case of $\hat{U}$ now easily follows from Lemma 7, Corollary 8 along the lines discussed in Section(iv) from the explicit expressions for $\rho^{(k)}(\alpha,\beta)$. The cases of $\hat{U}'$, $\hat{U}''$ can be treated along the same lines as $\hat{U}$: i.e. it is easy to find for $\rho^{(k)}(\alpha,\beta)$ an explicit expression which has essentially the same structure as the expression for $\rho^{(k)}$ in the case of $\hat{U}$.

We do not give the details of this part of the proof, not only because of the similarity between the proof for the case of $\hat{U}$ and the case of $\hat{U}'$, $\hat{U}''$ but also because in Appendix B a detailed proof of a more difficult similar result is given.

The expression for $\rho^{(k)}$ in the cases $\hat{U}'$, $\hat{U}''$ can be easily found with the help of an auxiliary smooth foliation ξ which contains $\gamma_c^{(k)}(x)$ or $\gamma_e^{(k)}(x)$ respectively: the formal expression for $\rho^{(k)}$ in this case is the same found in Appendix B for the quantity that, there, is called $\rho^{(k)}$.

The proofs of this appendix follow ref. [3], [1].

Appendix B. Proof of Lemma 15

Let ξ be the smooth foliation of G_0 with lines parallel to $\gamma_\ell(G_0)$ along $\gamma_u(G_0)$ (if $\gamma_\ell(G_0)$ is too short we continue it a little keeping its slope smooth and still between R_+^{-1} and $(R_-^{-1} + \tau_0^{-1})$).

If $\gamma \in \xi$ we denote $\phi_\xi^h(r)$ the equation of $T^h\gamma$. If $z \in G_0$ we denote $(d\phi_\xi^h/dr)(T^h z)$ as the slope at $T^h z$ of the curve of $T^h\xi$ which contains $T^h z$. Our assumption on the slope of $\gamma_\ell(G_0)$ allows us to deduce that

$$Z^{-1} \leq \frac{\frac{d\phi_\xi^h}{dr}(T^h z)}{\frac{d\phi_\xi^h}{dr}(T^h z')} \leq Z\ , \qquad \forall z, z' \in G_0\ ;\ \forall h = 0,\ldots,k,$$

where, we remember, $Z = R_+(R_-^{-1} + \tau_0^{-1})$.

Assume, for simplicity, that $L = 1$. Then the sides $\gamma_r(G_0)$, $\gamma_\ell(G_0) \in \xi$ and $\gamma_u(G_0)$, $\gamma_d(G_0)$ are contained in k-expanding fibers (by assumption).

If $z \in G_0$ we define the intrinsic coordinates of z to be the r-arc lengths (α,β) of the arcs xz and $\bar{z}z$ along, respectively, $\gamma_\ell(G_0)$ and the k-expanding fiber through z.

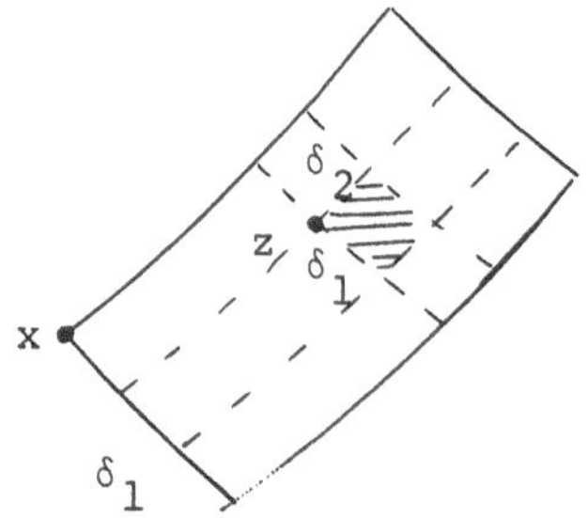

Figure 8

Let E be the set dashed in Figure 8 where the decreasing dotted lines represent close elements of ξ.

Let $|\cdot|$ denote the Euclidean length of an arc of a curve; then

$$r(\delta_1) = \frac{|\delta_1|}{\left[1 + \left(\frac{d\phi_\xi}{dr}(z)\right)^2\right]^{1/2}}, \quad z(\delta_2) = \frac{|\delta_2|}{\left[1 + \left(\frac{d\phi_e}{dr}(z)\right)^2\right]^{1/2}};$$

and the angle ω between δ_1 and δ_2 is

$$\sin\omega = \frac{(d\phi_e/dr)(z) - (d\phi_\xi/dr)(z)}{\left[1 + \left(\frac{d\phi_\xi}{dr}(z)\right)^2\right]^{1/2} \left[1 + \left(\frac{d\phi_e}{dr}(z)\right)^2\right]^{1/2}}$$

so that (remembering that, by definition, $-\cos\phi\, r(\delta) = p(\delta)$ for an infinitesimal arc $\delta \ni (r,\phi)$):

$$\rho^{(k)}(\alpha,\beta) = \frac{\nu(E)}{r(\bar{\delta}_1)\, r(\delta_2)} = -\frac{|\delta_1|\, |\delta_2|}{r(\bar{\delta}_1)\, r(\delta_2)} \sin\omega \cos\phi(z)$$

$$= -\cos\phi(\bar{z}) \frac{p(\delta_1)}{p(\bar{\delta}_1)} \left(\frac{d\phi_e}{dr}(z) - \frac{d\phi_\xi}{dr}(z)\right).$$

An identical construction can be performed on $T^h G_0$ around $T^h z$ using the system of intrinsic coordinates similar to the one used for G_0: i.e. defining the α-coordinate as the r-length of the image under T^h of the arc $\widehat{x\bar{z}}$ along $T^h\gamma_\ell(G_0)$ and the β-coordinate as the r-length of the image (under T^h) of the arc $\widehat{z\bar{z}}$ on the k-expanding fiber through z.

We find that, in this case, the Radon-Mykodim derivative of ν with respect to $d\alpha\, d\beta$ can be written as

$$(*) \qquad -\cos\phi(T^h \bar{z}) \frac{p(\delta_1)}{p(\bar{\delta}_1)} \left\{\frac{d\phi_e^h}{dr}(T^h z) - \frac{d\phi_\xi^h}{dr}(T^h z)\right\},$$

where δ_1 and $\bar{\delta}_1$ are infinitesimal arcs on the fibers of the foliation through, respectively, $T^h x$ and $T^h z$; the relation between δ_1 and $\bar{\delta}_1$ being that they are cut, on the fibers of $T^h\xi$ on which they lie, by the same two (k-h)-expanding fibers (as in Figure 8). We observe that the above infinitesimal arguments make sense beca because of the smoothness of the foliation $T^h\xi$ and of the (k-h)-expanding foliation.

We observe next that

$$-\frac{1}{\cos\phi(T^h z)}\frac{d\phi_\xi^h}{dr}(z) = \frac{k(T^h z)}{\cos\phi(T^h z)} + \cfrac{1}{\tau(T^{h-1}z) + \cfrac{1}{\frac{2k(T^{h-1}z)}{\cos\phi(T^{h-1}z)} + \dots \cfrac{1}{\tau(z) + \cfrac{1}{\frac{1}{-\cos\phi(z)}\left(\frac{d\phi_\xi}{dr}(z) - K(r)\right)}}}}$$

$$\frac{1}{\cos\phi(T^h z)}\frac{d\phi_e^h}{dr}(z) = \frac{k(T^h z)}{\cos\phi(T^h z)} + \cfrac{1}{\tau(T^h z) + \cfrac{1}{\frac{2k(T^{h+1}z)}{\cos\phi(T^{h+1}z)} + \dots \cfrac{1}{\tau(T^{k-1}z) + \cfrac{1}{\frac{2k(T^k z)}{\cos\phi(T^k z)}}}}}$$

We now consider the various elements that enter into the expression (*) and study how much they vary inside $T^h G_0$.

Clearly

$$\left|\frac{\cos\phi(T^h z)}{\cos\phi(T^h z')} - 1\right| \le \text{const.}\,(1+h^{b-4})^{-1}, \quad \forall z,z' \in G_0,\ h=0,\dots,k;$$

$$\left|\frac{K(T^h z)}{K(T^h z')} - 1\right| \le \text{const.}\,(1+h^b)^{-1}, \quad \forall z,z' \in G_0,\ h=0,\dots,k;$$

where the const. depends upon C but not on δ_0; if we were only interested in $k \geq k_0$ then the constant could be chosen C-independent (for the definition of k_0 see Section 5).

It is also not difficult to see that

$$\left|\frac{\tau(T^h z)}{\tau(T^h z')} - 1\right| \leq \text{const. } (1+h^{b-4})^{-1}, \quad \forall z,z' \in G_0,\ h=0,\dots,k-1.$$

This, in fact, immediately follows from the easy trigonometric estimates

$$\left|\frac{\partial \log \tau(x)}{\partial r}\right| \leq -\frac{\text{const.}}{\cos \phi(Tx)} \qquad \forall x \in M ;$$

$$\left|\frac{\partial \log \tau(x)}{\partial \phi}\right| \leq -\frac{\text{const.}}{\cos \phi(Tx)} \qquad \forall x \in M .$$

The above estimates on the maximum variations of the quantities τ, $\cos \phi$, k on $T^h G_0$ will be considered with the estimate of the logarithmic derivatives of the continued fraction expressing $d\phi^h/dr$ or $d\phi^h_e/dr$ (see above) considered in Appendix A.

We then easily obtain, using the continued fraction expressions for $d\phi^h_\xi/dr$ and $d\phi^h_e/dr$ that $h = 0,1,\dots,k$,

$$\left|\frac{\dfrac{1}{-\cos \phi(T^h z)} \dfrac{d\phi^h_\xi}{dr}(z)}{\dfrac{1}{-\cos \phi(T^h z')} \dfrac{d\phi^h_\xi}{dr}(z')} - 1\right| \leq \text{const. } \sum_{i=0}^{h} (1+\sigma)^{-2i}(1+(h-i)^{b-4})^{-1}$$

$$\leq \text{const.}/(1+h^{b-4}) ;$$

$$\left|\frac{\dfrac{1}{\cos \phi(T^h z)} \dfrac{d\phi^h_e}{dr}(z)}{\dfrac{1}{\cos \phi(T^h z')} \dfrac{d\phi^h_e}{dr}(z')} - 1\right| \leq \text{const. } \sum_{i=0}^{h} (1+\sigma)^{-2i}(1+(h+i)^{b-4})^{-1}$$

$$\leq \text{const.}/(1+h^{b-4}) ;$$

where σ can be taken τ_0/R_+ and the constant depends on C (but not on δ_0). It remains, therefore, to estimate the variation on T^hG_0 of $p(\delta_1)/p(\bar{\delta}_1)$.

As already remarked in the lectures

$$\frac{p(\delta_1)}{p(\bar{\delta}_1)} = \left\{ \prod_{i=0}^{k-h-1} \frac{\dfrac{p(T^{h+i}\ \delta_1)}{p(T^{h+i+1}\ \delta_1)}}{\dfrac{p(T^{h+i}\ \bar{\delta}_1)}{p(T^{h+i+1}\ \bar{\delta}_1)}} \right\} \frac{p(T^k\ \delta_1)}{p(T^k\ \bar{\delta}_1)}$$

and, assuming for simplicity $k(r) = R_e^{-1}$ = const. on the obstacle $\partial 0_\ell$,

$$= \left\{ \prod_{i=0}^{k-h-1} \frac{1 + \left(K(T^{h+1}\bar{z}) - \dfrac{d\phi_\xi^{h+i}}{dr}(T^{h+i}\ \bar{z})\right)\dfrac{\tau(T^{h+i}\ \bar{z})}{\cos\ \phi(T^{h+i}\ \bar{z})}}{1 + \left(K(T^{h+i}z) - \dfrac{d\phi_\xi^{h+i}}{dr}(T^{h+i}\ z)\right)\dfrac{\tau(T^{h+i}\ z)}{\cos\ \phi(T^{h+i}\ z)}} \right\}$$

$$\cdot \frac{R^{-1} + \dfrac{d\phi_\xi^{(k)}}{dr}(T^k\ z)}{R^{-1} + \dfrac{d\phi_\xi^{(k)}}{dr}(T^k\ \bar{z})}$$

where R^{-1} is the curvature of the obstacle on which T^kz, $T^k\bar{z}$ collide.

The just derived estimates on the variations of K, τ, cos ϕ, and $d\phi_\xi^h/dr$ on T^hG_0 allow us to immediately conc ude that the quantity appearing in the r.h.s. of the last equation differs from 1 by at most const.$\cdot(1+h^{b-4})^{-1}$, $h = 0,1,\ldots,k$ and the constant is δ_0-independent.

This completes the proof of the lemma when L = 1 and the obstacles are circular: the general case is a simple adaptation of the above arguments and we do not insist on the few obvious changes of symbols and of the numerical values of the constants (remember that L does not depend on C, δ_0 etc. but only on geometrical constants).

We notice that the above proof is essentially a repetation of the proof of the absolute continuity.

Appendix C. Proof of Lemma 17.

Let $C = 5R_+(R_-^{-1} + \tau_0^{-1}) = 5Z$. Let ε be a certain geometrical constant whose choice will be clear later. Consider the neighborhood $U = U(\varepsilon, C)$ of x_0.

Let $\frac{1}{2} > \eta > 0$ be a fixed positive number. Determine $q = q_\eta$ so that

$$\nu(U \setminus U \cap M_{q/4}) < \eta^2 \nu(U) .$$

Let (say) $\xi = 10^{-1} \arctan R_+^{1}$, and let $x, y \in M_{q/4} \cap U$ such that y is in the horizontal cone with vertex x and opening angle α such that $\tan \alpha = \xi$. Assume also that x,y are so close that it makes sense to speak of the quadrilateral G obtained by drawing through x,y the expanding and contracting germs. We also assume that the diameter of G is so small that Theorem 9 applies in a neighborhood of x containing G and, furthermore, the function $\theta_q(D(G))$ defined by Theorem 9 is such that

$$|\exp \pm \theta_q(D(G)) - 1| < \eta^2 .$$

Let Σ be the family of the (closed) quadrilaterals constructed as above. The family Σ covers the Lebesgue set of $M_{q/4} \cap U$ in the sense of Vitali.

Let $\{G_i\}_0^\infty$ be a sequence of pairwise disjoint sets $G_i \in \Sigma$, $i = 0,1,2,\ldots$, so that $G_i \subset U$, $\bigcup_i G_i \supset M_{q/4} \cap U$ (mod 0).

Then $\nu\left(\bigsqcup_{i=0}^{\infty} G_i \setminus M_{q/4} \cap U\right) < \eta^2 \nu(U)$.

Let I be the set of indices $i = 0,1,\ldots$ such that

(a) $\qquad \nu(G_i \cap (M_{q/4} \cap U)) > (1-\eta)\, \nu(G_i) , \quad i \in I$

then

(b) $\qquad \sum_{i \notin I} \nu(G_i) < \eta\, \nu(U) .$

We can choose a set J, finite and still satisfying (a), (b) above when J replaces I and, by relabeling, we can put $J = (1,2,\ldots,N)$.

Let $2d_\eta > 0$ be the smallest between the r-lengths of arcs joining the opposite sides of $G_0,\ldots,G_N$ and having slope bounded, in modulus, between R_+^{-1} and $(R_-^{-1}+\tau_0^{-1})$. Let $c_0(\eta)$ be the smallest between the measures $\nu(G_i)$, $i \in J$.

Let K_η be an integer such that for all $k \geq K_\eta$ the quadrilaterals $G_i^{(k)}$ obtained by drawing hrough x_i, y_i the k-germs of expanding and contracting fibers are so close to the respective G_i that $G_i^{(k)} \subset U$ and property (a) above holds when $G_i^{(k)}$ replaces G_i. This also means that

$$\nu\left(\bigsqcup_{i=1}^{N} G_i^{(k)} \setminus M_{q/4} \cap U\right) < 2\eta\, \nu(U) .$$

It is by virtue of Corollary 8 that the above K_η exists. Corollary 8 also implies that K_η can and will be so chosen that the smallest of the r-lengths of the segments of arcs joining opposite sides of $G_i^{(k)}$ and with absolute slope between R_+^{-1} and $(R_-^{-1} + \tau_0^{-1})$ is $> d_\eta$.

Consider $G_i^{(k)}$ and let $z \in M_q^{(k)^{\pm}} \cap G_i^{(k)}$, draw the k-contracting fiber $\gamma_e^{(k)}(z)$ or the k-expanding fiber $\gamma_c^{(k)}(z)$ through z. Then

$$r\left(\gamma_c^{(k)}(z) \cap G_i^{(k)}\right) > d_\eta$$

$$r\left(\gamma_e^{(k)}(z) \cap G_i^{(k)}\right) > d_\eta .$$

Let us open a side discussion in order to define like constants $C_i(\eta)$, $i = 1,2,\ldots,5$.

Consider a decreasing segment $\gamma \subset U$ with slope between $-(R_-^{-1} + \tau_0^{-1})$ and $-R_+^{-1}$ and such that $r(\gamma) > \frac{1}{GZ} d$ (say).

Draw the five equispaced horizontal lines as in Figure 10. Their spacing is $\Delta\phi \sim$ order of d_η.

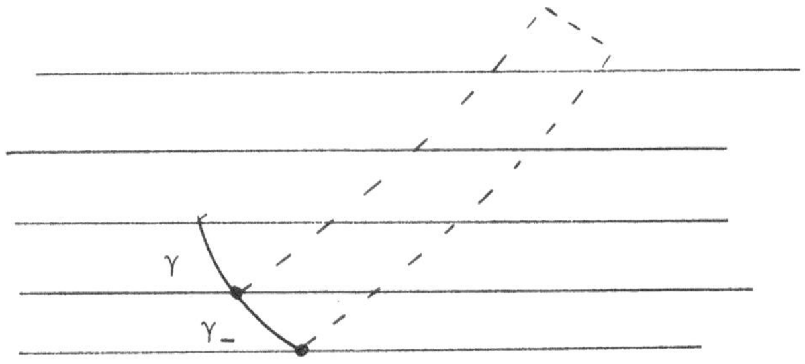

Figure 10

Apply Lemma 13 to the lower "half" γ_- of γ to construct a quadrilateral G for a $k \geq K_\eta$. The choice of C guarantees that the right side γ_r of G is above the uppermost line. Let D be the dashed region (i.e. the intersection of G with the strip between the fourth and the fifth line).

If we introduce on $\hat{G}^{(k)}$ the intrinsic coordinates mentioned in Lemma 13 and use that $\rho^{(k)}(\alpha,\beta)$ is bounded away from zero and infinity we see that $\exists c_1(\eta)$ such that

$$\nu(\hat{G}^{(k)} \cap D) > c_1(\eta) \qquad \forall k$$

where $c_1(\eta)$ is a suitable η-dependent (but k-independent) constant.

Also $\exists c_2(\eta)$, k-independent, such that

$$r(\hat{G}^{(k)} \cap \gamma_-) > c_2(\eta)\ r(\gamma_-) > 0 .$$

Similar constructions can be performed when γ_- is taken to be the right side of the quadrilateral G or when γ is assumed increasing (with the usual slope restriction) and is taken as the upper or lower side of G: we may, and shall, assume also that the constants $c_1(\eta)$, $c_2(\eta)$ are good also for the corresponding statements in these similar cases.

We consider next two segments $\gamma_+, \gamma_- \subset U$ such that

$\emptyset \neq \gamma_+ \cap \gamma_- = z \in U$. Let $\bar{\gamma}_+$, $\bar{\gamma}_-$ be the two longest (in the r-length sense) parts into which z cuts γ_+ and γ_- respectively. We assume that γ_+, γ_- are respectively increasing and decreasing with absolute slope between R_+^{-1} and $R_-^{-1} + \tau_0^{-1}$. Suppose also that $r(\bar{\gamma}_+)$, $r(\bar{\gamma}_-) >$ const. d_η where the constant is suitably small (say const. = 1/6 Z). Finally suppose that $r(\bar{\gamma}_+) = r(\bar{\gamma}_-)$.

We can apply the constructions allowed by Lemma 13 to $\bar{\gamma}_+$ or $\bar{\gamma}_-$ with C = 2 (say) and the same ε fixed at the beginning (and not yet exhibited) and arbitrary $k \geq K_\eta$. We obtain, in this way, two quadrilaterals G_+, G_-.

It is easy to see that there are geometrical constants $\varepsilon(x_0)$, $A(x_0)$ (η and k independent) so that

$$\varepsilon(x_0) \leq \frac{\nu(G_+ \cap G_-)}{\nu(G_\pm)} < \varepsilon(x_0)^{-1}$$

$$A(x_0) \leq \frac{\nu(G_+)}{\nu(G_-)} \leq A(x_0)^{-1}$$

Such inequalities follow from simple geometric considerations on the slopes of the sides of G_+, G_-. Then

$$\nu\left((G_+ \setminus \hat{G}_+^{(k)}) \cap G_+ \cap G_-\right) < \nu\left(G_+ \setminus \hat{G}_+^{(k)}\right) < \varepsilon\, \nu(G_+) .$$

Therefore

$$\nu(G_+ \cap G_-) - \nu(\hat{G}_+^{(k)} \cap G_+ \cap G_-) < \varepsilon\, \nu(G_+)$$

or

$$\nu(\hat{G}_+^{(k)} \cap (G_+ \cap G_-)) > \left(\varepsilon(x_0) - \varepsilon\right)\nu(G_+) ;$$

and similarly

$$\nu(\hat{G}_-^{(k)} \cap (G_+ \cap G_-)) > \left(\varepsilon(x_0) - \varepsilon\right) \nu(G_-) .$$

Hence

$$\nu\left((G_+ \setminus \hat{G}_+^{(k)}) \cap \hat{G}_-^{(k)} \cap (G_+ \cap G_-)\right) \leq \nu\left(G_+ \setminus G_+^{(k)}\right) < \varepsilon\, \nu(G_+) ,$$

or

$$\nu(\hat{G}_-^{(k)} \cap G_+ \cap G_-) - \nu(\hat{G}_+^{(k)} \cap \hat{G}_-^{(k)} \cap G_+ \cap G_-) < \varepsilon\, \nu(G_+)$$

or

$$\nu(\hat{G}_+^{(k)} \cap \hat{G}_-^{(k)}) \equiv \nu(\hat{G}_+^{(k)} \cap \hat{G}_-^{(k)} \cap G_+ \cap G_-) > (\varepsilon(x_0)-\varepsilon)\nu(G_-)-\varepsilon\nu(G_+)$$

$$> \left(\frac{\varepsilon(x_0)-\varepsilon}{A(x_0)} - \varepsilon\right)\nu(G_+)$$

it is now clear that ε should have been chosen so that

$\frac{\varepsilon(x_0)-\varepsilon}{A(x_0)} - \varepsilon > 0.$

With this choice of ε we also have

$$\nu(\hat{G}_+^{(k)} \cap \hat{G}_-^{(k)}) > c_3(\eta) > 0$$

$$\nu(\hat{G}_+^{(k)} \cap G_-) > c_3(\eta)$$

$$(\hat{G}_-^{(k)} \cap G_+) > c_3(\eta)$$

where $c_3(\eta)$ is a suitable k-independent number, $\forall k \geq K_\eta$ (one could take $c_3(\eta) = \frac{\varepsilon(x_0)}{A(x_0)} c_1(\eta))$.

Using intrinsic coordinates on $\hat{G}_+^{(k)}$, as in Lemma 13, we also find

$$r(\hat{G}_+ \cap \bar{\gamma}_+) > c_4(\eta)\; r(\bar{\gamma}_+)$$

$$r(\hat{G}_- \cap \bar{\gamma}_-) > c_4(\eta)\; r(\bar{\gamma}_-)$$

where $c_4(\eta)$ is a suitable k-independent number, $\forall k \geq K_\eta$. These inequalities imply the inequality that we describe in the next proposition.

Let E_1 be a subset of $\bar{\gamma}_+$ and $E_2 \subset \bar{\gamma}_-$; then if E_1, E_2 fill a large enough fraction of $\bar{\gamma}_+, \bar{\gamma}_-$, respectively, the set

$$\hat{E}_1 \times \hat{E}_2 = \{z \mid z=\gamma_e^{(k)}(x_2) \cap \gamma_c^{(k)}(x_1),\; x_1 \in \bar{\gamma}_+ \cap E_1 \cap \hat{G}_+^{(k)},$$
$$x_2 \in \bar{\gamma}_- \cap E_2 \cap \hat{G}_-^{(k)}\},$$

is such that

$$\nu\left((\hat{E}_1 \times \hat{E}_2) \cap (\hat{G}_+^{(k)} \cap \hat{G}_-^{(k)})\right) > c_5(\eta)$$

the fraction of $\bar{\gamma}_+$ and $\bar{\gamma}_-$ that has to be covered in order that the above inequality holds depends solely on η (and not on $k \geq K_\eta$). We may and shall assume that $c_5(\eta)$ measures the fraction of $\bar{\gamma}_+, \bar{\gamma}_-$ that can be left out of $\bar{\gamma}_+ \cap E_1$ or $\bar{\gamma}_- \cap E_2$ still keeping the last inequality valid.

Let $c_\eta = \min\limits_{i=0,\ldots,5} c_i(\eta)$.

We are now ready to complete the proof of Lemma 17.

Given $\eta > 0$ we choose $\bar{k}$ so large that all subsets $\Delta \subset U$ such that

$$\nu(\Delta) >> \frac{1}{2} c(\eta)$$

have the property

$$\nu(W^f_{\bar{k},p} \cap \Delta) > (1 - \sigma(\eta))\, \nu(\Delta)$$

where f is a continuous function on M, $W^f_{\bar{k},p}$ is the set of points where the Birchoff average $\bar{f}$ is attained within $\rho > 0$ after $\bar{k}$ steps at most; the function $\sigma(\eta)$ will be determined later.

Fix $K > \bar{K} + K_\eta$.

Consider the quadrilaterals $G_i^{(k)}$. Let

$$\hat{G}_i^{(k)} = \bigsqcup_{z \in M_{q/4} \cap G_i^{(k)}} \left(\gamma_c^{(k)}(z) \cap G_i^{(k)}\right).$$ Then

$$M_{q/4} \cap G_i^{(k)} \subset \hat{G}_i^{(k)} \subset G_i^{(k)} \cap M_{q/2}^{(k)^+} \subset G_i^{(k)}$$

and

$$\nu(\hat{G}_i^{(k)} \cap W^f_{\bar{k},\rho}) > (1 - \sigma(\eta))\, \nu(\hat{G}_i^{(k)}) \,.$$

So, using intrinsic coordinates as in Lemma 13, we can write this inequality as

$$\int_0^{\bar{\alpha}} d\alpha \int_{\alpha(\hat{G}_i^{(k)} \cap W^f_{\bar{k},\rho})} \rho^{(k)}(\alpha,\beta)\, d\beta > (1-\sigma(\eta)) \int_0^{\bar{\alpha}} d\alpha \int_0^{\beta(\alpha)} \rho^{(k)}(\alpha,\beta)\, \Lambda\beta \,,$$

where $\bar{\alpha}$, $\beta(\alpha)$ are the natural integraion bounds and $\alpha(\hat{G}_i^{(k)} \cap W^f_{\bar{k},\rho})$ denotes the r-measure of the intersection between $W^f_{\bar{k},\rho} \cap G_i^{(k)}$ and the arc $\gamma_c^{(k)}$ which has coordinate α .

This implies that the set N of the α's such that

$$\int_{\alpha(\hat{G}_i^{(k)} \cap W^f_{\bar{k},\rho})} \rho^{(k)}(\alpha,\beta)\, d\beta < \left(1 - \sqrt{\sigma(\eta)}\right) \int_{\alpha(\hat{G}_i^{(k)})} d\beta\, \rho^{(k)}(\alpha,\beta)$$

has the property

$$\int_N d\alpha \int_{\alpha(\hat{G}_i^{(k)})} \rho^{(k)}(\alpha,\beta)\, d\beta < \sqrt{\sigma(\eta)}\, \nu(\hat{G}_i^{(k)}) \ .$$

If

$$\int_{\alpha(\hat{G}_i^{(k)} \cap W^f_{\bar{k},\rho})} \rho^{(k)}(\alpha,\beta)\, d\beta \geq 1-\sqrt{\sigma(\eta)} \int_{\alpha(\hat{G}_i^{(k)})} d\beta\, \rho^{(k)}(\alpha,\beta)$$

it follows

$$\alpha(\hat{G}_i^{(k)}) - \alpha(\hat{G}_i^{(k)} \cap W^f_{\bar{k},\rho}) \geq 2\sqrt{\sigma}\, \alpha(G_i^{(k)})$$

since

$$\left|\frac{\rho^{(k)}(\alpha,\beta)}{\rho^{(k)}(\alpha',\beta')} - 1\right| < \eta^2 < \frac{1}{2} \text{ on } \hat{G}_i^{(k)} ,$$

by our construction of the $G^{(k)}$.

This means that the set V'^f of the points $z \in U$ through which it is possible to draw a k-contracting germ of fiber with r-length at least d_η and filled (in the r-length sense) by $W^f_{\bar{k},\rho}$ within $2\sqrt{\sigma(\eta)}$ has measure $\geq (1-(2\eta^2+2\sqrt{\sigma(\eta)}+2\eta))\nu(U)$.

Let x,x' be in $V^f_\eta = V'^f_\eta \cap W^f_{\bar{k},\rho}$ (then $\nu(V^f_\eta) > (1 - 2\sigma(\eta) - 2\eta^2 - 2\sqrt{\sigma(\eta)} - 2\eta)\ \nu(U))$;
we show that there is a number N_η and a chain $\gamma_0,\gamma_1,\dots,\gamma_{N_\eta}$ of alternatingly k-increasing and k-decreasing semgnets of k-fibers

such that $x \in \gamma_0$, $y \in \gamma_{N_\eta}$, $\gamma_i \cap \gamma_{i+1} \in W^f_{\bar{k},\rho}$, $i = 0,1,\ldots,N_\eta-1$.

Let γ_0' = k-contracting fiber segment hrough x which is filled within $2\sqrt{\sigma}$ by $W^f_{\bar{k},\rho}$.

Draw the five equispaced parallel lines, as in the above discussion, relative to the arc γ_0'.

Notice that the lower half $\gamma_0'^{-}$ of γ_0' must be filled within $c_1\sqrt{\sigma(\eta)}$ by $W^f_{\bar{k},\rho}$ (where c_1 is a suitable geometric constant which could be taken as 4Z).

Next we apply Lemma 13 to the segment $\gamma_0'^{-}$ with ε and C chosen as in the preliminary discussion and $K > K_\eta + \bar{K}$. Notice, also, that $r(\gamma_{0^-}) >$ const. d_η (with const. = 1/3Z (say)). Also, by our choice of $\bar{K}$:

$$\nu(\hat{G} \cap D \cap W^f_{\bar{k},\rho}) > (1 - \sigma(\eta))\ \nu(\hat{G} \cap D)$$

where D is the dashed region in Figure 10. On the other hand the relation

$$\int_{\gamma'\cap\hat{G}^{(k)}} d\alpha \int_{\alpha(\hat{G}^{(k)}\cap D\cap W^f_{\bar{k},\rho})} \rho^{(k)}(\alpha,\beta)\ d\beta > (1-\sigma(\eta)) \int_{\gamma_0'\cap\hat{G}} d\alpha \int_{\alpha(\hat{G}\cap D)} \rho^{(k)}(\alpha,\beta)\ d\beta$$

which is a rewriting of the preceding equation using the natural coordaintes (α,β) on $\hat{G}^{(k)}$ and symbols consistent with the preceding notations. This means that

$$\alpha(\hat{G} \cap D \cap W^f_{\bar{k},\rho}) > (1 - \sqrt{\sigma}\ B^2_\eta)\ \alpha(\hat{G} \cap D)$$

except for a set J of α's such that

$$\int_J d\alpha \int_{\alpha(\hat{G} \cap D)} \rho^{(k)}(\alpha,\beta)\ d\beta < \sigma^{1-\omega/2}\ \nu(\hat{G} \cap D)$$

where $0 < \omega < 1$ and B_η is the bound on $\rho^{(k)}(\alpha,\beta)/\rho^{(k)}(\alpha',\beta')$ on G.

So if $\bar{W} \subset \hat{G}^{(k)} \cap D \cap W^f_{\bar{k},\rho}$ is the set of points with intrinsic coordinates (α,β) the k-expanding segment through (α,β) is, inside D, filled within $\sqrt{\sigma(\eta)}^{\omega}\, B_\eta^2$ by points of $W^f_{\bar{k},\rho}$ has measure

$$\nu(\bar{W} \cap \hat{G} \cap D) \equiv \nu(\bar{W}) > \left(1 - \sqrt{\sigma(\eta)}^{2-\omega}\right) \nu(\hat{G} \cap D)$$

which implies that, if $E \subset \hat{\gamma}_0'$ is such that $z \in E \Rightarrow$ the k-expanding fiber $\gamma_e^{(k)}(z)$ has nonempty intersection with $\bar{W}$, we must have

$$r(\hat{\gamma}_0'^{-} \setminus E) < \text{const. } B_\eta^2 \sqrt{\sigma(\eta)}^{2-\omega} r(\hat{\gamma}_0'^{-})$$

where the const. is a suitable geometric constant (which is an upper bound for the ratio of the r-lengths of the arcs obtained by intersecting D with two k-expanding fibers of $\hat{G}^{(k)}$).

Therefore

$$r(E) > \left(1 - B_\eta^2 \sqrt{\sigma(\eta)}^{2-\omega} \text{const.}\right) r(\hat{\gamma}_0^-)$$

$$> \left(1 - B_\eta^2 \sqrt{\sigma(\eta)}^{2-\omega} \text{const.}\right) c_4(\eta)\, r(\gamma_0^{-1}) .$$

Remembering that $\gamma_0'^{-}$ was filled within $c_1\sqrt{\sigma(\eta)}$ by $W^f_{\bar{k},\rho}$ we realize that (if ω and $\sigma(\eta)$ so chosen that

$$\frac{1}{4}\, c_4(\eta) > c_1 \sqrt{\sigma(\eta)}$$

$$B_\eta^2 \sqrt{\sigma(\eta)}^{2-\omega} \text{const.} < \frac{1}{2}$$

$$B_\eta^2 \sqrt{\sigma(\eta)}^{\omega} < c_1 \sqrt{\sigma(\eta)}$$

there will be a point $x_1 \in \hat{\gamma}_0' \cap \hat{G}^{(k)}$ whose k-expanding fiber has intersection with D filled within $B_\eta^2\sqrt{\sigma(\eta)}^{\omega}$ by points of $W^f_{\bar{k},\rho} \cap D$.

Let γ_1' be the expanding fiber through x_1 cut at the intersection with the uppermost of the horizontal lines and let $\gamma_1'^{+}$ be $\gamma_1' \cap D$. The third property required on $\sigma(\eta)$ allows us to deduce that γ_1^+ is filled within $c_1\sqrt{\sigma(\eta)}$ by points of $W^f_{\bar{k},\rho}$ and $r(\gamma_1^-) \geq \frac{1}{2Z}\, d_\eta$. Hence we can iterate the construction.

A similar construction can be performed starting from x'.

In this way we easily obtain a chain $\gamma_0,\dots,\gamma_n$ and $x \in \gamma_0$, $x' \in \gamma_n$, $\gamma_i \cap \gamma_{i+1} \in W^f_{\bar{k},\rho}$ with only one possible exception occurring at some $i = i_0$; furthermore, the lines γ_i are alternatingly k-expanding and k-contracting segments (this sequence of arcs is obtained from the sequence of segments, constructed in the above interactive scheme, $\gamma_0',\gamma_1',\dots,\gamma_N'$ by cutting off each γ_i' the pieces that "stick out of the chain").

The exception occurs, possibly, where the chain starting in x and one starting in y meet.

Let us consider the first step in the above construction before continuing the proof. It is clear that

$$\nu(\hat{G}^{(k)} \cap W^f_{\bar{k},\rho}) > (1-\sigma(\eta))\ \nu(\hat{G}^{(k)})$$

by our choice of $\bar{K}$. Which means that the set $\bar{\bar{W}} \subset \hat{G}^{(k)}$ consisting of the points $(\alpha,\beta) \in \hat{G}^{(k)}$ such that the k-expanding fiber through them is filled within $\sqrt{\sigma(\eta)}^{\,\omega}\ B^2_\eta$ by $W^f_{\bar{k},\rho}$ has measure exceeding $(1 - \sqrt{\sigma(\eta)}^{\,2-\omega})\ \nu(\hat{G}^{(k)}) > (1 - \sqrt{\sigma}^{\,2-\omega})\ c_1(\eta) > \frac{1}{2}\ c_1(\eta)$ if $(1 - \sqrt{\sigma(\eta)}^{\,2-\omega}) > 1/2$.

So the set $\bar{\bar{W}}$ has the property that

$$\nu(\bar{\bar{W}} \cap W^f_{\bar{k},\rho}) > (1-\sigma(\eta))\ \nu(\bar{\bar{W}}))$$

Also

$$r(\bar{\bar{W}} \cap \hat{\gamma}_0^-) > (1 - \sqrt{\sigma(\eta)}^{\,2-\omega}\ B^2_\eta \text{ const.})\ r(\gamma_0^-) > \frac{c_4(\eta)}{2}\ r(\gamma_0^-)$$

if $1 - \sqrt{\sigma(\eta)}^{\,2-\omega}\ B^2_\eta$ const. $> 1/2$.

So we could replace in the above construction of $\gamma_1',\gamma_2',\dots$ $\hat{G}^{(k)}$ by $\bar{\bar{W}}$: this is the reason why we have used the generous factor 1/4 (instead of 1/2) in the first equation defining $\sigma(\eta)$.

Using this refinement of the above arguments we can construct the chain $\gamma_0,\gamma_1,\dots,\gamma_N$ so that also the property that γ_i is filled by $W^f_{\bar{k},\rho}$ within const. $\sqrt{\sigma(\eta)}^{\,\omega} B^2_\eta$ holds (the geometrical constant coming from the fact that γ'_i has to be chopped at one extreme to give γ_i).

We shall henceforth assume that the segments γ_i are chosen to also have this last property.

Consider γ_{i_0} and γ_{i_0+1}. We notice that we can always assume without loss of generality that $\gamma_{i_0} \cap \gamma_{i_0+1}$ is just a point (and not a segment): this comes from the fact that every time a new element of the chain is built it is chosen out of infinitely many possibilities.

The point Z divides on each of the segments γ_{i_0} and γ_{i_0+1} two segments. Let $\bar{\gamma}_{i_0}$, $\bar{\gamma}_{i_0+1}$ be the longest between them (in the r-length sense).

By construction $r(\bar{\gamma}_{i_0})$, $r(\bar{\gamma}_{i_0+1}) \geq 1/4Z\, d_\eta$. Furthermore $\bar{\gamma}_{i_0}$, $\bar{\gamma}_{i_0+1}$ are filled, in the r-arc length sense, within $\frac{c_1}{2} \sqrt{\sigma(\eta)}$, at least, by points $W_{\bar{k},\rho}$.

We can cut the longest between $\bar{\gamma}_{i_0}$ and $\bar{\gamma}_{i_0+1}$ to obtain two segments with equal arc length which we shall still call $\bar{\gamma}_{i_0}, \bar{\gamma}_{i_0+1}$, but which are now filled only within const. $\sqrt{\sigma(\eta)}$ by $W^f_{\bar{k},\rho}$ where the constant coujld be taken to be $c_1/4\,Z$.

By choosing $\sigma(\eta)$ smaller, if necessary, we may assume that the sets $E_1 = \bar{\gamma}_{i_0} \cap W^f_{\bar{k},\rho}$ and $E_2 = \bar{\bar{\gamma}}_{i_0+1} \cap W^f_{\bar{k},\rho}$ fill "enough" of $\bar{\gamma}_{i_0}$ and $\bar{\gamma}_{i_0+1}$ respectively that the set $\hat{E}_1 \times \hat{E}_2$ has measure $> c_5(\eta)$ (we use here the same symbol $\hat{E}_1 \times \hat{E}_2$ introduced in the preliminaries to our proof).

Then $W^f_{\bar{k},\rho} \cap (\hat{E}_1 \times \hat{E}_2)$ must be nonempty and if $\bar{z} \in W^f_{\bar{k},\rho} \cap (\hat{E}_1 \times \hat{E}_2)$ the k-fibers $\gamma_c^{(k)}(\bar{z})$ and $\gamma_e^{(k)}(\bar{z})$ will intersect E_2 and E_1, respectively, in $W^f_{\bar{k},\rho}$. It is clear that we have in this way constructed a sequence $\tilde{\gamma}_0, \tilde{\gamma}_1, \ldots, \tilde{\gamma}_{N+2}$ of alternatingly k-expanding and k-contracting fibers which have successive intersections in

$W^f_{\bar{k},\rho}$ and such that $x \in \tilde{\gamma}_0$, $x' \in \tilde{\gamma}_{N+2}$.

This completes the proof since we can choose

$$V^f = \bigcup_{n=1}^{\infty} V^f_{1/n} .$$

SPECTRAL INVARIANTS AND SMOOTH ERGODIC THEORY

D. A. Lind

Department of Mathematics, University of California, Berkeley 94720

1. Introduction

My purpose here is to briefly discuss the influence of spectral theory on the study of measure preserving transformations, and then to indicate how spectral invariants may play a role in deciding whether a given measure preserving transformation is isomorphic to a smooth one. As part of this discussion, I give a simple construction of a diffeomorphism of the two dimensional torus which has an everywhere discontinuous eigenfunction. I also hope to stimulate interest in the recent work of Anosov and Katok on constructing diffeomorphisms with prescribed ergodic behavior.

2. Spectral Theory of Measure Preserving Transformations

In the following X will denote a measure space with a σ-algebra of measurable subsets on which a probability measure μ is defined. To avoid unpleasant and needless pathology, we require X to be a so-called Lebesgue space, namely that X be measure isomorphic to the unit interval equipped with Lebesgue measure. For an axiomatic treatment of Lebesgue spaces, see Rohlin [17]. All "naturally occurring" measure spaces are Lebesgue.

A one-to-one and onto transformation T of one measure space to another is said to be measurable if $T^{-1}(E)$ is measurable whenever E is. T is called measure preserving if $T^{-1}(E)$ and E have the same measure. The objects of study in ergodic theory are measure preserving transformations of a space onto itself. Two such transformations, T acting on X and T' acting on X', are essentially the same from this point of view if there is a measure preserving transformation $\Phi: X \to X'$ such that $\Phi(Tx) = T'(\Phi x)$ for almost every $x \in X$. The basic problem of ergodic theory is to find ways of detecting whether two transformations, specified in perhaps very different ways, are isomorphic. Historically, spectral invariants have played a key role in this study.

The idea of using spectral theory to study measure preserving transformations goes back to Koopman [13]. He observed that a (measure preserving) transformation T of X induces a unitary operator U_T on $L^2(X)$ by the formula $(U_T f)(x) = f(Tx)$. Isomorphic transformations induce unitarily equivalent operators. Thus properties of the induced operators which are unitary invariants are the same for isomorphic transformations. These so-called spectral invariants allow us to prove that some transformations are not isomorphic.

The importance of these invariants becomes clear when one realizes that for over two decades, until Kolmogorov introduced entropy in 1958, they were the only useful ones known.

A transformation T is ergodic if whenever $T(E) = E$, then $\mu(E) = 0$ or 1. This is easily equivalent to the property that the only functions in $L^2(X)$ invariant under U_T are constant. This means that T is ergodic if and only if 1 is a simple eigenvalue of U_T, so that ergodicity is a spectral invariant.

In the same spirit, the point spectrum of U_T is a spectral invariant. Recall that a complex number λ is in the point spectrum P_T of U_T if there is a nonzero eigenfunction $f \in L^2(X)$ such that $U_T f(x) = \lambda f(x)$. P_T is a subgroup of the multiplicative unit circle group $\mathbb{T}$.

A good example to keep in mind is translation on the n-dimensional torus group $\mathbb{T}^n = \mathbb{T} \times \cdots \times \mathbb{T}$ with Haar measure. If $\lambda_i \in \mathbb{T}$, then translation on $\mathbb{T}^n$ by $(\lambda_1, \ldots, \lambda_n)$ is a measure preserving transformation whose point spectrum is the subgroup of $\mathbb{T}$ generated by $\lambda_1,\ldots,\lambda_n$. This translation is ergodic if whenever integers m_j are such that $\lambda_1^{m_1} \ldots \lambda_n^{m_n} = 1$, then all the $m_j = 0$. In this case the λ_i are said to be independent over the rationals. The eigenfunctions for this translation are just the multidimensional exponentials (the characters of $\mathbb{T}^n$), and since these span $L^2(\mathbb{T}^n)$, the point spectrum comprises the entire spectrum.

Koopman's idea is completely successful in this type of situation. Say that an ergodic transformation T has discrete spectrum if the eigenfunctions of U_T span $L^2(X)$.

Discrete Spectrum Theorem. If S and T are transformations with discrete spectrum, then S is measure isomorphic to T if and only if U_S is unitarily equivalent to U_T. Furthermore, this occurs if and only if $P_S = P_T$.

It can also be shown that any countable subgroup of $\mathbb{T}$ is the point spectrum of a transformation with discrete spectrum, and that every discrete spectrum transformation is isomorphic to a translation on a compact abelian group. For further details, see Halmos [9].

These facts give a very satisfying classification of transformations with discrete spectrum. However, as soon as the transformation has any stronger mixing properties such as weak mixing (see [6] for a definition), the point spectrum reduces to $\{1\}$ and so is useless for classification. Indeed, all Bernoulli shifts have unitarily equivalent induced operators and so are indistinguishable from the spectral point of view. Since there are nonisomorphic Bernoulli shifts, spectral invariants are useful but far from complete.

3. Smooth Transformations

Let M be a compact, connected, smooth (i.e., C^{∞}) manifold, and suppose that T is a smooth diffeomorphism of M which preserves a smooth measure μ, so that μ is given by a smooth positive density function on M. The manifold M now becomes the measure space we called X in Section 2, and the σ-algebra is the μ-completion of the Borel σ-algebra on M generated by the compact sets. A measure preserving transformation is defined to be smooth if it is measure isomorphic to such a diffeomorphism. Certain smooth systems, arising from the Hamiltonian flow on the energy surface of a mechanical system, were the original motivating examples of ergodic theory. An immediate question is: Are all ergodic transformations smooth? Or has ergodic theory, by throwing away the differentiable structure of the original examples, greatly enlarged the class of objects it studies?

The only definite result in this direction is that smooth transformations have finite entropy. A good source for the definition of entropy and related results is Billingsley [6]. Kouchnirenko [14] first proved this fact by using the isoperimetric inequality (see Arnold and Avez [4]). Bowen [7] has given a neat proof that the entropy of T is bounded above by $\max\{0, \dim M \cdot \sup_{x\in M}\| dT\, T_x M\|\}$, where the norm on the derivative is taken with respect to a suitable Riemannian metric on M.

The example in the previous section of translation T on $\mathbf{T}^n$ by the element $(\lambda_1,\ldots,\lambda_n)$ is certainly smooth.

The rank of P_T, written $\operatorname{rk} P_T$, is defined to be the largest number of elements in P_T which are independent over the rationals. Since P_T for the torus translation is generated by $\{\lambda_1,\ldots,\lambda_n\}$, it is easy to check that $\operatorname{rk} P_T \le n$, where n is the dimension of the manifold $\mathbf{T}^n$.

Up until about five years ago, in all smooth cases which could be checked, it turned out that

(1) $$\operatorname{rk} P_T \leq \dim M .$$

Indeed, Kolmogorov observed this in his address to the 1954 International Congress of Mathematicians [12], and it was conjectured to be generally true by Arnold and Avez [4; Appendix 16].

If the eigenfunctions of the diffeomorphism T are assumed to have certain smoothness properties, then finite bounds for $\operatorname{rk} P_T$ are known. We give below a simple proof that (1) holds if the eigenfunctions are once differentiable. Avez [5] has shown that it holds under the assumption of mere continuity of the eigenfunctions, and under the same hypothesis Arnold and Avez [4, Appendix 16] have shown that $\operatorname{rk} P_T \leq \dim H_1(M;\mathbb{R})$, the bound being the first Betti number of M.

Unfortunately, as was already known to Kolmogorov in 1953 [11], there are smooth systems with everywhere discontinuous eigenfunctions. Since examples seem to be available only in Russian, we give below a construction of this phenomenon on $\mathbb{T}^2$. Such badly behaved eigenfunctions obstructed further progress on the problem. Then about five years ago Anosov and Katok [3] published some constructions of diffeomorphisms which demolished the conjecture (1), but they left many questions open. We will describe their main result, and point out some areas which remain murky.

Let us first prove (1) for the point spectrum of well behaved eigenfunctions. Let D_T be the subgroup of P_T consisting of eigenvalues corresponding to once differentiable eigenfunctions.

<u>Theorem</u>. If T is an ergodic diffeomorphism of a manifold M, then $\operatorname{rk} D_T \leq \dim M$.

<u>Proof</u>: Let $\{\lambda_1,\ldots,\lambda_r\} \subset D_T$ be independent over the rationals.

Choose differentiable eigenfunctions f_j for λ_j. Since T is ergodic, $|f_j|$ is constant almost everywhere, hence everywhere by continuity. Normalize the eigenfunctions so that $|f_j| \equiv 1$ on M. Define a map $F: M \to \mathbb{T}^r$ by $F(x) = (f_1(x),\ldots,f_r(x))$. Notice that $F(Tx) = (\lambda_1,\ldots,\lambda_r)\cdot F(x)$. Hence invariance of M under T implies that the image $F(M)$ is invariant under translation by $(\lambda_1,\ldots,\lambda_r)$. Rational independence of the λ_i implies that the powers of $(\lambda_1,\ldots,\lambda_r)$ are dense in $\mathbb{T}^r$. Since $F(M)$ is compact and invariant under a dense set of translations, it must be all of $\mathbb{T}^r$. Now F is a differentiable onto mapping, so that the dimension of the range is bounded by the dimension of the domain. This gives $r \leq \dim M$, which proves the result.

4. Discontinuous Eigenfunctions

Our construction uses measure preserving flows, so we begin with some preliminary material. A measure preserving flow $\{T_t: t \in \mathbb{R}\}$ on a measure space X is a collection of measure preserving transformations satisfying the group property $T_s T_t = T_{s+t}$ together with the measurability assumption that $(x,t) \mapsto T_t x$ is a jointly measurable transformation from $X \times \mathbb{R}$ to X. A function $\psi: X \to \mathbb{R}$ is an eigenfunction for $\{T_t\}$ with eigenvalue $\lambda \in \mathbb{R}$ if $\psi(T_t x) = \exp(i\lambda t)\psi(x)$. This function is therefore an eigenfunction for each T_{t_0} with eigenvalue $\exp(i\lambda t_0)$.

We can build such flows from transformations by using the following suspension technique first introduced by von Neumann [15]. Suppose B is a measure space and $T: B \to B$ is measure preserving. Let $f: B \to (0,\infty)$ be measurable. There is a natural measure on the the region $X = \{(x,y): x \in B,\ 0 \leq y \leq f(x)\}$ under the graph of f obtained by restricting to X the product of the measure on B with Lebesgue measure on $\mathbb{R}$. The flow $\{S_t\}$ on X defined by moving a point (x,y) vertically at unit speed until $y = f(x)$, identifying $(x,f(x))$ with $(Tx,0)$, and continuing to flow vertically at $(Tx,0)$, is measurable and measure preserving. $\{S_t\}$ is said to be the flow built under the function f with base transformation T. Ambrose and Kakutani [2] have shown that all measure preserving flows can be put into this form.

In this construction, if B is a manifold, T a diffeomorphism of B, and f a smooth positive function on B, then X becomes a manifold M by identifying $(x,f(x))$ with $(Tx,0)$, i.e., by gluing the top and bottom edges of X together using f. Also, $\{S_t\}$ is a smooth flow on M.

A complex valued function on a manifold is everywhere discontinuous if on every open subset it does not agree almost everywhere with a continuous function. The function has an essential discontinuity at a point if it is not equal almost everywhere to a function continuous at the point.

The construction of an everywhere discontinuous eigenfunction for a smooth ergodic transformation proceeds as follows. Using a function constructed via harmonic analysis, we produce a measurable but not continuous conjugacy Φ between an irrational flow $\{S_t\}$ on $\mathbb{T}^2$ and a smooth flow $\{T_t\}$ on a manifold M diffeomorphic to $\mathbb{T}^2$. Then an exponential eigenfunction for $\{S_t\}$ carries over under Φ to everywhere discontinuous eigenfunction for $\{T_t\}$. A proper choice of t_0 then gives the ergodic diffeomorphism $T = T_{t_0}$ desired.

We now switch to additive notation on $\mathbb{T}$, it being considered as $[0,1)$ with addition modulo 1. For the sake of clarity, suppose first that we can find a function $\phi: \mathbb{T} \to [0,1)$ and $\omega \in \mathbb{T}$ such that $\phi(x+\omega) - \phi(x)$ is smooth, while $\exp(2\pi i\, \phi(x))$ has an essential discontinuity at 0. We will later modify the argument to handle unbounded ϕ, and then construct such a function. Put $g(\xi) = \phi(\xi+\omega)-$

$-\phi(\xi)+1$, a smooth positive function on $\mathbb{T}$. Let $\{S_t\}$ acting on X be the flow built under the constant function 1 with translation on $\mathbb{T}$ by ω as base transformation, and $\{T_t\}$ acting on M be the smooth flow built under g with the same base transformation as $\{S_t\}$. A typical point in X is denoted by (x,y), and one in M by (ξ,η). There is a natural map $x = \beta(\xi)$ from the base of $\{T_t\}$ to that of $\{S_t\}$.

We will define a conjugacy Φ: $M \to X$ using ϕ. Basically, Φ restricted to the base of $\{T_t\}$ is $\phi \circ \beta$, and then Φ extends to all of M by using the flow. More precisely,

$$\Phi(\xi,\eta) = \begin{cases} (\beta(\xi),\eta+\phi(\beta(\xi))), & \text{if } 0 \leq \eta \leq 1-\phi(\beta(\xi)), \\ (\beta(\xi)+\omega,\eta+\phi(\beta(\xi))-1), & \text{if } 1-\phi(\beta(\xi)) \leq \eta \leq g(\xi). \end{cases}$$

It is easy to check that Φ maps orbits to orbits and preserves the direction and speed of the flow. An easy argument using Fubini's theorem shows that Φ is measurable and measure preserving. Hence Φ is a conjugacy between $\{T_t\}$ and $\{S_t\}$.

Let χ be the eigenfunction for $\{S_t\}$ defined on X by $\chi(x,y) = \exp(2\pi iy)$. Then $\psi = \chi\circ\Phi$ is an eigenfunction for $\{T_t\}$ with eigenvalue 1. Now $\psi(\xi,0) = \exp(2\pi i\, \phi(\beta(\xi)))$ has an essential discontinuity at $\beta = 0$. But since $\psi \circ T_t = [\exp(2\pi it)]\psi$, this discontinuity can be translated by $\{T_t\}$ along the orbit of $(0,0)$, which is dense in M. Thus in every open subset of M the eigenfunction has an essential discontinuity, so that it is everywhere discontinuous.

Unhappily, it is not easy to find a bounded discontinuous function ϕ on $\mathbb{T}$ such that $\phi(x+\omega) - \phi(x)$ is smooth. Below we will produce a necessarily unbounded ϕ and an irrational $\omega \in T$ for which $\phi(x+\omega) - \phi(x)$ is smooth, bounded in absolute value by 1/4, and such that $\exp(2\pi i\, \phi(x))$ has an essential discontinuity at 0.

Let us indicate how to modify the definition of Φ if ϕ is unbounded. Basically, we use more of the orbit of $\{S_t\}$ to wrap the graph of ϕ.

For any real number a, let [a] denote the greatest integer in a, and {a} the fractional part of a. We then define

$$\Phi(\xi,\eta) = \begin{cases} (\xi+[\phi(\beta(\xi))]\omega,\ \eta+\{\phi(\beta(\xi))\})\ , & \text{if } 0 \le \eta \le 1-\{\phi(\beta(\xi))\}, \\ (\xi+[\phi(\beta(\xi))]\omega+\omega,\eta+\{\phi(\beta(\xi))\}-1) & \text{if } 1-\{\phi(\beta(\xi))\}\le \eta\le g(\xi). \end{cases}$$

The same arguments as before show that Φ is a conjugacy between $\{T_t\}$ and $\{S_t\}$. Φ reduces to the previous conjugacy if $0 \le \phi \le 1$. We also have as before that $\psi = \chi\circ\Phi$ is an everywhere discontinuous eigenfunction for $\{T_t\}$.

To obtain an ergodic diffeomorphism, choose t_0 such that $\exp(2\pi i t_0)$ and $\exp(2\pi i t_0\omega)$ are independent over the rationals. Then S_{t_0} is isomorphic to translation on $\mathbf{T}^2$ by $(\exp(2\pi i t_0),$ $\exp(2\pi i t_0\omega))$, and so is ergodic. Since T_{t_0} is isomorphic to S_{t_0}, $T = T_{t_0}$ is the required diffeomorphism.

The problem of finding a suitable ϕ and ω was considered by Furstenberg [8] in a somewhat different context. Our solution is parallel to his. Write the Fourier series of ϕ as

$$\phi(x) \sim \sum_{-\infty}^{\infty} c_n e^{2\pi inx} .$$

If $h(x) = \phi(x+\omega) - \phi(x)$, then

$$h(x) \sim \sum_{-\infty}^{\infty} c_n (e^{2\pi in\omega}-1)e^{2\pi inx} .$$

If ω is abnormally well approximable by rationals, the factor $e^{2\pi in\omega} - 1$ becomes small often enough to force smoothness of h while at the same time allowing the c_n's to be large enough to force an essential discontinuity of f at 0.

The Fourier series we consider will be lacunary. For any sequence of integers $n_k \to \infty$, with $n_{-k} = n_k$, the function

$$\phi(x) = \sum_{k\neq 0} \frac{1}{|k|} \exp\,(2\pi i n_k x)$$

is in $L^2(\mathbb{T})$. However, ϕ has an essential discontinuity at 0 since the Cesaro sums of the Fourier series of ϕ diverge at 0. By replacing ϕ with an appropriate multiple, we can guarantee that $\exp\,(2\pi i\,\phi(x))$ also has an essential discontinuity at 0. We will find n_k, ω such that

$$n_k^s|\exp(2\pi i n_k \omega) - 1| \to 0 \quad \text{as} \quad k \to \infty \tag{2}$$

for all $s \geq 0$. This forces h to have derivatives of all orders. Let $n_1 = 5$, $n_{k+1} = n_k^{k+1}$, and put $\omega = \sum_1^\infty 1/n_k$. Then if $k > s+1$,

$$n_k^s\{n_k\omega\} = n_k^{s+1} \sum_{j=k+1}^{\infty} \frac{1}{n_j} \leq n_k^{s+1} \left(\frac{2}{n_k^{k+1}}\right) < \frac{2}{n_k}\ ,$$

which proves (2) and also that $|h| < 1/4$. This yields the desired ϕ and ω.

We remark that such a ϕ cannot be bounded. The proof uses lacunarity of the Fourier series of ϕ and the method of Riesz products. Let

$$P_N(x) = \prod_{k=1}^{N} \left(1 + \cos\, 2\pi n_k x\right)\ .$$

Since the sequence $\{n_k\}$ is lacunary, the Fourier coefficient of P_N at 0 is 1, and at each $\pm n_k$ is $\frac{1}{2}$. Also, $P_N \geq 0$ since each factor is nonnegative. Suppose there were a C with $|\phi| \leq C$. Then we would have

$$\int_0^1 \phi(x)\ P_N(x)\ dx = \sum_{k=1}^{N} \frac{1}{k} \to \infty \quad \text{as} \quad N \to \infty\ ,$$

while

$$\left|\int_0^1 \phi(x)\ P_N(x)\ dx\right| \leq C \int_0^1 \left|P_N(x)\right|\ dx = C \int_0^1 P_N(x)\ dx = C\ .$$

This contradiction shows that ϕ must be unbounded.

We finish by briefly discussing the recent work of Anosov and Katok, and pointing out a few of the many still uncharted portions of the field.

What Anosov and Katok have done in [3] is to construct, for any given $n = 0,1,2,\ldots,\infty$, a diffeomorphism whose point spectrum has rank n on any manifold which admits a measure preserving action of the circle group (such as the two dimensional disk). However, the specific eigenvalues for these diffeomorphism are abnormally well approximable by rationals, and have quite a specific form. Thus while these examples destroy the original conjecture (1), arbitrary rank is a long way from arbitrary spectrum.

Generally, very little is known about available ergodic behavior for diffeomorphisms of a given manifold. For instance, do the various counterexamples constructed by Ornstein have smooth versions? There are some known Bernoulli diffeomorphisms, such as toral automorphisms (Katznelson [10]), and smooth Bernoulli flows, such as geodesic flow on the unit tangent bundle of a manifold of negative curvature (Ornstein and Weiss [16]). However, can any manifold support such Bernoulli diffeomorphisms or flows? A negative answer would mean that there are topological restrictions on the energy surface of a Bernoulli mechanical system, and this could have physical consequences.

References

[1] Abraham, R., Foundations of mechanics, Benjamin, New York (1967).

[2] Ambrose, W., and Kakutani, S., Structure and continuity of measurable flows, Duke Math. J. 9 (1942) 25-42.

[3] Anosov, D. V., and Katok, A. B., New examples in smooth ergodic theory. Ergodic diffeomorphisms, Trans. Moscow Math. Soc. 23 (1970) 1-35.

[4] Arnold, V. I., and Avez, A., Ergodic problems of classical mechanics, Benjamin, New York, 1958.

[5] Avez, A., Spectre discrete des systêmes ergodiques classiques, C. R. Acad. Sci. Paris 264 (1967) 49-52.

[6] Billingsley, P., Ergodic theory and information, Wiley, New York, 1965.

[7] Bowen, R., Entropy for group automorphisms and homogeneous spaces, Trans. Amer. Math. Soc. 153 (1971) 401-414.

[8] Furstenberg, H., Strict ergodicity and transformations of the torus, Amer. Jour. Math. 83 (1961) 573-601.

[9] Halmos, P. R., Lectures on ergodic theory, Publ. Math. Soc. Japan, no. 3, Math. Soc. Japan, Tokyo (1959).

[10] Katznelson, Y., Ergodic automorphism of T^n are Bernoulli, Israel Math. Jour. 10 (1971), 186-195.

[11] Kolmogorov, A. N., On dynamical systems with an integral invariant on the torus, Dokl. Akad. Nauk SSSR (N.S.) 93 (1953) 763-766 (in Russian).

[12] Kolmogorov, A. N., General theory of dynamical systems and classical mechanics, Proc. 1954 International Congress Math., North Holland: Amsterdam, 1957, pp. 315-333. (Russian; an English translation appears in Appendix D of [1]).

[13] Koopman, B. O., Hamiltonian systems and transformations in Hilbert spaces, Proc. Nat. Acad. Sci. 17 (1931) 315-318.

[14] Kouchnirenko, A. G., An estimate from above for the entropy of a classical system, Dokl. Akad. Nauk SSSR 161 (1965) 37-38.

[15] von Neumann, J., Zur Operatorenmethode in der klassichen Mechanik, Ann. of Math. 33 (1932) 587-642.

[16] Ornstein, D., and Weiss, B., Geodesic flows are Bernoullian, Israel J. Math. 14 (1973) 184-198.

[17] Rohlin, V. A., On the fundamental ideas of measure theory, Amer. Math. Soc. Transl., Ser. 1, Vol. 10, 1-54.

III. NONLINEAR DIFFERENTIAL EQUATIONS

NONLINEAR WAVE EQUATIONS

Martin Kruskal

Department of Astrophysical Sciences, Princeton University, N.J. 08540

1. Prefatory Remarks

I wish to take this opportunity to thank Jürgen Moser and Battelle Seattle Research Center for the opportunity provided me to deliver a series of lectures during the summer 1974 on the subject of nonlinear dispersive wave equations, as well as to attend the other lectures and participate in the discussions that took place. However, writing up these notes presented the difficulty that much or most of the material I covered was not being presented for the first time, but has been written up before by me and my colleagues once or even many times before in various forms. Sometimes one feels it worthwhile to recast, modify, or enlarge upon what has previously been presented, but other times one has nothing further to write on a given subject, even though oral exposition to a virgin audience is still worthwhile. Thus it turned out that these notes are very very far from a balanced representation of my lectures; a much better idea of the material covered can be obtained from the references to the work of our group (mainly Clifford Gardner, John Greene, Robert Miura, Norman Zabusky, and myself) given in the bibliography, except that I omitted certain topics because they were being covered by various other lecturers (e.g. the Fermi-Pasta-Ulam problem, lectured on by Zabusky, and the method of exact solution of the KdV and other equations by reduction to a sequence of linear problems, lectured on by Alan Newell). Thus these notes cover a somewhat varied set of topics which I felt still deserved putting in a new light or at least giving more emphasis than they have received in our publications, together with enough background and explanatory material to make them self-contained and give them, as I hope, a substantial degree of coherence. I refer the reader specifically to [Kruskal 1974] for some material I lectured on but do not include here.

I thank Robert Miura for making available to me his notes on my lectures.

2. Introductory Survey

My topic is nonlinear wave equations, and by a wave equation I mean first of all a partial differential equation (p.d.e.), like the classical linear wave equation

(1) $$u_{tt} - u_{xx} = 0 .$$

(Subscripts denote partial differentiation.) But our attention will be restricted only to a special class of wave equations called <u>conservative</u>, or <u>nondissipative</u>, or <u>reversible</u>. Without attempting to delimit them precisely, we can say that they are p.d.e.'s that conserve some quantities which may be interpreted as energy, or entropy, rather than dissipating them, and that in fact are invariant with respect to the reversal of time, so that the initial value problems for them are no better (nor worse) posed than the "final value" problems. These properties presuppose that one of the independent variables (t) be singled out to represent time.

Thus we will not be considering equations such as the classical heat equation

(2) $$u_t - u_{xx} = 0 ,$$

which is nonconservative, dissipative, and irreversible (the initial value problem is well posed, if reasonable boundary conditions are prescribed, but not the final value problem).

Among the conservative nonlinear p.d.e.'s there are a few very special ones that in the last few years have been discovered to possess some quite remarkable properties. They have solitary wave solutions which, when brought to interact with one another (in a fully nonlinear way!), emerge completely unchanged in size and shape. They possess not just several conservation laws, like any run of the mill conservative p.d.e., but infinitely many conservation laws (of a local nature). They are each associated with a linear ordinary (time-independent) eigenvalue equation whose spectrum remains invariant under the evolution in t of coefficient functions, which involve the dependent variable of the original p.d.e. They represent completely integrable systems, as described by Jürgen Moser elsewhere in these notes. They can be solved (exactly) by reduction to a sequence of linear problems.

What are these special equations? The earliest known is the Korteweg-deVries [1895] (KdV) equation

(3) $$u_t + uu_x + u_{xxx} = 0 .$$

Next is the "modified" or higher KdV equation

(4) $$v_t + v^2 v_x + v_{xxx} = 0 ,$$

which is closely related to the "standard" KdV equation (3), as we shall see. Then there is the so-called "sine-Gordon" equation [Rubinstein 1970],

(5) $$\phi_{tt} - \phi_{xx} + \sin \phi = 0 ,$$

which is a nonlinear version of the well known Klein-Gordon equation. Also the "nonlinear Schrödinger", or Zakharov-Shabat [1972], equation

(6) $$i \psi_t + \psi_{xx} + \psi^2 \psi^* = 0 ,$$

where $i = \sqrt{-1}$, asterisk denotes complex conjugate, and ψ is a complex variable (unlike the dependent variables in all the other cases, which are real). Finally, we have the recently discovered equation of Harry Dym [unpublished],

$$r_t = (r^{-1/2})_{xxx} , \tag{7}$$

again closely related to (3). Actually, each of these equations is merely the first (nontrivial) one in a sequence of increasingly higher order equations with closely linked properties. In very recent work, generalizations to other equations and even systems of equations have been found, as described by Alan Newell elsewhere in these notes.

Except for (7), all the equations above are quasilinear, i.e. linear in their highest derivative terms (where u_t, e.g., counts as a "highest derivative" term even though a third derivative term u_{xxx} is present, as long as no derivative term which would surely dominate it, such as u_{tt} or u_{xt}, is present). They are called "dispersive" wave equations, because the pure sinusoidal wave solutions of the corresponding linear equations (obtained by treating the dependent variable as small and neglecting higher order terms) disperse rather than travel together at a common phase velocity. Thus the linear equation corresponding to the KdV equation is

$$u_t + u_{xxx} = 0 , \tag{8}$$

which has sinusoidal solutions

$$u = \text{Re } [\text{const. } \exp (i\omega t + ikx)] \tag{9}$$

for k, ω satisfying the dispersion equation

$$\omega - k^3 = 0 . \tag{10}$$

Since the phase velocity

$$\frac{\omega}{k} = k^2 \tag{11}$$

depends on k (as does also the group velocity $d\omega/dk = 3k^2$), if a general localized initial function $u(x,0)$ is analyzed into sinusoidal modes, the solution of the linear problem at a later time will contain the same modes but at different relative positions, with different phase relations, and the solution $u(x,t)$ will be found to have "dispersed", i.e. spread out over a much wider interval than at $t = 0$, with essentially only one or a few discrete wavelengths apparent in any small local interval, though different ones in different intervals.

All this is in contrast to the standard linear wave equation (1), which is nondispersive, since its dispersion equation is

$$\omega^2 - k^2 = 0 \tag{12}$$

or

$$\omega = \pm k , \tag{13}$$

with two phase velocities

$$\frac{\omega}{k} = \pm 1 \tag{14}$$

(both) independent of k. Thus the general solution of (1) is

$$u(x,t) = f(x-t) + g(x+t) , \tag{15}$$

a superposition of two steady progressive waves travelling to the right and left without change of shape at velocities $\pm$ 1.

Note that despite the resemblance, the linear version of (5), the Klein-Gordon equation

$$\phi_{tt} - \phi_{xx} + \phi = 0 , \tag{16}$$

is dispersive because of the nonderivative term, its dispersion relation being

(17) $$\omega^2 = k^2 + 1$$

so that ω depends nonlinearly on k.

For the most part these very special equations have physical significance, and arise naturally in the study of physical problems. Thus the KdV equation was derived already in the preceding century as an approximate description of small amplitude water waves in a shallow channel (compared to the wavelengths of the waves). It was thereafter largely neglected until quite recently, although an excellent case could have been made for the importance of studying it on purely mathematical grounds. The KdV equation is, after all, the simplest nonclassical p.d.e.

The simplest p.d.e.'s have two independent variables. First order equations are well understood and are solvable by the method of characteristics. Second order equations are classified into elliptic, parabolic, and hyperbolic and are the backbone of classical mathematical physics. The simplest third order equation has a third derivative with respect to only one of the variables (x, say), and the minimum required to keep it from reducing to an ordinary differential equation is a first derivative with respect to the other variable (t). This gives the linearized KdV equation (8). But linear equations are in a certain sense trivial, at least those with constant coefficients. We might introduce a variable coefficient, most plausibly and simply x itself (this is in fact what Tricomi does to obtain his famous model equation with change of type from elliptic to hyperbolic), but spatial invariance has its own virtues and the term uu_x , nonlinear in the simplest way (quadratic), forms a natural and much precedented accompaniment to u_t , the sum of the two constituting the familiar convective (or substantial) derivative of fluid dynamics, if u is interpreted as a fluid velocity. (It might seem that u^2 would be a simpler term to add, but uu_x provides an extra symmetry, Galilean invariance, as will be used later.)

In this way, one should, or could, or might have been led to study the KdV equation on its mathematical merits, but in fact neither that consideration nor its physical relevance sufficed to turn attention to it for a long time, till it resurfaced in modern times in connection with the Fermi-Pasta-Ulam [1955] problem. So much for historical necessity!

Now in fact it turns out that the KdV equation arises in a large number of physical situations. Besides water waves, it has been shown to apply to magnetohydrodynamic waves, to ion-acoustic plasma waves, anharmonic lattice vibrations, and several other situations. How is it possible that so ubiquitous an equation managed for so long to evade attention?

Well, actually there is something strange about the KdV equation: it is first order with respect to t differentiation. All the familiar fundamental equations of physics are of even order, mostly second order (if they involve t at all; the potential equation doesn't) -- with perhaps two notable exceptions. The reason is that the fundamental processes of physics are time-reversible, so the equations describing those processes should be invariant under the replacement of t by $-t$. This invariance is enjoyed by even derivatives but not odd ones. (It is less apparent why the square of the first time derivative should never occur in a fundamental equation, but it seems not to. Quasilinearity would tend to militate against such a term, but why should fundamental equations be quasilinear? And even quasilinearity would be retained if the second time derivative were also present. Of course, many fundamental equations are outright linear.)

The two exceptions merely emphasize the point. The heat (or

diffusion) equation (2) describes an irreversible process, and in fact is not considered truly fundamental in physics, since all situations to which it is known to apply have been analyzed into finer constituents satisfying reversible equations, so that (2) is always viewed as ultimately an approximation. The Schrödinger equation

$$i\psi_t + \psi_{xx} - u\psi = 0 \tag{18}$$

looks first-order in t but is really second order, since it amounts to two coupled first order equations, ψ being complex; it is even time-reversible if one replaces i by $-i$ as well as t by $-t$ (as why shouldn't one, since there is no intrinsic way to specify one root of -1 in preference to the other?).

3. Derivation of KdV Equation

Like the heat equation, the KdV equation arises always as an approximation to a higher order equation (or system of equations). The approximation scheme employed is (some version of) the multi-scale method, which has only been formalized systematically in modern times and thus led to widespread applications of the KdV equation.

As a simple illustration of its derivation, consider the following prototypical weakly nonlinear, weakly dispersive wave equation

$$y_{tt} = y_{xx}(1 + \varepsilon y_x) + \alpha y_{xxxx} , \tag{19}$$

where ε and α, measuring the nonlinearity and the dispersiveness respectively, are two comparably small parameters. (The linear equation, with $\varepsilon = 0$, has dispersion relation $\omega^2 = k^2 - \alpha k^4$, so that ω/k is approximately constant for k not too large. If the relation is taken seriously even for k large, we should have $\alpha \leq 0$ to ensure well-posedness in the sense of Hadamard; otherwise ω can be negative imaginary and arbitrarily large, small wave-length perturbations can grow arbitrarily fast, and smooth initial data can become singular in arbitrarily short times or even immediately.)

Parenthetically, let me mention that (19) is actually just the lowest significant-order equation governing the evolution of the Fermi-Pasta-Ulam weakly nonlinear discrete mass lattice, governed by the equation

$$\text{(20)} \qquad \ddot{y}_n = h^{-2}(y_{n+1}-2y_n+y_{n-1}) \; [1+ \tfrac{1}{2}\, \varepsilon h^{-1}(y_{n+1}-y_{n-1})] \; ,$$

if the displacement function y_n and its time derivative $\dot{y}_n$ are assumed to discretely approximate smooth functions of $x = nh$, $y(x,t)$ and $y_t(x,t)$, and are Taylor expanded to fourth order in the (small) interparticle distance h. Here $\alpha = h^2/12$, since the terms $y_{xx} + \alpha y_{xxxx}$ together arise as the two leading nonvanishing terms of the expansion of $h^{-2}[y(x+h,t) - 2y(x,t) + y(x-h,t)]$, the elastic restoring force term for the lattice. [To be sure, this α is positive, but the Taylor expansion carried to the next nonvanishing term would give the linear equation $y_{tt} = y_{xx} + (h^2/12)y_{xxxx} + (h^4/360)y_{xxxxxx}$, with dispersion relation $\omega^2 = k^2-\alpha k^4+ \frac{2}{5}\alpha^2 k^6$, so that ω^2/k^2 is positive definite and well-posedness is restored.]

To lowest order $(\varepsilon = \alpha = 0)$, (19) becomes the linear wave equation whose solution is a superposition of right-progressing and left-progressing steady (shape-unchanging) "waves" (arbitrary functions), like (15). Let us seek a solution of the full equation (19) which is approximately just a right-progressing wave, not exactly steady of course, but at least only slowly varying. If we change variables from x,t to ξ,t, where $\xi = x-t$, then y (informally using the same symbol for what is mathematically a different function of the new arguments) should now vary only slowly with t.

The change of variables is effected by the replacements $\partial/\partial x \to \partial/\partial\xi$, $\partial/\partial t \to \partial/\partial t - \partial/\partial\xi$, so $y_{tt} \to y_{tt} - 2y_{\xi t} + y_{\xi\xi}$ and (19) becomes (cancelling $y_{\xi\xi}$)

(21) $$y_{tt} - 2y_{\xi t} = \varepsilon y_\xi y_{\xi\xi} + \alpha y_{\xi\xi\xi\xi} \, .$$

Since y now (i.e. expressed as a function of the new independent variables) varies slowly, by assumption, we set $\tau = \varepsilon t$, whence

(22) $$\varepsilon^2 y_{\tau\tau} - 2\varepsilon y_{\xi\tau} = \varepsilon y_\xi y_{\xi\xi} + \alpha y_{\xi\xi\xi\xi} \, .$$

Dropping the relatively negligible ε^2 term, we arrive at essentially the KdV equation; indeed if $u = \frac{1}{2} y_\xi$, then

(23) $$u_\tau + uu_\xi + \left(\frac{\alpha}{2\varepsilon}\right) u_{\xi\xi\xi} = 0 \, .$$

The remaining constant coefficient $\alpha/2\varepsilon$ could be transformed away by properly rescaling any two of the three variables u, τ, ξ. (If the $\alpha^2 y_{xxxxxx}$ term had been retained originally for the sake of well-posedness, it would have been dropped as higher order along with the term $\varepsilon^2 y_{\tau\tau}$, so the result would have been the same.)

4. Validity of the KdV Equation

At this point it is unfortunately necessary to divagate from straightforward exposition in order to indulge in a bit of polemical disputation. This is because in the last couple of years the validity of the KdV equation, as derived above and similarly elsewhere, has been challenged by T. Brooke Benjamin and his colleagues. In the following I shall be referring to and quoting from the article, "Model equations for long waves in nonlinear dispersive systems," by T. B. Benjamin, J. L. Bona, and J. J. Mahony [1972], which throws down the glove in unmistakable terms.

In order to exhibit the relevant arguments of Benjamin et al. and what I consider their essential flaws, it is preferable first to recast the derivation of the KdV equation given above into what may seem to be a simpler and more transparent form, though I believe its individual steps are not quite so well motivated. It has the advantage of never transforming the independent variables (which is, however, conducive to the misinterpretation of Benjamin et al., as will be seen).

Starting again with (19), we introduce right off

$$u \equiv y_x - y_t \; , \tag{24}$$

$$v \equiv y_x + y_t \; , \tag{25}$$

which are the so-called Riemann invariants of the lowest order (linear) approximation to (19), $y_{tt} = y_{xx}$. [This v, it should be said, has nothing to do with the modified KdV equation (4).] Noting that $y_x = (u + v)/2$, we calculate expressions for the time evolution of u and v, using (19) to eliminate y_{tt}. Thus

$$u_t = -u_x - \frac{u_x+v_x}{2}\,\varepsilon\,\frac{u+v}{2} - \alpha\,\frac{u_{xxx} + v_{xxx}}{2} \; , \tag{26}$$

$$v_t = v_x + \frac{u_x+v_x}{2}\,\varepsilon\,\frac{u+v}{2} + \alpha\,\frac{u_{xxx} + v_{xxx}}{2} \; . \tag{27}$$

If we assume that the initial disturbance in y was confined to a limited region, the right-going and left-going waves, represented by u and v respectively, soon separate and propagate each into regions where the other is absent, so that we can set v = 0 in (26) and obtain for u

$$u_t + u_x + \frac{\varepsilon}{4} uu_x + \frac{\alpha}{2} u_{xxx} = 0 \tag{28}$$

(and similarly of course for v). Thus the u and v equations have decoupled. [It should, however, not be thought that v = 0 is essential to obtain (28) -- the latter remains valid as a leading order approximation under not very severe assumptions on the form of v.]

Now (28) is nothing but the KdV equation, except written in the original independent variables x, t instead of being subjected to a Galilean transformation to eliminate the u_x term (and possibly some rescaling thereafter to adjust or normalize the coefficients). Furthermore, it is precisely in the form of equation (2.16) of Benjamin et al., aside from their use of capital letters, the presence of some inessential numerical coefficients, and the fact that their α is finite whereas mine is of order ε (and equal to $\alpha^2\varepsilon$ in their notation).

Benjamin et al. do not like some of the mathematical properties of the KdV equation, especially the behavior for large k of the dispersion function $\omega = \omega(k)$ of the linear version obtained by omitting the uu_x term. They therefore replace one x differentiation in the last term by a (negative) t differentiation, on the grounds that $u_x \approx - u_t$ by virtue of (28) itself, and so presumably $u_{xxx} \approx - u_{xxt}$. Accordingly they advocate the evolution equation

$$u_t + u_x + \frac{\varepsilon}{4} uu_x - \frac{\alpha}{2} u_{xxt} = 0 , \tag{29}$$

which, they say, though admittedly an approximation, with error $O(\varepsilon^2)$, is as valid as (28) but better behaved mathematically (for instance with respect to existence proofs). [Note the misprint just above their equation (2.18): the reference to (2.15) should obviously be to (2.16) instead.]

Before criticizing their espousal of (29), I must point out that the choice between (28) and (29) is no minor or side issue but the heart of their article. Indeed, just above their equation (2.18) they call the equivalence of what are written here as (28) and (29): "a point of central importance to our discussion". And they go on shortly to say, "It is worth further emphasis that, as an approximate model for long waves of small amplitude, [the two equations have] essentially the same formal justification."

Now for one thing we can dispute the supposed mathematical advantages of (29) over (28). Taking the equations in linearized form, introducing (9), and thinking of α as finite now (perhaps via rescaling), it comes to a comparison between the dispersion relations

$$\omega = -k + \frac{\alpha}{2} k^3 , \tag{30}$$

$$\omega = -k/(1+ \frac{\alpha}{2} k^2) , \tag{31}$$

for (28) and (29) respectively. The former has phase and group velocities which can take either sign, are unbounded in magnitude, etc.; as discussed by Benjamin et al. under the heading, "Shortcomings of the KdV equation," they find these properties disagreeable on various mathematical and computational grounds. The latter, however, is beautifully behaved in these respects. But, and it is a big but, this is the case only if $\alpha > 0$ (excluding the trivial case $\alpha = 0$). For $\alpha < 0$, (31) is if anything even worse behaved than (30), since then ω becomes infinite already for finite k. And (19) must have $\alpha < 0$ to be well posed, as explained parenthetically following it. It is true that Benjamin et al. purport to have $\alpha > 0$ in their case, which differs from (19), but if their approach were legitimate it should be expected to work in either case. Nevertheless, this is a relatively minor issue.

Another objection that might be raised is that (28), as derived above, is not an approximation but is exact, whereas (29) is at best only an approximation. Still, as mentioned before, in many cases (28) is also merely an approximation. Anyway, this too is not really an issue.

A more embarrassing criticism is to ask why Benjamin et al. stop with one such replacement. Why not replace uu_x by $-uu_t$? Or u_{xxx} in (28) by u_{xtt} or $-u_{ttt}$? This would be no less justifiable than what they do, but such a proposal begins to appear as a kind of reductio ad absurdum, since it strikes at the fundamental mathematical nature of the p.d.e. (e.g., more initial conditions at $t = 0$ would be needed to make a well posed problem).

What all this suggests is a crucial question which was inevitable right from the start. How can it be claimed that two distinct evolution equations are both satisfactory leading order approximations to the same given original "exact" equation? Since there is no suggestion that the equations could be equivalent (i.e. have the same solutions), and they obviously aren't, at most one of them can be correct. Properly performed complete systematic asymptotic reduction procedures lead to unique reduced equations--there can be no ambiguity. This is illustrated by the first (preferred) derivation given above, leading to the KdV equation (23). But the second derivation à la Benjamin et al. seems to offer the choice between (28) and (29), among many other possibilities. How can this be? It might be mentioned that the same question arises in the recent work of Neil Berger [1974] giving a rigorous derivation of the KdV equation. Again he can derive distinct forms of the equation with equal validity.

The obvious answer is that (28) and (29), etc., are not completely reduced leading order approximations. This is obvious because they explicitly contain ε and α. The point is obscured in the paper by Benjamin et al. because they emphasize what they call "the tidy form" of the equations, namely the equations they start right out with [their equations (1.1), (1.2)],

$$u_t + u_x + uu_x + u_{xxx} = 0 \ , \tag{32}$$

$$u_t + u_x + uu_x - u_{xxt} = 0 \ . \tag{33}$$

These are misleading because ε must be understood to occur implicitly in the definitions of the dependent and independent variables -- see their equation (2.17) and the sentence following it. What they have done is rescale the variables in (28) and (29) by certain powers of ε (and numbers) to obtain (32) and (33).

The essential significance of the KdV equation in the present context is that it is the <u>unique leading order reduced equation approximation.</u> In (28), we are to treat u, x, and t as finite, i.e. u as a finite-valued function varying with x and t on a finite scale, approximately a function only of the combination x-t. To eliminate the explicit appearance of the small parameters, the evident procedure is to make a Galilean transformation to get rid of the u_x term, and then (because u is now slowly varying in the new time variable) rescale the time by factor ε. This produces the KdV equation; of course, it is merely the procedure of the first derivation recapitulated. If we do the same procedure on (29), the Galilean transformation which eliminates u_x reintroduces u_{xxx}:

$$u_t + \frac{\varepsilon}{4} uu_x - \frac{\alpha}{2} (u_{xxt} - u_{xxx}) = 0 \ . \tag{34}$$

Since now u is slowly varying in t, we drop the αu_{xxt} term as being of higher order, rescale the time, and again end up with the ubiquitous and inescapable KdV equation.

In short, the KdV equation really holds a unique and privileged place, and the equation proposed and studied by Benjamin et al. is merely an approximation to it. If one is willing, however, to countenance extra terms to improve the mathematical properties of the equation, there are other ways to do so. Thus R. Témam [1969] has studied the KdV equation (3) as the limit of the well posed higher order evolution equation

(35) $$u_t + uu_x + u_{xxx} + \varepsilon u_{xxxx} = 0$$

as ε approaches zero from above.

5. Conservation Laws

A (differential) conservation law says that a divergence vanishes. In our case (time and one space variable) this means it takes the form

(36) $$T_t + X_x = 0 ,$$

where T and X are functionals of u (and conceivably functions of t and x). Integrating over an interval gives

(37) $$\frac{\partial}{\partial t} \int_a^b dx\, T + [X]_a^b = 0 ,$$

whence T may be considered the density of a quantity which is conserved in the sense that the rate of change of the total amount of it in any interval [a,b] can be interpreted as the net effect due to the fluxes through the endpoints, each flux being -X evaluated locally. Conservation laws are of widespread occurrence and frequently have both mathematical and physical significance.

The KdV equation is itself a conservation law, as seen by rewriting it in the form

$$(u)_t + (\tfrac{1}{2} u^2 + u_{xx})_x = 0 . \tag{38}$$

However, it implies more conservation laws as well. Multiplying it by u gives what can be rewritten

$$(\tfrac{1}{2} u^2)_t + (\tfrac{1}{3} u^3 - \tfrac{1}{2} u_x^2 + uu_{xx})_x = 0 . \tag{39}$$

A slightly more complicated manipulation [Whitham 1965] gives

$$(\tfrac{1}{3} u^3 - u_x^2)_t + (\tfrac{1}{4}u^4 - 2uu_x^2 + u^2 u_{xx} + u_{xx}^2 - 2u_x u_{xxx})_x = 0 . \tag{40}$$

Zabusky, Miura, and I found, by brute force calculations with undetermined coefficients, that this process could be continued indefinitely; as high a power as we started with (up to u^{10}, specifically), there was always a conservation law with that power occurring in the conserved density T, with T and X each being a polynomial in u and its x derivatives, and with each such polynomial having all terms of equal rank [the rank of a monomial $u^{a_0} u_x^{a_1} u_{xx}^{a_2} \ldots$ is defined as the number of factors $u, u_x, \ldots$ plus half the number of x-differentations, i.e. $a_0 + a_1 + a_2 + \ldots + \frac{1}{2}(a_1 + 2a_2 + 3a_3 + \ldots)$, this being the relevant combination because in the KdV equation uu_x combines with u_{xxx}, so that an extra factor u balances two extra x differentiations].

We further found that each such conservation law is essentially unique, viz., unique within two trivial transformations. The first is multiplication by a constant. The second is addition of an arbitrary x derivative to T and corresponding modification of X, for if $T_t + X_x = 0$ then also $(T+Q_x)_t + (X-Q_t)_x = 0$. Incidentally, we generally utilize this freedom, in effect the freedom to integrate by parts, to put each polynomial conserved density into a canonical form such that each monomial in it has its highest order x derivative, if any, occurring at least squared. A conserved density which is a perfect x derivative is trivial, and usually of no significance.

Our group eventually even obtained an explicit formula for the n-th conservation law, assuming it existed, but were nevertheless unable to parlay this result straightforwardly into a proof (of existence) by induction. The breakthrough came from an inspired observation of Miura's.

Miura applied the same brute-force approach to the modified KdV equation (4), and found that it too possessed a seemingly endless sequence of conservation laws,

$$(v)_t + (\tfrac{1}{3} v^3 + v_{xx})_x = 0 , \tag{41}$$

$$(\tfrac{1}{2} v^2)_t + (\tfrac{1}{4} v^3 - \tfrac{1}{2} v_x^2 + vv_{xx})_x = 0 , \tag{42}$$

$$(\tfrac{1}{4} v^4 - \tfrac{3}{2} v_x^2)_t + (\tfrac{1}{6} v^6 - 3v^2 v_x^2 + v^3 v_{xx} + \tfrac{3}{2} v_{xx}^2 - 3v_x v_{xxx})_x = 0 , \tag{43}$$

and so on. (To avoid giving a wrong impression, let me emphasize that this is a rare and remarkable property of a p.d.e. -- we have for instance proved that none of the still higher forms of the KdV equation, whether obtained from it by further raising the power of the dependent variable in the coefficient or by taking a derivative of higher order than the third, has more than three independent polynomial conserved densities. See [ ].)

Now Miura observed that the (slightly strange-seeming) transformation

$$u = v^2 + \sqrt{-6}\, v_x \tag{44}$$

takes conserved densities of the KdV dquation into those of the modified KdV equation. (The fact that this transformation is complex, whereas up to now the equations and their solutions have been thought of as real, should not be allowed to disturb one. Think of it as purely formal at this point. When we change a coefficient by a negative factor shortly, everything will in fact become real.) Thus the conserved density u itself goes into v^2 (the term $\sqrt{-6}\, v_x$ being dropped as a perfect derivative), which is (twice) the second conserved density for the v equation. Similarly $\frac{1}{2} u^2 = \frac{1}{2} v^4 - 6\, v^2 v_x - 3\, v_x^2$, which after dropping the perfect derivative te~m $6\, v^2 v_x$ is (twice) the third conserved density for the v equation. This process continues all the way up. The omission of the first conserved density, v itself, in this process has a particular significance to be explained shortly.

In fact, not merely the conservation laws but the differential equations themselves are related by Miura's transformation. For it is easy to verify that if $u = v^2 + \sqrt{-6}\, v_x$, then

$$(45) \qquad u_t + uu_x + u_{xxx} = (2v + \sqrt{-6}\, \tfrac{\partial}{\partial x})\ (v_t + v^2 v_x + v_{xxx}) \ .$$

Thus if v satisfies the modified KdV equation, u satisfies the (original) KdV equation. Miura's transformation is very reminiscent of the famous transformation of Hopf [1950] and Cole [1951], which takes the quadratically nonlinear Burgers' equation

$$(46) \qquad u_t + uu_x - u_{xx} = 0$$

into the corresponding linear heat equation (2), except that it takes the quadratically nonlinear KdV equation into a similar but cubically nonlinear equation.

We now generalize Miura's transformation by introducing a parameter, without losing the quadraticity of the coefficient of v_x in the v equation, by replacing v everywhere by v+k, k a constant.

Then $u = (v+k)^2 + \sqrt{-6}\, v_x$ implies that $u_t + uu_x + u_{xxx}$ $= (2v + 2k + \sqrt{-6}\, \partial/\partial x)\, [v_t + (v^2 + 2kv + k^2) v_x + v_{xxx}]$. To simplify the new v equation (obtained by setting the quantity in square brackets equal to zero) we make a Galilean transformation to get rid of the term $k^2 v_x$. With t and v unchanged, we replace $x \to x + k^2 t$, so that $\partial/\partial x$ is unchanged and $\partial/\partial t \to \partial/\partial t - k^2\, \partial/\partial x$. In order to keep the left side unchanged we replace $u \to u + k^2$, in effect exploiting the Galilean invariance of the KdV equation. This leads us to the result that if $u = v^2 + 2kv + \sqrt{-6}\, v_x$, then

$$u_t + uu_x + u_{xxx} = (2v + 2k + \sqrt{-6}\, \frac{\partial}{\partial x})\ [v_t + (v^2+2kv) v_x + v_{xxx}] .$$

We now choose to normalize the coefficient of the vv_x term (at the expense of the $v^2 v_x$ term) and so let $v = w/2k$. (Approaching our final result we want a new name for the much-transformed v.) To simplify the form of the coefficient of $\partial/\partial x$ in the resulting parenthetical operator we also set $k = \sqrt{6}/2\varepsilon$. The final result (essentially suggested by Clifford Gardner) is: If

$$u = w + i\varepsilon w_x + \frac{\varepsilon^2}{6} w^2 , \tag{47}$$

then

$$u_t + uu_x + u_{xxx} = (1+i\varepsilon \frac{\partial}{\partial x} + \frac{\varepsilon^2}{3} w)\ [w_t+(w+ \frac{\varepsilon^2}{6} w^2) w_x+ w_{xxx}] . \tag{48}$$

The reason for these successive transformations was to get the new variable, w, in a form which reduces to u in a limiting case, viz. $\varepsilon = 0$. It (or more precisely $\varepsilon w/\sqrt{6}$) reduces to v in the opposite limiting case $\varepsilon \to \infty$. (The reason for having chosen the coefficient of $\partial/\partial x$ to simplify is that it is the dominant one of the small terms as $\varepsilon \to 0$, since only ε^2 occurs elsewhere.)

Since w is merely a simply transformed version of v, the

"remodified" KdV equation

$$w_t + (w + \frac{\varepsilon^2}{6} w^2) w_x + w_{xxx} = 0 \quad , \tag{49}$$

like (4), has a sequence of conservation laws, which are obtainable from those for the KdV equation by the transformation (47). As before, there is an anomalous one not so obtained, viz. (49) itself rewritten:

$$(w)_t + (\frac{1}{2} w^2 + \frac{\varepsilon^2}{18} w^3 + w_{xx})_x = 0 \ . \tag{50}$$

There are now two crucial points. First, we note that (47), viewed in a formal way for $\varepsilon \to 0$, can be inverted to give w as a series of nonnegative powers of ε with coefficients which are polynomials in u and its derivatives, thus

$$\begin{aligned} w = u - i\varepsilon w_x - \varepsilon^2[u_{xx} + \frac{1}{6} u^2] + i\varepsilon^3[u_{xxx} + \frac{2}{3} uu_x] \\ + \varepsilon^4[u_{xxxx} + \frac{2}{3} uu_{xx} + \frac{5}{6} u_x^2 + \frac{1}{18} u^3] + \dots \ . \end{aligned} \tag{51}$$

Second, since (49) is (formally) a consequence of the KdV equation, because the operator $1 + i\varepsilon \frac{\partial}{\partial x} + \frac{\varepsilon^2}{3} w$ in (48) can be formally inverted [its inverse is $1 - i\varepsilon \frac{\partial}{\partial x} - \varepsilon^2(\frac{1}{3} w + \frac{\partial^2}{\partial x^2}) + i\varepsilon^3(\frac{1}{3} w_x + \frac{2}{3} w \frac{\partial}{\partial x} + \frac{\partial^3}{\partial x^3}) + \dots$], and the KdV equation is independent of ε, it follows that order by order in ε, (50) provides a distinct conservation law for the KdV equation. Thus we construct an infinite sequence of conservation laws!

Actually, every second conservation law so obtained is trivial. To see this, we split w into real and imaginary parts, $w = r + is$, insert this into (47), and separate real and imaginary parts to obtain

(52) $$u = r - \varepsilon s_x + \frac{\varepsilon^2}{6}(r^2 - s^2) \ ,$$

(53) $$0 = s + \varepsilon r_x + \frac{\varepsilon^2}{3} rs \ .$$

Solving the latter gives

(54) $$\begin{aligned} s &= -\varepsilon r_x / (1 + \frac{\varepsilon^2}{3} r) \\ &= [-\frac{3}{\varepsilon} \ln (1 + \frac{\varepsilon^2}{3} r)]_x \ , \end{aligned}$$

a perfect derivative. Hence the odd powers of ε, which, obviously, constitute collectively the imaginary part of w, are all trivial conserved densities.

Substituting (54) into (52) gives

(55) $$u = r + \varepsilon^2 r_{xx}/(1 + \frac{\varepsilon^2}{3} r) - \frac{\varepsilon^4}{2} r_x^2 (1 + \frac{\varepsilon^2}{3} r)^2 + \frac{\varepsilon^2}{6} r^2 \ ,$$

which generates r directly by inversion. To see that none of these conserved densities (the even powers of ε in w) is trivial, we note that the terms with u alone, without derivatives of u, which will be generated are just those generated by inverting (43) with r_x and r_{xx} set equal to zero, namely

(56) $$u = r + \frac{\varepsilon^2}{6} r^2 \ .$$

But this quadratic equation has the solution

(57) $$r = \frac{3}{\varepsilon^2} \left(-1 \pm \sqrt{1 + \frac{2}{3}\varepsilon^2 u}\,\right),$$

or, since the plus sign is clearly the appropriate one,

(58) $$r = \frac{3}{\varepsilon^2} \sum_{n=1}^{\infty} \binom{1/2}{n} \left(\frac{2}{3}\varepsilon^2 u\right)^n .$$

Evidently every nonnegative even power of ε, and correspondingly every positive integral power of u, actually occurs in this expansion, since the binomial coefficient never vanishes. Because the derivative of any polynomial in u, u_x, u_{xx}, ... obviously cannot contain a monomial which is a pure power of u, we see that there is a nontrivial (i.e. non-perfect-derivative) conserved density of each positive rank.

6. Schrödinger Equation

At this point it is convenient to eliminate the peculiar factor -6 by the replacements $u \to -6u$, $v \to \sqrt{-6}\, v$. Miura's transformation becomes: If

(59) $$u = v^2 + v_x ,$$

then

(60) $$u_t - 6uu_x + u_{xxx} = (2v + \frac{\partial}{\partial x})\ (v_t - 6v^2 v_x + v_{xxx}) .$$

Accordingly we will take the KdV and modified KdV equations in the forms

(61) $$u_t - 6uu_x + u_{xxx} = 0 ,$$

(62) $$v_t - 6v^2 v_x + v_{xxx} = 0 .$$

Now observe that Miura's transformation goes easily in only one direction. Given v we obtain u immediately from (59); and if v satisfies (62) then u is immediately seen to satisfy (61), by virtue of (60). The converse is less clear. Given u, we must solve the Riccati equation (59) for v, and the solution is not unique; and if u satisfies (61), it does not necessarily follow that v satisfies (62), since the operator $(2v + \partial/\partial x)$ in (60) has no inverse -- since it annihilates the function $\exp\ (-2 \int^x v\, dx)$. [It, or rather its analog in w, $(1 + i\varepsilon\ \partial/\partial x + \frac{\varepsilon^2}{3} w)$, as in (48), was usefully found invertible only because we were then working formally in series of nonnegative integral powers of ε. The function $\exp\ (\frac{i}{\varepsilon} x + \frac{i\varepsilon}{3} \int^x w\, dx)$ that is annihilated by this operator has no such expansion.]

The handy way to solve the Riccati equation (59) for v, given u, is by means of the linearizing transformation

(63) $$v = \frac{\psi_x}{\psi} ,$$

which transforms (59) into

(64) $$\psi_{xx} = u\psi .$$

Let us suppose the conceptually simplest boundary value problems for (61) and (62), that u and v are required to be periodic functions of x with a given period ℓ. (Obviously if v is periodic, u must be too, and with the same period.) It is natural to seek to have ψ similarly periodic. If v were known, ψ could be obtained from (63), as $\psi = \exp \int^x v\,dx$, and periodicity of ψ would require $\int_0^\ell v\,dx = 0$. It is tempting to hope to impose this as an extra condition to pick one special solution v out of the one-parameter family of solutions of (59). Unfortunately, however, this cannot generally be achieved, because (64) does not generally possess even a single periodic solution ψ (other than the unacceptable $\psi \equiv 0$); e.g., if $u > 0$ everywhere, then if $\psi \not\equiv 0$ were periodic it would have a positive maximum or a negative minimum, at neither of which (64) could hold.

It would actually be quite natural to abandon the imposition of periodicity on ψ, and instead, due to the homogeneity of its relations with u and v, merely require it to be "exponential-periodic" with period ℓ, that is to be the product of some exponential exp ax (a being constant) with a function of period ℓ. Such a ψ satisfying (64) can always be found, according to Floquet theory. It is perhaps fortunate that we thought of a different remedy first, since it opened up to us a vast new world.

That remedy was to utilize the Galilean invariance of the KdV equation to introduce a constant into Miura's transformation, much as done before for the proof of existence of infinitely many conservation laws. Because of the factor -6 introduced, if we make the replacement $u \to u - \lambda$ and wish to keep t and (61) invariant, we must replace $x \to x + 6\lambda t$.

We conclude from (59) and (60) that if

$$u - \lambda = v^2 + v_x \,, \tag{65}$$

then

$$u_t - 6uu_x + u_{xxx} = (2v + \frac{\partial}{\partial x}) \; [v_t - 6(\lambda+v^2)v_x + v_{xxx}] \,. \tag{66}$$

If we now linearize (65) as before by (63), we obtain

$$\psi_{xx} + (\lambda - u)\psi = 0 \,. \tag{67}$$

This is the well known so-called "time-independent Schrödinger equation," with u in the role of the potential. [The time referred to here, say τ, has nothing to do with the time t of our evolution equations but is represented by the residue of Fourier-analyzing out the quantum-mechanical time dependence, the term $\lambda\psi$, coming from an original term $i\psi_\tau$ in the time-dependent Schrödinger equation, cf. (18).] We shall here, as is customary, call it simply the "Schrödinger equation."

Now that we have the disposable parameter λ, we can choose it (given any periodic u) to permit a periodic solution ψ of (67); this is a standard Sturm-Liouville eigenvalue problem, and it is known that (67) possesses an infinite discrete spectrum of real eigenvalues $\lambda_1 < \lambda_2 \leq \lambda_3 < \lambda_4 \leq \lambda_5 < \lambda_6 \leq \ldots$ with $\lambda_n \to \infty$ as $n \to \infty$. Typically the λ_n are all distinct and each determines a real periodic eigenfunction ψ_n which is unique up to scaling (multiplication by a constant).

These last considerations were based on viewing (67) as an ordinary differential equation, since t enters only implicitly as a parameter and there are no differentiations with respect to it. In fact, however, u is supposed to evolve with time according to (61), hence the λ_n and ψ_n all evolve. To determine how they do so, we impose on v the "remodified" evolution equation

(68) $$v_t - 6(\lambda + v^2)v_x + v_{xxx} = 0 ,$$

where λ is some constant which is one of the eigenvalues λ_n of (67) at $t = 0$, say. It is important to note that we do not assume that λ remains an eigenvalue, since λ is taken to be a constant and the eigenvalues are presumed to vary with time. By (63) and (67) we have (65), hence (66), and are thus assured that u does evolve according to the KdV equation (61).

The evolution of ψ is easily obtained by first integrating (63) to obtain

(69) $$\psi = \exp \int^x v\, dx .$$

We observe by the way that the condition of periodicity on ψ, namely $\int_0^\ell v\, dx = 0$, which is satisfied at $t = 0$ by virtue of the way λ was chosen (as one of the λ_n), remains true because v is a conserved density of (68) whose fluxes at the endpoints of the interval $[0,\ell]$ are equal, by the periodicity of v, and hence cancel in their contributions (one is inward, the other outward), so that the total amount in the one-period interval cannot change with time. Thus we know that ψ remains periodic. (Note that we have again made special use of the anomalous first conservation law of the v equation.) Differentiating (69) and using (68) gives

$$(70)\qquad \psi_t = \psi \int^x v_t \, dx$$

$$= \psi \int^x [6(\lambda+v^2)v_x - v_{xxx}] \, dx$$

$$= \psi\left(6\lambda v + 2v^3 - v_{xx} + \text{const.}\right).$$

The constant of integration may be omitted since its presence merely corresponds to the freedom to rescale ψ arbitrarily at each time, i.e. to multiply ψ by an arbitrary function of t. (Alternatively, we could impose a normalization condition on ψ, such as $\int_0^\ell \psi^2 \, dx = 1$, and use this to determine the "constant" as a function of time.) Eliminating v by (63) and simplifying quickly gives for ψ the autonomous evolution equation

$$(71)\qquad \psi_t - \left(6\lambda + 3\,\frac{\psi_{xx}}{\psi}\right)\psi_x + \psi_{xxx} = 0\ ,$$

or, linear but not autonomous,

$$(72)\qquad \psi_t - 3(\lambda + u)\psi_x + \psi_{xxx} = 0\ .$$

To interpose a mathematical observation, it might be supposed that since this linear evolution equation has periodic coefficients, it would follow that an initially periodic ψ would necessarily evolve so as to remain periodic. This would obviate the relevance of the argument given just before to establish the periodicity of ψ. However, (71) does not determine the evolution of ψ in a local way, even though it is a local (i.e. differential) equation, because it is of "superparabolic" type. Three boundary conditions are undoubtedly needed to make the initial value problem well posed. Here the three conditions are taken to be $[\psi]_0^\ell = 0$, $[\psi_x]_0^\ell = 0$, $[\psi_{xx}]_0^\ell = 0$, in accord with the established periodicity of ψ.

To tie things up, we need merely observe that at any time t, not merely at $t = 0$, we have (63) and (65), hence also (67). Thus the constant λ remains an eigenvalue of (67) for all time; or turning the statement around, every eigenvalue of (67) remains constant for all time! The argument presented here is the most intuitive and least calculational that I know, though it is a bit subtle in its logic.

Actually, of more real significance than the constancy of the eigenvalues of (67) under evolution of u in accord with (61), is the fact that ψ evolves according to a p.d.e. of its own, viz. (71). Under any deformation of the potential u in the Schrödinger equation one can obtain the corresponding deformation of ψ by standard perturbation theory, but this deformation is generally global in character, so that an integrodifferential equation for ψ would result. Only for very special deformations of u is the deformation of ψ expressible locally.

The systematic search for such special evolution equations for u was carried out by Lax [1968] and A. Lenard (see [ ]). The lowest order p.d.e. found is $u_t + u_x = 0$, which is trivial since it states that u is a function of only the single variable (x-t), and hence merely translates rigidly as t varies, whereupon ψ simply translates along with it. The next order p.d.e. found is the third order KdV equation (61). Then comes the fifth order p.d.e. (to be derived presently)

(73) $$u_t + 3u^2 u_x - 2u_x u_{xx} - uu_{xxx} + \frac{1}{10} u_{xxxxx} = 0 ,$$

a seventh order one, and so on.

To see how this goes, we follow Lenard, approximately. Consider the equation

(74) $$\psi_{xx} - h\psi = 0 ,$$

where h and ψ are assumed to be periodic with period ℓ. Here h will later be replaced by $u-\lambda$, to make this into the Schrödinger equation, or by λr, to make this into a Sturm Liouville equation we shall also touch on.

We now perturb h and calculate the corresponding perturbation of ψ. Denoting perturbed quantities by a circumflex, we have

(75) $$\hat{\psi}_{xx} - h\hat{\psi} - \hat{h}\psi = 0 .$$

This is an inhomogeneous linear equation for $\hat{\psi}$; it may be solved by quadratures by setting

(76) $$\hat{\psi} = \alpha\psi$$

in it and using (74) to obtain

(77) $$\alpha_{xx}\psi + 2\alpha_x\psi_x - \hat{h}\psi = 0 ,$$

which is a first order equation for α_x. Multiplying by the integrating factor ψ leads to

(78) $$\alpha_x\psi^2 = \int^x \hat{h}\,\psi^2\,dx .$$

Periodicity of α and hence $\hat{\psi}$ (assuming that of ψ, h, and $\hat{h}$) requires

(79) $$\int_0^\ell \hat{h}\psi^2\,dx = 0 ,$$

but this will be guaranteed anyway by what follows.

If α and hence $\hat{\psi}$ are to come out locally determined, $\hat{h}\psi^2$ must be integrable; we therefore set it equal to a perfect derivative, and since $\hat{h}\psi^2$ is of second degree in ψ and all our relations are homogeneous in ψ and its derivatives, we assume this to have the form

(80) $$\hat{h}\psi^2 = (A\psi^2 + B\psi\psi_x + C\psi_x^2)_x .$$

Carrying out the indicated differentiations, eliminating ψ_{xx} by (74), and then separately equating the ψ^2, $\psi\psi_x$, and ψ_x^2 terms gives

$$\hat{h} = A_x + Bh , \tag{81}$$

$$0 = 2A + B_x + 2Ch , \tag{82}$$

$$0 = B + C_x . \tag{83}$$

Eliminating B by (83) and then A by (82) yields

$$\hat{h} = \frac{1}{2} C_{xxx} - 2C_x h - Ch_x . \tag{84}$$

We can now choose any function C and obtain $\hat{h}$ from this. We also obtain B and A from (83) and (82), so from (78) and (80),

$$\begin{aligned} \alpha_x &= A + \frac{B\psi_x}{\psi} + \frac{C\psi_x^2}{\psi} + \frac{k}{\psi^2} \\ &= \frac{1}{2} C_{xx} - Ch - \frac{C_x\psi_x}{\psi} + \frac{C\psi_x^2}{\psi^2} + \frac{k}{\psi^2} . \end{aligned} \tag{85}$$

If we choose the constant of integration $k = 0$, this becomes automatically integrable! Since $h = \psi_{xx}/\psi$, the integral is

$$\alpha = \frac{1}{2} C_x - \frac{C\psi_x}{\psi} , \tag{86}$$

where we have omitted the new constant of integration since its effect on $\hat{\psi} = \alpha\psi$ is a trivial manifestation in the perturbation of the freedom to scale the exact full ψ. Thus we conclude that

$$\hat{\psi} = \frac{1}{2} C_x\psi - C\psi_x . \tag{87}$$

As a first example, if we choose $C = -1$ (a multiplicative constant in C is unimportant, the constant being chosen merely for convenience) we obtain

$$\hat{h} = h_x , \quad \hat{\psi} = \psi_x . \tag{88}$$

Interpreting the perturbation as a time derivative, we get the trivial case of (uniform) translation with time mentioned previously. A second example is provided by $C = -2\lambda h - 6\lambda^2$:

(89) $$\hat{h} = -\lambda h_{xxx} + 6\lambda(h+\lambda)h_x ,$$

(90) $$\hat{\psi} = -\lambda h_x\psi + 2\lambda(h+3\lambda)\psi_x .$$

Thus we recover the KdV equation (set $h = u-\lambda$, $\hat{h} = \lambda u_t$) and (71), the evolution equation for ψ (set $\hat{\psi} = \lambda\psi_t$ and eliminate $h = u-\lambda$ $= \psi_{xx}/\psi$). A third example, studied in recent work by Harry Dym and myself [unpublished], results from choosing $C = 2h^{-1/2}$ [to make the last two terms of (84) cancel], whereby

(91) $$\hat{h} = (h^{-1/2})_{xxx} ,$$

(92) $$\hat{\psi} = (h^{-1/2})_x\psi - 2h^{-1/2}\psi_x .$$

With $h = \lambda r$, $\hat{h} = {}^{-1/2}r_t$, we obtain what I call Dym's equation, (7). (Higher order equations of Dym's type can also be obtained by the method about to be illustrated.)

Returning to the selection $h = u-\lambda$, we may ask what possibilities there are for C to make $\hat{h}$ given by (84) come out to be a function of u and its x-derivatives, to be interpreted as u_t. Actually, we can tolerate a power of λ as a common factor on the right, since it can be absorbed into t -- or $\hat{h}$ can be interpreted as u_t multiplied by a power of λ. Since λ occurs linearly on the right side of (84) (now that $h = u-\lambda$), it is natural to think of working with a finite series of integral powers of λ for C, chosen to cancel out power by power except at one end (only) of the series, where the term left "sticking out" can serve to provide the evolution equation. It is plausible to ask first for a formal infinite series for C which satisfies (84) with $\hat{h}$ omitted on the left. Such a series can then be truncated at any order to provide an evolution equation.

There is a question whether the series should be in increasing or decreasing powers of λ, i.e. whether we should think of λ as small or large. If the former, we will have a third order equation to integrate at every step, and this seems impossible to carry out explicitly. In the latter case, however, we have merely a quadrature at every step; indeed, our equation can be written for recursive generation as

(93) $$C = -\frac{1}{2}\lambda^{-1}\int^{x}\left[\frac{1}{2}C_{xxx} - 2C_{x}u - Cu_{x}\right]dx .$$

Starting with a constant (of integration) on the right, say 1, we obtain successively

(94) $$C = 1 - \frac{1}{2}\lambda^{-1}\int^{x}[-u_{x}]\,dx + O(\lambda^{-2})$$
$$= 1 + \frac{1}{2}\lambda^{-1}u + O(\lambda^{-2}) ,$$

(95) $$C = 1 + \frac{1}{2}\lambda^{-1}u - \frac{1}{4}\lambda^{-2}\int^{x}\left[\frac{1}{2}u_{xxx} - 3u_{x}u\right]dx + O(\lambda^{-3})$$
$$= 1 + \frac{1}{2}\lambda^{-1}u - \frac{1}{8}\lambda^{-2}(u_{xx} - 3u^{2}) + O(\lambda^{-3}) ,$$

and so on: to our surprise we are able to integrate explicitly at every step, apparently indefinitely. The reason for this, pointed out to me by Ira Bernstein in a different but remarkably similar situation, is that the homogeneous version of equation (84) for C, though linear, can be integrated explicitly by treating it nonlinearly; if we multiply by C we can integrate to obtain

(96) $$\lambda = \frac{1}{2}CC_{xx} - \frac{1}{4}C_{x}^{2} - C^{2}(u - \lambda) ,$$

where the constant of integration (on the left) has been chosen so that $C = 1 + O(\lambda^{-1})$ in accord with the previous choice. Solving for C from the dominant term on the right, $C^{2}\lambda$, gives

(97) $$C = [1 - \lambda^{-1}(\frac{1}{2} CC_{xx} - \frac{1}{4} C_x^2 - C^2u)]^{1/2} ,$$

which after formal expansion in nonpositive integral powers of λ is suitable for recursive generation of C. No further integration is required (they have already been performed all at once, as it were), and the result must evidently be identical with that obtained by the previous linear recursion -- which shows that at every step in the latter algorithm the integration will be explicitly possible.

Now if we truncate (95) at the first term, taking $C = 1$, (84) (now inhomogeneous) gives

(98) $$\hat{h} = - u_x ,$$

which with $\hat{h} = u_t$ is the simple translational evolution again. The next higher truncation is $C = 1 + \frac{1}{2} \lambda^{-1} u$, whence

(99) $$\hat{h} = \frac{1}{4} \lambda^{-1}u_{xxx} - \lambda^{-1}u_x(u-\lambda) - (1 + \frac{1}{2} \lambda^{-1}u)u_x$$

$$= \lambda^{-1}[\frac{1}{4} u_{xxx} - \frac{3}{2} uu_x] ,$$

which with $\hat{h} = - \frac{1}{4} \lambda^{-1} u_t$ gives the KdV equation (61). Note how the λ^0 terms (necessarily) cancelled. The next choice is $C = 1 + \frac{1}{2} \lambda^{-1}u - \frac{1}{8} \lambda^{-2}(u_{xx} - 3u^2)$, which produces (after automatic cancellation of the λ^0 and λ^{-1} terms)

(100) $$\hat{h} = \frac{5}{8} \lambda^{-2}[- \frac{1}{10} u_{xxxxx} + uu_{xxx} + 2u_xu_{xx} - 3u^2u_x] ,$$

which leads to the evolution equation (73) given in advance.

The evolution equation for ψ, (72), is linear but contains λ; it is hardly surprising that all the eigenfunctions do not evolve according to a single common linear equation. But it is easy to obtain a common nonlinear equation: merely use (67) to eliminate λ in (72) to get

(101) $$\psi_t - (6u - 3 \frac{\psi_{xx}}{\psi})\psi_x + \psi_{xxx} = 0 .$$

This is not very interesting, but becomes more so when multiplied by 2ψ and rewritten in the form

$$(\psi^2)_t - 6u(\psi^2)_x + (\psi^2)_{xxx} = 0 \; ; \tag{102}$$

the squares of the eigenfunctions do all evolve according to a single common linear equation! (This equation will play a significant role later.) Moreover, another nonsurprise, ψ^2 is a conserved density, though this is less apparent from (102) than from (71), which after multiplication by 2ψ gives

$$(\psi^2)_t + (-6\,\lambda\psi^2 - 4\psi_x^2 + 2\psi\psi_{xx})_x = 0 \, . \tag{103}$$

Of course this conservation law does not really give information about a solution u of the KdV equation, since ψ can be any periodic solution of (67) with eigenvalue λ, and there is no obvious independent way to set its scale (to normalize it).

This dfficulty may be obviated by the following consideration. Since the KdV equation (61) results from (99) by setting $\hat{h} = -\frac{1}{4}\lambda^{-1}u_t$, and $\hat{h} = (u-\lambda)^{\wedge} = \hat{u}$, we interpret $\hat{\psi}$ in (87) as $-\frac{1}{4}\lambda^{-1}\psi_t$, while $C = 1 + \frac{1}{2}\lambda^{-1}u$. [See lines above and below (99).] Both sides of (87) have the integrating factor ψ^{-3}, so that

$$(\psi^{-2})_t = (4\lambda C\psi^{-2})_x \, . \tag{104}$$

I.e. ψ^{-2} is also a conserved density. The constancy in time of $\left(\int_0^\ell \psi^2\,dx\right)\left(\int_0^\ell \psi^{-2}\,dx\right)$ is now genuinely informative, since the expression as a whole is invariant under rescaling of ψ. Both $\psi^2\int_0^\ell \psi^{-2}\,dx$ and $\psi^{-2}\int_0^\ell \psi^2\,dx$ are genuine conserved densities for the KdV equation, though not local functionals of u. (If ψ vanishes somewhere, $\int \psi^{-2}\,dx$ may be interpreted by taking the path of integration a little off the real axis, in case the functions involved are analytic, or equivalently but more generally by taking the Hadamard finite part.)

Another way to generate the polynomial conserved densities (due to C. Gardner, and in fact the earliest way we knew), which is simplified by using the preceding results, is to solve (67) asymptotically for $\lambda \to \infty$ by the WKB method. We write (formally)

(105) $$\psi = A \exp i \lambda^{1/2} \int^x B\, dx ,$$

where A and B [which have nothing to do with A, B, C introduced in (80)] are series in nonnegative integral powers of $\lambda^{-1/2}$ (actually of λ^{-1}, it will turn out), the A being included for convenience, though it gives no greater generality (since it can be put logarithmically into the exponent and then differentially under the integral and thus absorbed into B). Substituting (105) into (67) gives

(106) $$A_{xx} + 2A_x i\lambda^{1/2}B + A(-\lambda B^2 + i\lambda^{1/2}B_x + \lambda - u) = 0 .$$

We use the extra freedom provided by A to impose an extra condition, that the (formally, and actually if not yet so established) real and imaginary parts of (106) hold separately. The imaginary part gives $2A_xB + AB_x = 0$, or

(107) $$B = A^{-2}$$

if we arbitrarily fix the amplitude of (105). With this the real part of (106) becomes $A_{xx} + A(-\lambda A^{-4} + \lambda - u) = 0$, from which A can be found recursively when it is solved for in the way appropriate for large λ,

(108) $$A = [1 + \lambda^{-1}(\frac{A_{xx}}{A} - u)]^{-1/4} ,$$

and the right side is expanded in powers of λ^{-1}. It is evident that the coefficients of these powers are polynomials in u and its derivatives.

Actually, it is not so much A we want as A^2, since that is a conserved density. The reason is that A obtained as just described is evidently real, hence so is B, whence the real and imaginary parts of ψ are $A \cos (\lambda^{1/2} \int^x B\, dx)$ and $A \sin (\lambda^{1/2} \int^x B\, dx)$. Since λ is (being taken to be) real, the real and imaginary parts of ψ separately satisfy (67), so that their squares are conserved densities. Adding, we see that A^2 is a conserved density. Clearly, the coefficient of each power of λ^{-1} in A^2 is by itself a conserved density, and is a polynomial in u and its derivatives.

Also A^{-2} is a conserved density. This is not so obvious from the fact that ψ^{-2} is, as one might have expected, but follows from (107). For, the periodicity condition on ψ as given by (105) is

(109) $$\lambda^{1/2} \int_0^{\ell} B\, dx = 2\pi N ,$$

where N is a (large) positive integer characterizing a pair of eigenfunctions (by being half the number of their zeros in the period ℓ). This shows that $\int_0^{\ell} B\, dx$ is constant, which implies (in the present context) that B is a conserved density. So, of course, is the coefficient of each power of λ^{-1} in its expansion. Naturally we do not obtain truly different polynomial conserved densities this way, but merely the same ones as before, modulo perfect derivatives.

7. Solitons

When the KdV equation is solved on the infinite interval $-\infty < x < \infty$ with essentially localized initial data, i.e. $u(x,0)$ vanishing outside a finite interval or at least approaching zero sufficiently rapidly as $|x| \to \infty$, it is found that the asymptotic solution for $t \to \infty$ consists of a certain number (possibly zero) of distinct widely spaced solitary waves travelling to the right at distinct velocities, as well as a dispersing "hash" left behind and decaying in amplitude. The solitary steady progressing wave solutions of (61) can be found by assuming for u the form

$$u(x,t) = U(x-ct) , \tag{110}$$

c being the velocity of the wave. This gives for U the ordinary differential equation

$$- cU' - 6UU' + U''' = 0 . \tag{111}$$

Integrating and assuming that $U \to 0$ as $|x| \to \infty$ gives

$$- cU - 3U^2 + U'' = 0 . \tag{112}$$

(In view of Galilean invariance this is not essentially more restrictive than assuming $U \to$ constant.) Multiplying by $2U'$ and again integrating gives

$$- cU^2 - 2U^3 + U'^2 = 0 . \tag{113}$$

An elementary quadrature yields finally

$$U = - \frac{c}{2} \operatorname{sech}^2 \frac{\sqrt{c}}{2} (x - ct - x_0) . \tag{114}$$

Thus a solitary wave can travel only to the right $(c > 0)$ -- for $c < 0$ (114) still makes sense, in that U is real, but it becomes oscillatory and fails to vanish at ∞ (is no longer solitary). It takes always the form of a dip, a decrease in u compared to the circumambient value but over only an essentially finite region, in view of the fast (exponential) decay of the hpyerbolic secant. (In the original form of the KdV equation, without the coefficient -6, the solitary waves are positive, increases in u.) We note that all solitary waves have the same shape up to translation and scaling of the variables, and that the speed of the wave (c) is proportional to its amplitude (c/2) and inversely proportional to the square of its width $(2/\sqrt{c})$.

The remarkable fact is now that if we let the initial condition evolve backwards in time, we find for $t \to -\infty$ exactly the same number of solitary waves emerging from the hash, and with the same set of values of c. It is hard to resist identifying each of the solitary waves that appears at large positive t with its spitting image at large negative t. A somewhat metaphysical entity, which manifests itself as a solitary wave in the absence of disturbing influences (other waves or hash), may now be posited: this entity is called a soliton [Zabusky-Kruskal 1965].

If we eliminate the irrelevant emphasis on an initial state, on $t = 0$, we can describe the situation as follows. At large negative t let there be a number of solitary waves of distinct amplitudes and corresponding velocities travelling inward from infinity on the negative x-axis. During some essentially finite interval of time they interact, with each other and with whatever hash may be present. Eventually they emerge and separate, going off to infinity on the positive x-axis with their original shapes and velocities. During the interaction it is hard to know what "they" are -- there may be nothing present looking like the original waves, and if there are "waves" present they are certainly not solitary. We shall see that there is a very reasonable way to interpret the solitons during interaction and say just how they are distorted.

In the absence of hash we can obtain exact descriptions of the interactions of solitons by a simple variational procedure based on the conserved densities. For the KdV equation in the form (61), the corresponding constants of motion are

(115) $$I_1 = \int_{-\infty}^{\infty} u \, dx \; ,$$

(116) $$I_2 = \int_{-\infty}^{\infty} u^2 \, dx \; ,$$

(117) $$I_3 = \int_{-\infty}^{\infty} (u^3 + \frac{1}{2} u_x^2) \, dx \; ,$$

(118) $$I_4 = \int_{-\infty} (u^4 + 2uu_x^2 + \frac{1}{5} u_{xx}^2) \, dx \; ,$$

and so on. If we make I_3 stationary, for a prescribed value of I_2 (it is best to omit I_1, though we would include it if were were working on a finite interval with periodic functions, as earlier), we obtain the Euler variational equation

(119) $$2\alpha u + 3u^2 - u_{xx} = 0 \; ,$$

where α is a Lagrange multiplier. A function u satisfying (119) at one time continues to satisfy it as it evolves under the KdV equation -- for instance, if u minimizes I_3 for given I_2, it continues to do so because I_2 and I_3 are constants of motion. Thus we have found an ordinary differential equation (119), satisfied by a solution of the p.d.e. (61).

So far this isn't very interesting, because a solution of (119) is a soliton, as we see by comparison with (112). ($\alpha = \frac{1}{2}$ c.) But now let us do the same thing one level up. If we vary I_4 with I_2 and I_3 constrained to given values, we obtain

(120) $$2\beta u + \gamma(3u^2 - u_{xx}) + 4u^3 - 2u_x^2 - 4uu_{xx} + \frac{2}{5} u_{xxxx} = 0 \; .$$

Again, if (120) is satisfied at one time, it remains satisfied as u evolves. Now among the many solutions of (120) it is possible to find solitons. To see this one can use (119) to eliminate higher derivatives of u in favor of lower ones, $u_{xxxx} = (2\alpha u + 3u^2)_{xx}$, etc.

Eventually one will want also $u_x^2 = 2\alpha u^2 + 2u^3$ [cf. (113)]. The higher powers of u cancel automatically and (120) reduces to

$$(121) \qquad (2\beta - 2\alpha\gamma + \frac{8}{5}\alpha^2)u = 0 ,$$

whereupon the factor u may be omitted. For suitable values of β and γ this may have one or even two positive real solutions α. Suppose there are two. Then either one of two solitons satisfies (120). Hence so does a function consisting, loosely speaking, of both those solitons but "infinitely far apart." In the course of time the two solitons, moving at different velocities, will come closer together (if the faster is the one to the left) and eventually they will interact. At every time, however, (120) will remain satisfied and govern whatever shape their interaction takes.

Since we have an infinite sequence of conservation laws, the same argument can be carried out for any number of simultaneously interacting solitons. We obtain in each case a single higher order but still ordinary differential equation governing all stages of the interaction.

This is all very well, but still does not provide an interpretation of what the solitons are during the interaction. To that end we can apply a different variational argument. Let us minimize I_2 under the constraint that u permit n eigenfunctions $\psi^{(1)},\dots,\psi^{(n)}$ with prescribed real eigenvalues $\lambda^{(1)},\dots,\lambda^{(n)}$. The constraint is that a $\psi^{(m)}$ exist satisfying (67) for each $\lambda^{(m)}$,

$$(122) \qquad \psi_{xx}^{(m)} + (\lambda^{(m)} - u)\psi^{(m)} = 0 , \qquad m = 1,\dots,n.$$

This is to be satisfied on the infinite interval, and we assume that the $\psi^{(m)}$ are square-integrable functions and normalize them,

$$(123) \qquad \int_{-\infty}^{\infty} \psi^{(m)^2}\, dx = 1 , \qquad m = 1,\dots,n.$$

(We take the $\psi^{(m)}$ real; the $\lambda^{(m)}$ are evidently all negative.) With appropriate Lagrange multipliers $\alpha^{(m)}(x)$ and $\beta^{(m)}$, we require that

$$(124) \quad \int_{-\infty}^{\infty} \left\{ \frac{1}{2} u^2 + \sum_{m=1}^{n} [\alpha^{(m)} (\psi_{xx}^{(m)} + (\lambda^{(m)} - u)\psi^{(m)}) + \beta^{(m)} \psi^{(m)2}] \right\} dx ,$$

be stationary with respect not only to variation of u but also each $\psi^{(m)}$. This gives the variational conditions

$$(125) \qquad u - \sum_{m} \alpha^{(m)} \psi^{(m)} = 0 ,$$

$$(126) \qquad \alpha_{xx}^{(m)} + \alpha^{(m)}(\lambda^{(m)} - u) + 2\beta^{(m)}\psi^{(m)} = 0 , \quad m=1,\ldots,n.$$

Multiplying the latter by $\psi^{(m)}$ and using (122) (after double integration by parts) and (123) gives

$$(127) \qquad 2\beta^{(m)} = 0 , \qquad m = 1,\ldots,n .$$

Therefore (126) becomes for $\alpha^{(m)}$ exactly the same equation as (122) is for $\psi^{(m)}$. We infer that $\alpha^{(m)}$ equals $\psi^{(m)}$, up to a constant multiplier:

$$(128) \qquad \alpha^{(m)}(x) = k^{(m)} \psi^{(m)}(x) .$$

(The second independent solution blows up as $|x| \to \infty$ in such a way as to make the terms in (124) not individually integrable, contrary to implicit assumption. If we were in the periodic case, the second independent solution would normally not be periodic and excluded on that ground.) From (125) we therefore obtain

$$(129) \qquad u = \sum_{m=1}^{n} k^{(m)} \psi^{(m)2} .$$

That is, the minimizing u is a linear superposition of the squares of its n eigenfunctions with the prescribed n eigenvalues.

As with the previous variational principles, since the conditions are constants of motion (specifically, here, I_2 and the $\lambda^{(m)}$), the conclusion continues to hold as u evolves. It is not a priori evident that the $k^{(m)}$ do not vary with t, but this in fact follows from (102); if we take the $k^{(m)}$ constant, the weighted sum of (102) applied to each $\psi^{(m)}$ gives back (61) and justifies the constancy of the $k^{(m)}$ a posteriori. Assuming that u separates into solitary waves for large $|t|$ (which is proved in [ ]), the $k^{(m)}$ can be evaluated as if $n = 1$ (since at most one term in the sum is nonnegligible at any one place). Integrating (129) over a limited but effectively infinite interval, with u having the single-soliton form (114), gives $-2\sqrt{c^{(m)}} = k^{(m)}$ by elementary quadrature. To relate $c^{(m)}$ to $\lambda^{(m)}$, a simple way is to observe that for $|x| \to \infty$, (112) becomes approximately $u'' \approx cu$ so that $u \sim \exp(-\sqrt{c}|x|)$, while (67) becomes $\psi_{xx} \approx -\lambda\psi$ so that $\psi \sim \exp(-\sqrt{-\lambda}\,|x|)$; since $u \sim \psi^{(m)^2}$ here, we have $\sqrt{c} = 2\sqrt{-\lambda}$ or

$$k^{(m)} = -4\sqrt{-\lambda^{(m)}}\,. \tag{130}$$

The fact that $k^{(m)}\psi^{(m)^2}$ is a solitary wave when it is the only nonnegligible term in the sum, and the validity of (129) throughout interaction, provide strong and almost compelling justification for viewing (129) as a decomposition of u as a linear superposition of solitons at all times, with the solitons of course distorted by interaction when they are not well separated. Indeed, it is plausible to generalize further and call $-4\sqrt{-\lambda^{(m)}}\,\psi^{(m)^2}$ a soliton for any decently behaved u, not merely one made up of nothing but its solitons as given by (129). The right side of (129) then represents the soliton part of u, and the remainder provides an exact definition of the so-called hash even while it is interacting with the solitons.

As a final remark, it should be pointed out that a soliton

itself can be further analyzed into a superposition of poles in the complex plane. This is described briefly in [Kruskal 1974] and has been worked out in detail by W. R. Thickstun [to be published].

7. References

The main work of our group is distilled in a sequence of papers referred to in the text simply by the corresponding roman numeral. Those that have appeared by now are:

Miura, R. M., Korteweg-deVries equation and generalizations. I. A remarkable explicit nonlinear transformation, J. Math. Phys. 9, 1(1968), 1202-1204.

Miura, R. M., Gardner, C. S., and Kruskal, M. D., Korteweg-deVries equation and generalizations. II. Existence of conservation laws and constants of motion. J. Math. Phys. 9 (1968), 1204-1209.

Su, C. H., and Gardner, C. S., Korteweg-deVries equation and generalizations. III. Derivation of the Korteweg-deVries equation and Burgers' equation. J. Math. Phys 10 (1969), 536-539.

Kruskal, M. D., Miura, R. M., Gardner, C. S., and Zabusky, N.J., Korteweg-deVries equation and generalizations. V. Uniqueness and nonexistence of polynomial conservation laws, J. Math. Phys., 11 (1970), 952-960.

Gardner, C. S., Greene, J. M., Kruskal, M.D., and Miura, R.M., Korteweg-deVries equation and generalizations. VI. Methods for exact solution, Comm. Pure Appl. Math. 27 (1974), 97-133.

The initial discovery of solitons, and of the method of exact solution of the KdV equation, were announced respectively in:

Zabusky, N. J., and Kruskal, M. D., Interaction of "solitons" in a collisionless plasma and the recurrence of initial states, Phys. Rev. Letters 15 (1965), 240-243.

Gardner, C. S., Greene, J. M., Kurskal, M. D., and Miura, R.M., Method for solving the Korteweg-deVries equation, Phys. Rev. Letters 19 (1967), 1095-1097.

The remaining works referred to in the text are:

Benjamin, T. B., Bona, J. L., and Mahony, J. J., Model equations for long waves in nonlinear dispersive systems, Phil. Trans. Roy. Soc. London A, 272 (1972), 47-78.

Berger, N., Estimates for the derivatives of the velocity and pressure in shallow water flow and approximate shallow water equations, SIAM J. Appl. Math. 27 (1974), 256-280; see especially p. 276.

Cole, J. D., On a quasi-linear parabolic equation occurring in aerodynamics, Quart. Appl. Math., 9 (1951), 225-236.

Fermi, E., Pasta, J. R., and Ulam, S. M., Studies of nonlinear problems, part I, Report LA-1940, Los Alamos Scientific Laboratory, Los Alamos, N. M., (1955).

Hopf, E., The partial differential equation $u_t + uu_x = \mu u_{xx}$, Comm. Pure Appl. Math. 3 (1950), 201-230.

Korteweg, D. J., and deVries, G., On the change of form of long waves advancing in a rectangular canal, and on a new type of long stationary waves, Phil. Mag. 39 (1895), 422-443.

Kruskal, M. D., The Korteweg-deVries equation and related evolution equations, Amer. Math. Soc. Lectures in Appl. Math. 15 (1974), 61-83.

Lax, P. D., Integrals of nonlinear equations of evolution and solitary waves, Comm. Pure Appl. Math. 21 (1968), 467-490.

Rubinstein, J., Sine-Gordon equation, J. Math. Phys. 11 (1970) 258-266.

Témam, R., Sur un problème nonlinéaire J. Math. Pures Appl., 48 (1969), 159.

Thickstun, W. R., A system of particles equivalent to solitons, to be published.

Whitham, G. B., Non-linear dispersive waves, Proc. Roy. Soc. A, 283 (1965), 238-261.

Zakharov, V. E., and Shabat, A. B., Exact theory of two-dimensional self-focussing and one-dimensional self-modulating of waves in nonlinear media, English translation in Soviet Physics JETP 34 (1972), 62-69.

INTEGRABLE SYSTEMS OF NONLINEAR EVOLUTION EQUATIONS

Hermann Flaschka

Department of Mathematics, University of Arizona, Tuscon, Arizona

Alan C. Newell

Department of Mathematics, Clarkson College of Technology, Potsdam NY

Abstract

The method, developed by Gardner, Greene, Kruskal and Miura, for solving nonlinear evolution equations is looked at from three perspectives; (i) as a nonlinear Fourier transform, (ii) as a completely integrable Hamiltonian system, and (iii) from the operator theoretic standpoint. The interrelation between the three points of view is discussed and, in particular, the ubiquitous role of the "transmission coefficient" is brought out. The spectral representation of Lax's equation is studied in an abstract setting.

Table of Contents

Chapter

Introduction and General Discussion

Our goal in this paper is to discuss some of the exciting progress which has been made in the field of nonlinear partial differential equations, and in particular evolution equations, over the past decade. An evolution equation describes how a particular quantity evolves in time from a given initial state. To this date, the progress has resulted in a method (which has been labelled IST-Inverse Scattering Transform [2]) whereby the initial value problem for certain classes of these equations can be solved exactly. Furthermore, the long time behavior of the general solution may be analyzed and discussed in much the same way as one analyzes solutions to linear partial differential equations [5,24].

The pioneering work was done by Gardner, Greene, Kruskal and Miura [GGKM: 15, 16] who developed the method and applied it successfully to the Korteweg deVries (KdV) and related equations. The principal step is to build from the unknown variable of the evolution equation an eigenvalue problem whose spectrum remains invariant as the unknown variable changes with time. For the KdV equation, the appropriate eigenvalue problem is the Schrödinger equation in which the dependent variable in the KdV equation plays the role of the potential. The method consists of (1) mapping the potential at the initial time into the scattering data associated with the eigenvalue problem, (2) following the time evolution of the scattering data, and (3) inverting the mapping at any later time. The simplicity of the expressions describing the time evolution of the scattering data is one of the principal keys to success. Just as in linear problems, the equation becomes separable in the new transform variables. Seen from this perspective, the method may be considered to be an extension of the ideas of the Fourier transform to nonlinear problems. The concept of normal modes carries over but the superposition principle is not linear.

A second perspective provides a link with mechanics. It is well known that there are relatively few finite (or infinite) dimensional integrable systems. However, here we suddenly have class after class of infinite dimensional integrable system and it is natural to attempt to put each system in a Hamiltonian framework. Such a step is possible provided the system is, in a sense which we describe below, conservative. In these cases, the mapping described in the previous paragraph is a canonical transformation. For the s pecial class of flows discussed in this paper, the scattering data act as active-angle coordinates and the integration of Hamilton's equations in the new coordinate frame is trivial.

A third perspective concerns the spectral theory of operators and may be appropriately termed the Lax approach. In a very elegant work, Lax [30] was able to write the KdV evolution equation as a statement concerning the unitary equivalence of two operators. The expression of this statement in the spectral representation provides the time evolution of the scattering data.

In this article our aim is to introduce and discuss some of the merits of those various perspectives. Along the way we give some new results. We begin in the first chapter by outlining some important historical highlights in the development of the theory with some comments of our own added. In the second chapter we discuss in some detail the work of Ablowitz, Kaup, Newell and Segur (AKNS: 1a,b,2) who strongly advocate (not, however, in any sense to the exclusion of the other points of view) the first perspective mentioned above. Most, but not all, of the results contained in this chapter are also contained in AKNS [2]. We point out how each evolution equation may be characterized by its <u>dispersion relation</u> (which we define as the frequency-wave number relation of its linearized version). Some of the most interesting equations arise when the dispersion relation is singular (with pole singularities). The existence of motion invariants is not crucial to the success of the method although in most applications to this date there has been an infinite set of conserved quantities.

In the third and fourth chapters we elaborate on the Hamiltonian structure and on Lax's approach. There is one unifying idea which connects the different parts of the discussion; namely, the central role of $a(\zeta)$, the transmission coefficient (one* of the scattering functions). We shall show (1) how it can be used to generate all Hamiltonian flows solvable by the AKNS method, (2) how it leads to nonlinear Parseval relations and (3) how, even in a completely abstract setup, the analog of $a(\zeta)$ can be identified and used in a discussion of motion invariants.

We begin, in Chapter 3, by showing how an asymptotic expansion of $\ln a(\zeta)$ generates all (polynomial) conserved quantities both as functionals of the original variables of the evolution equation as well as of the scattering variables. Each such functional can be used as a Hamiltonian for a flow in which all others are conserved. These ideas are already present in the works of Zakharov, Shabat and Faddeev [44,46]; however, the systematic applciation to all the equations contained in the AKNS framework is new. Also new, and of some importance, is the realization that $\ln a$ itself, (estimated at a particular $\hat{\zeta}$), can be used as the Hamiltonian for nonlinear equations whose dispersion relations have pole singularities (at $\hat{\zeta}$); the derivation of this class of equations via the use of $\ln a(\zeta)$ is considerably simpler than by AKNS' original method. A significant feature of these equations is their structure. They arise as a coupled set of partial differential equations in which the quadratic

* The one whose zeros provide the discrete eigenvalues of the underlying eigenvalue problem.

products of solutions of the underlying eigenvalue problem appear as dependent variables. It is thus natural to anticipate that in these cases the quadratic products have a direct physical meaning and that the eigenvalue problem itself may have a physical interpretation (this conjecture has been suggested and shown to be true in certain cases by McLaughlin and Corones [33]). On account of the direct relation between the system Hamiltonian and $\ln a(\zeta)$ in these circumstances, it is suggested that $a(\zeta)$ may also play the role of a partition function. A third contribution of this chapter is the interpretation of the scattering representations of the Hamiltonians as nonlinear analogs of Parseval's relations. A connection is thereby established between the Fourier and Hamiltonian perspectives, but so far this connection has not been explored at any deeper level. Some of the results of this chapter were obtained independently and simultaneously by our colleagues, McLaughlin [32], and Kaup and Levermore [21]. McLaughlin follows an approach similar to that of Zakharov and Faddeev and our own and derives the Hamiltonian representation for the Toda lattice, the nonlinear Schrödinger and sine-Gordon equations. Kaup and Levermore take a somewhat different tack. Using the same convenient shortcuts outlined in Section 2.4, they explicitly obtain relations between infinitesimal changes in the potentials and infinitesimal changes in the scattering data. Then, assuming that the equations of motion are in Hamiltonian form, they write down what the variation of the Hamiltonian must be, transform these expressions in terms of the scattering data and read off the action-angle variables.

In Chapter 4, we develop yet another -- the operator-theoretic -- approach to the study of constants of the motion and it is here that the reason for the significance of the function $\ln a(\zeta)$ appears perhaps most clearly. It is seen that $\ln a(\zeta)$ is related to a suitably defined version of the trace (in finite dimensions, the trace is the sum of the eigenvalues; for further discussion, see [11]) of the Zakharov and Shabat [45] eigenvalue problem, given in (2.1). In view of the invariance of the spectrum, one might anticipate that $a(\zeta)$ -- being in a sense an "average" of the spectrum -- should remain constant as the coefficients of the system vary. This is indeed the case for most of the AKNS equations; the details are worked out in the beginning of the chapter. It is possible, however, that the coefficients of the system can change in such a way so that the spectrum is the only invariant of the motion. This is illustrated in Chapter 2 and again in Section 4.4 with a class of equations which admit solution by IST but for which $\ln a(\zeta)$ is time dependent. The best physical example is the system known as the self induced transparency problem [3] in which a significant portion of an initial pulse can be lost by resonant absorption.

These illustrations point up the distinction between the invariance of the spectrum and the existence of an infinite number of conserved quantities. The fact that the latter is not a necessary consequence of the theory is easily seen to be compatible with the framework in which Lax [30] discusses integrable equations. In Chapter 4, we study Lax's abstract equation $L_t = [B,L]$ and show (this appears to be a new deduction) that this equation becomes separable when transformed to scattering variables. Furthermore, general scattering theory provides a function analogous to $a(\zeta)$ and in terms of this we give abstract criteria for the existence of motion invariants which, in concrete cases, reduce to conditions already known.

One of the necessary features of the Lax approach is the skew adjointness of B. Without this property, the eigenfunctions do not evolve by unitary transformations and it is possible to lose control of their size as time evolves. Such a catastrophe does occur in the general AKNS system and at particular points in space-time the mapping from original variables to scattering data cannot be inverted. We have found that the eigenfunctions develop a singular behavior at precisely these locations. However, by somewhat restricting the AKNS system we can develop a Lax type formalism using the notion of J-spaces. We have not developed these ideas in any great depth and it may be very well possible that all systems which fall into the J-space category may be transformable to self adjoint systems.

In this article, we will not dwell on four other very important features of the general theory. The first of these is the nature of the general solution and in particular the "soliton", an entity we predict (as others have before) is destined to have a profound impact in both practical and theoretical mathematical physics. For discussions see [16,37,41,5]. The second topic relates to the question: given an evolution equation, how can one find out whether it is integrable, and if so, what its appropriate eigenvalue problem is? The answer to this question should help provide insight into the transformation theory underpinning the whole structure. For discussions we refer to [2,8,29,34,43]. A third area of much interest is the extension of the method to equations in higher dimensions. Some ideas are sketched very briefly in Chapter 4. Last, we have ignored altogether the completely analogous theory which can be applied to differential-difference equations. The first step in this area was taken by Flaschka who solved the Toda lattice equations (reported elsewhere in this volume). For further references we suggest articles [20, 7,4,36].

Suspecting that some readers may approach the conclusion of this lengthy introduction with sighs of relief rather than gasps of anticipation, we pause at this point to express our gratitude to the Battelle Institute which hosted the conference at which most of this work was done. In particular, we thank Jurgen Moser whose leadership during the conference helped provide an atmosphere which made for a most memorable summer.

Our thanks also to Bob Miura who contributed much in the formative stage of many of the ideas. Portions of this work were supported by National Science Foundation Grants GP-32839X and GP-42739.

1. Historical Development

The first equation which was solved by this method was the Korteweg deVries (KdV) equation

$$q_t + qq_x + q_{xxx} = 0 , \tag{1.1}$$

with the initial data $q(x,0)$ given and $q(x,t) \to 0$ sufficiently rapidly as $|x| \to \infty$. Motivated by a transformation relating solutions of (1.1) and (1.2), the "modified KdV" equation

$$p_t - 6\zeta^2 p_x + p^2 p_x + p_{xxx} = 0 , \tag{1.2}$$

(traditionally, the modified KdV equation is (1.2) with $\zeta^2 = 0$), GGKM were led to consider the Schrodinger equation

$$v_{xx} + (\zeta^2 + \frac{1}{6} q(x,t))v = 0 \tag{1.3}$$

in which the unknown variable in (1.1) plays the role of potential. Their first remarkable discovery was that as $q(x,t)$ evolves according to (1.1), the spectrum of (1.3) remains invariant. For $q(x,t)$ real and bounded, the eigenvalues are the set of all real $\zeta = k$ (the continuous spectrum) and a finite number of distinct imaginary numbers $\zeta = iK_n$, $n = 1,\dots N$ (the discrete spectrum). The corresponding generalized and proper eigenfunctions have the following asymptotic behavior as $x \to \pm\infty$:

$$\begin{aligned} v(x,t;k) &\to e^{-ikx} + R(k,t)e^{ikx} , & x \to +\infty \\ &\to T(k,t)\, e^{-ikx} , & x \to -\infty \end{aligned} \tag{1.4}$$

which relations define the reflection coefficient $R(k,t)$ and transmission coefficient $T(k,t)$; for $\zeta = iK_n$,

$$\begin{aligned} v_n(x,t) &\to C_n(t)e^{-K_n x} , & x \to +\infty \\ &\to D_n(t)\, e^{K_n x} , & x \to -\infty , \end{aligned} \tag{1.5}$$

and for convenience v_n is normalized $\int_{-\infty}^{\infty} |v_n|^2 \, dx = 1$. Essentially, then, the association of (1.1) with (1.5) generates a mapping between $q(x,t)$ and what we shall call the <u>scattering data</u> S,

$$q(x,t) \to S\left[\left\{iK_n, C_n(t)\right\}_{n=1}^{N}, T(k,t), R(k,t)\right], \tag{1.6}$$

a composition of the spectrum of (1.3) and coefficients representing the asymptotic behavior of the corresponding eigenfunctions.

<u>Comment</u>. As noted above, the motivation for (1.3) was provided by finding a transformation [35], suggested by a comparison of the two sets of conservation laws, between (1.1) and (1.2). The transformation is in the form of a Ricatti equation for p which when linearized gives (1.3). The observation of the interrelation between the appropriate eigenvalue problem and transformation theory was another of GGKM's great discoveries. Indeed, it was just such a transformation suggested by Kruskal [28] that led AKNS to the appropriate eigenvalue problem with which to study the sine-Gordon equation. More information on these transformations may be found in [2,29,34,43].

We have already noted that as $q(x,t)$ evolves according to (1.1), the spectrum remains invariant. However, the eigenfunctions do change according to

$$v_t = \left(\tfrac{1}{6} q_x + c(\zeta)\right)v + \left(4\zeta^2 - \tfrac{1}{3} q\right)v_x, \tag{1.7}$$

where $c = 0$ for $\zeta = iK_n$, and $c = 4ik^3$ for $\zeta = k$. Cross differentiation of (1.7) and (1.3) will give that $(\zeta^2)_t = 0$ if (1.1) holds. Now, the second remarkable discovery of GGKM was the observation that while the determination of $v(x,t,\zeta)$ at later times requires a knowledge of $q(x,t)$, the determination of the asymptotic behavior of $v(\pm\infty,t,\zeta)$ does not! Indeed as $x \to \pm\infty$, (1.7) is linear and the time evolution of the scattering data is given simply:

$$K_{n_t} = 0$$

$$T_t(k,t) = 0 \tag{1.8}$$

$$R_t(k,t) = 8ik^3 R(k,t)$$

$$C^2_{n_t} = 8K_n^3 C_n^2 \ .$$

The third observation is that the mapping (1.6) can be inverted, a result of Gel'fand and Levitan [18,19], Marchenko [31] and Kay and Moses [24]. Define

$$B(p,t) = \sum_{n=1}^{\infty} C_n^2(t)\, e^{-K_n p} + \frac{1}{2\pi}\int_{-\infty}^{\infty} R(k,t)\, e^{ikp}\, dK \tag{1.9}$$

and solve the linear integral equation for $y > x$,

$$K(x,y,t) + B(x+y,t) + \int_x^{\infty} K(x,z,t)\, B(y+z,t)\, dz = 0 \ , \tag{1.10}$$

subject to the condition

$$K(x,z) \to 0 \ , \qquad z \to \infty. \tag{1.11}$$

Then

$$q(x,t) = 12 \frac{d}{dx} K(x,x,t) \ . \tag{1.12}$$

Comment. We may summarize the procedure as follows. Map the initial data into the scattering data (1.6), follow its time evolution by (1.8) and invert (1.10). The close analogy with the recipe for solving linear problems suggests that one may think of the method in terms of a nonlinear extension of Fourier transforms. Furthermore, the fact that the reflection coefficient evolves according to the dispersion relation (of double the wave number) of the linearized KdV equation is no coincidence. In general, for any equation which leaves the spectrum of the Schrödinger operator invariant, $R_t(k,t) = -2\Omega(k)R(k,t)$ where $\Omega(k)$ * is its dispersion relation.

* More precisely, $\Omega(k) = ik\, C_0(k^2)$ where $C_0(k^2)$ is the phase velocity of the wave number $2k$ of the linearized equation.

In addition, the general solution may be though of as a composition of normal modes corresponding to the different solutions associated with the discrete and continuous spectrum. The superposition principle is, however, nonlinear.

The form of (1.8) suggests that the transformation (1.6) is a canonical transformation into action-angle variables. This interpretation was first given by Gardner [17] (he cast the periodic KdV equation in a Hamiltonian framework and showed that the constants of motion were in involution) and Zakharov and Faddeev [44] for the infinite region $-\infty < x < \infty$. The latter authors showed that the action variables are proportional to $\{K_n^2\}_{n=1}^N$ and $\ln(1-|R(k,t)|^2)$ and that the angle variables are proportional to Arg $R(k,t)$-Arg $T(k,t)$ and $\ln C_N^2$. The role of the Hamiltonian is played by the functional $\int (\frac{1}{2} q_x^2 - q^3)\, dx$, and each motion invariant of the KdV flow may be used as a Hamiltonian to generate a countably infinite sequence of flows all integrable by the means described above. Each flow is characterized by its dispersion relation. In Chapter 3, we will extend the Hamiltonian framework to include flows whose dispersion relations are singular.

Comment. It is now possible to indicate the importance of the observation that the systems (1.3) and (1.7) are linear as $x \to \pm\infty$. It is the existence of a known reference potential (in this case $q = 0$) to which the perturbed potential tends as $x \to \pm\infty$, which allows (1.7) to be solved as a linear equation. It is the lack of suitable angle variables that has frustrated, at least to date, the resolution of (1.1) over a finite region $a \le x \le b$ with periodic boundary conditions.* The constants of the motion only provide half the required information.

* We have just learned of some progress on this question by McKean and von Moerbecke.

<u>Further Remark</u>. A comparison with the Hopf-Cole [9] procedure, (see also Forsythe [14]), for solving Burgers' equation

$$u_t - 2uu_x - u_{xx} = 0 \tag{1.13}$$

may be helpful. Consider the equation pair analogous to (1.3) and (1.7),

$$v_x = u(x,t)v \tag{1.13a}$$

$$v_t = A(x,t)v \;. \tag{1.13b}$$

For this simple illustration we consider v, u and A to be just scalars although considerable generalization is possible. Cross differentiating (1.13a) and (1.13b) we get $u_t = A_x$ which is Burgers' equation when $A = u^2 + u_x$. The strategy for solution then runs as follows: given $u(x,0)$, solve for v at the initial time and attempt to follow its evolution using (1.13b). If we are able to accomplish this then $u(x,t)$ may be recovered at later times by again using (1.13a). But, and here is the rub, we must be able to follow the evolution of $v(x,t)$ for all values of x. The success of this transformation thus lies solely in the special nature of A, namely, $Av = (u^2+u_x)v = v_{xx}$, whereupon (1.13b) becomes the heat equation. In solving the KdV equation, we only had to follow the asymptotic behavior of $v(x,t)$, $x \to \pm\infty$. The price we had to pay, of course, is that we had to add an extra dimension characterized by the eigenvalue ζ^2.

An elegant perspective on the relation between (1.1) and (1.3) was provided by Lax [30] at an early stage in the development of the theory. Lax observed that the requirement, that all operators of a family $\{L(t)\}$ be unitarily equivalent, namely,

$$U^* L(t) = L(0)\, U^* \tag{1.14}$$

leads to the operator differential equation

$$L_t = [B,L], \tag{1.15}$$

where

$$U_t = B\,U\,. \tag{1.16}$$

B is only determined up to an operator commuting with L. The spectrum remains invariant. Indeed if λ is an eigenvalue of L(0) with eigenfunction v(0), then λ is also an eigenvalue of L(t) with eigenfunction v(t) = U v(0). One may recover the GGKM result concerning the constancy of eigenvalues by choosing

$$L(t) = -\frac{\partial^2}{\partial x^2} - \frac{1}{6}\,q(x,t) \tag{1.17}$$

and

$$B(t) = -4\frac{\partial^3}{\partial x^3} - q\frac{\partial}{\partial x} - \frac{1}{2}\,q_x\,. \tag{1.18}$$

Comment. The fact that there is a family of B's allows one to define corresponding families of unitary operators. As discussed in Chapter 4, two such intertwining operators (the Moller wave operators) are related by the scattering operator whose evolution in time may be computed in principle. The spectral representation of this equation leads to the equations for the evolution of the scattering data (1.8).

Comment. However, the Moller wave operators are defined in such a way that their corresponding B's are nonlocal. Note that the "B" given in (1.7) with $c(\zeta)$ proportional to ζ^3 is not local when looked at from x-space. ζ^2 acting on v(x,t) does give a local operator in x-space (by using (1.3)), but ζ^3 does not. Specific examples are given in (4.4).

Comment. For reasons already explained in the introduction, it is important that the evolution of the eigenfunctions be unitary, as it allows some control on the size of v(t).

Comment. Equation (1.15) only guarantees that the spectrum of L(t) is time invariant. The existence of other motion invariants depends on the form of B. This point will be discussed again in Sections 2.4, 4.3 and 4.4.

The greatest value of the Lax approach in the early stages of the inverse method was that it provided a straightforward computational way of verifying that the spectrum of some linear differential operator remains constant when its coefficients change in some special way, and it also gives a simple prescription for then finding the evolution of the scattering data. Thus, the original, sometimes rather involved, arguments of GGKM can be replaced by routine computations. The difficulty, of course, lies in guessing the correct L and B that will give rise to a specific equation. Such ingenuity was provided in a very significant paper by Zakharov and Shabat [45] who were able to cast the nonlinear Schrodinger equation

$$q_t = i(q_{xx} + 2q^2q^*) \tag{1.19}$$

into the form (1.15). In order to carry out the inverse scattering procedure, however, they transformed their self adjoint L into a non-selfadjoint operator of the form

$$\begin{pmatrix} i\frac{\partial}{\partial x} & -iq \\ -iq^* & -i\frac{\partial}{\partial x} \end{pmatrix} \tag{1.20}$$

Shortly thereafter, Wadati [42] found an L and a B which would serve to express the modified KdV equation (1.2) in Lax's form and he too transformed to (1.20) for the purpose of inversion. In both examples the choices of L and B appeared to be ad hoc.

In a series of papers, AKNS [1a,b,2] provided a systematic procedure for finding all the evolution equations which could be solved using a particular, but nonselfadjoint, operator. The operator they use is a generalization of the one given by Zakharov and Shabat,

$$L = \begin{pmatrix} i\frac{\partial}{\partial x} & -i\,q \\ i\,r & -i\frac{\partial}{\partial x} \end{pmatrix} . \tag{1.21}$$

Beginning with L, AKNS define the scattering data, which are related to the asymptotic behavior of the eigenfunctions of $Lv = \zeta v$. Next, the time evolution of the eigenfunctions are computed for completely arbitrary time evolutions of both $q(x,t)$ and $r(x,t)$. (The expressions may be written in the form $v_t = Bv$ where B is ζ-dependent; but at this stage B is horribly complex and involves q_t and r_t, the time derivatives of the coefficients $q(x,t)$ and $r(x,t)$). In particular, the time evolution of the scattering data may be found and in general leads to complicated expressions involving integrals over products of q_t and r_t with quadratic products of solutions to $Lv = \zeta v$. AKNS then ask what it would take to make these expressions simple. From these choices, they are able to identify the class of evolution equations which are integrable via the operator L. Although it is not necessary for the purposes of solution, one may then compute B, a posteriori, as a functional of q and r by using the evolution equation to replace q_t and r_t.

Comment. Apart from the selfadjoint and unitary properties, the AKNS and the Lax approaches are similar in structure but differ in strategy in that the former approach puts less emphasis on the equation $L_t = [B,L]$ but rather attempts to go directly to its spectral representation.

Comment. We show in Chapter 4.2 that the self adjoint and unitary properties allow the application of some deep results in operator theory. To date no corresponding analytic machinery has been worked out for general nonselfadjoint operators such as (1.21). We suggest a framework for such a theory in Section 4.5 which is applicable to a subclass of (1.21). There is reason to anticipate that this subclass of nonselfadjoint operators can be converted to selfadjoint operators by ζ-dependent transformations. However, bound states of the former operator do not necessarily transform to bound states of the latter. This suggests that it may be more relevant to think of the soliton (the solution of (1.1) corresponding to a discrete eigenvalue i.e., a bound state of (1.3)) as being related to poles of the scattering matrix rather than to the bound states.

In the following chapter, we present a slightly modified version of the AKNS analysis in some detail. For completeness, we outline the results for the direct and inverse scattering problem in Sections 2.2 and 2.3 but it is to Section 2.4, the evolution of the scattering data, that we primarily direct the reader's attention. It is this step that is distinctly novel in the AKNS approach.

2. Review of the Work of AKNS

2.1 Introduction

Consider the eigenvalue problem

$$(2.1)\qquad \begin{aligned} v_{1x} + i\zeta v_1 &= q(x,t)\,v_2 \\ v_{2x} - i\zeta v_2 &= r(x,t)\,v_1 \,. \end{aligned} \qquad -\infty < x < \infty$$

AKNS were able to show that as the potentials q(x,t) and r(x,t) (assumed to decay at $|x| = \infty$ for all time) evolve in time according to any member of the class of flows given by

$$(2.2)\qquad \begin{pmatrix} r_t \\ -q_t \end{pmatrix} + 2\Omega(L^+)\begin{pmatrix} r \\ q \end{pmatrix} = 0$$

the spectrum (the set of eigenvalues ζ of (2.1) corresponding to bounded v_1, v_2) of (2.1) remains invariant. Furthermore (2.2) is a completely integrable system. Each equation pair (or flow) is characterized by the function $\Omega(\zeta)$ which is directly related to the dispersion relation of the linearized version of (2.2). $\Omega(\zeta)$ may be singular but can have at most poles and in these situations certain additional restrictions must be placed on the initial data. The operator L^+ is given by

$$(2.3)\qquad L^+ = \frac{1}{2i}\begin{pmatrix} \frac{\partial}{\partial x} - 2r\int_{-\infty}^{x} dy\; q & 2r\int_{-\infty}^{x} dy\; r \\ -2q\int_{-\infty}^{x} dy\; q & -\frac{\partial}{\partial x} + 2q\int_{-\infty}^{x} dy\; r \end{pmatrix}$$

For example when $\Omega(\zeta) = -4i\zeta^3$ and $r = -q$ we obtain the modified KdV equation (1.2); for $\Omega(\zeta) = 2i\zeta^2$ and $r = -q^*$, we obtain the nonlinear Schrodinger equation (1.22); for $\Omega = \frac{i}{4\zeta}$ and $r = -q = \frac{u_x}{2}$ we find the sine-Gordon equation,

(2.4) $$q_t = -\frac{1}{2}\sin\left[-2\int_{-\infty}^{x} q\,dy\right],$$

for which the additional restriction is that $\int_{-\infty}^{\infty} q\,dy = n\pi$, $n = 0, 1, 2, \ldots$.

Our goal in this chapter is to show how the connection between (2.1) and (2.2) may be described in three stages. First, assuming that $q(x,t)$ and $r(x,t)$ are known we define the scattering data S, namely, the spectrum and the asymptotic behavior of the eigenfunctions of (2.1) as $x \to \pm\infty$. Second, we give the formula from which $q(x,t)$ and $r(x,t)$ may be recovered from the scattering data. Third, and this is the most emphasized section of this chapter, we will attempt to obtain for arbitrary flows $q(x,t)$, $r(x,t)$ expressions for the time evolution of the scattering data. In general, these expressions are complicated and not simply integrable in time. However, the demand that they be simply integrable will identify a certain class of flows which in the simplest case may be written as (2.2).

2.2 The Direct Scattering Problem

For ζ real, we define $\phi(x,t,\zeta)$, $\bar{\phi}(x,t,\zeta)$, $\psi(x,t,\zeta)$ and $\bar{\psi}(x,t,\zeta)$ to be solutions of (2.1) whose boundary conditions are given in the following table:

TABLE 1

	$x = -\infty$	$x = +\infty$
ϕ	$\begin{pmatrix}1\\0\end{pmatrix}e^{-i\zeta x}$	$\begin{pmatrix}a(\zeta,t)e^{-i\zeta x}\\b(\zeta,t)e^{i\zeta x}\end{pmatrix}$
$\bar{\phi}$	$\begin{pmatrix}0\\-1\end{pmatrix}e^{i\zeta x}$	$\begin{matrix}\bar{b}(\zeta,t)e^{-i\zeta x}\\-\bar{a}(\zeta,t)e^{i\zeta x}\end{matrix}$
ψ	$\begin{pmatrix}\bar{b}(\zeta,t)e^{-i\zeta x}\\a(\zeta,t)e^{i\zeta x}\end{pmatrix}$	$\begin{pmatrix}0\\1\end{pmatrix}e^{i\zeta x}$
$\bar{\psi}$	$\begin{pmatrix}\bar{a}(\zeta,t)e^{-i\zeta x}\\-b(\zeta,t)e^{i\zeta x}\end{pmatrix}$	$\begin{pmatrix}1\\0\end{pmatrix}e^{-i\zeta x}$

These functions are related by

(2.5a) $$\phi = a\bar{\psi} + b\psi\ , \qquad \bar{\phi} = -\bar{a}\psi + \bar{b}\bar{\psi}\ ,$$

and

(2.5b) $$\psi = \bar{b}\phi - a\bar{\phi}\ , \qquad \bar{\psi} = \bar{a}\phi + b\bar{\phi}\ ,$$

where

(2.6) $$a\bar{a} + b\bar{b} = 1\ .$$

We list the following results which are all proved in [2]:

(a) If $q(x,t)$ and $r(x,t)$ decay sufficiently rapidly as $|x| \to \infty$, (we will assume $\int_{-\infty}^{\infty} |x^n|\,|q|\,dx$, $\int_{-\infty}^{\infty} |x^n|\,|r|\,dx$ exists for all n), then $\phi e^{i\zeta x}$, $\psi e^{-i\bar{\zeta} x}$ are analytic for $\mathrm{Im}\,\zeta \geq 0$, $\bar{\phi}e^{-i\zeta x}$, $\bar{\psi}e^{i\zeta x}$ are analytic for $\mathrm{Im}\,\zeta \leq 0$.

(b) In particular, $a(\zeta,t) = \phi_1\psi_2 - \phi_2\psi_1$ is analytic and may be extended to $\mathrm{Im}\,\zeta \geq 0$. Its zeros in the upper half plane $\{\zeta_k\}_{k=1}^{N}$ are the discrete eigenvalues of (2.1) with $\mathrm{Im}\,\zeta_k > 0$. At each eigenvalue $\phi(\zeta_k,t) = b_k(t)\psi(\zeta_k,t)$. Similarly the zeros $\{\bar{\zeta}_k\}_{k=1}^{\bar{N}}$ of $\bar{a}(\zeta,t) = \bar{\phi}_1\bar{\psi}_2 - \bar{\phi}_2\bar{\psi}_1$ are the discrete eigenvalues of (2.1) in the lower half plane: $\mathrm{Im}\,\bar{\zeta}_k < 0$. At these points $\bar{\phi}(\bar{\zeta}_k,t) = \bar{b}_k(t)\bar{\psi}(\bar{\zeta}_k,t)$. The analyticity of $a(\zeta,t)$ and $\bar{a}(\zeta,t)$ in their respective half planes, together with the fact that both tend to unity as $\zeta \to \infty$, $\mathrm{Im}\,\zeta \gtrless 0$ respectively, ensures that N and $\bar{N}$ are finite. We will <u>assume</u> that neither a nor $\bar{a}$ has multiple zeros nor zeros on the real ζ axis.

(c) If q and r are on compact support, the regions of analyticity of $(\phi e^{i\zeta x}, \psi e^{-i\zeta x}, a(\zeta,t))$ and $(\bar{\phi}e^{-i\zeta x}, \bar{\psi}e^{i\zeta x}, \bar{a})$ can be extended into the entire lower and upper half planes respectively so that each set is analytic at all finite ζ. Furthermore both $b(\zeta,t)$ and $\bar{b}(\zeta,t)$, hitherto only defined on the real axis, can also be analytically extended everywhere and in particular $b(\zeta_k,t)=b_k(t)$, $\bar{b}(\bar{\zeta}_k,t) = \bar{b}_k(t)$.

(d) We call the set $S\{(\zeta_k,b_k)_{k=1}^{N}, (\bar{\zeta}_k,\bar{b}_k)_{k=1}^{\bar{N}}, a(\zeta,t), b(\zeta,t), \bar{a}(\zeta,t), \bar{b}(\zeta,t)\}$ the scattering data.

(e) Whenever r is linearly related to q or q^*, simplifications occur. First, consider $r = \alpha q$ where α is any nonzero, complex scalar constant. In this case $\bar{\psi}(\zeta,x) = S\psi(-\zeta,x)$ and $\bar{\phi}(\zeta,x) = -\frac{1}{\alpha}S\phi(-\zeta,x)$ where $S = \begin{pmatrix}0 & 1\\ \alpha & 0\end{pmatrix}$. Consequently $\bar{a}(\zeta) = a(-\zeta)$, $\bar{b}(\zeta) = -\frac{1}{\alpha}\, b(-\zeta)$ and the zeros of a and $\bar{a}$ are paired such that $\bar{N}= N$ and $\bar{\zeta}_k = -\zeta_k$, and $\bar{b}_k = -\frac{1}{\alpha}\, b_k$. When* $r = \alpha q^*$, α real, we have $\bar{\psi}(\zeta,x) = S\psi^*(\zeta^*,x)$, $\bar{\phi}(\zeta,x) = -\frac{1}{\alpha}\, S\phi^*(\zeta^*,x)$ which gives $\bar{a}(\zeta) = a^*(\zeta^*)$, $\bar{b}(\zeta) = -\frac{1}{\alpha}\, b^*(\zeta^*)$, $\bar{N} = N$, $\bar{\zeta}_k = \zeta_k^*$ and $\bar{b}_k = -\frac{1}{\alpha}\, b_k^*$. When r, q and α are real we can deduce that if ζ_k is an eigenvalue in the upper half plane such that $\operatorname{Re}\zeta_k \neq 0$, then $-\zeta_k^*$ is also an eigenvalue in the upper half plane. It sometimes transpires that the velocities of the solitons corresponding to ζ_k and $-\zeta_k^*$ are the same and the two solitons together form a composite state which is manifested as the well known 0π pulse or breather in the context of the sine-Gordon equation.

(f) A very useful relation for $\ln a(\zeta)$ expressed as a functional of q and r may be found by integrating the identity

$$i(\phi_1\psi_2+\phi_2\psi_1) - ia = \frac{\partial}{\partial x}\left(-\frac{\partial\phi_1}{\partial\zeta}\,\psi_2 + \frac{\partial\phi_2}{\partial\zeta}\,\psi_1\right) - ia\,,$$

between $-R$ and R and allowing $R \to +\infty$. We find, using Table 1,

$$\text{(2.7a)}\qquad \frac{\partial}{\partial\zeta}\ln a(\zeta) = -i\int_{-\infty}^{\infty}\left(\frac{\phi_1\psi_2+\phi_2\psi_1}{a} - 1\right)dx\,,\qquad \operatorname{Im}\zeta > 0$$

or

$$\text{(2.7b)}\qquad \ln a(\zeta) = -i\int_{i\infty}^{\zeta}d\zeta\int_{-\infty}^{\infty}\left(\frac{\phi_1\psi_2+\phi_2\psi_1}{a} -1\right)dx,\qquad \operatorname{Im}\zeta > 0$$

We use this result to prove the trace formula in Chapter 4, equation (4.6).

---

* refers to complex conjugate.

(g) Here we derive a most important asymptotic relation for $\ln a(\zeta,t)$, which quantity will turn out to play the central role in the analysis of the next two chapters. Define c by the principal branch of the expression $e^c = \phi_1 e^{i\zeta x}$ and note from Table 1 that as $x \to -\infty$, $c \to 0$ whereas as $x \to +\infty$, $c \to \ln a$. By (2.1), c satisfies the equation (here we follow closely the ideas of Zakharov and Shabat [45])

$$2i\zeta \frac{dc}{dx} = -qr + q\frac{d}{dx}\frac{1}{q}\frac{dc}{dx} + \left(\frac{dc}{dx}\right)^2 \tag{2.8}$$

from which we can derive

$$\frac{dc}{dx} \sim \sum_{n=1}^{\infty} \frac{1}{(2i\zeta)^n}\phi_n ,$$

where

$$\begin{aligned} \phi_1 &= -qr , \\ \phi_{n+1} &= q\frac{d}{dx}\left(\frac{1}{q}\phi_n\right) + \sum_{j+k=n}\phi_j\phi_k , \qquad n = 1,2,\ldots . \end{aligned} \tag{2.9}$$

Integrating (2.8) from $-\infty$ to ∞, we find

$$\ln a \sim \sum_{n=1}^{\infty} \frac{1}{\zeta^n}\frac{1}{(2i)^n}\int_{-\infty}^{\infty}\phi_n\, dx = -\sum_{n=1}^{\infty}\frac{1}{\zeta^n}C_n \tag{2.10}$$

where we have defined $C_n = \frac{-1}{(2i)^n}\int_{-\infty}^{\infty}\phi_n\, dx$. These quantities will turn out to be motion invariants for the class of flows given by (2.2). The first three are

$$\begin{aligned} C_1 &= \frac{1}{2i}\int_{-\infty}^{\infty} qr\, dx, \quad C_2 = \frac{1}{(2i)^2}\int_{-\infty}^{\infty} qr_x\, dx , \\ C_3 &= \frac{1}{(2i)^3}\int_{-\infty}^{\infty}(qr_{xx} - q^2r^2)\, dx . \end{aligned} \tag{2.11}$$

Similarly, by finding the asymptotic expansion for $\bar{c}$, where $e^{\bar{c}} = \bar{\psi}_1 e^{i\zeta x}$, we find

(2.12) $$\ln \bar{a} \sim \sum_{1}^{\infty} \frac{1}{\zeta^n} C_n \quad \text{as} \quad \zeta \to \infty , \quad \text{Im}\ \zeta < 0.$$

2.3 The Inverse Scattering Formulae

In order to recover the potentials $q(x,t)$, $r(x,t)$ we may use one of two sets of scattering data. We will omit writing in the explicit t dependence of all quantities.

(a) Given

(2.13) $$S_+ = S_+\left\{ \left(\zeta_k , \frac{b_k}{a_k'}\right)_{k=1}^{N}, \left(\bar{\zeta}_k, \frac{\bar{b}_k}{\bar{a}_k'}\right)_{k=1}^{\bar{N}}, \frac{b(\xi)}{a(\xi)}, \frac{\bar{b}(\xi)}{\bar{a}(\xi)}, \xi \text{ real} \right\}$$

define

(2.14) $$F(z) = i \sum \frac{b_k}{a_k'} e^{i\zeta_k z} + \frac{1}{2\pi} \int_{-\infty}^{\infty} \frac{b(\xi)}{a(\xi)} e^{i\xi z} \, d\xi ,$$

and

(2.15) $$\bar{F}(z) = i \sum \frac{\bar{b}_k}{\bar{a}_k'} e^{-i\bar{\zeta}_k z} + \frac{1}{2\pi} \int_{-\infty}^{\infty} \frac{\bar{b}(\xi)}{\bar{a}(\xi)} e^{-i\xi z} \, d\xi .$$

Then solve for $K(x,y)$, $\bar{K}(x,y)$ where

(2.16) $$\bar{K}(x,y) + \binom{0}{1} F(x+y) + \int_x^{\infty} K(x,s)\, F(s+y)\, ds = 0 , \qquad y > x$$

(2.17) $$K(x,y) - \binom{1}{0} \bar{F}(x+y) - \int_x^{\infty} \bar{K}(x,s)\, \bar{F}(s+y)\, ds = 0 , \qquad y > x$$

and find $q(x)$ and $r(x)$ from

(2.18) $$K_1(x,x) = -\frac{1}{2} q(x) , \qquad \bar{K}_2(x,x) = -\frac{1}{2} r(x) .$$

(b) Given

(2.19) $$S_- = S_-\left\{ \left(\zeta_k, \frac{1}{b_k a_k'}\right)_{k=1}^{N}, \left(\bar{\zeta}_k, \frac{1}{\bar{b}_k \bar{a}_k'}\right)_{k=1}^{\bar{N}}, \frac{\bar{b}(\xi)}{a(\xi)}, \frac{b(\xi)}{\bar{a}(\xi)}, \xi \text{ real} \right\}$$

define

$$(2.20) \qquad G(z) = -i \sum \frac{1}{b_k a_k'} e^{-i\zeta_k z} + \frac{1}{2\pi} \int_{-\infty}^{\infty} \frac{\bar{b}(\xi)}{a(\xi)} e^{-i\xi z} d\xi$$

and

$$(2.21) \qquad \bar{G}(z) = i \sum \frac{1}{\bar{b}_k \bar{a}_k'} e^{i\zeta_k z} + \frac{1}{2\pi} \int_{-\infty}^{\infty} \frac{b(\xi)}{\bar{a}(\xi)} e^{i\xi z} d\xi .$$

Then, solve for $L(x,y)$, $\bar{L}(x,y)$ where

$$(2.22) \qquad \bar{L}(x,y) + \binom{1}{0} G(x+y) - \int_{-\infty}^{x} L(x,s)\, G(s+y)\, ds = 0 , \qquad x > y$$

$$(2.23) \qquad L(x,y) + \binom{0}{1} \bar{G}(x+y) + \int_{-\infty}^{x} \bar{L}(x,s)\, \bar{G}(s+y)\, ds = 0 , \qquad x > y$$

and find $q(x)$ and $r(x)$ from

$$(2.24) \qquad \bar{L}_1(x,x) = \frac{1}{2} q(x) , \qquad L_2(x,x) = -\frac{1}{2} r(x) .$$

Comment. If we imagine (2.1) as the system of ordinary differential equations with frequency parameter ζ, which is derived by a Fourier transform with respect to a time τ of a system of wave equations, then S_+ corresponds to the scattering, by the potentials q and r, of a delta function pulse initiated from $x = +\infty$ at $\tau = -\infty$. S_- corresponds to the scattering of a right moving pulse from $x = -\infty$.

Most of the results stated here were first given by Shabat [38,39,40] and derived independently by AKNS [2].

2.4 Evolution of the Scattering Data

As the potentials $q(x,t)$, $r(x,t)$ change in time according to some evolution equation, the scattering data will also change. The original approach taken by GGKM was to substitute (1.3) into (1.1) and determine the resultant evolution of the eigenfunctions. This approach works well when one has a specific equation and when one can isolate the potential in the eigenvalue problem. Lax's approach was to find a local B (depending on derivatives) such that the evolution equation could be expressed as $L_t = [B,L]$. The significant step taken by AKNS was to allow the operator B to depend on the eigenvalue ζ explicitly. Instead of finding the operator B which corresponded to a specific equation and from there determining the time evolution of the scattering data dS/dt, the strategy of AKNS was to determine directly the quantity dS/dt in terms of functionals of q_t and r_t. The demand that these expressions are readily integrable leads immediately to a condition from which one may derive, a posteriori, the most general class of evolution equations which can be solved using (2.1) as the eigenvalue problem. The appropriate B may also be found a posteriori.

In this article we follow the AKNS strategy but use a more direct computational approach. The system (2.1) may be written in matrix form

$$\Phi_x = P\Phi \tag{2.25}$$

where

$$\Phi = \begin{pmatrix} \phi_1 & \bar{\phi}_1 \\ \phi_2 & \bar{\phi}_2 \end{pmatrix}, \quad P = \begin{pmatrix} -i\zeta & q \\ r & i\zeta \end{pmatrix}.$$

Consider making arbitrary infinitesimal variations δq and δr in q constrained only by the requirement that $\delta q, \delta r \to 0$ as $|x| \to \infty$ and the demand that the discrete spectrum of (2.25) remain invariant. As yet, no demands are made that these variations follow some particular flow path. We find

$$\delta\Phi = \Phi \int_{-\infty}^{x} \Phi^{-1} \begin{pmatrix} 0 & \delta q \\ \delta r & 0 \end{pmatrix} \Phi \, dx , \tag{2.26}$$

where $\Phi^{-1} = \begin{pmatrix} -\bar{\phi}_2 & \bar{\phi}_1 \\ \phi_2 & -\phi_1 \end{pmatrix}$. Considerable use is made of the fact that the Wronskian of the two solutions ϕ and $\bar{\phi}$ is -1. Taking the limit $x \to +\infty$ and using Table 1, we find the corresponding changes in the scattering data to be

$$(2.27)\qquad \begin{pmatrix} \delta a & \delta\bar{b} \\ \delta b & -\delta\bar{a} \end{pmatrix} = \begin{pmatrix} a & \bar{b} \\ b & -\bar{a} \end{pmatrix} \begin{pmatrix} I(\phi,\bar{\phi}) & I(\bar{\phi},\bar{\phi}) \\ -I(\phi,\phi) & -I(\phi,\bar{\phi}) \end{pmatrix}$$

where

$$(2.28)\qquad I(u,v) = \int_{-\infty} (-\delta q u_2 v_2 + \delta r u_1 v_1)\, dx .$$

We use the relations (2.5a,b) and show

$$(2.29)\qquad \delta a = -I(\phi,\psi)\ ,\ \delta b = I(\phi,\bar{\psi})\ ,\ \delta\bar{a} = -I(\bar{\phi},\bar{\psi})\ ,\ \delta\bar{b} = -I(\bar{\phi},\psi).$$

If the variations are consistent with a flow characterized by the time parameter t, then

$$(2.30)\ a_t = -\hat{I}(\phi,\psi),\ b_t = \hat{I}(\phi,\bar{\psi}),\ \bar{a}_t = -\hat{I}(\bar{\phi},\bar{\psi}),\ \bar{b}_t = -\hat{I}(\bar{\phi},\psi)\ ,$$

where $\hat{I}(u,v) = \int_{-\infty}^{\infty} (-q_t u_2 v_2 + r_t u_1 v_1)\, dx$. In particular the time evolution of the scattering data S_- is

$$(2.31)\qquad \left(\frac{\bar{b}}{a}\right)_t = \frac{\bar{b}}{a}\,\frac{\hat{I}(\psi,\psi)}{a\bar{b}}\ ,\qquad \left(\frac{b}{\bar{a}}\right)_t = \left(\frac{b}{\bar{a}}\right)\frac{\hat{I}(\psi,\psi)}{\bar{a}b} .$$

To this point, we have not made any demands on q_t and r_t and for any q_t and r_t we could integrate (2.31) locally (step by step) in time. However, certain choices of $\hat{I}(\psi,\psi)$ and $\hat{I}(\bar{\psi},\bar{\psi})$ render (2.31) immediately integrable. Here we will make the simplest choice,

$$(2.32)\qquad \hat{I}(\psi,\psi) = 2\Omega(\zeta)a\bar{b}\ ,\qquad \hat{I}(\bar{\psi},\bar{\psi}) = -2\bar{\Omega}(\zeta)\bar{a}b\ ,$$

which make (2.31) linear equations

(2.33a) $$\frac{\partial}{\partial t}\ln\frac{\bar{b}}{a} = 2\Omega(\zeta)\ , \qquad \frac{\partial}{\partial t}\ln\frac{b}{\bar{a}} = -2\bar{\Omega}(\zeta)\ ,$$

with analytic extensions*

(2.33b) $$\frac{\partial}{\partial t}\ln\frac{1}{b_k a_k'} = 2\Omega(\zeta_k)\ ; \qquad \frac{\partial}{\partial t}\ln\frac{1}{\bar{b}_k \bar{a}_k'} = -2\bar{\Omega}(\bar{\zeta}_k)\ .$$

In the cases where the Hamiltonian formulation may be applied, ln b is an angle variable. It will turn out that the system is only conservative when $\Omega = \bar{\Omega}$ for which the two cases (a) Ω an entire function of ζ, and (b) Ω meromorphic, are discussed separately. Finally in case (c) we show that some non-conservative systems are equally tractable by the method even though there are no motion invariants. In particular, we note that the condition given in (4.19) for the existence of constant traces is not satisfied.

<u>Case (a)</u> $\Omega = \bar{\Omega}$, Ω entire. Noting that we may write $a\bar{b} = -\psi_1\psi_2\Big|_{-\infty}^{\infty} = -\int_{-\infty}^{\infty}(q\psi_2^2 + r\psi_1^2)\,dx$, the condition $I(\psi,\psi) = 2\Omega(\zeta)a\bar{b}$ may be written

$$\int_{-\infty}^{\infty}\left[\begin{pmatrix} r_t \\ -q_t \end{pmatrix} + 2\Omega(\zeta)\begin{pmatrix} r \\ q \end{pmatrix}\right]\cdot\begin{pmatrix} \psi_1^2 \\ \psi_2^2 \end{pmatrix} dx = 0\ .$$

Using (2.1) and Table 1 extensively we find that

(2.34) $$L\begin{pmatrix} \psi_1^2 \\ \psi_2^2 \end{pmatrix} = \zeta\begin{pmatrix} \psi_1^2 \\ \psi_2^2 \end{pmatrix}$$

with

(2.35) $$L = \frac{1}{2i}\begin{pmatrix} -\frac{\partial}{\partial x} - 2q\int_x^{\infty} dy\, r & -2q\int_x^{\infty} dy\, q \\ 2r\int_x^{\infty} dy\, r & \frac{\partial}{\partial x} + 2r\int_x^{\infty} dy\, q \end{pmatrix}$$

*The residue of the analytic extension of $\frac{\bar{b}}{a}$ at $\zeta = \zeta_k$ is $\frac{1}{b_k a_k'}$ where $a_k' = ((d/d\zeta)a)_{\zeta_k}$.

If $\Omega(\zeta)$ is entire, then $\Omega(\zeta)\begin{pmatrix}\psi_1^2\\ \psi_2^2\end{pmatrix} = \Omega(L)\begin{pmatrix}\psi_1^2\\ \psi_2^2\end{pmatrix}$. Integration by parts and changing the order of integration in (2.33) gives

$$\int_{-\infty}^{\infty}\left[\begin{pmatrix} r_t\\ -q_t\end{pmatrix} + 2\Omega(L^+)\begin{pmatrix} r\\ q\end{pmatrix}\right]\cdot\begin{pmatrix}\psi_1^2\\ \psi_2^2\end{pmatrix}\,dx = 0\ , \tag{2.36}$$

with L^+ the adjoint operator to L,

$$L^+ = \frac{1}{2i}\begin{pmatrix} \frac{\partial}{\partial x} - 2r\int_{-\infty}^{x} dyq & 2r\int_{-\infty}^{x} dyr \\ -2q\int_{-\infty}^{x} dyq & -\frac{\partial}{\partial x} + 2q\int_{-\infty}^{x} dyr \end{pmatrix}. \tag{2.37}$$

The choice

$$\begin{pmatrix} r_t\\ -q_t\end{pmatrix} + 2\Omega(L^+)\begin{pmatrix} r\\ q\end{pmatrix} = 0\ , \tag{2.38}$$

which is the general evolution equation, satisfies both equations (2.32). By (2.38) it follows that

$$\hat{I}(\psi,\bar{\psi}) = -2\Omega(\zeta)b\bar{b}\ ,\quad \hat{I}(\phi,\phi) = -2\Omega(\zeta)ab\ ,\quad \hat{I}(\bar{\phi},\bar{\phi}) = 2\Omega(\zeta)\bar{a}\bar{b}. \tag{2.39}$$

Using (2.30) and (2.39), we may also obtain the time evolution of the scattering data S_+:

$$\frac{\partial}{\partial t}\frac{b}{a} = \frac{\hat{I}(\phi,\phi)}{ab}\frac{b}{a} = -2\Omega\frac{b}{a}\ ,\quad \frac{\partial}{\partial t}\frac{\bar{b}}{\bar{a}} = \frac{\hat{I}(\bar{\phi},\bar{\phi})}{\bar{a}\,\bar{b}}\frac{\bar{b}}{\bar{a}} = 2\Omega\frac{\bar{b}}{\bar{a}} \tag{2.40a}$$

with analytic extensions,*

$$\frac{\partial}{\partial t}\frac{b_k}{a_k'} = -2\Omega(\zeta_k)\frac{b_k}{a_k'}\ ,\qquad \frac{\partial}{\partial t}\frac{\bar{b}_k}{\bar{a}_k'} = 2\Omega(\zeta_k)\frac{\bar{b}_k}{\bar{a}_k'}\ . \tag{2.40b}$$

From (2.39) and (2.5a,b),

*Again we refer to the residue of the analytic extension of b/a at $\zeta = \zeta_k$.

(2.41a) $$a_t = -\hat{I}(\phi,\psi) = -a\hat{I}(\bar{\psi},\psi) - b\hat{I}(\psi,\psi) = 0 ,$$

(2.41b) $$\bar{a}_t = -\hat{I}(\bar{\phi},\bar{\psi}) = \bar{a}\hat{I}(\psi,\bar{\psi}) - \bar{b}I(\bar{\psi},\bar{\psi}) = 0 .$$

These relations immediately furnish motion invariants $\{\zeta_k\}_{k=1}^{N}$, $\{\bar{\zeta}_k\}_{k=1}^{N}$ and from the asymptotic expansions (2.10), $\{C_n\}_{n=1}^{\infty}$. Furthermore we have a continuum of invariants, namely the function ln a itself, a fact we will stress in much detail in the following chapter.

Case (b): If $\Omega = \Omega_1/\Omega_2$ where Ω_1 and Ω_2 are entire, we find

(2.43) $$\Omega_2(L^+)\begin{pmatrix} r_t \\ -q_t \end{pmatrix} + 2\Omega_1(L^+)\begin{pmatrix} r \\ q \end{pmatrix} = 0 ,$$

or

(2.43) $$\begin{pmatrix} r_t \\ -q_t \end{pmatrix} + 2\Omega_2^{-1}(L^+)\Omega_1(L^+)\begin{pmatrix} r \\ q \end{pmatrix} = 0 .$$

Again it may be shown that $a(\zeta)$ and $\bar{a}(\zeta)$ are conserved by (2.43). However, the requirement that $\Omega_2^{-1}(L^+)\Omega_1(L^+)\binom{r}{q}$ be defined requires some restrictions on the initial data, a matter discussed in some detail in AKNS [2]. If $\Omega(\zeta)$ has a pole of order M at $\hat{\zeta}$, Im $\hat{\zeta} > 0$, we require that $a(\hat{\zeta})$ and its first M-1 derivatives be nonvanishing at $\hat{\zeta}$. If $\Omega(\zeta)$ has a pole of order M on the real axis $\zeta = \hat{\zeta}$, we require that $b(\hat{\zeta})$, $\bar{b}(\hat{\zeta})$ and their first M-1 derivatives vanish at this pole. For example, for the sine-Gordon equation (2.4), $\Omega(\zeta) = i/4\zeta$ and the condition that $b(0) = 0$ is equivalent to the boundary condition $\int_{-\infty}^{\infty} q\, dy = n\pi$, $n = 0,\pm1,\ldots,2,\ldots$. For boundary conditions where $\int_{-\infty}^{\infty} q\, dy = (2n+1)\pi/2$, $a(0) = 0$ and the present analysis is not applicable although this case is of physical interest. (In all probability the solution is primarily a π pulse which travels at the speed of light and seeks to remove the unstable equilibrium at $x = +\infty$ as rapidly as possible).

In both cases (a) and (b) considerations of stability generally mean that $\Omega(\zeta)$ must be purely imaginary when ζ is real.

Case (c): $\Omega(\zeta) \neq \bar{\Omega}(\zeta)$. In this case we cannot deduce (2.38) from (2.38), as both equations (2.32) would not be satisfied. Instead we use the completeness of the states $\begin{pmatrix} \psi_1^2 \\ \psi_2^2 \end{pmatrix}$, $\begin{pmatrix} \bar{\psi}_1^2 \\ \bar{\psi}_2^2 \end{pmatrix}$ (for simplicity in this presentation we assume no bound states, a more complete analysis will be given by AKNS shortly: the completeness proof was given by Kaup [22]), generated by the operator L and the adjoint states $\begin{pmatrix} \phi_2^2 \\ -\phi_1^2 \end{pmatrix}$ $\begin{pmatrix} \bar{\phi}_2^2 \\ -\bar{\phi}_1^2 \end{pmatrix}$ generated by L^+. Noting that equation (2.32) gives the projection of $\begin{pmatrix} r_t \\ -q_t \end{pmatrix}$ into the adjoint states $\begin{pmatrix} \phi_2^2 \\ -\phi_1^2 \end{pmatrix}$ and $\begin{pmatrix} \bar{\phi}_2^2 \\ -\bar{\phi}_1^2 \end{pmatrix}$, we have

$$(2.44)\qquad \begin{pmatrix} r_t \\ -q_t \end{pmatrix} = \frac{2}{\pi}\int_{-\infty}^{\infty} \Omega(\zeta)\frac{\bar{b}}{a}\begin{pmatrix} \phi_2^2 \\ -\phi_1^2 \end{pmatrix} d\zeta + \frac{2}{\pi}\int_{-\infty}^{\infty} \bar{\Omega}(\zeta)\,\frac{b}{\bar{a}}\begin{pmatrix} \bar{\phi}_2^2 \\ -\bar{\phi}_1^2 \end{pmatrix} d\zeta\ ,$$

where we have used

$$(2.45)\qquad \int_{-\infty}^{\infty} \begin{pmatrix} \psi_1^2 \\ \psi_2^2 \end{pmatrix}_{\zeta} \begin{pmatrix} \phi_2^2 \\ -\phi_1^2 \end{pmatrix}_{\zeta'} dx = \pi a^2 \delta(\zeta-\zeta')\ , \qquad \zeta,\zeta' \text{ real.}$$

Equation (2.44) may be simplified by noting that

$$(2.46)\qquad \int_{-\infty}^{\infty} g(\eta)\begin{pmatrix} \phi_2\bar{\phi}_2 \\ -\phi_1\bar{\phi}_1 \end{pmatrix} d\eta = \frac{i}{2\pi}\int_{-\infty}^{\infty} d\zeta\, \frac{\bar{b}}{a}\begin{pmatrix} \phi_2^2 \\ -\phi_1^2 \end{pmatrix}\left(\int_{C_U} \frac{g(\eta)}{\zeta-\eta}\, d\eta\right)$$

$$+ \frac{i}{2\pi}\int_{-\infty}^{\infty} d\zeta\, \frac{b}{\bar{a}}\begin{pmatrix} \bar{\phi}_2^2 \\ -\bar{\phi}_1^2 \end{pmatrix}\left(\int_{C_A} \frac{g(\eta)}{\zeta-\eta}\, d\eta\right),$$

when C_U and C_A are contours along the real η axis indenting under and over the pole at $\eta = \zeta$ respectively. Thus, if we choose,

$$(2.47)\quad \Omega(\zeta) = -\frac{i}{4}\int_{C_U} \frac{g(\eta)}{\zeta-\eta}\, d\eta \;, \qquad \bar{\Omega}(\zeta) = -\Omega^*(\zeta) = -\frac{i}{4}\int_{C_A} \frac{g(\eta)}{\zeta-\eta}\, d\eta \;,$$

together with the consistent choice $r = \alpha q^*$, α real ($g(\eta)$ real), we find (2.44) is

$$(2.46)\qquad q_t = \int_{-\infty}^{\infty} g(\eta)\; \phi_1\bar{\phi}_1\, d\eta \;.$$

The auxiliary equations are ($r = \alpha q^*$, $\bar{\phi}_1 = -\frac{1}{\alpha}\phi_2^*$, $\bar{\phi}_2 = -\phi_1^*$ for real ζ),

$$(2.47)\qquad \lambda_x + 2i\zeta\lambda = 2qN \;,$$

$$(2.48)\qquad N_x = \alpha(q\lambda^* + q^*\lambda) \;,$$

when $\lambda = 2\phi_1\bar{\phi}_1$ and $N = \phi_1\bar{\phi}_2 + \bar{\phi}_1\phi_2$. When $\alpha = -1$, equations (2.46) through (2.48) are the Maxwell-Bloch equations describing the propagation of pulses through an inhomogeneously broadened and resonant medium with broadening distribution $g(\eta)$, electric field envelope $2q$, polarization λ and excitation number density N. For further details we refer the reader to [3].

When $\alpha = +1$, the quantity N is negative definite for all x and thus no permanent pulses can occur. This dovetails with the mathematics of the problem since in this case the original eigenvalue problem (2.1) is selfadjoint and therefore there are no bound states.

The principal reason for introducing this example here is to illustrate that we still may have integrable systems <u>without any motion invariants</u> other than the spectrum of (2.1). We may compute $\hat{I}(\phi,\psi) = b\hat{I}(\psi,\psi) + a\hat{I}(\psi,\bar{\psi})$ and hence from (2.30) find

$$\frac{\partial}{\partial t} \ln a = \frac{i}{4} P \int_{-\infty}^{\infty} g(\eta) \frac{2b\bar{b}(\eta)}{\zeta-\eta} d\eta \tag{2.49}$$

when P refers to the Cauchy Principal Value. From (2.10), we note that none of the $\{C_k\}_{k=1}^{\infty}$ are conserved. Thus the invariance of the spectrum and the existence of an infinite sequence of conserved quantities are separate questions and the latter property is not crucial to the success of IST. In Chapter 4, Sections 4.3 and 4.4, we again study the question of the existence of motion invariants in a more abstract setting.

3. The Hamiltonian Formulation

3.1 Introduction

The goal of this chapter is to express the evolution equations in the framework of Hamiltonian mechanics. Our aim is to show how the equation (2.2) can be written in the Hamiltonian form

$$q_t = \frac{\delta H_\Omega}{\delta r}, \qquad r_t = -\frac{\delta H_\Omega}{\delta q}$$

and so our first task will be to show the second term in (2.2) is a gradient. Here H_Ω is a functional of q and r which may be constructed directly from (2.1) from a knowledge of $\Omega(\zeta)$, the dispersion relation, alone. The functional ln a, which must be a constant of the motion,* plays the central role of the analysis. First, given $\Omega(\zeta)$, we may use ln a directly to generate H_Ω. Second, since ln a may be expressed both as a function of q and r by (2.7) and (2.10), and of the scattering data S (Section 3.3), the equality of these expressions provides the expression for the Hamiltonian in terms of the action-angle variables. The equations of motion in the new variables are precisely (2.40), (2.41) and so we may interpret the transforms from (r,q) to the scattering data as a canonical transformation to action-angle variables. As a byproduct, the expression for ln a in the two sets of variables provides interesting nonlinear analogues of Parseval's relations in linear Fourier analysis.

We treat three cases. First, when r and q are prescribed independently, it turns out that they play the role of conjugate variables. The underlying symplectic form which is preserved is $\int_{-\infty}^{\infty} (dq \wedge dr)\, dx$. Second, when r is directly proportional to q, we show that the equations (2.2) arise most naturally in the form

* As in cases (a) and (b) but not (c), Section 2.4.

$$q_t = \frac{\partial}{\partial x}\frac{\delta H}{\delta q} ,$$

which corresponds to the underlying symplectic form $\omega(\delta_1 q, \delta_2 q) = = \int_{-\infty}^{\infty} (\delta_1 q \int_{-\infty}^{x} \delta_2 q \, dy) \, dx$. Finally, we treat the case when $r = -1$, for which the system (2.1) becomes the Schrodinger equation with eigenvalue ζ^2. We reproduce the results of Zakharov and Faddeev [44] and in addition provide the equations and the corresponding Hamiltonian framework when the dispersion relation $\Omega(\zeta)$ is singular.

3.2 The Variables r and q Have Independent Variation

Our first aim is to show that the term $-2\Omega(L^+)\binom{r}{q}$ of equation (2.2) can be expressed as a gradient. For this purpose, we begin from equation (2.29) which is

$$\delta \ln a = \int_{-\infty}^{\infty} \left(\delta q \left(\frac{\phi_2 \psi_2}{a}\right) - \delta r \left(\frac{\phi_1 \psi_1}{a}\right)\right) dx . \tag{3.1}$$

But the quadratic products of solutions to (2.1) satisfy the third order system

$$\frac{1}{2}(\phi_1\psi_2 + \phi_2\psi_1)_x = q\phi_2\psi_2 + r\phi_1\psi_1 , \tag{3.2}$$

$$(\phi_1\psi_1)_x + 2i\zeta\phi_1\psi_1 = q(\phi_1\psi_2 + \phi_2\psi_1) , \tag{3.3}$$

$$(\phi_2\psi_2)_x - 2i\zeta\phi_2\psi_2 = r(\phi_1\psi_2 + \phi_2\psi_1) , \tag{3.4}$$

which system, after solving (3.2) for

$$\phi_1\psi_2 + \phi_2\psi_1 - a = \int_{-\infty}^{x} (2q\phi_2\psi_2 + 2r\phi_1\psi_1) \, dx , \tag{3.5}$$

may be written

$$(3.6) \qquad \zeta\begin{pmatrix} \dfrac{\phi_2\psi_2}{a} \\ -\dfrac{\phi_1\psi_1}{\alpha} \end{pmatrix} = L^+\begin{pmatrix} \dfrac{\phi_2\psi_2}{a} \\ -\dfrac{\phi_1\psi_1}{a} \end{pmatrix} - \frac{1}{2i}\binom{r}{q}$$

where L^+ is defined by (2.3). Equation (3.6) is of great value as it immediately provides the relation between the vector $\binom{r}{q}$ and the gradient of ln a (see (3.1)). To see this, we treat ζ large, Im $\zeta > 0$ and solve (3.6) iteratively to find

$$(3.7) \qquad \begin{pmatrix} \dfrac{\phi_2\psi_2}{a} \\ -\dfrac{\phi_1\psi_1}{a} \end{pmatrix} = -\sum_{m=0} \frac{1}{2i\zeta^{m+1}} L^{+m}\binom{r}{q} .$$

Then from (3.1)

$$(3.8) \qquad \delta \ln a = -\frac{1}{2i}\sum_{m=0} \frac{1}{\zeta^{m+1}} \int_{-\infty}^{\infty} \begin{pmatrix}\delta q\\ \delta r\end{pmatrix}\cdot L^{+m}\binom{r}{q}\, dx$$

which by comparison with (2.10) gives

$$(3.9) \qquad 2i\delta C_{m+1} = \int_{-\infty}^{\infty} \begin{pmatrix}\delta q\\ \delta r\end{pmatrix}\cdot L^{+m}\binom{r}{q} , \qquad m = 0,1,\ldots,$$

or

$$(3.10) \qquad 2i \text{ grad}_{q,r} C_{m+1} = L^{+m}\binom{r}{q} .$$

The gradient operator is considered to be the variational (Frechet) derivative with respect to q and r respectively. From (3.10), we deduce that the functionals $\{C_n\}_{n=1}^{\infty}$ and linear combinations thereof play the role of Hamiltonians for all those flows (2.2) for which the dispersion relation is a positive power series for ζ. For example, if $\Omega(\zeta) = \zeta^m$, then (2.2) is

$$(3.11) \qquad \begin{pmatrix} r_t \\ -q_t \end{pmatrix} = -2L^{+m}\binom{r}{q}$$

$$= -4i \text{ grad}_{q,r} C_{m+1} .$$

Thus if we set

(3.12) $$H_m = -4iC_{m+1}, \qquad m = 0,1,2,\ldots,$$

we have

(3.13) $$r_t = \frac{\delta H_m}{\delta q}, \qquad q_t = -\frac{\delta H_m}{\delta r},$$

which is the standard Hamiltonian form with q and r as conjugate variables. Therefore each C_m, $m = 1,2,\ldots,$ generates a flow and the flow with this particular Hamiltonian conserves all of the $\{C_m\}_{m=1}^{\infty}$ and moreover the total function $a(\zeta)$ for all ζ. The two-form $\int_{-\infty}^{\infty} (dq \wedge dr)\, dx$ is also preserved.

Introducing the Poisson bracket,

(3.14) $$\langle f,g\rangle = \int_{-\infty}^{\infty} \left(\frac{\delta f}{\delta q}\frac{\delta g}{\delta r} - \frac{\delta f}{\delta r}\frac{\delta g}{\delta q}\right) dx,$$

we may write the time evolution of any functional of q and r, $I[q,r]$, as

(3.15) $$\frac{dI}{dt} = \int_{-\infty}^{\infty} \left(\frac{\delta I}{\delta q} q_t + \frac{\delta I}{\delta r} r_t\right) dx = \langle H_m, I\rangle$$

as q and r evolve according to (3.13). If I is any motion invariant,

(3.16) $$\langle H_m, I\rangle = 0,$$

and in particular

(3.17) $$\langle H_m, H_n\rangle = 0, \quad \text{all } m,n \geq 0,$$

which shows that all the constants of motion are in involution.

It is also true that for each m,

(3.18) $$\langle H_m, \ln a(\hat{\zeta}_1)\rangle = 0, \quad \operatorname{Im} \hat{\zeta}_1 > 0, \quad a(\hat{\zeta}_1) \neq 0,$$

and also

(3.19) $$\langle H_m, \frac{\partial^n}{\partial \zeta^n} \ln a(\hat{\zeta}_1)\rangle = 0,$$

which suggests that the function $\ln a(\hat{\zeta}_1)$ itself and its derivatives can play the role of Hamiltonians provided $\hat{\zeta}_1$ is not an eigenvalue of (2.1). We now show that these flows all correspond to dispersion relations $\Omega(\zeta)$ which have poles to some order at $\zeta = \hat{\zeta}_1$.

To see this, observe from (3.6) that

$$(3.20)\qquad (L^+ - \zeta)^{-1} \binom{r}{q} = 2i \begin{pmatrix} \dfrac{\phi_2\psi_2}{a} \\ -\dfrac{\phi_1\psi_1}{a} \end{pmatrix}$$

Then if $\Omega(\zeta) = 1/(\zeta - \hat{\zeta}_1)$,

$$(3.21)\qquad \begin{pmatrix} r_t \\ -q_t \end{pmatrix} = -2(L^+ - \hat{\zeta}_1)^{-1} \binom{r}{q} = -4i \begin{pmatrix} \dfrac{\phi_2\psi_2}{a} \\ -\dfrac{\phi_1\psi_1}{a} \end{pmatrix}_{\hat{\zeta}_1}$$

which by (3.1) is,

$$(3.22)\qquad = -\, 4i \,\mathrm{grad}_{q,r} \ln a(\hat{\zeta}_1) \;.$$

Furthermore, if $\Omega(\zeta) = 1/(\zeta - \hat{\zeta}_1)^n$,

$$(3.23)\qquad \begin{pmatrix} r_t \\ -q_t \end{pmatrix} = -2(L^+ - \hat{\zeta}_1)^{-n} \binom{r}{q} = -\frac{4i}{(n-1)!} \frac{\partial^{n-1}}{\partial \zeta^{n-1}} \begin{pmatrix} \dfrac{\phi_2\psi_2}{a} \\ -\dfrac{\phi_1\psi_1}{a} \end{pmatrix}_{\hat{\zeta}_1}$$

$$(3.24)\qquad = -\, 4i \,\mathrm{grad} \frac{1}{(n-1)!} \frac{\partial^{n-1}}{\partial \zeta^{n-1}} \ln a(\hat{\zeta}_1) \;,$$

where the second step is obtained by differentiating (3.20) $n-1$ times and setting $\zeta = \hat{\zeta}_1$. The equations (3.21) and (3.23) must be augmented by the generalized Bloch equations (3.2), (3.3), (3.4) which give the quadratic products of solutions to (2.1) as functionals of q and r. In fact, in physical situations the equations arise most naturally in this format [3]. If Im $\hat{\zeta}_1 < 0$, then the relevant Hamiltonians are $+4i \ln \bar{a}(\hat{\zeta}_1)$ and $4i \frac{1}{(n-1)!} \frac{\partial^{n-1}}{\partial \zeta^{n-1}} \ln \bar{a}(\hat{\zeta}_1)$. If $\hat{\zeta}$ is real, then either $-4i \ln a(\hat{\zeta}_1)$ or $4i \ln \bar{a}(\hat{\zeta}_1)$ may be the Hamiltonian but in order for $(L^+-\hat{\zeta}_1)^{-1}\binom{r}{q}$ to exist we must demand that $b(\hat{\zeta}_1) = \bar{b}(\hat{\zeta}_1) = 0$. This step is consistent with equating the Hamiltonians $-4i \ln a(\hat{\zeta}_1)$ and $4i \ln a(\hat{\zeta}_1)$ when one recalls that for real ζ, $a\bar{a} + b\bar{b} = 1$. This requirement is analogous to the criterion one must place on the Fourier transform of the initial condition $r(x,0)$ of a quantity $r(x,t)$ which evolves according to the linear equation $r_{xt} - 2i\hat{\zeta}r_t = r$, if $r(x,t)$ is to remain analytic analytic at later times. When $\hat{\zeta}_1$ is real the flow corresponding to $\Omega(\zeta) = 1/(\zeta-\hat{\zeta}_1)$ is given by

$$
\begin{aligned}
r_t &= 4i(\phi_2\bar{\phi}_2)_{\hat{\zeta}_1} , \\
q_t &= 4i(\phi_1\bar{\phi}_1)_{\hat{\zeta}_1} ,
\end{aligned}
\tag{3.25}
$$

which system reduces to the sine Gordon equation (2.4) when $\hat{\zeta}_1 = 0$ (for details see Section 3.6).

We may now write down the general result. If

$$
\Omega(\zeta) = \sum_j \sum_{n=1}^{N_j} a_{-n}^{(j)} \frac{1}{(\zeta-\hat{\zeta}_1)^n} + \sum_{n=0}^{\infty} a_n \zeta^n
\tag{3.26}
$$

is the dispersion relation of the lineraized evolution equation, then the corresponding Hamiltonian for this system is

$$
\begin{aligned}
H_\Omega[q,r] = &- 4i \sum_j \sum_{n=1}^{N_j} a_{-n}^{(j)} \frac{1}{(n-1)!} \frac{\partial^{n-1}}{\partial \zeta^{n-1}} (\ln a)^+_{\hat{\zeta}_j} \\
&+ 4i \sum_{n=0}^{\infty} \frac{a_n}{(n+1)!} \frac{\partial^{n+1}}{\partial (\frac{1}{\zeta})^{n+1}} (\ln a)\Big|_{\zeta\to\infty}
\end{aligned}
\tag{3.27}
$$

The operator $\frac{1}{(n+1)!} \frac{\partial^{n+1}}{\partial (\frac{1}{\zeta})^{n+1}}$ acting on ln a means the $1/\zeta^{n+1}$ coefficient in the asymptotic expanion (2.10) for ln a. The quantity $(\ln a)^+$ has the following interpretation

(3.28) $(\ln a)^+ = \ln a, \operatorname{Im} \hat{\zeta}_j > 0$ and $a(\hat{\zeta}_j), \ldots \frac{\partial^{N_j-1}}{\partial\zeta^{N_j-1}} a(\hat{\zeta}_j)$ nonzero,

$= \ln a = -\ln\bar{a}, \operatorname{Im}\hat{\zeta}_1 = 0$ and $b(\hat{\zeta}_j)$, $\bar{b}(\hat{\zeta}_j)$ and their

first $(N_j - 1)$ derivatives are zero at $\hat{\zeta}_j$

$= -\ln\bar{a}, \operatorname{Im}\hat{\zeta}_j < 0$ and $\bar{a}(\hat{\zeta}_j), \ldots \frac{\partial^{N_j-1}}{\partial\zeta^{N_j-1}} a(\hat{\zeta}_j)$ nonzero.

A direct proof that all the constants of motion are in involution i.e., $\langle \ln a(\hat{\zeta}_1), \ln a(\hat{\zeta}_2)\rangle = 0$ may be given by use of the identity

$$\left(\tfrac{1}{2}(\phi_1\phi_2+\phi_2\psi_1)_{\zeta_1}(\phi_1\psi_2+\phi_2\psi_1)_{\zeta_2} - (\phi_1\psi_1)_{\zeta_1}(\phi_2\psi_2)_{\zeta_2} - (\phi_1\psi_1)_{\zeta_2}(\phi_2\psi_2)_{\zeta_1}\right)_x$$

(3.29)

$$= \quad 2i(\zeta_1-\zeta_2)\left((\phi_1\psi_1)_{\zeta_1}(\phi_2\psi_2)_{\zeta_2} - (\phi_1\psi_1)_{\zeta_2}(\phi_1\psi_1)_{\zeta_1}\right)$$

where the subscripts ζ_1,ζ_2 refer to the dependence of the quantities in brackets on ζ. Integration with respect to x and using (3.1) yields

(3.30a) $$2i(\zeta_1-\zeta_2)\int_{-\infty}^{\infty}\left(\frac{\delta\ln a(\zeta_1)}{\delta r}\frac{\delta\ln a(\zeta_2)}{\delta q} - \frac{\delta\ln a(\zeta_1)}{\delta q}\frac{\delta\ln a(\zeta_2)}{\delta r}\right)dx = 0$$

or

(3.30b) $$\langle \ln a(\zeta_1), \ln a(\zeta_2)\rangle = 0 \; .$$

Differentiation of (3.29) with respect to ζ_1 (or ζ_2) and use of inductive arguments will give furthermore that for all m,n,

(3.31) $$\left\langle \frac{\partial^m \ln a(\zeta_1)}{\partial\zeta^m}, \frac{\partial^n \ln a(\zeta_2)}{\partial\zeta^n}\right\rangle = 0 \; .$$

3.3 Representation of the Hamiltonian in Action-Angle Variables

Expressions for ln a and ln $\bar{a}$ in terms of the scattering data S_+ and S_- were derived in AKNS [2]. Here we briefly review these derivations. Since $a(\zeta)$ is analytic for $\text{Im}\,\zeta \geq 0$, with a finite number of zeros, and with asymptotic behavior $a(\zeta) \to 1$, $\zeta \to \infty$, $\text{Im}\,\zeta > 0$, the function $f(\zeta)$, defined as

$$f(\zeta) = a(\zeta) \prod_{1}^{N} \frac{\zeta-\zeta_k^*}{\zeta-\zeta_k}, \tag{3.32}$$

is analytic, $\text{Im}\,\zeta > 0$, and has neither zeros nor poles in the upper half plane. We assume that all zeros of $a(\zeta)$ and $\bar{a}(\zeta)$ are simple and that none lie on the real axis. Similarly $\bar{f}(\zeta)$, where

$$\bar{f}(\zeta) = \bar{a}(\zeta) \prod_{1}^{\bar{N}} \frac{\zeta-\bar{\zeta}_k^*}{\zeta-\bar{\zeta}_k}, \tag{3.33}$$

is analytic, $\text{Im}\,\zeta < 0$, and has neither zeros nor poles in the lower half plane. Thus, for $\text{Im}\,\zeta > 0$,

$$\ln f(\zeta) = \frac{1}{2\pi i} \int_{-\infty}^{\infty} \frac{\ln f(\xi)\, d\xi}{\xi - \zeta}$$

and

$$0 = \frac{1}{2\pi i} \int_{-\infty}^{\infty} \frac{\ln \bar{f}(\xi)\, d\xi}{\xi - \zeta}.$$

Adding and using (3.32), (3.33),

$$\ln a(\zeta) = \sum_{k=1}^{N} \ln \frac{\zeta-\zeta_k}{\zeta-\zeta_k^*} + \frac{1}{2\pi i} \int_{-\infty}^{\infty} \frac{d\xi}{\xi-\zeta} \left[\ln a\bar{a} + \sum_{k=1}^{N} \ln \frac{\zeta-\zeta_k^*}{\zeta-\zeta_k} + \sum_{k=1}^{\bar{N}} \ln \frac{\zeta-\bar{\zeta}_k^*}{\zeta-\bar{\zeta}_k} \right]. \tag{3.34}$$

A useful result, used in (4.25), is also obtained by subtracting these two expressions. Similarly for $\text{Im}\,\zeta < 0$,

$$\ln \bar{a}(\zeta) = \sum_{k=1}^{N} \ln \frac{\zeta-\zeta_k}{\zeta-\zeta_k^*} - \frac{1}{2\pi i} \int_{-\infty}^{\infty} \frac{d\xi}{\xi-\zeta} \left[\ln a\bar{a} + \sum_{1}^{N} \ln \frac{\zeta-\zeta_k^*}{\zeta-\zeta_k} + \sum_{1}^{N} \ln \frac{\zeta-\bar{\zeta}_k^*}{\zeta-\bar{\zeta}_k} \right]. \tag{3.35}$$

In both the expressions (3.34) and (3.35) $\ln a\bar{a}$ is complex and must change by $2\pi i(N - \bar{N})$ as ξ goes from $-\infty$ to $+\infty$. This variation is to offset the logarithm terms, the principal branch of which we choose so as to make $\log \frac{\xi-\zeta_k^*}{\xi-\zeta_k}$ real as $\xi \to +\infty$. Then as $\xi \to -\infty$, there is a contribution of $2\pi i$ from $\ln \frac{\zeta-\zeta_k^*}{\zeta-\zeta_k}$ and $-2\pi i$ from $\log \frac{\zeta-\bar{\zeta}_k^*}{\zeta-\bar{\zeta}_k}$. Recall that $\text{Im}\ \zeta_k > 0$, $k = 1,\ldots,N$, and $\text{Im}\ \bar{\zeta}_k < 0$, $k = 1,\ldots,\bar{N}$. Thus the integrals in (3.34), (3.35) converge but cannot be decomposed into three separate integrals.

If ζ is real, we can also find an expression for $\ln a$ by adding the contributions of

$$\int_c \frac{\ln f(\xi)\, d\xi}{\xi - \zeta} \quad \text{and} \quad \int_{\bar{c}} \frac{\ln \bar{f}(\xi)\, d\xi}{\xi - \zeta}$$

where c and $\bar{c}$ are continuous along the real ξ axis indented over and below $\xi = \zeta$ respectively. In this case

$$\ln a = \frac{1}{2}\sum_1^N \ln \frac{\zeta-\zeta_k}{\zeta-\zeta_k^*} + \frac{1}{2}\sum_1^{\bar{N}} \ln \frac{\zeta-\bar{\zeta}_k^*}{\zeta-\bar{\zeta}_k}$$

$$(3.36) \qquad - \frac{1}{2\pi i} P \int \frac{d\xi}{\xi-\zeta}\left[\ln a\bar{a} + \sum_{k=1}^{N} \ln \frac{\xi-\zeta_k^*}{\xi-\zeta_k} + \sum_1^{\bar{N}} \ln \frac{\xi-\bar{\zeta}_k^*}{\xi-\bar{\zeta}_k}\right]$$

where P is the Cauchy principal value.

Equations (3.34), (3.35) and (3.36) give expressions for $\ln a$ in terms of half of the scattering data, namely, $\{\zeta_k\}_{k=1}^N$, $\{\bar{\zeta}_k\}_{k=1}^{\bar{N}}$ and $\ln a\bar{a}(\xi)$. These quantities then serve as the action variables which each flow will conserve and the remaining scattering data will serve as the angle variables whose time evolution is given by the variation of $\ln a$ with respect to the action variables. We now compute these variations.

From (3.34), we note that

$$\frac{\delta \ln a}{\delta \zeta_k} = -\frac{1}{\zeta-\zeta_k} + \frac{1}{2\pi i}\int_{-\infty}^{\infty} \frac{d\xi}{\xi-\zeta}\,\frac{1}{\xi-\zeta_k}$$

$$(3.37) \qquad = -\frac{1}{\zeta-\zeta_k},$$

by closing the contour in the lower half plane. Also

$$\frac{\delta \ln a}{\delta \zeta_k} = \frac{1}{2\pi i}\int_{-\infty}^{\infty} \frac{d\xi}{\xi-\zeta}\,\frac{1}{\xi-\zeta_k}$$

$$(3.38) \qquad = \frac{1}{\zeta-\bar{\zeta}_k}$$

by closing the contour in the upper half plane. In both these cases, the derivative is the ordinary one. Taking the variational derivative of ln a with respect to - ln $a\bar{a}$, we obtain

$$(3.39) \qquad \frac{\delta \ln a}{\delta(-\ln a\bar{a})} = \frac{1}{2\pi i}\,\frac{1}{\zeta-\xi}\,.$$

The function ln $a\bar{a}$ is complex and both its real and imaginary parts (i.e. two pieces of information) may be arbitrarily prescribed except for the constraints that (a) its total variation from $-\infty$ to $+\infty$ is $2\pi i(N - \bar{N})$, (b) when $\Omega(\zeta)$ is singular, certain restrictions must be placed on the initial data.

We define the action variables to be

$$(3.40) \qquad R_k = 2i\zeta_k\,, \quad \bar{R}_k = 2i\bar{\zeta}_k\,, \quad R(\xi) = \frac{1}{\pi}\ln a\,\bar{a}\,,$$

and the corresponding angle variables to be

$$(3.41) \qquad Q_k = \ln b_k\,, \quad \bar{Q}_k = \ln \bar{b}_k\,, \quad Q(\xi) = \ln b(\xi)\,.$$

We may verify that these choices do indeed preserve the symplectic form by checking the various Poisson brackets. By repeatedly using relations analogous to (3.29) we find

$$\langle R_k, Q_j \rangle = \delta_{kj} , \quad \langle \bar{R}_k, \bar{Q}_j \rangle = \delta_{kj} , \quad \langle R(\xi), Q(\xi) \rangle = \delta(\xi - \xi')$$

and all other brackets are zero.

Using (3.26), (3.27), (3.37), (3.38), (3.39) and the fact that (3.34), (3.35), (3.36) give us expressions for the Hamiltonian in terms of the active variables, we compute the motion of the new coordinates.

$$\begin{aligned}
\frac{dQ_k}{dt} &= \frac{d}{dt} \ln b_k = \frac{\partial H}{\partial R_k} = -2\Omega(\zeta_k) , \\
\frac{d\bar{Q}_k}{dt} &= \frac{d}{dt} \ln \bar{b}_k = \frac{\partial H}{\partial \bar{R}_k} = 2\Omega(\bar{\zeta}_k) , \\
\frac{dQ(\xi)}{dt} &= \frac{d}{dt} \ln b = \frac{\delta H}{\delta Q_k} = -2\Omega(\xi) , \\
\frac{dR_k}{dt} &= \frac{d}{dt} 2i\zeta_k = -\frac{\partial H}{\partial Q_k} = 0 , \\
\frac{d\bar{R}_k}{dt} &= \frac{d}{dt} 2i\bar{\zeta}_k = -\frac{\partial H}{\partial \bar{Q}_k} = 0 , \\
\frac{dR(\xi)}{dt} &= \frac{d}{dt} \frac{1}{\pi} \ln a\bar{a} = -\frac{\delta H}{\delta Q(\xi)} = 0 .
\end{aligned} \tag{3.42}$$

These relations agree precisely with (2.40), (2.41).

Once the variables ζ_k, $\bar{\zeta}_k$, $\frac{1}{\pi}$ ln $a\bar{a}$, b_k, $\bar{b}_k$ and $b(\xi)$ are given, their time evolution may be followed. At any time they provide sufficient information from which we can recover the original potentials q and r. (At least formally; however there are still open questions. For example for arbitrary scattering data, are q and r always bounded etc.?) Given $b(\xi)$ and $a\bar{a}(\xi)$, we know $\bar{b}(\xi)$. From (3.34) and (3.35), a and $\bar{a}$ may be found in their respective half planes.

Our conclusion, then, is that the inverse scattering transform is a canonical transformation which for a special class of Hamiltonians (the "conserved" quantities) leads to the set of trivially integrable relations (3.42). Furthermore, we expect that more general systems (with Hamiltonians outside the special class) can be handled by the same canonical map as long as the corresponding relations (3.42) are still integrable. Two particular interesting avenues for further investigation spring to mind. One, the formulation is well suited to examining systems whose Hamiltonians are perturbations to perfect Hamiltonians. Two, it would be interest interesting to construct systems for which the discrete eigenvalues are made to move in a prescribed fashion. The dispersion relation describes how the b's move. What quantity would naturally describe the motion of the ζ_k's?

3.4 Integral Relations Between q,r and the Action Variables

The expressions (2.10) and the asymptotic expansions of ln a as as given in (3.34) can be equated to find relations (in fact, moment equations) between integrals over products of q and r and their derivatives to corresponding expressions in the action variables in close analogy to expressions obtained by Zakharov and Faddeev [44]. We can find expressions for C_m in terms of the asymptotic expansion of (3.34) as a power series in ζ^{-1}. We will simplify this discussion by examining two limiting cases where ln $a\bar{a}$ is real and single valued.

First, let q and r be small in the sense $\int_{-\infty}^{\infty} |q|dx \int_{-\infty}^{\infty} |r|\, dx << 1$. Then there are no bound states, a and $\bar{a}$ are approximately unity and $b(\zeta)$ and $\bar{b}(\zeta)$ are the ordinary Fourier transforms $\tilde{r}(2\zeta)$, $-\tilde{q}(-2\zeta)$ of $r(x,t)$ and $q(x,t)$ respectively. Then, from (3.34)

$$\ln a \sim -\frac{1}{2\pi i}\sum_{m=1}^{\infty}\frac{1}{\zeta^m}\int_{-\infty}^{\infty}\xi^{m-1}\, b\bar{b}\, d\xi. \tag{3.43}$$

But, applying the linear limit to (2.10), we find

$$\ln a \sim -\sum_{m=1}^{\infty}\frac{1}{\zeta^m} C_m$$

where

$$C_m = \frac{1}{(2i)^m}\int q\,\frac{\partial^{m-1}}{\partial x^{m-1}}\, r\, dx$$

$$= \frac{1}{2i}\int_{-\infty}^{\infty} q\left(\frac{1}{2i}\frac{\partial}{\partial x}\right)^{m-1} r\, dx\,, \qquad m = 1,2,\ldots\,.$$

Therefore

$$\int_{-\infty}^{\infty} q\left(\frac{1}{2i}\frac{\partial}{\partial x}\right)^{m-1} r\, dx = \frac{1}{\pi}\int_{-\infty}^{\infty}\xi^{m-1}\, b(\xi)\,\bar{b}(\xi)\, d\xi, \tag{3.44}$$

which the reader will recognize as Parseval's relations.

Second, if $r = -q^*$, then $\bar{a}(\zeta) = a^*(\zeta^*)$, $\bar{b}(\xi) = b^*(\xi)$, $N = \bar{N}$, $\bar{\zeta}_k = \zeta_k^*$, $k = 1,\ldots,N$ and the logarithm terms in the integral of (3.34) cancel. Thus equating powers of ζ^{-m} between (2.10) and (3.34) we obtain

$$C_m = -\sum_{k=1}^{N}\frac{(-\zeta_k)^m - (-\bar{\zeta}_k)^m}{m} + \frac{1}{2\pi i}\int_{-\infty}^{\infty}\xi^{m-1}\ln aa^*\, d\xi. \tag{3.45}$$

The integral (3.45) exists for all m as aa^* decays exponentially as $|\zeta| \to \infty$, ζ real if the condition $\int |x^n||q| < \infty$, for each n, is satisfied. The expressions (3.45) constitute a system of moment equations for one set of motion invariants $\{C_m\}_{m=1}^{\infty}$ in terms of another, i.e., the action variables. Clearly if the latter are prescribed, the former are given by (3.45). It is an open question as to whether the converse holds, although if one views (3.46) as moment equations it seems likely. McLaughlin [32] has shown that the reverse holds for the analogous moment equations derived for the Toda lattice problem. In the framework of Hamiltonian mechanics, it is somewhat natural to think of the constants $\{C_n\}_{n=1}^{\infty}$ (and consequently the action variables $\{\zeta_k\}_{k=1}^{N}$ and $\ln aa^*$) as providing "half" of the information required to recover q and r.

3.5 Special Examples

We will examine two cases. In both we will take $r = -q^*$. If q is complex, we can regard q and r as being prescribed independently. First let $\Omega(\zeta) = -2i\zeta^2$; then the flow (2.2) becomes the nonlinear Schrodinger equation

$$q_t + i(q_{xx} + 2q^2q^*) = 0 \tag{3.46}$$

The Hamiltonian for the system is from (3.26), (3.27), and (2.10)

$$H = i\int_{-\infty}^{\infty} \left(q_x q_x^* - q^2 q^{*2}\right) dx$$

i.e. proportional to the $1/\zeta^3$ component in the expansion (2.10) for ln a. One may verify that

$$q_t = -\frac{\delta H}{\delta(-q)} = \frac{\delta H}{\delta q^*} \tag{3.47}$$

gives (3.46). In terms of the action variables the Hamiltonian is

$$H = -8\sum_{k=1}^{k} \frac{\zeta_k^3 - \zeta_k^{*3}}{3} + \frac{4i}{\pi}\int_{-\infty}^{\infty} \xi^2 \ln aa^*\, d\xi. \tag{3.48}$$

Then,

$$\frac{\partial}{\partial t} \ln b_k = \frac{\partial H}{\partial(2i\zeta_k)} = 4i\zeta_k^2$$

(3.49)

$$\frac{\partial}{\partial t} \ln b = \frac{\partial H}{\partial \frac{1}{\pi} \ln aa^*} = 4i\xi^2$$

and of course

(3.50) $$\frac{\partial \zeta_k}{\partial t} = \frac{\partial aa^*}{\partial t} = 0 \ .$$

Second, let $\Omega(\zeta) = i/4\zeta$. Since $\Omega(\zeta)$ has a singularity on the real axis, we must take the initial data such that $b(\xi = 0, t = 0) = \bar{b}(\xi = 0, t = 0) = 0$. The Hamiltonian for this system is

(3.51) $$H = \ln a(0)$$

and the equations are

$$r_t = -q_t^* = -(\phi_2\bar{\phi}_2) = (\phi_2\phi_1^*)_{\zeta=0} \ ,$$

(3.52)

$$q_t = -(\phi_1\bar{\phi}_1) = -(\phi_1\phi_2^*)_{\zeta=0} \ ,$$

since $\bar{\phi}_1 = \phi_2^*$, $\bar{\phi}_2 = -\phi_1^*$. A special case of the system is found if q is initially real for then it stays real. If we set $q = -u_x/2$, then

(3.53) $$q_t = -\frac{u_{xt}}{2} = -(\phi_1\phi_2)_{\zeta=0} \ .$$

But $\phi_{1x} = -\frac{u_x}{2}\phi_2$, $\phi_{2x} = \frac{u_x}{2}\phi_1$, when $\zeta = 0$ and thus $\phi_1 = \cos u/2$, $\phi_2 = \sin u/2$ and (3.74) becomes the sine-Gordon equation (2,4),

(3.54) $$u_{xt} = \sin u \ .$$

The criterion that $b(0) = 0$ stipulates that $u(+\infty) = \pm 2n\pi$, or that $\int_{-\infty}^{\infty} q \, dy = \pm n\pi$. However, since r and q do not have independent variation, the quantity $\ln a(0)$ cannot be the Hamiltonian. Indeed in the special case when $r = -q$, $a(0) = 1$. Nevertheless, we can derive the flow equations (2.2) for the cases $r = \pm q$ by treating r and q as independent and looking at a special subclass of of the resulting flows.

It would be of value to have a Hamiltonian for the particular class $r = -q$. It will turn out to be generated by $(\frac{1}{\zeta} \ln a)$ which, in the special case where $\zeta = 0$, becomes $\frac{\partial \ln a}{\partial \zeta}$.

3.6 Hamiltonian Framework for the System (2.1) when $r = -q$

From (2.32), we know that the choices

(3.55) $$\hat{I}(\psi,\psi) = 2\Omega(\zeta)a\bar{b} \ , \quad \hat{I}(\bar{\psi},\bar{\psi}) = -2\Omega(\zeta)\bar{a}b$$

lead directly to the flow (2.2). Here we present a somewhat different derivation taking advantage of the relation $r = -q$. The form of the resulting evolution equation will be slightly different from that presented in AKNS Appendix 2, but will have the advantage of being directly convertible to Hamiltonian form. We write the first equation in (3.55) in the form (use $a\bar{b} = -\psi_1\psi_2\Big|_{-\infty}^{\infty} = +\int_{-\infty}^{\infty} q(\psi_1^2-\psi_2^2)\,dx$),

(3.56) $$\int_{-\infty}^{\infty} \left(-q_t(\psi_1^2+\psi_2^2) - 2\Omega(\zeta)\, q(\psi_1^2-\psi_2^2)\right) dx = 0 \ .$$

But,

(3.57a) $$(\psi_1\psi_2)_x = -q(\psi_1^2-\psi_2^2)$$

(3.57b) $$(\psi_1^2+\psi_2^2)_x + 2i\zeta(\psi_1^2-\psi_2^2) = 0$$

(3.57c) $$(\psi_1^2-\psi_2^2)_x - 2i\zeta(\psi_1^2+\psi_2^2) = 4q\psi_1\psi_2 \ .$$

Define $Q = -\int_{\infty}^{x} q_t\, dy$ and note that (using (3.57b)),

(3.58) $$\int_{-\infty}^{\infty} -q_t(\psi_1^2+\psi_2^2)\, dx = [Q(\psi_1^2+\psi_2^2)]_{-\infty}^{\infty} + 2i\zeta \int_{-\infty}^{\infty} Q(\psi_1^2-\psi_2^2)\, dx \ .$$

We demand that $Q(-\infty) = 0$ which implies $\frac{\partial}{\partial t}\int_{-\infty}^{\infty} q\, dy = 0$, or that $\int_{-\infty}^{\infty} q\, dy$ is a conserved quantity. Then (3.56) becomes

$$\int_{-\infty}^{\infty} \left(Q + i\,\frac{\Omega(\zeta)}{\zeta}\, q\right) (\psi_1^2-\psi_2^2)\, dx = 0 \;. \tag{3.59}$$

When $r = -q$, the dispersion relation $\Omega(\zeta)$ must be an odd function of ζ in order that the two equations (2.2) for r and q be compatible. Set $\Omega(\zeta) = \zeta F(\zeta^2)$. For reasons of stability we will require F to be pure imaginary when ζ is real. From (3.57) and Table 1, we may show that

$$L_F(\psi_1^2-\psi_2^2) = \zeta^2(\psi_1^2-\psi_2^2) \;, \quad L_F = -\frac{1}{4}\frac{\partial^2}{\partial x^2} - q^2 + q_x \int_x^{\infty} dy\, q \;. \tag{3.60}$$

Assuming F is entire (if not, write F as F_1/F_2 and apply F_2 to Q), we have $F(\zeta^2)(\psi_1^2-\psi_2^2) = F(L_F)(\psi_1^2-\psi_2^2)$. Therefore,

$$\begin{aligned}\int_{-\infty}^{\infty} (Q+iF(\zeta^2)q)(\psi_1^2-\psi_2^2)\, dx &= \int_{-\infty}^{\infty} (Q+iqF(L_F))(\psi_1^2-\psi_2^2)\, dx \\ &= \int_{-\infty}^{\infty} (Q+iF(L_F^+)q)(\psi_1^2-\psi_2^2)\, dx\end{aligned} \tag{3.61}$$

with

$$L_F^+ = -\frac{1}{4}\frac{\partial^2}{\partial x^2} - q^2 + q\int_{-\infty}^{x} dy\, q_x \;. \tag{3.62}$$

We may satisfy (3.59) with the choice,

$$-\int_{\infty}^{x} q_t\, dy + iF(L_F^+)q = 0 \;, \tag{3.63}$$

or equivalently,

$$q_t = i\,\frac{\partial}{\partial x}\,(F(L_F^+)q) \;.$$

For example, if $F(L_F^+) = -4iL_F^+$, then $L_F^+q = -\frac{1}{4}(q_{xx}+2q^3)$ and $q_t = -q_{xxx} - 6q^2q_x$, the modified KdV equation. In order to find the Hamiltonian for the system (3.64) we note from (2.29) that when $r = -q$,

(3.65) $$\delta \ln a = \int_{-\infty}^{\infty} \delta q \left(\frac{\phi_2 \psi_2}{a} + \frac{\phi_1 \psi_1}{a}\right) dx .$$

From the generalized Bloch equations (3.2), (3.3) and (3.4) we may show

(3.66) $$\left(L_F^{+} - \zeta^2\right)\left(\frac{\phi_2 \psi_2}{a} + \frac{\phi_1 \psi_1}{a}\right) = + i\zeta q \quad ,$$

which is the direct analogue of equation (3.6) and serves to relate q with the gradient of ln a. Solving (3.66) iteratively for large ζ, we find

$$\frac{\phi_2 \psi_2}{a} + \frac{\phi_1 \psi_1}{a} = - i\zeta \sum_{m=0}^{\infty} \frac{(L_F^{+})^m q}{\zeta^{2m+2}} .$$

Thus,

$$\delta \ln a = - i \sum_{0}^{\infty} \frac{1}{\zeta^{2m+1}} \int_{-\infty}^{\infty} \delta q \, L_F^{+m} q \, dx$$

$$= \sum_{n=1}^{\infty} - \frac{1}{\zeta^n} \delta C_n \qquad \text{from (2.10).}$$

For n even, $C_n \equiv 0$ and for $n = 2m+1$,

(3.67) $$\frac{\delta C_{2m+1}}{\delta q} = i L_F^{+m} q .$$

The Hamiltonian for the system with

(3.68) $$\Omega(\zeta) = \sum_{m=0}^{\infty} a_{2m+1} \zeta^{2m+1}$$

is

(3.69) $$H_\Omega = \sum_{m=0}^{\infty} a_{2m+1} C_{2m+1}$$

and

(3.70) $$q_t = \frac{\partial}{\partial x} \frac{\delta H_\Omega}{\delta q} .$$

For example, if $\Omega(\zeta) = -4i\zeta^3$, $H = -4iC_3 = \frac{1}{2}\int_{-\infty}^{\infty} (q_x^2 - q^4)\, dx$, whence $\delta H/\delta q = -q_{xx} - 2q^3$ and $q_t + 6q^2 q_x + q_{xxx} = 0$ which is (1.2).

A Hamiltonian system with 2n degrees of freedom may be written in the form $dz/dt = B\,\mathrm{grad}_z\,H$ when B is a skew-symmetric matrix, $B^T = -B$, and the corresponding quadratic form which is preserved as $(dz)^T B^{-1} (dz)$. More generally, in systems with infinite degrees of freedom B may be a skew-symmetric operator, in the case above, $\frac{\partial}{\partial x}$. The inverse of the differential operator is the integration operator and the natural symplectic form for these systems is

$$\omega(\delta_1 q, \delta_2 q)\, dx = \int_{-\infty}^{\infty} \left(\delta_1 q \int_{-\infty}^{x} dy\, \delta_2 q - \delta_2 q \int_{-\infty}^{x} dy\, \delta_1 q\right) dx$$

as originally suggested by Zakharov and Faddeev [44] in a similar context.

From the properties (e) given in Section1.2 we know that for q real, $-\zeta_k^*$ is an eigenvalue if ζ_k is. Also $\bar{a}(\xi) = a^*(-\xi) = a(\xi)$ when ξ is real. Therefore, the expression (3.34) for ln a may be written

$$(3.71)\quad \ln a = \sum_k \left(\ln\frac{\zeta-\zeta_k}{\zeta-\zeta_k^*} + \ln\frac{\zeta+\zeta_k^*}{\zeta+\zeta_k}\right) + \sum_k \ln\frac{\zeta-i\eta_k}{\zeta+i\eta_k} + \frac{\zeta}{\pi i}\int_0^{\infty} \frac{\ln aa^* d\xi}{\xi^2-\zeta^2}$$

where the first summation is over all complex ζ_k, $\mathrm{Im}\,\zeta_k > 0$ and the second over all discrete eigenvalues lying on the positive imaginary ζ-axis. Thus

$$(3.72)\quad C_{2m+1} = 2\sum \frac{\zeta_k^{2m+1} - \zeta_k^{*2m+1}}{2m+1} + 2\sum \frac{(i\eta_k)^{2m+1}}{2m+1} + \frac{1}{\pi i}\int_0^{\infty} \xi^{2m} \ln aa^*\, d\xi\ .$$

A check of the Poisson brackets (defined in (3.82)) reveals that the appropriate action variables are

$$-\ln \zeta_k \quad (\text{and } -\ln (i\eta_k)) \quad \text{and} \quad \frac{-1}{2\pi\xi} \ln aa^*$$

with the corresponding angle variables

$$\ln b_k \quad \text{and} \quad \mathrm{Arg}\, b(\xi)$$

respectively.

If $F(\zeta^2)$ is singular, we use (3.66) and recognize that

(3.73) $$i\left(L_F^+ - \zeta^2\right)^{-1} q = \left(\frac{\phi_2\psi_2 + \phi_1\psi_1}{a\zeta}\right)$$

and in general (differentiate (n-1) times with respect to ζ^2)

(3.74) $$i\left(L_F^+ - \zeta^2\right)^{-n} q = \frac{1}{(n-1)!} \frac{\partial^{n-1}}{\partial(\zeta^2)^{n-1}} \left(\frac{\phi_2\psi_2+\phi_1\psi_1}{a\zeta}\right)$$

(3.75) $$= \frac{\delta}{\delta q} \frac{1}{(n-1)} \frac{\partial^{n-1}}{\partial(\zeta^2)^{n-1}} \left(\frac{\ln a}{\zeta}\right)$$

by (3.65). Thus if $F(\zeta^2) = 1/(\zeta^2-\hat{\zeta}^2)^n$, the corresponding Hamiltonian is $\frac{1}{(n-1)} \frac{\partial^{n-1}}{\partial(\zeta^2)^{n-1}} \left(\frac{1}{\zeta} \ln a\right)_{\hat{\zeta}}$. Again, if the poles $\pm\hat{\zeta}$ are real, we must demand $b(\pm\hat{\zeta})$ and its first (n-1) derivatives to be zero. If $\hat{\zeta}$ is complex, the demand is that $a(\hat{\zeta})$ and its first (n-1) derivatives be nonvanishing at $\hat{\zeta}$.

The time rate of change of the angle variables is given by (for simplicity take $\Omega = \zeta/(\zeta^2-\hat{\zeta}^2)$),

$$\frac{\partial}{\partial t} \ln b_k = \frac{\partial H}{\partial(-\ln \zeta_k)} = -2 \frac{\zeta_k}{\zeta_k^2-\hat{\zeta}^2} = -2\Omega(\zeta_k),$$

$$\frac{\partial}{\partial t} \operatorname{Arg} b = \frac{\delta H}{\delta\left(\frac{-1}{2\pi\xi} \ln aa^*\right)} = 2\frac{\xi}{\xi^2-\hat{\zeta}^2} = 2i\Omega(\xi),$$

which are precisely (2.40), (2.41).

As an example, take $\Omega = i/4\zeta$, whence

$$F = \frac{i}{4\zeta^2} \quad \text{and} \quad H = \frac{i}{4}\left(\frac{\ln a}{\zeta}\right)_{\zeta=0} = \frac{i}{4}\left(\frac{\partial}{\partial\zeta} \ln a\right)_{\zeta=0}.$$

Then using (3.65) with ζ real ($\psi_2 = -a\bar{\phi}_2$, $\psi_1 = -a\bar{\phi}_1$ when $\zeta = 0$)

(3.76) $$q_t = -i \frac{\partial}{\partial x} \left(\frac{\phi_2\bar{\phi}_2+\phi_1\bar{\phi}_1}{\zeta}\right)_{\zeta=0}.$$

But, from the equivalent equation to (3.57b) in terms of $\phi_1\bar{\phi}$ and $\phi_2\bar{\phi}_2$,

$$\frac{\partial}{\partial x} \frac{\phi_2\bar{\phi}_2+\phi_1\bar{\phi}_1}{\zeta} = -2i(\phi_1\bar{\phi}_1-\phi_2\bar{\phi}_2).$$

Therefore, if $q = -u_x/2$,

$$q_t = -\frac{u_{xt}}{2} = -\frac{1}{2}\left(\phi_1\bar{\phi}_1-\phi_2\bar{\phi}_2\right)_{\zeta=0}$$

which is (see (3.53))

$$u_{xt} = \sin u \, . \tag{3.77}$$

The Hamiltonian may be written in terms of u (and q) directly:

$$H = \frac{i}{4\zeta}\left(\frac{\partial}{\partial\zeta}\ln a\right)_{\zeta=0}$$

$$= -\frac{1}{4}\int_{-\infty}^{\infty}(\phi_1\bar{\phi}_2+\bar{\phi}_1\phi_2+1)\,dx \qquad \text{from (2.7)}$$

$$= -\frac{1}{4}\int_{-\infty}^{\infty}(1-\cos u)\,dx\ , \quad u = -2\int_{-\infty}^{x} q\,dy\ . \tag{3.78}$$

The criterion that $b(0) = 0$ is reflected in the boundary condition that $u \to \pm 2n\pi$ as $x \to \pm\infty$. As a second example, consider the equation

$$q_t + 6q^2q_x + q_{xxx} + \frac{\alpha}{2}\sin\left(-2\int_{-\infty}^{x} q\,dy\right) = 0 \tag{3.79}$$

a combination of the sine-Gordon equation (2.4) and the modified KdV equation (1.2) recently suggested by Konno, Kameyana and Samuki [25] for investigating the weak dislocation effects on wave propagation in an anharmonic crystal. The dispersion relation corresponds to

$$\Omega(\zeta) = -4i\zeta^3 + \frac{i\alpha}{4\zeta} \tag{3.80}$$

and the corresponding Hamiltonian is

$$H = \frac{1}{2}\int_{-\infty}^{\infty}(q_x^2-q^4)\,dx - \frac{\alpha}{4}\int_{-\infty}^{\infty}\left(1-\cos\left(-2\int_{-\infty}^{x} q\,dy\right)\right) \tag{3.81}$$

Some interesting phenomena occur in these cases where the two effects (dislocation potential and nonlinear potential) balance at eigenvalues $\zeta = (\alpha/16)^{1/4}$, whence $\Omega(\zeta) = 0$.

As before, we may write the rate of change of the functional $I[q]$ of q in terms of a Poisson bracket,

$$\frac{dI}{dt} = \int_{-\infty}^{\infty} \frac{\delta I}{\delta q} q_t \, dx = \int_{-\infty}^{\infty} \frac{\delta I}{\delta q} \frac{\partial}{\partial x} \frac{\delta H}{\delta q} \, dx$$

$$= \frac{1}{2} \int_{-\infty}^{\infty} \left(\frac{\partial I}{\partial q} \frac{\partial}{\partial x} \frac{\delta H}{\delta q} - \frac{\delta H}{\delta q} \frac{\partial}{\partial x} \frac{\delta I}{\delta q}\right) dx$$

(3.82) $$= \langle H, I\rangle \, .$$

If I is any motion invariant, $\{C_{2n+1}\}_{n=0}^{\infty}$, $\ln a(\hat{\zeta})$ or its derivatives, then

$$\langle H, I\rangle = 0 \, .$$

The case of $r = +q$ may be handled in a completely analogous fashion. The Hamiltonians for the two better known examples

$$q_t - 6q^2 q_x + q_{xxx} = 0$$

and

$$q_t = -2 \sinh \left(-2 \int_{-\infty}^{x} q \, dy\right)$$

are $\frac{1}{2}\int_{-\infty}^{\infty} (q_x^2+q^4)\, dx$ and $-\frac{1}{4}\int_{-\infty}^{\infty} (1- \cosh 2 \int_{-\infty}^{x} q \, dy)\, dx$ respectively. If q is real, no solitons exist for either equation.

3.7 The Hamiltonian Framework for the System (2.1) when $r = -1$

When $r = -1$, the system (2.1) is equivalent to the Schrodinger equation

(3.80) $$v_{2xx} + (\zeta^2+q)v_2 = 0 \, .$$

Following the same procedure as outlined previously, we define the solutions to (3.80), $\phi, \psi, \bar{\phi}$ and $\bar{\psi}$ with the following asymptotic properties given in Table 2.

Table 2

	$x = -\infty$	$x = +\infty$
ϕ	$\begin{pmatrix} 2i\zeta \\ 1 \end{pmatrix} e^{-i\zeta x}$	$\begin{pmatrix} 2i\zeta a\, e^{-i\zeta x} \\ ae^{-i\zeta x} + be^{i\zeta x} \end{pmatrix}$
$\bar{\phi}$	$\begin{pmatrix} 0 \\ 1 \end{pmatrix} e^{i\zeta x}$	$\begin{pmatrix} 2i\zeta\bar{b}\, e^{-i\zeta x} \\ \bar{a}e^{i\zeta x} + \bar{b}\, e^{-i\zeta x} \end{pmatrix}$
ψ	$\begin{pmatrix} -2i\zeta\bar{b}e^{-i\zeta x} \\ ae^{i\zeta x} - \bar{b}\, e^{-i\zeta x} \end{pmatrix}$	$\begin{pmatrix} 0 \\ 1 \end{pmatrix} e^{i\zeta x}$
$\bar{\psi}$	$\begin{pmatrix} 2i\zeta\bar{a}\, e^{-i\zeta x} \\ \bar{a}e^{-i\zeta x} - b\, e^{i\zeta x} \end{pmatrix}$	$\begin{pmatrix} 2i\zeta \\ 1 \end{pmatrix} e^{-i\zeta x}$

The solutions bear the relations

$$\phi = a\bar{\psi} + b\psi\ , \qquad \bar{\phi} = \bar{a}\psi + \bar{b}\psi\ ,$$

(3.81)

$$\psi = a\bar{\phi} - \bar{b}\phi\ , \qquad \bar{\psi} = \bar{a}\phi - b\phi\ ,$$

where $\bar{b}(\zeta) = b(-\zeta) = b^*(\zeta)$ and $\bar{a}(\zeta) = a(-\zeta) = a^*(\zeta^*)$ and where (because the Wronskian $W(\phi,\bar{\phi}) = 2i\zeta$) $a\bar{a} - b\bar{b} = 1$. The quantities $1/a$ and b/a are the transmission and reflection coefficients T and R defined in the Introduction, (1.4).

Following the method of Section 2.2 we may find the changes in the scattering data due to infinitesimal changes in the potentials q and r. Defining

(3.81)
$$J(u,v) = \int_{-\infty}^{\infty} -\delta quv\, dx$$

we find

$$\delta a = \frac{-1}{2i\zeta} J(\phi_2,\psi_2)\ , \qquad \delta b = \frac{1}{2i\zeta} J(\phi_2,\bar{\psi}_2)\ ,$$

(3.82)

$$\delta \bar{a} = \frac{1}{2i\zeta} J(\bar{\phi}_2,\bar{\psi}_2)\ , \qquad \delta \bar{b} = \frac{-1}{2i\zeta} J(\bar{\phi}_2,\psi_2)\ .$$

In particular,

$$(3.83)\qquad \begin{aligned} a_t &= -\frac{1}{2i\zeta}\hat{J}(\phi_2,\psi_2)\,, & b_t &= \frac{1}{2i\zeta}\hat{J}(\phi_2,\bar{\psi}_2)\,,\\ \bar{a}_t &= \frac{1}{2i\zeta}J(\bar{\phi}_2,\bar{\psi}_2)\,, & \bar{b}_t &= \frac{-1}{2i\zeta}\hat{J}(\bar{\phi}_2,\psi_2)\,, \end{aligned}$$

with $\hat{J}(u,v) = \int_{-\infty}^{\infty} -q_t\, uv\, dx$. The change in time of the reflection coefficient $R = b/a$ is given by

$$(3.84)\qquad R_t = R\,\frac{\hat{J}(\phi_2,\phi_2)}{2i\zeta ab}\,.$$

Choosing

$$(3.85)\qquad \hat{J}(\phi_2,\phi_2) = -2\Omega(\zeta)(2i\zeta ab)$$

gives us that

$$(3.86)\qquad R(t) = R(0)\, e^{-2\Omega(\zeta)t}\,.$$

Using the identity

$$(3.87)\qquad 4\zeta^2 ab = \int_{-\infty}^{\infty} q_x\phi_2^2\,,$$

we find (3.85) is

$$(3.88)\qquad \int_{-\infty}^{\infty} \left(q_t + C_0(\zeta^2)q_x\right)\phi_2^2\, dx = 0$$

with $i\zeta C_0(\zeta^2) = \Omega(\zeta)$. In order that the evolution equation obtained from (3.88) be compatible with that found from the expression for $(\bar{b}/\bar{a})_t$, it is necessary that C_0 be a function of ζ^2. After some manipulation (see AKNS [2] Appendix 1 for details) we find

$$(3.89)\qquad \zeta^2\phi_2^2 = L_s\phi_2^2\,,\qquad L_s = -\frac{1}{4}\frac{\partial^2}{\partial x^2} - q + \frac{1}{2}\int_{-\infty}^{x} dy\, q_x$$

whereupon we find that (3.88) may be satisfied by

$$(3.90)\qquad q_t + C_0(L_s^+)q_x = 0\,,\qquad L_s^+ = -\frac{1}{4}\frac{\partial^2}{\partial x^2}q + \frac{1}{2}q_x\int_{x}^{\infty} dy\,.$$

Equation (3.90) may be rewritten as

$$(3.91)\qquad q_t + \frac{\partial}{\partial x}C_0(L_F^+)q = 0\,,$$

where

$$(3.92)\qquad L_F^{+} = -\frac{1}{4}\frac{\partial^2}{\partial x^2} - q - \frac{1}{2}\int_x^{\infty} dy\, q_x \,.$$

For example, if $C_0(\zeta^2) = -4\zeta^2$, (3.91) is

$$q_t + \frac{\partial}{\partial x}(q_{xx}+3q^2) = 0$$

which is the KdV equation (1.1).

Using the generalized Bloch equations (3.2), (3.3), (3.4) with $r = -1$, we can show

$$(3.93)\qquad (L_F^{+}-\zeta^2)\ \frac{\phi_2\psi_2}{a} - 1 = \frac{q}{2} \,.$$

Equation (3.93) is the analogue of (3.6) and (3.66) and provides the link between q and the gradient of ln a. Solving (3.93) iteratively for large ζ^2,

$$(3.94)\qquad \frac{\phi_2\psi_2}{a} = 1 - \frac{1}{2}\sum_0^{\infty}\frac{1}{(\zeta^2)^{n+1}}\, L_F^{+m}\, q \,.$$

Therefore from (2.10), (3.82) and (3.94),

$$(3.95)\qquad \delta \ln a = -\sum_{\substack{n=1\\ n\ \text{odd}}}^{\infty}\frac{1}{\zeta^n}\,\delta C_n = \frac{1}{2i\zeta}\int_{-\infty}^{\infty} +\delta q\Big(1-\frac{1}{2}\sum_0^{\infty}\frac{1}{(\zeta^2)^{m+1}}\, L_F^{+m}\, q\Big)\, dx.$$

Therefore

$$(3.96)\qquad \frac{\delta C_{2m+3}}{\delta q} = \frac{1}{4i}\, L_F^{+m}\, q \,, \qquad m = 0,1,\ldots\,.$$

Note that $C_1 = \frac{-1}{2i}\int_{-\infty}^{\infty} q$, and $\delta C_1/\delta q = -1/2i$. Thus if

$$(3.97)\qquad \Omega(\zeta) = i\zeta \sum_{m=0} a_{2m+1}(\zeta^2)^m \,,$$

the Hamiltonian is

$$(3.98)\qquad H_\Omega = -4i \sum_{m=0} a_{2m+1}\, C_{2m+3}$$

and

$$(3.99)\qquad q_t = \frac{\partial}{\partial x}\frac{\delta H}{\delta q} \,.$$

For example, $\Omega = -4i\zeta^3$, $H = 16iC_5 = \frac{1}{2}\int_{-\infty}^{\infty} (q_x^2 - 2q^3)\, dx$ and

$$q_t = \frac{\partial}{\partial x}\frac{\delta H}{\delta q} = \frac{\partial}{\partial x}(-q_{xx} - 3q^2) = -q_{xxx} - 6qq_x \ .$$

Making use of the properties of a, b and R, we may write

(3.100) $$\ln a = \sum \ln \frac{\zeta - i\eta_k}{\zeta + i\eta_k} - \frac{\zeta}{2\pi i}\int_{-\infty}^{\infty} \frac{1}{\xi^2 - \zeta^2} \ln(1 - RR^*)\, d\zeta$$

and

(3.101) $$C_{2m+1} = 2\sum \frac{(i\eta_k)^{2m+1}}{2m+1} - \frac{1}{2\pi i}\int_{-\infty}^{\infty} \xi^{2m} \ln(1 - RR^*)\, d\xi \ .$$

Note that for the KdV equation, m = 2 and

$$H = 16iC_5 = -\frac{32}{5}\sum \eta_k^5 - \frac{8}{\pi}\int_{-\infty}^{\infty} \xi^4 \ln(1 - |R|^2)\, d\xi \ ,$$

which is precisely the expression given by Zakharov and Faddeev. The appropriate <u>action variables</u> are $-2\eta_k^2$ and $\frac{-\xi}{\pi}\ln(1-|R|^2)$ and the corresponding angle variables are $\ln b_k$ and Arg b respectively, where b_k is the residue of the analytic extension of b to $\zeta = i\eta_k$. One may verify,

(3.102) $$\frac{\partial}{\partial t}\ln b_k = \frac{\delta H}{\delta(-2\eta_k^2)} = -2\Omega(i\eta_k)$$

and

(3.103) $$\frac{\partial}{\partial t}\text{Arg}\, b = \frac{\delta H}{\delta\left(\frac{-\xi}{\pi}\ln(1-|R|^2)\right)} = 2i\Omega(\xi)$$

which are (3.84), (3.85).

Observe that the trivial constant of the motion $\int q\, dx$ is not included (see (3.96)) among the class of allowable Hamiltonians. As a Hamiltonian it would correspond to a stationary flow in x-space x-space but the corresponding flow in action-angle coordinates would not be. This indicates that we must always restrict ourselves to manifolds of constant $\int q\, dx$. In general, such restrictions would seem to be necessary when the skew symmetric operator B ($q_t = B\frac{\delta H}{\delta q}$; in this case $B = \frac{\partial}{\partial x}$) has a nontrivial null space.

Finally, if C_0 is singular, we use

$$(3.104)\qquad (L_F^{+}-\zeta^2)^{-1}\, q = 2\left(\frac{\phi_2\psi_2}{a} - 1\right)$$

and more generally,

$$(3.105)\qquad (L_F^{+}-\zeta^2)^{-n} = 2\,\frac{1}{(n-1)!}\,\frac{\partial^{n-1}}{\partial(\zeta^2)^{n-1}}\left(\frac{\phi_2\psi_2}{a} - 1\right)$$

Again, if C_0 has a pole of order N at real $\hat{\zeta}^2$, we must demand that $b(\hat{\zeta}^2)$ and its first (N-1) derivatives vanish at $\hat{\zeta}^2$. We must always insist that $\pm\hat{\zeta}$ is never an eigenvalue of (3.80) and (2.1).

For example, if $\Omega(\zeta) = i\zeta/(\zeta^2-\hat{\zeta}^2)$, Im $\hat{\zeta} > 0$, we have

$$(3.106)\qquad q_t = -2\,\frac{\partial}{\partial x}\left(\frac{\phi_2\psi_2}{a}\right)_{\zeta=\hat{\zeta}}$$

and the Hamiltonain for this system is

$$(3.107)\qquad H_\Omega = (-4i\zeta\,\ln a)_{\zeta=\hat{\zeta}}\ .$$

For a pole of order N, $\Omega(\zeta) = i\zeta/(\zeta^2-\hat{\zeta}^2)^N$, Im $\hat{\zeta} > 0$,

$$(3.108)\qquad q_t = -2\,\frac{\partial}{\partial x}\,\frac{1}{(N-1)!}\,\frac{\partial^{N-1}}{\partial(\zeta^2)^{N-1}}\left(\frac{\phi_2\psi_2}{a} - 1\right)$$

and

$$(3.109)\qquad H_\Omega = \frac{1}{(N-1)!}\,\frac{\partial^{N-1}}{\partial(\zeta^2)^{N-1}}\,(-4i\zeta\,\ln a)_{\zeta=\hat{\zeta}}\ .$$

We are very grateful to Dave McLaughlin for many illuminating discussions and key observations on the material of this Chapter.

4. The Lax Approach

4.1 Introduction

In the last chapter, we derived a great many results from a study of the transmission coefficient $a(\zeta)$. Here we offer an operator-theoretic explanation for this ubiquitous role of $a(\zeta)$, directly for the Zakharov-Shabat system, as well as in abstract terms.

Scattering theory is a special way of comparing two operators L_0 and L. The scattering operator is one measure of the "difference" between L_0 and L. As the scattering operator in its entirety is sometimes a very complicated object, other "coarser" measures have been devised to describe the same kind of "difference" between operators. Abstractly, the coefficient $a(\zeta)$ is a scalar function on the spectrum which measures the "shift" from the spectrum of L_0 to that of L. Inasmuch as scattering theory requires that L_0 and L have the same continuous spectrum, this "shift" is a subtle concept, and refers to a redistribution of some "weighting" of the spectrum, rather than to a displacement of the spectrum as a whole.

Our disucssion of these ideas comes in two parts. First, we give a brief computation relating $a(\zeta)$ for the Zakharov-Shabat operator L to the function

$$\text{Trace}\,[(L-\zeta)^{-1} - (L_0-\zeta)^{-1}]\,, \tag{4.1}$$

L_0 corresponding to $q = r = 0$. The expression (4.1) is a generalization, to the singular operator L, of a familiar concept. The trace (= sum of eigenvalues) of a matrix, or the trace $\int_a^b K(x,x)\,dx$ of an integral operator with kernel $K(x,y)$, are much used in linear algebra and Fredholm theory. Formula (4.1) is reminiscent of these quantities; but the ordinary trace is associated with a single operator, while (4.1) requires a reference operator L_0 (to subtract off a divergent integral, see below). Hence, (4.1) is one bridge between spectral theory and perturbation theory.

Elsehwere in these proceedings [11], it was explained how the trace-functional provides a unified way of obtaining constants of the motion for nonlinear equations associated with regular linear eigenvalue problems.* The difference between using regular and singular boundary value problems for the study of nonlinear equations is that in the latter class an infinite number of conserved quantities need not exist.

This becomes particularly clear in the second half of this chapter, where we transform the general Lax equation $L_t = [B,L]$ to scattering variables, and study the relation between (4.1) and an abstract version of $a(\zeta)$.

Throughout, we draw heavily on work done primarily by Soviet mathematicians over the last two decades. The abstract theory of the spectral shift was developed by M. G. Krein [26,27] and related to scattering theory by Krein and Birman [6]. The study of trace formulas (4.1) for differential operators was initiated by Gel'fand and Levitan [19] and continued by other mathematicians; of decisive importance for us is the application of (4.1) to the KdV equation by Zakharov and Faddeev [44]. These works are our sources for this chapter; we shall not give specific references below. What is new here aside from computing the trace formula for the general AKNS system is the use made of the abstract theory in the discussion of Lax's equation.

* E.g., matrices, or differential operators on a finite interval.

4.2 The Trace for the AKNS System

Set

$$L = \begin{pmatrix} i\frac{d}{dx} & -iq \\ ir & -i\frac{d}{dx} \end{pmatrix} .$$

The Green function of $L-\zeta$ is, as usual, the solution G of

$$(L_x-\zeta)\ G(x,y) = \delta(x-y)I .$$

It is easily verified that (ζ is not written; but Im $\zeta > 0$)

$$G(x,y,\zeta) = \frac{i}{a}\begin{pmatrix} \phi_1(x)\psi_2(y) & \phi_1(x)\psi_1(y) \\ \phi_2(x)\psi_2(y) & \phi_2(x)\psi_1(y) \end{pmatrix}, \quad x < y ,$$

(4.2) $$= \frac{i}{a}\begin{pmatrix} \psi_1(x)\phi_2(y) & \psi_1(x)\phi_1(y) \\ \psi_2(x)\phi_2(y) & \psi_2(x)\phi_1(y) \end{pmatrix}, \quad x > y .$$

(The jump $G(y+0,y) - G(y-0,y)$ is found to be

$$i\begin{pmatrix} -1 & 0 \\ 0 & 1 \end{pmatrix}$$

upon use of the Wronskian relationship, $\phi_1\psi_2-\phi_2\psi_1 = a$).

Now we consider the integral operator

(4.3) $$D(\zeta) \equiv (L-\zeta)^{-1} - (L_0-\zeta)^{-1} , \qquad \text{Im } \zeta > 0 ,$$

with kernel $\qquad g(x,y) \equiv G(x,y) - G_0(x,y).$

The trace of $D(\zeta)$, provided it exists, is obtained by summing the diagonal entries of g,

(4.4) $$R(\zeta) \equiv \text{Tr } D(\zeta) = \int_{-\infty}^{\infty} \text{Tr } g(x,x)\ dx .$$

The integral (4.4) may exist and yet not be the trace of $D(\zeta)$ in the technical sense, but for rapidly decaying q and r this difficulty does not arise.

Now, using (4.2) and Table 1,

$$(4.5)\qquad \int_{-\infty}^{\infty} \operatorname{Tr} g(x,x)\, dx = \int_{-\infty}^{\infty} \left[\frac{\psi_1(x)\phi_2(x)+\psi_2(x)\phi_1(x)}{a} - 1\right] dx ,$$

and according to equation (2.7), this equals $-\frac{\partial}{\partial\zeta} \ln a(\zeta)$. Hence follows the desired formula

$$(4.6)\qquad R(\zeta) = -\frac{d}{d\zeta} \ln a(\zeta) .$$

Remark 1. In Fredholm theory, one shows that the resolvent kernel satisfies

$$\int R(x,x)\, dx = -\frac{\Delta'(\lambda)}{\Delta(\lambda)} ,$$

where Δ is the Fredholm determinant. One may thus think of $a(\zeta)$ as a Fredholm determinant of L relative to L_0; the name "perturbation determinant" [3] is common. Abstractly, if L differs from L_0 only by an n-dimensional perturbation, then $R(\zeta)$ is the log derivative of $n\times n$ determinant. Compare also equations (4.13, 4.14) below.

Remark 2. We saw in Chapter 3 that the constants of motion of AKNS evolution equations were obtained as coefficients C_n in the expansion of $\ln a(\zeta)$ for large ζ. Correspondingly we may now expand $R(\zeta)$ in the same way; in a purely formal sense, the resulting coefficients would be proportional to $\operatorname{tr}(L^n - L_0^n)$. Since $L^n - L_0^n$ is an unbounded operator, this expression has no independent meaning, but it does point up the fundamental similarity between the integrals of the AKNS equations, and those of a superficially very different system like the finite Toda lattice [11].

Remark 3. Formally, one can derive from (4.6) the fundamental variational equation (3.1) for ln a which formed the basis for Chapter 3.

Integrating (4.6) we get

$$\ln a(\zeta) = \operatorname{Tr}[\ln(L-\zeta) - \ln(L_0-\zeta)]$$

so that

$$\begin{aligned}\delta \ln a(\zeta) &= \mathrm{Tr}[\ln(L+\delta L-\zeta)-\ln(L-\zeta)]] \\ &= \mathrm{Tr}\,[\ln(L-\zeta)[1+(L-\zeta)^{-1}\delta L]-\ln(L-\zeta)] \\ &= \mathrm{Tr}\,\ln[1+(L-\zeta)^{-1}\delta L + \ldots] \\ &= \mathrm{Tr}(L-\zeta)^{-1}\,\delta L\,.\end{aligned}$$

But, $\delta L = \begin{pmatrix} 0 & -i\delta q \\ i\delta r & 0 \end{pmatrix}$, and therefore,

$$\delta \ln a(\zeta) = \int_{-\infty}^{\infty} \mathrm{Tr} \begin{pmatrix} G_{11} & G_{12} \\ G_{21} & G_{22} \end{pmatrix} \cdot \begin{pmatrix} 0 & -i\delta q \\ i\delta r & 0 \end{pmatrix} dx$$

$$(4.7) \qquad = \int_{-\infty}^{\infty} (iG_{12}\delta r - iG_{21}\delta q)\, dx = \int_{-\infty}^{\infty} \left(\frac{\phi_2\psi_2}{a}\,\delta q - \frac{\phi_1\psi_1}{a}\,\delta r\right) dx$$

which is (3.1).

4.3 Some Abstract Considerations

We have mentioned repeatedly, that Lax's representation $L_t = [B,L]$ leads immediately to the constancy of the spectrum of the Schrodinger operator as the potential evolves according to KdV. It has apparently not been noticed before, that this same representation has in it the seeds of the inverse scattering solution method. Indeed, we shall show that the equation $L_t = [B,L]$, when transformed to scattering variables, becomes (in a certain sense) diagonal (separable). Moreover, some general results of operator theory suggest an abstract analogue of $a(\zeta)$, and thereby provide criteria for the existence of infinitely many motion invariants. It will be seen that the existence of these is not characteristic of equations of Lax type without further conditions on the operators $B(t)$. (A more systematic investigation of the ideas sketched here is in progress, and results will be presented elsewhere.)

We begin by reviewing some facts from spectral theory. The basic concept we need is that of the "spectral representation" of a self-adjoint operator. Consider, for example, the self adjoint case $(r = q^*)$ of the Zaharov-Shabat operator (1.21),

(4.8) $$L = \begin{pmatrix} i\partial & -iq \\ iq^* & -i\partial \end{pmatrix} .$$

In the notation of Table 1, we define normalized* improper eigenfunctions:

(4.9) $$m(x,\zeta) \equiv \frac{1}{a(\zeta)}\,\phi(x,\zeta) , \qquad p(x,\zeta) \equiv \frac{1}{a(\zeta)}\,\psi(x,\zeta)$$

(m and p for "minus" and "plus" in the exponents of the asymptotic behavior). It has already been remarked that (4.8) has no point spectrum, and by computing the jump of Green's function across the spectrum $(-\infty,\infty)$ one finds the formula for expanding elements of $L_2^{(2)}(\mathbb{R})$ ** in eigenfunctions of L: If $f(x) = \begin{pmatrix} f_1(x) \\ f_2)x) \end{pmatrix} \in L_2^{(2)}(\mathbb{R})$, define

(4.10) $$F(\zeta) = \int_{-\infty}^{\infty} f(y)\; m^*(y,\zeta)\; dy , \quad G(\zeta) = \int_{-\infty}^{\infty} f(y)\; p^*(y,\zeta)\; dy ;$$

then

(4.11) $$f(x) = \frac{1}{2\pi}\int_{-\infty}^{\infty} d\zeta \left\{F(\zeta)m(x,\zeta) + G(\zeta)p(x,\zeta)\right\} .$$

Furthermore,

(4.12) $$\int_{-\infty}^{\infty} |f_1(x)|^2 + |f_2(x)|^2\; dx = \int_{-\infty}^{\infty}\left(|F(\zeta)|^2 + |G(\zeta)|^2\right) d\left(\frac{\zeta}{2\pi}\right) .$$

In abstract terms, we have associated with each element f(x) of $L_2^{(2)}(\mathbb{R})$ a vector-valued function $\begin{pmatrix} F(\zeta) \\ G(\zeta) \end{pmatrix}$ defined on the spectrum of L. Let us denote this function by $\hat{f}(\zeta)$. This $\hat{f}$ belongs to a

---

* $$\frac{1}{2\pi}\int_{-\infty}^{\infty} m(x,\zeta)\; m^*(x,\zeta')\; dx = \delta(\zeta-\zeta') .$$

** superscript (2) indicates that the functions in question are 2-vectors.

Hilbert space $\mathcal{L} \equiv L_2(\Lambda,K,dm)$, consisting of functions defined on the spectrum of L (here, $\Lambda = R$), taking values in an auxiliary vector space K (here, dim K = 2), and square-integrable with respect to a measure dm on Λ (here, $dm = \frac{1}{2\pi} d\zeta$). Moreover, (4.12) shows that the mapping $\sigma\colon f \to \hat{f}$ from $L_2^{(2)}(\mathbb{R})$ to $\mathcal{L}$ preserves norm. Standard theory asserts that it is, infact, unitary:(*) $\sigma\sigma^*$ is the identity on $\mathcal{L}$, $\sigma^*\sigma$ is the identity on $L_2^{(2)}(\mathbb{R})$. Finally, we note that the operator L, when transferred to the space $\mathcal{L}$, is extremely simple: (4.11) shows that if $\sigma\colon f(x) \mapsto \begin{pmatrix} F(\zeta) \\ G(\zeta) \end{pmatrix}$, then $\sigma\colon (Lf)(x) \mapsto \zeta\cdot\begin{pmatrix} F(\zeta) \\ G(\zeta) \end{pmatrix}$. We express this by writing $\hat{L} \equiv \sigma L \sigma^* = \zeta I$.

This concrete example is abstracted as follows. Let L be a selfadjoint operator on a Hilbert space H; L is assumed to have an absolutely continuous spectrum Λ of uniform multiplicity. Then there exists an auxiliary Hilbert space K, and a measure dm on Λ , such that H is isomorphic to $\mathcal{L} \equiv L_2(\Lambda,K,dm)$ (K-valued functions on Λ, square integrable with respect to dm). That is, there is a map : $H \xrightarrow{\text{onto}} \mathcal{L}$, sending $f \in H$ to $\hat{f} \in \mathcal{L}$, such that

$$\|f\|_H^2 = \int_\Lambda \|\hat{f}(\zeta)\|_K^2 \, dm(\zeta) .$$

Moreover, the image under σ of Lf is $\zeta\hat{f}(\zeta)$.

Remark. The auxiliary space K may be large; if $L = -\Delta$ on $\mathbb{R}^3$, for example, then $\dim K = \infty$.

Let A be an operator on H. We say A is diagonal in the spectral representation of L , if the image under σ of Af is $\hat{A}(\zeta)\hat{f}(\zeta)$, each $\hat{A}(\zeta)$ being an operator on K. That is, A acts separately for each ζ in the spectral representation. Any bounded selfadjoint or unitary A which commutes with L has this property; if A is unbounded, it may commute with L and yet fail to be diagonal.

(*) * denotes the adjoint operator.

Let us now fix a selfadjoint operator L_0 , to be thought of as "unperturbed." Its spectral representation map to $\mathcal{L}$ will be denoted by σ_0, and images under σ_0 by "tilde": $\sigma_0 : f \to \tilde{f}$. Another L is said to be unitarily equivalent to L_0 , if there exists a unitary operator U which <u>intertwines</u> L and L_0:

(4.13) $$LU = UL_0 .$$

L can then be spectrally represented on the same space as L_0, and in fact

(4.14) $$\sigma = \sigma_0 U^* ;$$

the operator

(4.15) $$\hat{L} = \sigma_0 U^* L U \sigma_0^*$$

is again multiplication by ζ (we use ^ in connection with perturbed operators). There may be different intertwining operators implementing the unitary equivalence between L and L_0. If $LU = UL_0$, and $LV = VL_0$, then

(4.16) $$V = UN ,$$

for a unitary N which is diagonal in the spectral representation of L_0 (indeed, $NL_0 = U^*VL_0 = U^*LV = L_0U^*V = L_0N$, so N and L_0 commute).

As an illustration, take L_0 to be (4.8) with $q = 0$. Given $f \in L_2^{(2)}(\mathbb{R})$, define $F(\zeta)$, $G(\zeta)$ by (4.10), but use the unperturbed eigenfunctions m_0, p_0; then, invert by (4.11), with the <u>perturbed</u> eigenfunctions m, p. The result will, of course, be different from f -- call it U_-f. The operator U_- so defined* intertwines L_0 and L, and any other intertwining operator differs only in the normalization of the generalized eigenfunctions. That normalization is described by the operator N.

* U_- is one of the Møller wave operators; see Ch. 4.4.

In scattering theory, two particular intertwining operators are singled out by physical considerations. These are the Møller wave operators, defined (when they exist) by

$$U_{\pm} = \underset{s \to \pm\infty}{\text{strong-lim}}\, e^{isL} e^{-isL_0} . \tag{4.17}$$

The normalizing operator (4.16) relating these is called the <u>scattering operator</u>, S:

$$U_- = U_+ S . \tag{4.18}$$

In accordance with the general fact mentioned after (4.16), S is unitary, and in the L_0 spectral representation it is described by a function $\tilde{S}(\zeta)$. $\tilde{S}(\zeta)$ is, for each ζ, an operator on K; in our recurring example $\tilde{S}(\zeta)$ is a 2×2 matrix. Matrix representations of $\tilde{S}$ are often useful, and $\tilde{S}(\zeta)$ is commonly called the scattering matrix.

Various hypotheses ensure the existence of the limits (4.17); we shall refer to the following:

(4.19) <u>If</u> $\zeta \notin \Lambda$, <u>the operator</u> $(L-\zeta)^{-1} - (L_0-\zeta)^{-1}$ <u>is trace class</u>.

According to Krein, (4.19) also entails the existence of a function $\Xi(\lambda)$, $\lambda \in \mathbb{R}$, with support on Λ, such that

$$\operatorname{Tr}\left[(L-Z)^{-1}-(L_0-Z)^{-1}\right] = -\int_{-\infty}^{\infty} \frac{\Xi(\lambda)}{(\lambda-Z)^2}\, d\lambda, \qquad z \notin \Lambda. \tag{4.20}$$

Ξ is called the <u>spectral displacement function</u>.

<u>Remark</u>. Equation (4.20) can be generalized considerably; for a large class of functions h,

$$\operatorname{Tr}\left[h(L)-h(L_0)\right] = \int_{-\infty}^{\infty} \Xi(\lambda)\, dh(\lambda) . \tag{4.21}$$

Formula (4.20) corresponds to $h(\lambda) = (\lambda-Z)^{-1}$.

There is a deep and important connection between the scattering matrix and Ξ. For each $\zeta \in \Lambda$, it is possible to define the determinant of $\tilde{S}(\zeta)$ (even though $\tilde{S}(\zeta)$ may be an infinite matrix), and

(4.22) $$\det \tilde{S}(\zeta) = e^{-2i\pi\Xi(\zeta)} .$$

Let us now apply these ideas to Lax's equation. We are given a fixed ("unperturbed") operator L_0, and a family $L(t)$ of unitarily equivalent "perturbations". ($L(0)$ may or may not be equal to L_0.) Thus,

$$L(t)U(t) = U(t)L_0 ,$$

each $U(t)$ being unitary, whence

(4.23) $$L_t(t) = [B(t),L(t)],$$

each $B(t)$ being skew-adjoint. From now on, we omit the variable t; fixed quantities (usually related to L_0) will be distinguished by a subscript 0. The assumptions on L_0 are those made earlier; in particular, it has no point spectrum. We ignore precise smoothness hypotheses; our purpose here is to outline the formal structure of equation (4.23).

Suppose now there is a second intertwining family $V(t)$, so that we have

$$LV = VL_0 , \quad L_t = [C,L] , \quad V_t = CV ,$$

as well as

$$LU = UL_0 , \quad L_t = [B,L] , \quad U_t = BV .$$

By (4.16), there is a normalizing family $N(t)$, $U = VN$, with $\tilde{N}$ diagonal. Now

$$\begin{aligned} N_t = (V^*U)_t &= V_t^*U + V^*U_t = V^*C^*U + V^*BU \\ &= -V^*CU + V^*BU = V^*(B-C)VV^*U \\ &= V^*(B-C)VN . \end{aligned}$$

Since $\sigma_0^*\sigma_0 = I$, we can transfer this relation to the spectral representation of L_0, as follows:

(4.24) $$\sigma_0 N_t \sigma_0^* = (\sigma_0 V^* (B-C) V \sigma_0^*)(\sigma_0 N \sigma_0^*) .$$

$\sigma \equiv \sigma_0 V^*$ maps to the spectral representation of L; we rewrite (4.24) as an equation in $\mathcal{L}$:

(4.25) $$\tilde{N}_t = \sigma (B-C) \sigma^* \tilde{N} .$$

If B-C were a bounded operator, we could conclude that $\sigma(B-C)\sigma^*$ is diagonal (since B-C commutes with L, as is seen from $[B-C,L] = L_t - L_t = 0$). Typically, however, B-C is unbounded, so we assume that $\sigma(B-C)\sigma^*$ is diagonal, acting in $\mathcal{L}$ via an operator function $\widehat{B-C}\ (\zeta)$. Then, (4.25) becomes separable:

(4.26) $$\tilde{N}_t(\zeta) = \widehat{B-C}\ (\zeta)\ \tilde{N}(\zeta) .$$

The point of all this is that under assumption (4.19), a family L(t) of unitarily equivalent operators has associated with it two intrinsically defined intertwining families: the Møller wave operators $U_{\pm}$, which now depend on the parameter t, and satisfy differential equations $U_{\pm t} = B_{\pm} U_{\pm}$. We now apply the earlier argument, taking $B = B_-$, $C = B_+$, $U = U_-$, $V = U_+$. Then N is the scattering operator S (function of t), and (4.26) becomes the evolution equation for the scattering matrix,

(4.27) $$\tilde{S}_t(\zeta) = \mathcal{B}(\zeta)\ \tilde{S}(\zeta) ,$$

where $\mathcal{B} \equiv \sigma_0 U_+^* (B_- - B_+) U_+ \sigma_0^*$.

It thus becomes apparent that Lax's equation $L_t = [B,L]$ leads (under some reasonable hypotheses) to a family (parametrized by ζ) of possibly simpler equations, (4.27). The equivalence of Lax's equation and (4.27), on the other hand, is not implied: that is essentially the inverse-scattering problem. An abstract discussion of the inverse scattering solution method would undoubtedly require much more structure than we have imposed. We can, however, ask whether there exist motion invariants of Lax's equation. They could not be associated with proper eigenvalues, since these have been hypothesized away. In view of our discussion at the beginning of this chapter, another likely source of invariants is the function $R(Z) = \text{Tr}\,[(L-Z)^{-1} - (L_0-Z)^{-1}]$; one may expect the coefficients in a Taylor or asymptotic expansion to be independent of t _if the same is true of_ R(Z). When is R(Z) constant? A _sufficient_ condition, according to Krein's formula (4.20), is: $\Xi(\zeta,t)$ _is independent of_ t.* Equivalently, by (4.22), one may require that $\det \tilde{S}(\zeta)$ be independent of t. To characterize this last condition in terms of the equation satisfied by $\tilde{S}$, we note that $\log\det \tilde{S}(\zeta) = \text{Tr}\log \tilde{S}(\zeta)$, and differentiate:

$$\frac{d}{dt}\log\det \tilde{S}(\zeta) = \frac{d}{dt}\text{Tr}\log\tilde{S}(\zeta) = \text{Tr}\,\tilde{S}_t(\zeta)\,\tilde{S}(\zeta)^* = \text{Tr}\,\mathcal{B}(\zeta)\ .$$

Hence:

(4.28) _If_ $\text{Tr}\,\mathcal{B}(\zeta) \equiv 0$, _then_ R(z) _is independent of_ t.

This criterion, and the form (4.27) of the $\tilde{S}$ equation, are conceptually pleasing because they involve only invariantly defined quantities, such as $U_{\pm}$ and $B_{\pm}$. They do have potential drawbacks, which reveal themselves in specific computations. For instance, the entries of $\mathcal{B}$ depend on those of $\tilde{S}$, so that (4.27) is nonlinear in appearance, even though the entries of $\tilde{S}$ satisfy linear equations (see the example in Section 4.4). Furthermore, it turns out that $\Xi \equiv 0$ in some useful** examples; in that case, $\det \tilde{S}(\zeta) \equiv 1$, and criterion (4.28) becomes meaningless.

A more refined criterion for the existence of constants of motion is based on the following idea: _if the operators_ $\tilde{S}(\zeta,t)$ _are unitarily equivalent_ (for fixed ζ, variable t), _their eigenvalues_ $\lambda_j(\zeta)$ _are invariant functions of_ ζ _which can be used to define_

* provided Ξ is not identically 0; otherwise, the criterion is useless. See below.

**for example, the original Zakharov-Shabat selfadjoint operator; see [45].

constants of motion (by averages, moments, or analytic continuation). Thus, we try to cast the $\tilde{S}$ equation in the form

$$\tilde{S}_t(\zeta) = [T(\zeta), S(\zeta)] , \tag{4.29}$$

with $T(\zeta)$ skew-adjoint. Here is one way to do this: we have

$$S_t = U_{+_t}^* U_- + U_+^* U_{-_t} = -U_+^* B_+ U_- + U_+^* B_- U_- .$$

Let B be a third skew-adjoint family satisfying $L_t = [B,L]$. Adding and subtracting $U_+^* B U_-$, we get

$$\begin{aligned} S_t &= U_+^*(B-B_+)U_- - U_+^*(B-B_-)U_- \\ &= U_+^*(B-B_+)U_+ S - SU_-^*(B-B_-)U_- . \end{aligned}$$

Let $T_{\pm} = U_{\pm}^*(B-B_{\pm})U_{\pm}$. Thus,

$$S_t = T_+ S - ST_- . \tag{4.30}$$

If it is possible to choose B so that $T_+ = T_- = T$, then (4.30) becomes $S_t = [T,S]$, since $T_{\pm}$ are clearly skew, this is the desired representation. Equation (4.29) is a direct consequence.

The examples of Section 4.4 will show (4.29) to be much simpler than (4.27); the disadvantage (with the present formulation, at any rate) is clear: there is no invariant definition of the B which makes $T_+ = T_-$ (if any such B exists). A more detailed theory remains to be worked out.

It is an interesting exercise in spectral theory to compute the various formulas presented above for the operator (4.8). This will be done in a moment. First, a few comments.

1. It is not known at present to what extent the inverse scattering solution method is applicable to higher dimensional problems -- for instance, with $L = -\Delta+q$. In that case, for each ζ, $\tilde{S}(\zeta)$ is an integral operator acting on functions defined on the unit sphere. An attempt at discovering nontrivial equations of the form $L_t = [B,L]$ via (4.29) would involve, first, finding Lax-type

evolutions (4.29) of these integral operators; but then the linearization (4.29) -- which in one space dimension is given by the simplest differential equations -- would still be a flow on an infinite-dimensional space. The spherical harmonics span the auxiliary space K in this case, and (4.29) may well become an infinite system of coupled differential equations in this basis. How much of a simplification this would be remains to be seen.

2. For nonlinear equations described in Lax's setup, it is actually unusual for the soliton part of the solution to be connected with proper eigenvalues. This is the case for KdV and its associated Hamiltonian flows; however, matters are different for the nonlinear Schrodinger equation, the modified KdV, and others. These were originally related to eigenvalue problems in the manner of Lax, i.e., they were written as $L_t = [B,L]$ with selfadjoint L (see [42, 45]). Only in the second step was theeigenvalue problem for L transformed to the nonselfadjoint system investigated by Zaharov and Shabat. When one looks back at the selfadjoint operators, one finds finds that they have no discrete spectrum, and in fact the "eigenvalues" which give rise to solitons lie off the real axis and correspond to expinentially divergent eigenfunctions. Our classification of constants of the motion as being associated with either the discrete or the continuous spectrum was therefore incomplete; a thorough abstract discussion would require consideration of $\tilde{S}(\zeta)$ as a function of complex ζ. The exceptional complex points may be poles of the scattering matrix, or zeros of a certain entry (such as $a(\zeta)$). In either case, something must be continued off the real axis, and -- judgingby the example set by Lax-Phillips theory-- a much more elaborate theory would be required to encompass analytic properties of scattering coefficients.

4.4 Example: The Self-Adjoint Zaharov-Shabat System

We continue the discussion of the operator (4.8),

$$L = \begin{pmatrix} i\partial & -iq \\ iq^* & -i\partial \end{pmatrix} .$$

A. The Wave Operators

It has already been pointed out that different intertwining operators are determined by different normalizations of the perturbed eigenfunctions. The Møller operators are determined by particular choices of these, once a choice of the unperturbed eigenfunctions has been made. For the latter, we pick $m_0 \equiv \begin{pmatrix} e^{-i\zeta x} \\ 0 \end{pmatrix}$ $p_0 \equiv \begin{pmatrix} 0 \\ e^{i\zeta x} \end{pmatrix}$. The U_--eigenfunctions are then found from the Lippmann-Schwinger equation:

$$\text{(4.31)} \quad \begin{aligned} f(x,\zeta) &= m_0(x,\zeta) - \int_{-\infty}^{\infty} G_0(x,y,\zeta+i0) \begin{pmatrix} 0 & -iq(y) \\ iq^*(y) & 0 \end{pmatrix} f(y,\zeta)\, dy , \\ g(x,\ ) &= p_0(x,\zeta) - \int_{-\infty}^{\infty} G_0(x,y,\zeta+i0) \begin{pmatrix} 0 & -iq(y) \\ iq^*(y) & 0 \end{pmatrix} g(y,\zeta)\, dy . \end{aligned}$$

Here, $G_0(x,y,\zeta+i0)$ is the limit of the Green function (4.2), with $q = 0$, as ζ approaches the real axis from above. The solutions f,g turn out to be precisely the m,p as defined in (4.9).* The U_+-eigenfunctions are defined by equations similar to (4.31), but with Green function $G_0(x,y,\zeta-i0)$. They can be shown to be $\bar{\psi}/a^*$ and $-\bar{\phi}/a^*$, respectively. Let us denote these functions by m_+ and p_+.

B. The Scattering Matrix

For $f \in L_2^{(2)}(\mathbb{R})$, let $(\sigma f)(\zeta) = \begin{pmatrix} F(\zeta) \\ G(\zeta) \end{pmatrix}$ (the L_0-representer). Then $U_- f$ is given by

---

*One knows that f,g are linear combinations of m,p; consideration of the asymptotic behavior of f,g as defined by (4.31) shows them to be precisely **m**,p.

$$(4.32)\qquad U_-f(x) = \frac{1}{2\pi}\int_{-\infty}^{\infty} F(\zeta)\, m(x,\zeta) + G(\zeta)\, p(x,\zeta)\, d\zeta \ .$$

By (2.5), $m = \frac{1}{a} m_+ + \frac{b}{a} p_+$, $p = -\frac{b^*}{a} m_+ + \frac{1}{a} p_+$.

Inserting this into (4.32), we get

$$(4.33)\qquad U_-f(x) = \frac{1}{2\pi}\int_{-\infty}^{\infty} \left(\frac{1}{a} F - \frac{b^*}{a} G\right)m_+ + \left(\frac{b}{a} F + \frac{1}{a} G\right)p_+ \, d\zeta \ .$$

Now, the right side of (4.33) is U_+g , where g is that function whose L_0-representer is

$$\begin{pmatrix} \frac{1}{a} & -\frac{b^*}{a} \\ \frac{b}{a} & \frac{1}{a} \end{pmatrix} \begin{pmatrix} F \\ G \end{pmatrix}$$

On the other hand, $g = U_+^* U_- f = Sf$. Hence, the matrix-function of ζ appearing in (4.34) is precisely $\tilde{S}(\zeta)$.

C. <u>The Spectral Shift Function</u>.

One finds that $\det \tilde{S}(\zeta) = a^*(\zeta)/a(\zeta)$. If we set $\arg a = \eta$, then

$$(4.35)\qquad \det \tilde{S} = e^{-2i\eta} \ .$$

The general theory indicates that η should be related to the trace-difference $R(z)$. Indeed, from (4.6) we have

$$(4.36)\qquad R(z) = -\frac{d}{dz} \ln a(z) \ , \qquad \operatorname{Re} z > 0.$$

On the other hand, the analyticity of $\ln a(\zeta)$ in $\operatorname{Re} z > 0$ allows $\ln a$ to be expressed in terms of the values of η on the real axis*:

$$(4.37)\qquad \ln a(z) = \frac{1}{\pi}\int_{-\infty}^{\infty} \frac{\eta(\zeta)}{\zeta - z}\, d\zeta \ .$$

* Simply use the ideas used in deriving (3.34) except subtract instead of add.

From (4.36) and (4.37), we find

$$(4.38) \qquad R(z) = -\frac{1}{\pi}\int_{-\infty}^{\infty}\frac{\eta(\zeta)}{(\zeta-z)^2}\,d\zeta \;.$$

Comparison of (4.38) and (4.20) shows that

$$\Xi(\zeta) = \frac{1}{\pi}\,\eta(\zeta) \;;$$

consequently, the explicit result (4.35) bears out Krein's abstract formula (4.22).

D. <u>The Condition</u> $\mathrm{Tr}\,\mathcal{B} \equiv 0$

We already know (cf. (4.27)) that $\tilde{S}(\zeta)$ must satisfy an equation of the form

$$\tilde{S}_t(\zeta) = \mathcal{B}(\zeta)\,\tilde{S}(\zeta) \;.$$

$\mathcal{B}$ is most easily computed from $\mathcal{B} = \tilde{S}\,\tilde{S}^*$. The result is

$$(4.39) \qquad \mathcal{B} = \frac{1}{aa^*}\begin{bmatrix} -a^*a_t+b_t^*b & -b_t^* \\ b_t & b^*b_t-a^*a_t \end{bmatrix}$$

One can verify that $\mathcal{B}^* = -\mathcal{B}$ (as it should be), by using the relation obtained from differentiation of the identity $1+bb^* = aa^*$. This substitution shows that $\mathrm{Tr}\,\mathcal{B} = (a_t^*a-a^*a_t)/aa^*$, so that

$\mathrm{Tr}\,\mathcal{B} = 0$ <u>iff</u> $a_t^*a = a^*a_t$ <u>iff</u> $\frac{a_t}{a} = \frac{a_t^*}{a^*}$ <u>iff</u> $(\ln a)_t = (\ln a^*)_t$ <u>iff</u> $(\ln \frac{a}{a^*})_t = 0$ iff $\eta_t = 0$. Thus,

$$\mathrm{Tr}\,\mathcal{B} = 0 \Leftrightarrow \eta_t = 0 \,.^{(*)}$$

In view of (4.37), $\eta_t = 0$ iff $\ln a(z)$ is independent of t, for $\mathrm{Re}\,z > 0$. That, of course, has been repeatedly mentioned as criterion for the existence of motion invariants.

Let us now consider a specific evolution of q, namely, case (a), Chapter 2.4. Recall that here $a(\zeta)$ is independent of t,

(*) In terms of (2.32), $\mathrm{Tr}\,\mathcal{B} = 2a\bar{a}\hat{I}(\psi,\bar{\psi})-a\bar{b}\hat{I}(\psi,\psi)+\bar{a}b\hat{I}(\bar{\psi},\bar{\psi})$, which is zero when $\Omega = \bar{\Omega}$.

while $b_t(\zeta) = -2\bar{\Omega}(\zeta)b(\zeta)$. Furthermore, the conditions $\bar{b}(\zeta) = b^*(\zeta)$, $\zeta \in \mathbb{R}$ (due to $r = q^*$), and $\Omega = \bar{\Omega}$ (this is case (a)), imply (cf. (2.33a)) that $\Omega^* = -\Omega$, i.e., Ω is purely imaginary for real ζ. Let us write $\Omega = iA_0$.

From (4.39), we find

$$\mathcal{B} = \frac{2iA_0}{|a|^2}\begin{bmatrix} |b|^2 & -b^* \\ -b & -|b|^2 \end{bmatrix}. \tag{4.40}$$

Equation (4.40) is time-dependent, so that the $\tilde{S}$ evolution, $\tilde{S}_t = \mathcal{B}\tilde{S}$, is a nonautonomous equation. If one takes the reasonable point of view, that the entries of $\tilde{S}$ are unknown, then this equation seems to be nonlinear; however, the separate equations for the entries are linear. The $\tilde{S}$ evolution in the form (4.29), on the other hand, clearly exhibits the trivial time-dependence of the scattering matrix. To explain this, we must refer to the original AKNS theory as expounded in [1,2]. It was observed there that an eigenfunction v of L evolves in time according to

$$v_t = \begin{pmatrix} A & E \\ C & D \end{pmatrix} v \equiv \mathcal{N}v , \tag{4.41}$$

A, C, D, E being certain functions of x, t, and ζ. These functions must satisfy a series of integrability relations (obtained by differentiating $v_t = \mathcal{N}v$ with respect to x, $Lv = \zeta v$ with respect to t, and equating mixed partials). If one wishes to preserve (in t) a certain asymptotic behavior of the eigenfunctions, $\mathcal{N}$ must be restricted further:

The eigenfunctions $m(x,\zeta,t)$ satisfy (4.41), with $A,E,C \to 0$ as $x \to \pm\infty$, $D \to 0$ as $x \to -\infty$, and $D \to -2iA_0$ as $x \to +\infty$, and D+A is independent of x.

The eigenfunctions $p(x,\zeta,t)$ satisfy (4.41), now with $A \to 2iA_0$ as $x \to +\infty$, $A \to 0$ as $x \to -\infty$, and $E,C,D \to 0$ as $x \to \pm\infty$.

Let us make an in-between choice for A,E,C,D; namely let $E,C\to 0$ as $|x|\to\infty$, $A\to iA_0$ as $|x|\to\infty$, $D=-A$. (The last relation is a consequence of the integrability conditions.) Then

$$\phi_t = \begin{pmatrix} A-iA_0 & E \\ C & -A-iA_0 \end{pmatrix}\phi \equiv N_1\phi \, , \tag{4.42}$$

$$\psi_t = \begin{pmatrix} A+iA_0 & E \\ C & -A+iA_0 \end{pmatrix}\psi \equiv N_2\psi \, . \tag{4.43}$$

We now express $B_-(t)$ by means of $N_{1,2}$. Let $\hat{t}$ and $f(x)$ be given, choose $f_0(x)$ so that $U_-(\hat{t})f_0 = f$; then $\frac{d}{dt} U_-(t)f_0\big|_{t=\hat{t}} = B_-(\hat{t})f$. Since $U_-(\hat{t})f_0 = \int (Fm + Gp)\, d\zeta$ where $\binom{F}{G} = \tilde{f}_0$, and m,p are evaluated at $t=\hat{t}$, we find

$$B_-(\hat{t}) = \int (Fm_t+Gp_t)\, d\zeta\big|_{t=\hat{t}} = \int FN_1m + GN_2p\, d\zeta. \tag{4.44}$$

Remember that N_1 and N_2 depend on x; hence (4.44) is quite different from a Fourier inversion (actually, $B_-(\hat{t})$ is a "pseudo-differential" operator).

Now define two more families of eigenfunctions by

$$v = e^{iA_0t}\phi \, , \qquad w = e^{-iA_0t}\psi \, . \tag{4.45}$$

One finds that both v and w satisfy $u_t = Nu$, where

$$N \equiv \begin{pmatrix} A & E \\ C & -A \end{pmatrix} ; \tag{4.46}$$

they differ in their initial values, and hence are independent. Define U(t) by (again, $\tilde{f}_0 = [{F \atop G}]$)

$$U(t)f_0 = \int (Fv + Gw)\, d\zeta \, . \tag{4.47}$$

We have just found an expression for $B_-(\hat{t})f$; let us compute $B(\hat{t})f$ for the same f. We have:

$$f = U_-(\hat{t})f_0 = \int Fm + Gp = \int (e^{-iA_0\hat{t}}F)v + (e^{iA_0\hat{t}}G)w ,$$

which is $U(\hat{t})f_1$ for a certain f_1. Now,

$$Bf = \frac{d}{dt} U(\hat{t})f_1\Big|_{t=\hat{t}} = \int \frac{1}{a}(e^{-iA_0\hat{t}}F)v_t + \frac{1}{a}(e^{iA_0\hat{t}}G)w_t$$

(4.48)
$$= \int FNm + GNp , \text{ (using (4.45, 46))}.$$

Subtraction of (4.44) from (4.48) gives

$$(B-B_-)f = \int F(N-N_1)m + G(N-N_2)p ,$$

or

(4.49)
$$(B-B_-)f = \int iA_0 Fm - iA_0 Gp .$$

This shows that the U_--representer of $B-B_-$ (i.e., $\hat{T}_- \equiv \sigma_0 U_-^*(B-B_-)U_-\sigma_-^*$ in the language of (4.30)) is $\begin{pmatrix} iA_0 & 0 \\ 0 & -iA_0 \end{pmatrix}$.

A similar argument leads to the conclusion:

$$\sigma_0 U_+^*(B-B_+)U_+\sigma_0^* \equiv \hat{T}_+ = \begin{pmatrix} iA_0 & 0 \\ 0 & -iA_0 \end{pmatrix} .$$

According to (4.30) then, with

(4.50)
$$T(\zeta) = \begin{pmatrix} iA_0(\zeta) & 0 \\ 0 & -iA_0(\zeta) \end{pmatrix},$$

we have

(4.29)
$$\tilde{S}_t(\zeta) = [T(\zeta),S(\zeta)] .$$

This is easily verified directly.

To summarize: For the conservative case (a) of Section 2.4, we have verified that the abstract equations (4.29,30) exhibit a manifestly linear and autonomous evaluation of $\tilde{S}$.

Remarks. 1. The matrix $T(\zeta)$ is clearly the carrier of the linearized dispersion relation. We do not know of an abstract expression of this fact.

2. If $A_0(\zeta)$ is a polynomial, then the operators B defined by (4.48) are local. For $A_0(\zeta) = -2\zeta^2$, for example, we have [2]

$$N = \begin{pmatrix} -2i\zeta^2 - iqq^* & 2q\zeta + iq_x \\ 2q^*\zeta - iq_x^* & 2i\zeta^2 + iqq^* \end{pmatrix},$$

and

$$B = \begin{pmatrix} 2i\partial^2 - iqq^* & -iq_x - 2iq\partial \\ iq_x^* + 2iq^*\partial & -2i\partial^2 + iqq^* \end{pmatrix}.$$

(The equation $L_t = [B,L]$ is $q_t = iq_{xx} - 2iq|q|^2$, the "stable" nonlinear Schrodinger equation.) The operators (4.48) are thus precisely the B's needed for a Lax-type representation with local operators. The operator T is therefore the difference (appropriately represented) between the Møller $B_\pm$ and the local B. It should be noted that both $B_\pm$ are integrodifferential operators.

3. One can show by direct computation that $\tilde{S}_t = [T,\tilde{S}]$ if and only if $a_t = 0$. The entries of T will be constant in t precisely in cases (a) and (b) of Section 2.4. The direct computation for case (c) is very heavy and will not be given here. Part of the difficulty is that neither E nor C $\to 0$ as $x \to +\infty$.

4. Concerning the problem of "interpolating" the correct B which will make $T_+ = T_-$ in (4.30), we offer two remarks.

If there exists a local B for which $L_t = [B,L]$, it should be the correct choice -- in any example in which L is local. An intriguing possibility is that these ideas may be applicable to periodic problems; the wave and scattering operators would have to be defined by stationary methods (Lippmann-Schwinger equation) over intervals of the spectrum. While the approach centering on formula (4.20) would be inapplicable, a counterpart of (4.30) should exist.

When (as in cases (b) or (c) of Section 2.4) a local B does not exist, we do now know a prescription for obtaining the simplest one of the many possible equations $S_t = T_+S - ST_-$ (cf. (4.30)).*
It is reasonable to expect that there should be a simplest such equation, with $T_\pm$ reflecting the different dispersion relations (see case (c)), the SIT problem). Again, we do not know whether there is a "best" interpolating B, or even why there should be one at all.

4.5 Relating the Lax and AKNS Approaches

The discussion, in Section 4.3 of the consequences of Lax's equation, raises the question: how much of the abstract theory applies to the nonselfadjoint AKNS system as well? On the fact of it, not very much, since the equations studied by AKNS cannot, with a few exceptions, be cast in Lax's framework with the same L (the operators B are not skew adjoint). One way to include these equations in Lax's framework would be to transform the eigenvalue problem (2.1) into a selfadjoint problem by a ζ-dependent change of variable; this would, in effect, undo the type of transformation by which Zakharov, Shabat and Wadati [42,45] went from the Lax equation to the nonselfadjoint problem. As mentioned in Section 4.3, the abstract theory would then have to deal with the complex scattering states.

There is another possibility, not yet investigated, but perhaps of enough interest to warrant a brief description. Lax's theory might be applied in spaces with indefinite metric.

Let H be the Hilbert space of vector functions $v = \begin{pmatrix} v_1 \\ v_2 \end{pmatrix}$ with

* In case (b), it is actually easy to see that B must be defined by an N in which $A = -D$ (see (4.46)). The question is whether this B is distinguished in any other way.

inner product,

$$(4.51) \qquad (v,w)_H = \int_{-\infty}^{\infty} (v_1 w_1^* + v_2 w_2^*) \, dx \ .$$

Let J be a (2×2) constant matrix such that $J^* = J$ and $J^2 = I$ and define on H the bilinear form,

$$(4.52) \qquad [v,w] = (Jv,w) \ .$$

Next, define (1) the operator A as being J-self adjoint if

$$(4.53) \qquad A^* = JAJ \ ,$$

(2) the operator B as being J-skew adjoint if

$$(4.54) \qquad B^* = -JBJ \ ,$$

(3) the operator C as being J-unitary if

$$(4.55) \qquad C^* JC = J \ ,$$

where (*) denotes H-adjoint.

Given a family L(t) of J-self adjoint operators, which are J-unitarily equivalent,

$$(4.56) \qquad L(t)U = UL(0)$$

where

$$(4.57) \qquad U^* JU = J \ .$$

Differentiation (4.57) and find

$$(4.58) \qquad U_t = - JU^{*^{-1}} U_t^* JU \ .$$

Define

$$(4.59) \qquad B = - JU^{*^{-1}} U_t^* J \ .$$

Then

$$\begin{aligned} B^* &= -JU_t U^{-1} J \\ &= J(JU^{*^{-1}} U_t^* JUU^{-1}) J \quad \text{on using (4.58)} \end{aligned}$$

$$= U^{*-1} U_t^* \quad (4.60)$$

Thus $-JB^*J = B$ and B is J-skew adjoint. Next, differentiate (4.56) and find

$$L_t U + LBU = BUL_0$$
$$= BLU$$

which immediately gives

$$L_t = [B,L] \quad (4.61)$$

To illustrate these properties, let us examine some of the better known equations which fall within the AKNS framework. First, for the dispersion relation $\Omega(\zeta) = -2i\zeta^2$, the equations (2.2) are

$$\begin{aligned} q_t - i(q_{xx} - 2q^2 r) &= 0 \\ r_t + i(r_{xx} - 2qr^2) &= 0 \end{aligned} \quad (4.62)$$

The operator L is

$$L = \begin{pmatrix} +iD & -iq \\ ir & -iD \end{pmatrix} \quad \text{with} \quad D = \frac{\partial}{\partial x} ,$$

and

$$B = 2i\begin{pmatrix} D^2 & 0 \\ 0 & -D^2 \end{pmatrix} - iqr\begin{pmatrix} 1 & 0 \\ 0 & -1 \end{pmatrix} + i\begin{pmatrix} 0 & q_x \\ r_x & 0 \end{pmatrix} + 2i\begin{pmatrix} 0 & -q \\ r & 0 \end{pmatrix} D + 2iD\begin{pmatrix} 0 & -q \\ r & 0 \end{pmatrix} .$$

Note that if $r = q^*$, L is selfadjoint and B is skew adjoint and (4.62) becomes the "stable" (or no soliton) nonlinear Schrodinger equation. On the other hand, if $r = -q^*$ it may easily be verified that with

$$J = \begin{pmatrix} 1 & 0 \\ 0 & -1 \end{pmatrix} \quad (4.63)$$

L is J-self adjoint and B is J skew adjoint. Indeed, the choice (4.63) for J is appropriate for all systems where $r = -q^*$.

If we take

(4.64) $$L = \begin{pmatrix} -D & q \\ -r & D \end{pmatrix},$$

then L is J-self adjoint with

(1) $$J = \begin{pmatrix} 0 & 1 \\ 1 & 0 \end{pmatrix} \quad \text{if } q \text{ and } r \text{ are real}$$

(2) $$J = \begin{pmatrix} 0 & i \\ -i & 0 \end{pmatrix} \quad \text{if } q \text{ and } r \text{ are imaginary}$$

As an illustration, we may take the modified KdV equation with $r = -q$, q real, for which the operator B is

$$B = -4\begin{pmatrix} D^3 & 0 \\ 0 & D^3 \end{pmatrix} + 3D\begin{pmatrix} -q^2 & q_x \\ -q_x & -q^2 \end{pmatrix} + 3\begin{pmatrix} -q^2 & q_x \\ -q_x & -q^2 \end{pmatrix} D$$

and it may be verified that $B^* = -JBJ$.

Furthermore, it may be verified in these examples that the evolution equation can be written in the Lax form $L_t = [B,L]$.

This approach will not work for arbitrary complex $q(x,t)$ and $r(x,t)$ but in any event (as we have already noted) in this case the eigenfunctions of L develop singular behavior in finite time. A possible conjecture is that membership in a J-space is a necessary and sufficient condition (along with certain integrability conditions on the initial potential) to guarantee that a system is integrable for all time.

The problem with these ideas at present is that not enough is known about operators in J-spaces to draw the kind of conclusions arrived at in the earlier sections of this chapter; on the other hand the very fact that so much can be said about the Zakharov and Shabat system indicates that there is a potentially rich and complete theory of certain classes of J-operators that still remains to be worked out.

References

1 Ablowitz, M. J., Kaup, D., Newell, A. C., and Segur, H., (1973), (a) Phys. Rev. Lett. 30 (1973) 1962, (b) Phys. Rev. Lett. 31 (1973) 125.

2 Ablowitz, M. J., Kaup, D., Newell, A. C., and Segur, H., Studies in Appl. Math., to appear (December 1974).

3 Ablowitz, M. J., Kaup, D., and Newell, A. C., J. Math. Phys. to appear (November 1974).

4 Ablowitz, M. J., and Ladik, J., to appear (1974).

5 Ablowitz, M. J., and Newell, A. C., J. Math. Phys. 14 (1973). 1277.

6 Birman, M. S., and Krein, M. G., Dokl. Akad. Nauk 144 (1962) 475-478.

7 Case, K. M., and Kac, M., J. Math. Phys. 14 (1973) 594.

8 Clairin, S., Ann. Toulouse 5,2 (1903) 437.

9 Cole, J. D., Quart. Appl. Math. 9 (1951) 225.

10 Faddeev, L. D., Dokl. Akad. Nauk SSSR 121 (1958) 63.

11 Flaschka, H., These Proceedings.

12 Flaschka, H., Prog. Theoretical Phys. 51 (1974) 703.

13 Flaschka, H., Phys. Rev. B 9 (1974) 1924.

14 Forstyhe, A. R., Dover Publishers, New York, Volume 6 (1959).

15 Gardner, C. S., Greene, J. M., Kruskal, M. D., Miura, R. M., Phys. Rev. Letters 19 (1967) 1095

16 Gardner, C. S., Greene, J. M., Kruskal, M. D., and Miura, R.M., Comm. Pure Appl. Math. 27 (1974) 97.

17 Gardner, C. S., Princeton Report, PPLAP45.

18 Gel'fand, I. M., and Levitan, B. M., Amer. Math. Soc. Transl. 2, 1 (1951) 253.

19 Gel'fand, I. M., and Levitan, B. M., Dokl. Akad. Nauk 88 (1953), 593-596.

20 Henon, M., Phys. Rev. B. 9, 1921

21 Kaup, D. J., and Levermore, C. D., to appear.

22 Kaup, D. J., to appear.

23 Kaup, D. J., to appear.

24 Kay, I., and Moses, H. E., Nuovo Cimento 3 2 (1956) 277.

25 Korno, K., Kameyana, W., and Sanuki, H., J. Phys. Soc. Japan, 37 (1974) 171.

26 Krein, M. G., Mat. Sb. 33 (153) 587-626

27 Krein, M. G., Dokl. Akad. Nauk, 143 (1962) 506.

28 Kruskal, M. D., "Nonlinear Wave motion," Editor: A. C. Newell, Lectures in Appl. Math. Vol. 15 (1974), AMS, Providence, R.I.

29 Lamb, G. L., to appear, J. Math. Phys.

30 Lax, P. D., Comm. Pure Appl. Math. 21 (1968) 467.

31 Marcenko, V. A., Dokl. Akad. Nauk SSSR 104 5 (1955) 696.

32 McLaughlin, D. W., to appear, J. Math. Phys.

33 McLaughlin, D. W. and Corones, J., to appear, Phys. Rev. A.

34 McLaughlin, D. W., and Scott, A. C., J. Math. Phys. 14, (1973) 1878.

35 Miura, R. M., J. Math. Phys. 9 (1968) 1202.

36 Moser, J., These Proceedings.

37 Scott, A. C., Chu, F. Y. F., and McLaughlin, D. W., Proc. IEEE, 61 (1973) 1449.

38 Shabat, A. B., Dokl. Akad. Nauk 205 (1972) 546

39 Shabat, A. B., Dokl, Akad. Nauk 211 (1973) 1310.

40 Shabat, A. B., Diff. Equations (Soviet Journal) 8 (1972) 164.

41 Segur, H., J. Fluid. Mech. 59 (1973) 721

42 Wadati, M., J. Phys. Soc., Japan 32 (1972) 1681.

43 Wahlquist, H., and Estabrook, F., to appear.

44 Zakharov, V. E., and Faddeev, L.D., Funk. Anal. Priloz. 5 (1971) 18-27.

45 Zakharov, V. E., and Shabat, A. B., JETP 34 (1972) 62.

DISCRETE AND PERIODIC ILLUSTRATIONS OF SOME ASPECTS OF THE INVERSE METHOD

H. Flaschka

Department of Mathematics, University of Arizona, Tuscon, Ariz 85721

Introduction

Since the original discovery, by Gardner, Greene, Kruskal, Miura and Zabusky, of the many interesting properties of the Korteweg-de Vries equation $u_t - 6uu_x + u_{xxx} = 0$, there have been several attempts at a clarification of the structure underlying these properties. The ultimate aim of these more abstract studies, of course, is this: to develop a practical way to tell when a non-linear evolution equation possesses a great number of conservation laws, when it has soliton solutions, and in particular, when it can be linearized by inverse scattering techniques. Any such criterion is still far in the future. Here, I want to illustrate the ideas presented in two pioneering "interpretive" papers:

(1) Lax's representation $L_t = [B,L]$ (see [10]), which explains why equations such as the Korteweg-de Vries equation should be related to a linear eigenvalue problem.

(2) The interpretation, due to Zaharov and Faddeev [15], of the constants of motion of the Korteweg-de Vries equation as regularized traces of a singular boundary value problem.

In Section 1, these two ideas will be illustrated by the so-called Toda lattice, a Hamiltonian system of 2N nonlinear ODEs. The whole discussion takes place in a finite dimensional setting, few new definitions are required, and there are none of the technical difficulties encountered with the Korteweg-de Vries

*Supported by National Science Foundation Grant GP-42739.

equation on the line. On the other hand, our results are only fragmentary, and really do no more than illustrate the ideas described above; a "theory of the Toda lattice" remains to be constructed.

In Section 2, we consider the Korteweg-de Vries equation with periodic boundary conditions. This problem has countably many degrees of freedom, but, after showing how the techniques of Section 1 carry over with comparatively little change, we focus our attention on special solutions described by a finite number of parameters, and sketch some of the new results in classical spectral theory that were motivated by the connection with nonlinear evolution problems, most of them originating in (or proved by means of) the ideas of Lax, Zaharov, and Faddeev.

The most complete part of Korteweg-de Vries theory is the inverse scattering method of solution; we do not deal with it here. Some abstract considerations, and an exposition of the canonical structure of the method, will be given in the paper by Alan Newell and myself.

1. Finite Systems

A. Lax's Representation [10].

Let $t \to L(t)$ be a smooth function whose values are symmetric $N \times N$ matrices. We ask: under what conditions on $t \to L(t)$ will the eigenvalues of $L(t)$ be independent of t? A sufficient (but not necessary) condition is that there exist a unitary matrix function, $t \to U(t)$, $U^{-1} = U^{*}$, such that

$$U^{-1}(t)L(t)U(t) = L(0) \quad \text{for all } t . \tag{1.1}$$

In finite dimensions, for the constancy of the eigenvalues it is necessary and sufficient that there exist a <u>similarity</u> transformation from $L(t)$ to $L(0)$; however, we want eventually to extend the arguments that follow to unbounded operator functions $L(t)$, and since similarity transformations in Hilbert space are poorly understood, we restrict our discussion to the situation described by (1.1).

To get a differential version of (1.1), we apply $\dot{} = d/dt$:

$$(1.2) \qquad - U^{-1}\dot{U}U^{-1}LU + U^{-1}\dot{L}U + U^{-1}L\dot{U} = 0 .$$

Now, a unitary function must satisfy a differential equation of the form

$$(1.3) \qquad \dot{U} = BU ,$$

where $B = - B^{*}$. Substitute (1.3) into (1.2) and simplify, to get

$$(1.4) \qquad \dot{L} = [B,L] = BL - LB .$$

Thus, if all $L(t)$ are unitarily equivalent to $L(0)$, then $L(t)$ satisfies a matrix differential equation of the form (1.4), with some skew symmetric $B(t)$.

The important thing is that the converse is also true. Suppose we give a skew-symmetric function $B(t)$, and look at (1.4) as a differential equation for $L(t)$. Defining $U(t)$ by (1.3) with initial condition $U(0) = I$, we verify easily that (1.2) holds, and hence also (1.1): the $L(t)$ are unitarily equivalent to $L(0)$.

That is the strategy of the "inverse method": if a particular system of differential equations can be cast in the form (1.4) by clever choice of L and B, then the eigenvalues of the matrices $L(t)$ will be "first integrals" or "constants of the motion". The existence of these is usually of interest, but, as we shall show, Lax's representation allows one to draw further conclusions as well.

Example 1. The simplest system of differential equations is surely the harmonic oscillator:

$$(1.5) \qquad \begin{aligned} \dot{a} &= -b \\ \dot{b} &= a . \end{aligned}$$

It has the first integral $I = a^2 + b^2$. This can be obtained from equation (1.4). Set

$$L(t) = \begin{bmatrix} b(t) & a(t) \\ a(t) & -b(t) \end{bmatrix}, \qquad B(t) = \begin{bmatrix} 0 & \frac{1}{2} \\ -\frac{1}{2} & 0 \end{bmatrix}.$$

Then

$$[B,L] = \begin{bmatrix} a(t) & -b(t) \\ -b(t) & -a(t) \end{bmatrix},$$

and the preceding discussion shows that if $\dot{L} = [B,L]$, i.e., if

$$(1.6) \qquad \begin{bmatrix} \dot{b} & \dot{a} \\ \dot{a} & \dot{b} \end{bmatrix} = \begin{bmatrix} a & -b \\ -b & -a \end{bmatrix},$$

then the eigenvalues $= \pm \sqrt{a^2+b^2}$ of L will be constant in t. Of course, (1.5) and (1.6) are identical restrictions on the functions a and b.

Example 2. One can make up somewhat more complicated equations with the L of Example 1. This is done by changing the entry of the matrix B:

$$B(t) = \begin{bmatrix} 0 & \frac{1}{2} g(t) \\ -\frac{1}{2} g(t) & 0 \end{bmatrix}.$$

Equation (1.4), $\dot{L} = [B,L]$, leads to the following restrictions on a and b:

$$(1.7) \qquad \begin{aligned} \dot{a} &= -gb , \\ \dot{b} &= ag . \end{aligned}$$

Thus, whenever a and b evolve according to a system of the form (1.7), the eigenvalues $\lambda = \pm \sqrt{a^2+b^2}$ of L are constant. It is, of course, easy to verify directly that a^2+b^2 is an integral of (1.7).

Observe that if g is a known function, (1.7) is a linear

system with possibly variable coefficients, while if g itself is an expression involving a and b, (1.7) is nonlinear. An illustration of this is: $g = a$, leading to

$$\dot{a} = -ab , \qquad \dot{b} = a^2 ,$$

which is a precursor of the Toda lattice. --

The usefulness of Lax's representation (1.4) lies precisely in the fact that it can encompass and yield information about nonlinear systems. The solution of such systems is not necessarily made easier, however. In Example 1, one could integrate $\dot{U} = BU$ (equation (1.3)) to find

$$U(t) = \begin{bmatrix} \cos\frac{t}{2} & -\sin\frac{t}{2} \\ \sin\frac{t}{2} & \cos\frac{t}{2} \end{bmatrix} ,$$

and one can then use (1.1) in the form $L(t) = U(t)L(0)U^{-1}(t)$ to express the solution of (1.5) in terms of its initial data; the customary formula results. When B itself depends on the unknown functions, however,

$$\dot{L} = [B,L] \qquad \text{and} \qquad \dot{U} = BU$$

from a coupled nonlinear system, which is at least as difficult as the original one, so that one would not try to obtain the solution $\dot{L} = [B,L]$ via (1.1). Situations in which $\dot{U} = BU$ can nonetheless be used to get a solution of (1.4) are described in [5].

B. The Toda Lattice

Take L and B of the form

$$L = \begin{bmatrix} b_1 & a_1 & 0 & 0 & \cdots & a_N \\ a_1 & b_2 & a_2 & 0 & \cdots & 0 \\ 0 & a_2 & b_3 & a_3 & \cdots & 0 \\ \vdots & & & & & \\ a_N & \cdot & \cdot & \cdot & \cdots & b_N \end{bmatrix}, \qquad B = \begin{bmatrix} 0 & a_1 & 0 & \cdots & -a_N \\ -a_1 & 0 & a_2 & \cdots & 0 \\ & & & & \\ \vdots & & & & \\ a_N & \cdot & \cdot & \cdot & 0 \end{bmatrix};$$

all entries are functions of t. The condition $\dot{L} = [B,L]$ turns out to impose the following requirement:

$$\dot{a}_n = a_n(b_{n+1} - b_n) , \qquad \dot{b}_n = 2(a_n^2 - a_{n-1}^2) , \tag{1.8}$$

(a periodicity convention is used: $a_{N+1} = a_1$, $a_0 = a_N$, etc.). Thus: <u>if the entries of</u> L <u>change according to</u> (1.8), <u>the eigenvalues of</u> L <u>are constant, i.e., they are first integrals of</u> (1.8).

The equations (1.8) are equivalent to those of a very interesting mass nonlinear spring chain invented by M. Toda. He considered springs whose stored energy, as a function of extension r, is given by

$$\phi(r) = e^{-r} + r - 1 . \tag{1.9}$$

A chain of unit mass is connected by such springs, with q_n denoting the displacement of mass n from equilibrium, and $p_n = \dot{q}_n$. N such masses are arranged in a circle, i.e., the periodicity conditions $q_{n+N} = q_N$, $p_{n+N} = p_n$ are imposed.

The equations of motion are Hamiltonian:

$$\dot{q}_n = \frac{\partial H}{\partial p_n} , \qquad \dot{p}_n = -\frac{\partial H}{\partial q_n} , \tag{1.10}$$

where $H = \sum_{n=1}^{N} \left\{ \frac{1}{2} p_n^2 + \phi(q_n - q_{n+1}) \right\}$. Without going into the details of the computation, we state that if $q_1,\ldots,p_N$ satisfy (1.10), and if $a_1,\ldots,b_N$ are defined by

$$a_n = \frac{1}{2} \exp\left(\frac{q_{n-1} - q_n}{2}\right) , \qquad b_n = -\frac{1}{2} p_{n-1} , \qquad n = 1,\ldots,N, \tag{1.11}$$

then $a_1,\ldots,b_N$ satisfy (1.8); the eigenvalues of L are integrals of the Toda equations [2,3].

It would be too much of a digression to describe those properties of the Toda lattice which first drew attention to it: the discovery of explicit solutions, the possibility of nonlinear superposition of normal modes, the existence of soliton solutions to the infinitely long lattice, the fact that the Korteweg-deVries and Boussinesq equations arise as continuum limits, etc. For these matters we refer to the recent comprehensive treatise by Toda [14].

Returning to the study of (1.10), we note first that the eigenvalues $\lambda_1,\ldots,\lambda_N$ of L(t) are horribly complicated functions of the canonical variables $q_1,\ldots,p_N$. Somewhat more pleasant to contemplate are certain symmetric functions of the λ_i (these are also important conceptually, see subsection C below):

(1.12) $$I_1 = \sum_j \lambda_j \ , \quad I_2 = \sum_{i,j} \lambda_i\lambda_j \ , \ \ldots$$

(1.13) $$J_1 = \sum_j \lambda_j \ , \quad J_2 = \sum_j \lambda_j^{\,2} \ , \ \ldots$$

The integrals in (1.12) are essentially the coefficients of the characteristic polynomials of L, while those in (1.13) are the traces of powers of L. Closed expressions for both sets of integrals are given in [8].

To continue our analysis of the Toda equation, we recall this definition [1]:

Definition. A Hamiltonian system (1.10) is completely integrable in a region $D \subset \mathbb{R}^{2N}$, if there exist on D functions $I_1,\ldots,I_N$ of $q_1,\ldots,p_N$, which are

(i) functionally independent in D,

(ii) in involution in D, i.e., the Poisson bracket

$$\left\{I_i,I_j\right\} = \sum_k \frac{\partial I_i}{\partial q_k}\frac{\partial I_j}{\partial p_k} - \frac{\partial I_i}{\partial p_k}\frac{\partial I_j}{\partial q_n}$$

vanishes identically in D.

According to a theorem of Liouville, the solution of a completely integrable system can (in principle) be obtained by quadratures. This is not necessarily easy to carry out for specific examples, and apparently not for the Toda lattice (although Dr. Rüssmann has shown me how to complete the integration if $N = 3$). A greater significance of complete integrability is geometrical. Under some additional conditions [1], the sets $I_i = c_i$, $i = 1,\ldots,N$, are N-dimensional tori for all values of $c_1,\ldots,c_N$, (with the possible exception of a $\leq(N-1)$-dimensional manifold)), and the Hamiltonian flows generated by the I_i,

$$\dot{q}_n = \frac{\partial I_i}{\partial p_n}, \qquad \dot{p}_n = -\frac{\partial I_i}{\partial q_n}, \tag{1.14}$$

commute. The solution curves of the N systems (1.14) issuing from a particular point $\overset{\circ}{q}_1,\ldots,\overset{\circ}{p}_N$ define (in general) N independent directions, and can be taken as a coordinate system of the torus $I_i = c_i$. By travelling along suitable pieces of solution curves of (1.14) for $i = 1,2,\ldots,N$ in succession, one can connect any point on the torus with any other point.

It appears, then, that the geometry of the phase space of completely integrable systems is well understood in a qualitative way. Moreover, to such systems one can apply the Kolmogorov-Arnol'd-Moser (KAM) theory [1], and deduce the existence of quasiperiodic solutions of perturbations of the original system. This fact may turn out to have importance in the theory of lattices; it is believed that the existence of solitons is connected with the complete integrability of the equations of motion, and the fact that computer solutions of nonintegrable lattices indicate soliton-like behavior may be a reflection of KAM perturbation theory.

These remarks may explain the use of the following

<u>Proposition 1.</u> <u>The periodic Toda lattice is completely integrable.</u>

<u>Proof:</u> We verify that the Poisson bracket of two simple, distinct eigenvalues μ, λ of L vanishes.

Let u be the normalized eigenvector for λ, $(u,u) = 1$. Now $\lambda = (Lu,u)$, so

$$\frac{\partial\lambda}{\partial p_j} = \left(\frac{\partial L}{\partial p_j} u,u\right) + \left(L \frac{\partial u}{\partial p_j},u\right) + \left(Lu, \frac{\partial u}{\partial p_j}\right)$$

But now $(L \frac{\partial u}{\partial p_j},u) = (\frac{\partial u}{\partial p_j},Lu) = \lambda(\frac{\partial u}{\partial p_j},u) = \frac{1}{2}\lambda \frac{\partial}{\partial p_j}(u,u) = 0$, and likewise, $(Lu, \partial u/\partial p_j) = 0$. Hence

$$\frac{\partial\lambda}{\partial p_j} = \left(\frac{\partial L}{\partial p_j} u,u\right) .$$

The dependence of L on p_j is extremely simple, in fact, $\partial L/\partial p_j$ has a $-1/2$ in the $j+1$, $j+1$ position, and zeroes elsewhere. Hence,

(1.15)
$$\frac{\partial\lambda}{\partial p_j} = -\frac{1}{2} u_{j+1}^2$$

In a similar way, one gets

(1.16)
$$\frac{\partial\lambda}{\partial q_j} = \frac{1}{2}\left\{a_{j+1}u_ju_{j+1} - a_ju_ju_{j-1}\right\} ,$$

where a_j was defined in (1.11).

For $\mu = (Lv,v)$, we get formulas analogous to (1.15), (1.16). Then

$$\begin{aligned}\{\lambda,\mu\} &= -\frac{1}{4}\sum_{j=1}^{N}\{u_{j+1}^2[a_{j+1}v_jv_{j+1} - a_jv_jv_{j-1}] \\ &\quad - v_{j+1}^2[a_{j+1}u_ju_{j+1} - a_ju_ju_{j-1}]\} \\ &= -\frac{1}{4}\sum_{j=1}^{N} u_jv_j[a_j(u_jv_{j+1} - v_ju_{j+1}) + a_{j-1}(v_ju_{j-1} - u_jv_{j-1})] \\ &\equiv \frac{1}{4}\sum_{j=1}^{N} u_jv_j[R_j + R_{j-1}]\end{aligned}$$

where $R_j = a_j(v_ju_{j+1} - u_jv_{j+1})$. On the other hand, multiplying the equation for u,

$$a_ju_{j+1} + a_{j-1}u_{j-1} + b_ju_j = \lambda u_j$$

by v_j, the corresponding equation for v by u_j, and subtracting, we find

$$u_j v_j = \frac{1}{\lambda-\mu} [R_j - R_{j-1}] .$$

Hence,

$$\{\lambda,\mu\} = \frac{1}{4(\lambda-\mu)} \sum_{j=1}^{N} [R_j^2 - R_{j-1}^2] ,$$

and by periodicity this sum collapses to zero.

Now it can be shown (we omit the argument) that except on manifolds in phase space of dimension $\leq$ N-1, the eigenvalues of L <u>are</u> simple. Away from the exceptional sets, then, there are N distinct eigenvalues in involution. It follows that the integrals (1.12) or the integrals (1.13), being functions of the λ_j , are also in involution. The independence of the integrals (1.12) was established by Henon [8].

Further insight into the relation of the invariant tori to the distribution of eigenvalues of L can be obtained by a more detailed study of the spectrum of L. To this end, it proves to be convenient to consider not L, but an infinite three term recursion relation

$$(1.17) \quad (\hat{L}u)_n = a_n u_{n+1} + a_{n-1} u_{n-1} + b_n u_n = \lambda u_n , \qquad -\infty < n < \infty ,$$

where $a_{n+N} = a_n$, $b_{n+N} = b_n$. The theory of (1.17) is quite similar to that of Hill's equation (the differential analog of (1.17), see Section 2.B), and there are undoubtedly counterparts of the results to be described in Section 2.D. It turns out that the spectrum of $\hat{L}$ (as operator on ℓ^2) is continuous, consisting of N closed intervals $[\alpha_n, \beta_n]$. N of these end points $(\beta_N, \alpha_{N-1}\ \beta_{N-2}, \alpha_{N-3}, \ldots)$ are eigenvalues of the earlier N×N matrix L. We have already indicated that the eigenvalues of L are integrals of the Toda equations. When some of these eigenvalues become double (due to particular choice of initial data $\mathring{q}_1, \ldots, \mathring{p}_N$), the number of independent integrals is apparently reduced, and, when as many eigenvalues as possible are double, the spectrum of $\hat{L}$ takes on this structure:

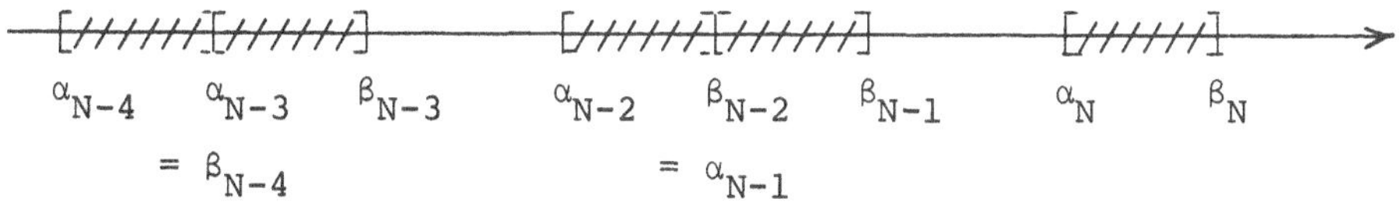

It turns out that as long as the other endpoints of the spectral intervals do not coalesce, the double eigenvalues $\beta_{N-2}, \beta_{N-4}, \ldots,$ can be varied independently. Further coalescence of spectral intervals, however, offers an obstruction to independent variation of the double eigenvalues, with consequent reduction of the number of degrees of freedom. The extreme case is:

Proposition 2. When the spectrum of $\hat{L}$ is connected, then $a_n = \frac{1}{2}$, and $b_n \equiv$ constant.

The proof will be sketched in the Appendix. The point of the result is, that for certain values of the integrals of the Toda lattice, the corresponding invariant torus is zero dimensional.

In terms of the physical variables, this means that the lattice is undergoing a uniform translation; if one agrees to keep the center of mass fixed, then the lattice is at rest. When the spectrum of $\hat{L}$ degenerates in this way, the eigenvalues λ_j of L are essentially the normal modes of the N-particle harmonic lattice. For this particular choice of λ_j, then, the lattice has no finite motion, whereas its infinitesimal (linearized) motion is entirely characterized by these numbers.

By analogy with known results for the Korteweg-de Vries equation, one would expect that initial data, for which the spectrum of $\hat{L}$ has two components, would be precisely those leading to Toda's elliptic-function solutions, but this remains to be shown.

All this is still far from a complete theory of the toda lattice. One would really like to find a set of N "angle" variables with which to coordinatize the invariant tori, and to investigate the nature of the possible motions: quasiperiodic, stable or unstable periodic, etc. This program has been carried out in a simplified example by Moser (see his paper in these proceedings), but the periodic problem remains unsolved.*

C. Traces

We conclude this paragraph with one apparently random remark, which will become important in Section 2.C. Above, it was observed that the expressions $J_p = \mathrm{Tr}\, L^p$, $p = 1,\dots,N$, being moments of the constant eigenvalues, are integrals of the Toda lattice. This is a rather trivial observation, but it is not quite so trivial that the constants of the Korteweg-de Vries equation can also be interpreted as traces of powers of an unbounded operator. This is a result first stated in [15], and will be discussed in Section 2. Here, we only observe that the J_p are coefficients in an expansion of an invariant function: Since $\mathrm{Tr}\,(L-\lambda)^{-1} = \sum \frac{1}{\lambda_j - \lambda}$, and since the λ_j are constant, the function $R(\lambda) = \mathrm{Tr}\,(L-\lambda)^{-1}$ is independent of t for any λ at which it is defined. $R(\lambda)$ generates the integrals J, since

$$R(\lambda) = \mathrm{Tr}\left\{-\frac{1}{\lambda}\sum_{j=0}^{\infty}\frac{1}{\lambda^j}L^j\right\} = -\frac{1}{\lambda}\sum_{j=0}^{\infty}\frac{1}{\lambda^j}\,\mathrm{Tr}\,L^j .$$

In more complicated examples, $R(\lambda)$ is replaced by a Fredholm determinant, or more generally by a coefficient of the scattering matrix.

2. The Periodic Korteweg-de Vries Equation

We shall now describe some properties of the KdeV equation $q_t - 6qq_x + q_{xxx} = 0$; of interest are solutions satisfying periodic boundary conditions $q(x,t) = q(x+\pi,t)$ and evolving from

* Recently, Kac and van Moerbecke (preprints) have devised a method (based on orthogonal polynomials) which applies simultaneously to the semiinfinite Toda lattice, as well as to Moser's finite example.

a specified initial function $q(x,0)$ (also π-periodic).

A. Lax's Representation

Define the one parameter families $L(t)$ and $B(t)$ of differential operators by

$$L(t) = -\frac{\partial^2}{\partial x^2} + q(x,t) ,$$

$$B(t) = -4\frac{\partial^3}{\partial x^3} + 3\left(q(x,t)\frac{\partial}{\partial x} + \frac{\partial}{\partial x} q(x,t)\right) . \tag{2.0}$$

A formal computation [10] shows that equation (1.4), $\dot{L} = [B,L]$, reduces to

$$q_t = 6qq_x - q_{xxx} . \tag{2.1}$$

If boundary conditions for L and B are chosen in a way that makes these operators self- and skew-adjoint, respectively, we can repeat the arguments of Section 1.A to deduce that the $L(t)$ are unitarily equivalent when the coefficient $q(x,t)$ changes according to (2.1). (Differentiability questions will not be dealt with, so our discussion will be a bit formal.) In particular, the eigenvalues (if any) of $L(t)$ will be constant if the coefficient q satisfies the KdeV equation; this was the original remarkable discovery of Gardner, Greene, Kruskal and Miura (see [7] for a comprehensive survey of, and references to, their work). Before discussing the significance of this fact for the behavior of solutions of (2.1), we give a brief review of the spectral theory of the operator L.

B. Hill's Equation

Being concerned, for the moment, only with the spectral properties of L, we omit the variable t and consider a fixed operator with π-periodic coefficient (Hill's operator)

$$L = -d^2/dx^2 + q(x) . \tag{2.2}$$

We want to impose boundary conditions which make L selfadjoint on some Hilbert space. This can be done in several ways:

$(I)_k$. The domain of L consists of a function on $L^2([0,\pi])$ for which

$$(2.5)_k \qquad y(\pi) = e^{ik}\, y(0)\ , \qquad\qquad y'(\pi) = e^{ik}\, y'(0)\ ,$$

and $Ly \in L^2([0,\pi])$. k is a fixed number in $(-\pi,\pi]$.

(II). The domain of L consists of all functions $y \in L^2(-\infty,\infty)$ for which $Ly \in L^2(-\infty,\infty)$.

The resulting boundary value problems are quite different. $(I)_k$ is regular, and the spectrum of L will consist of eigenvalues, the corresponding eigenfunctions being square integrable on L^2 $L^2([0,\pi])$. (II) is a singular boundary value problem, the spectrum is purely continuous, and there are no eigenfunctions which are square integrable over $(-\infty,\infty)$.

Nonetheless, all these spectra can be studied in a unified way. We let $y_0(x,\lambda)$, $y_1(x,\lambda)$ be solutions of $Ly = \lambda y$ which satisfy

$$y_0(0) = 1\ , \quad y_0'(0) = 0\ ; \qquad y_1(0) = 0\ , \quad y_1'(0) = 1\ .$$

Then the solution of the initial value problem

$$Ly = \lambda y\ , \qquad y(0) = \alpha\ , \quad y'(0) = \beta\ ,$$

can be found in terms of y_0 and y_1:

$$\begin{pmatrix} y(x) \\ y'(x) \end{pmatrix} = \begin{pmatrix} y_0(x,\lambda) & y_1(x,\lambda) \\ y_0'(x,\lambda) & y_1'(x,\lambda) \end{pmatrix} \begin{pmatrix} \alpha \\ \beta \end{pmatrix}$$

Solutions of $(I)_k$ will exist for some appropriate initial condition exactly when

$$\det \begin{pmatrix} y_0(\pi,\lambda)-e^{ik} & y_1(\pi,\lambda) \\ y_0'(\pi,\lambda) & y_1'(\pi,\lambda)-e^{ik} \end{pmatrix} = 0\ ,$$

i.e., when $e^{2ik} - \Delta(\lambda)\, e^{ik} + 1 = 0$, where $\Delta(\lambda) = y_0(\pi,\lambda)+y_1'(\pi,\lambda)$.

Thus, the eigenvalues of $(I)_k$ are the solutions $\lambda_j(k)$ of the transcendental equation

(2.4) $$\Delta(\lambda) = 2 \cos k .$$

One can show [12] that $\Delta(\lambda)$ has the asymptotic behavior $2 \cos \sqrt{\lambda}\, \pi$ as $|\lambda| \to \infty$; it has the following shape:

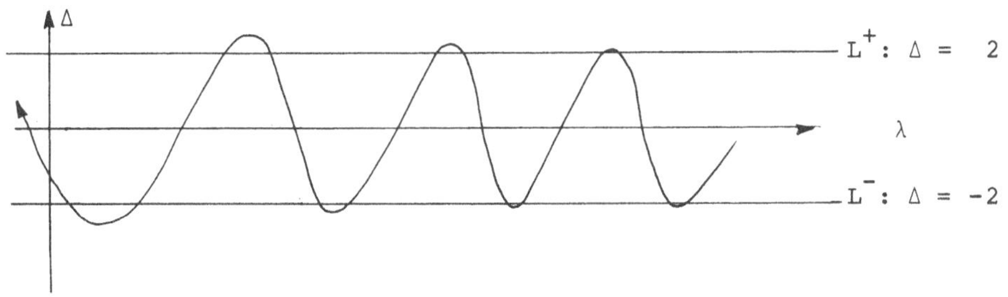

All intersections with L^+ and L^- are real and at most double. The intersections with L^+ give eigenvalues of $(I)_0$, with π-periodic eigenfunctions, those with L^- give eigenvalues of $(I)_\pi$, with 2π-periodic eigenfunctions. The spectrum of L with boundary conditions (II) turns out to be the set $\{\lambda \mid |\Delta(\lambda)| \leq 2\}$; it consists of intervals which are separated by regions where $|\Delta| > 2$. When λ belongs to one of the latter regions, all solutions of $Ly = \lambda y$ are unbounded for either $x \to +\infty$ or $x \to -\infty$. These intervals are referred to, variously, as "instability intervals", "forbidden bands", or "gaps". Gaps may degenerate to points, and coefficients q for which all but a finite number of gaps disappear have many special properties; these will be the subject of Section D.

C. Constants of the Motion of KdeV

We can now give a unified, spectral theoretic description of the constants of motion of the KdeV equation (2.1). We have observed that the eigenvalues of L stay constant for any boundary condition $(I)_k$. Hence the zeroes of $\Delta(\lambda)$, being eigenvalues of

$(I)_{\pi/2}$ (see (2.4)) are constant, but $\Delta(\lambda)$, being an entire function of order $\frac{1}{2}$ [12] is determined by its zeroes and its asymptotic behavior $\sim 2 \cos \sqrt{\lambda}\, \pi$, so that actually $\Delta(\lambda)$ is constant as the coefficient $q(x,t)$ changes according to (2.1). Recall that in Section 1.C we describeda scalar function of $L(t)$ and λ which remained constant in t, namely $\mathrm{Tr}\,(L-\lambda)^{-1}$, and that all integrals of the Toda lattice were derived as coefficients in an expansion of this function in powers of λ^{-1}. The situation here is quite similar.

By direction computation, or by appeal to Fredholm theory, one shows that the Green function $G(x,y,\lambda)$ of $L-\lambda I$ with periodic boundary conditions $(I)_0$ satisfies (for λ not one of the periodic eigenvalues $\lambda_0, \lambda_1, \lambda_2, \ldots$)

$$\int_0^\pi G(x,x,\lambda)\, dx = -\frac{d}{d\lambda} \ln[\Delta(\lambda) - 2] .$$

The left side is in fact nothing but

$$\sum_{i=0}^{\infty} \frac{1}{\lambda_i - \lambda} = \mathrm{Tr}\,(L-\lambda I)^{-1} .$$

The coefficients in an asymptotic expansion of $\ln \Delta(\lambda)$ would therefore be independent of t, and these coefficients are precisely the constants of the motion discovered by Gardner, Greene, Kruskal, and Miura [7].

Theorem. $\ln \Delta(-k)$ admits an asymptotic expansion

$$\ln \Delta(-k) \sim \pi\sqrt{k} - \ln 2 + \frac{1}{\sqrt{k}} \sum_{n=0}^{\infty} \frac{(-1)^n I_n(q)}{k^n}$$

as $k \to +\infty$; the coefficients $I_n(q)$ are defined as follows.

Let $M \equiv -d^3/dx^3 + 4q\, d/dx + 2q_x$, set $\beta_0 = 1$, and define β_n recursively by

(2.5) $$MB_n = 4 \frac{d}{dx} B_{n+1} .$$

It is possible [7] to choose B_{n+1} to be a polynomial in q and its derivatives. Then $I_n(q)$ is defined by

$$I_n(q) = \int_0^\pi B_{n+1}(q)\ (x)\ dx .$$

The proof is fairly straightforward; one uses integral representations of $y_0(\pi,\lambda)$ and $y_1'(\pi,\lambda)$ to compute asymptotic expansions for those functions and recursion relations for their coefficients. Then one expands $\Delta = y_0 + y_1'$, and then ln Δ. Some of the computations are done in [4], some in [7].

The idea of obtaining the constants of motion of the KdeV as coefficients in an asymptotic expansion is, of course, original with Gardner et al. [7]. However, the particular choice of Zaharov-Faddeev, to expand the trace of the resolvent, provides a unified way of understanding the occurrence of such constants of the motion in all known examples of the inverse method.

As a further application of the idea of Zaharov-Faddeev, consider the question of constants of the motion for any π-periodic evolution of q(x,t),

$$q_t = F(q)$$

which leaves the operators L(t) unitarily equivalent. The operators L(t) (with boundary conditions $(I)_0$ for example) have plenty of eigenvalues, all of which remain constant. Alternatively, one can argue that the discriminant $\Delta(\lambda,t)$ of L(t) is in fact constant in t, since the spectra of the L(t) are identical. Thus, the functionals $I_n(q)$ are constant for arbitrary evolutions of the type $\dot{L} = [B,L]$. * When one attempts to carry over this argument to

* An infinite number of such evolutions, all with the same L, has been found by Lax. These flows on the space of q's are actually commuting Hamiltonian flows, generated by the $I_n(q)$. This is discussed in some more detail in [6].

the KdeV on the whole line, one finds that some evolutions of the type $\dot{L} = [B,L]$, while leaving discrete eigenvalues invariant, may not preserve the familiar functionals $I_n(q)$. This is discussed in [5].

D. N-Gap Solutions

As far as I know, there is an analytical expression for just one kind of solution of the periodic KdeV, the "cnoidal" traveling wave,

$$q(x,t) = -6acn^2\left\{\sqrt{\frac{a-b}{12}}\left(x-\frac{a+b}{3}t\right);k\right\}, \qquad k = \frac{a}{a-b},$$

in which $cn(\xi,k)$ is the Jacobi elliptic function with modulus k. Much is already known, however, about a certain class of solutions, namely, those for which the operator $-\frac{d^2}{dx^2} + q(x,t)$ has precisely N finite spectral gaps (see the end of Section 2.B).

It has been known for some time that there are functions $q(x)$ for which the spectrum of $-d^2/dx^2 + q$ consists of exactly N+1 components. Theorems of Erdelyi and Ince (see [11]) assert this about Lame's equation

$$-y''(x) + N(N+1)k^2sn^2(x,k)y(x) = \lambda y(x);$$

for integer values of N, there are exactly N finite gaps.

Lax [11] conjectured that all N-gap coefficients q would be solutions of the variational problem:

(2.6) minimize $I_{N+1}(q)$, subject to $I_n(q)$ = constant, $n = 0,\dots,N$.

By the Lagrange method, this amounts to saying that q is a periodic solution of a nonlinear ODE of order 2N, of the form (with some coefficients c_n)

$$\text{grad } I_{N+1} + \sum_0^N c_n \text{ grad } I_n = 0\,. \tag{2.7}$$

This has now been proved; the fact that N-gap coefficients satisfy equation (2.7) was shown in [4], and the converse has been demonstrated by Lax (to appear) and Novikov [13]. In the latter paper, it is argued that equation (2.7) can be thought of as a Hamiltonian system in x, having, in general, quasiperiodic solutions q, but that these, too, give rise to N-gap spectra for the operator $-d^2/dx^2 + q$.

This result is of interest both in spectral theory and in KdeV theory. On the one hand, it characterizes a class of coefficients of Hill's operator with special properties, and includes some "inverse theorems" already known. One can, for example, compute that the gradients of I_0, I_1, I_2 are proportional to 1, $\frac{1}{2}q$, and $\frac{1}{8}(3q^2 - q'')$, and deduce that 0- and 1-gap coefficients must satisfy equations of the form

$$q = \text{const.} , \quad \text{resp.} \quad q'' = 3q^2 + Aq + B ,$$

(the solution of the second equation is essentially sn^2). These results are originally due to Borg and Hochstadt (see [9,12]). For KdeV theory, the significance of the result is that KdeV solutions starting from N-gap data must always remain solutions of (2.7), so that such special solutions wind around on low-dimensional manifolds, i.e., they have finitely many degrees of freedom.

There are also some very suggestive analogies between N-gap solutions of the periodic KdeV and N-soliton solutions of the KdeV on $-\infty < x < \infty$. It has been known for some time that N-soliton solutions satisfy the variational equation (2.7). Recently, Novikov [13] has defined a function b(k) associated with the spectral theory of Hill's equation, which, in a manner reminiscent of the reflection coefficient [7] in the infinite case, vanishes in an asymptotic sense on components of the spectrum not strictly separated by gaps.

Finally, it is known that N-soliton solutions can be represented in the form [7]

(2.8) $$q(x,t) = \sum_{1}^{N} c_i \psi_i^2(x,t) \ ,$$

where the ψ_i are the square integrable eigenfunctions of

$$- y''(x) + q(x,t)y(x) = \lambda y(x) \ .$$

Motivated by results of Lax and Kruskal, I have recently shown that any N-gap coefficient q, as well as any conserved density $\mathcal{B}_n$ (see (2.5)) can be written (for all t) as a linear combination of the form

(2.9) $$\sum_{0}^{N} c_i \psi_i^2 \ , \qquad c_i \text{ constants,}$$

where now the ψ_i are the eigenfunctions corresponding to the N+1 simple eigenvalues of the periodic problem $(I)_0$. In fact, it turns out that the moduli of the squares of any N+1 eigenfunctions of any problem $(I)_k$ (excepting double eigenfunctions for k = 0 or π) are linearly independent, while any more are dependent. However, the particular choice made in (2.9) is most suggestive of N-soliton solutions, as it tends to the representation (2.8) in the case of Lamé's equation when the period of sn^2 becomes infinite (modulus $k \to 1$).

I understand that other authors have obtained further results on N-gap solutions, but I am not yet familiar with these. Because of that, the sketchy exposition given above may soon have to be revised, augmented, or completed, but it may still give an idea of the fascinating interplay between spectral theory and the theory of nonlinear evolution equations which has enriched both fields.

Appendix

Our aim is to prove Proposition 2 of Section 1.B. Recall that with the periodic Toda lattice, we had associated an infinite, three-term recurrence relation

$$(\hat{L}u)_n = a_n u_{n+1} + a_{n-1} u_{n-1} + b_n u_n \ ,$$

where $a_{n+N} = a_n$, $b_{n+N} = b_n$. The spectrum of $\hat{L}$ (as operator on ℓ^2) is a union of intervals. We prove now that <u>if the spectrum is connected, then</u> $a_n \equiv \frac{1}{2}$, $b_n \equiv$ <u>constant</u>. This result is the discrete analog of Borg's theorem about Hill's equation (that if the spectrum is connected, then $q(x) \equiv$ const., see Section 2.D). The theorem lends credence to our earlier conjectures about the periodic lattice (end of Section 1.B), and suggests that the properties of the periodic lattice will be quite similar to that of PKdeV.*

To begin, we define solutions $\phi = \{\phi_n\}$, $\psi = \{\psi_n\}$ of

(A.1) $$\hat{L}u = \lambda u \ ,$$

by

(A.2) $$\phi_0 = 1 \ , \quad \phi_1 = 0 \ ;$$

(A.3) $$\psi_0 = 0 \ , \quad \psi_1 = 1 \ .$$

One sees that these are polynomials in λ, with

(A.4) $$\phi_n(\lambda) = - \frac{a_0}{a_1 \cdots a_{n-1}} \lambda^{n-2} + \ \dots \ ,$$

(A.5) $$\psi_n(\lambda) = \frac{1}{a_1 \cdot \ldots \cdot a_{n-1}} \lambda^{n-1} + \ \dots \ .$$

---

* However, the solitons of ∞KdeV can move to the right only, whereas those of the ∞ Toda lattice can move in either direction. I suspect that periodic Toda "solitons" move to one side or the other depending on whether the spectral gap is located at the right or left end of the spectrum.

As in Section 2.B, one obtains a criterion for the existence of a solution u of (A.1) for which

$$u_N = \rho u_0 , \quad u_{N+1} = \rho u_1 .$$

Namely, $\rho = \rho(\lambda)$ must solve

$$\rho^2 - \Delta\rho + 1 = 0 ,$$

where

$$\Delta(\lambda) = \phi_N(\lambda) + \psi_{N+1}(\lambda) . \tag{A.6}$$

$\Delta(\lambda)$ is a polynomial of degree N,

$$\Delta(\lambda) = 2^N\lambda^N + \ldots . \tag{A.7}$$

As in the theory of Hill's equation (Section 2.B), the spectrum σ of $\hat{L}$ is the set $\{\lambda \mid |\Delta(\lambda)| \leq 2\}$. If σ is connected, then all roots of $\Delta(\lambda) = \pm 2$ are double (with the exception of the smallest, α, and the largest, β). If we add a suitable constant to all b_n, we can translate $\sigma = [\alpha,\beta]$ so that it becomes an interval $[-A,A]$. Assume this done.

Lemma 1. $A = 1$.

Proof: $2^{-N}\Delta$ is a polynomial, with leading coefficient 1, which assumes its maximum value, 2^{-N+1}, N+1 times on $[-A,A]$, with alternating signs. Hence, it must be the Nth Chebyshev polynomial on $[-A,A]$, but this is well known to have maximum value $A^N 2^{-N+1}$. Hence $A = 1$.

From now on, we may confine our attention to $\lambda \in [-1,1]$. We set $\lambda = \cos\theta$, and define

$$T_n = \cos n\theta ,$$

$$S_n = \frac{\sin n\theta}{\sin\theta} .$$

S_n solves

$$\frac{1}{2} u_{n-1} + \frac{1}{2} u_{n+1} = \lambda u_n ,$$

and satisfies the initial condition (A.3). We will show that the solution ψ_n , for connected spectrum, must equal S_n; the desired conclusion will follow.

Denote by $\lambda_0,\ldots,\lambda_N$ the points cos $j\pi/N$, at which Δ (now known to equal $2T_N$) assumes its extreme values $\pm$ 2. For these λ, all solutions of (A.1) have period N or 2N. Hence,

$$\psi_N(\lambda_j) = \psi_0(\lambda_j) = 0 , \qquad j = 1,\ldots,N-1$$

and

$$\phi_{N+1}(\lambda_j) = \phi_1(\lambda_j) = 0 , \qquad j = 1,\ldots,N-1.$$

By (A.2.3), both these polynomials are of degree N-1, and therefore these are their only zeroes. Furthermore, we remember from (1.11) that $a_1 \cdots a_N = 2^{-N}$, and then see from (A.2,3), that

$$\phi_{N+1}(\lambda) = - a_N 2^N \lambda^{N-1} + \ldots ,$$

$$\psi_N(\lambda) = a_N 2^N \lambda^{N-1} + \ldots ,$$

so that

(A.8)
$$\phi_{N+1}(\lambda) = - \psi_N(\lambda) .$$

The zeroes of ψ_N coincide with those of S_N , and comparison of leading coefficients shows that

(A.9)
$$\psi_N(\lambda) = 2a_N S_N(\lambda)$$

Next, the discrete analog of the constancy of the Wronskian shows that

(A.10)
$$\phi_N\psi_{N+1} - \psi_N\phi_{N+1} = 1 ;$$

by means of (A.6) and (A.8), this converts into

$$\phi_N^2 - \Delta\phi_N + 1 - \psi_N^2 = 0 .$$

Solving this, we get

(A.11) $$\phi_N = \frac{1}{2}\,\Delta \pm \sqrt{\frac{1}{4}\,\Delta^2 - 1 + \psi_N^2}\ .$$

The expression under the radical reduces to

(A.12) $$\left(\frac{4a_N^2}{1-\lambda^2} - 1\right) \sin^2\theta .$$

Since ϕ_N is real, (A.12) must be nonnegative, for all $\lambda \in [-1,1]$.

Now we use a crucial trick, first applied by Hochstadt in the continuous case [9]: $\Delta(\lambda)$ is invariant under the translation $a_n \to a_{n+k}$, $b_n \to b_{n+k}$, for any k. Hence, (A.12) must be valid for any coefficient a:

(A.13) $$\left(\frac{4a_n^2}{1-\lambda^2} - 1\right) \sin^2\theta \geq 0\ , \quad \lambda \in [-1,1]\ , \quad \text{for } n = 1,\dots,N.$$

Because $a_1 \cdots a_N = 2^{-N}$, this is possible only when $a_n \equiv \frac{1}{2}$.

This fact, combined with (A.9), gives $\psi_N = S_N$. But then we can find ϕ_N from (A.11), and from (A.10), that $\psi_{N+1} = S_{N+1}$.

Finally, we note that (remember $a_n \equiv \frac{1}{2}$)

$$\psi_{N-1}(\lambda) = -\psi_{N+1} + 2(\lambda - b_N)\psi_N\ .$$

For $\lambda = \lambda_j$, $j = 1,\dots,N-1$, we get

$$\psi_{N-1}(\lambda_j) = -S_{N+1}(\lambda_j) = \pm 1\ ;$$

again, a comparison of leading coefficients shows that $\psi_{N-1} = S_{N-1}$.

Now we know that

$$\frac{1}{2}\,S_{N-1} + \frac{1}{2}\,S_{N+1} = \lambda S_N$$

and upon replacing S by ψ, we deduce that $b_N = 0$. The translation trick $a_n \to a_{n+k}$ shows that in fact $b_n \equiv 0$, which proves the proposition.

We now remark that the shift of the spectrum [-A,A] would have been unnecessary, had we agreed beforehand to keep the center of mass fixed. Under this assumption, Lemma 1 is equivalent to the following interesting fact: <u>the eigenvalues of the</u> $N \times N$ <u>matrix</u> L <u>coincide with the normal frequencies of the periodic, N-particle harmonic lattice precisely when the Toda lattice is at rest</u>.

References

[1] Arnol'd, V. I., Avez, A. A., Ergodic Problems of Classical Mechanics, W. A. Benjamin, New York, 1968.

[2] Flaschka, H., "On the Toda lattice I," Phys. Rev. B<u>9</u> (1974) 1924-1925.

[3] Flaschka, H., "On the Toda Lattice II," Prog. Theor. Phys. <u>51</u> (1974) 703-716.

[4] Flaschka, H., "On the Inverse Problem for Hill's Operator," to appear, Arch. Rat. Mech. Anal.

[5] Flaschka, H., and Newell, A. C., paper in these Proceedings.

[6] Gardner, C. S., "Korteweg-de Vries Equation and Generalization IV." J. Math. Phys. <u>12</u> (1971) 1548-1551.

[7] Gardner, C. S., Green, J. M., Kruskal, M. D., and Miura, R. M., "Korteweg-de Vries equation and Generalizations, VI," Comm. Pure Appl. Math. <u>27</u> (1974) 97-133.

[8] Hénon, M., "Integrals of the Toda Lattice," Phys. Rev. B <u>9</u> (1974) 1421-1423.

[9] Hochstadt, H., "On the Determination of Hill's Equation from its Spectrum," Arch. Rat. Mech. Anal. <u>19</u> (1965) 353-362.

[10] Lax, P. D., "Integrals of Nonlinear Equations of Evolution and Solitary Waves," Comm. Pure Appl. Math. <u>21</u> (1968) 467-490.

[11] Lax, P. D., "Periodic Solutions of the KdeV Equations," in Lectures in Appl. Math., Vol. 15, AMS, Providence, R.I. 1974.

[12] Magnus, W., and Winkler, S., <u>Hill's Equation</u>, Interscience-Wiley, Tracts in Pure and Appl. Math., No. 20, New York, 1966.

[13] Novikov, S., "The Periodic Problem for the KdeV Equation, I", (in Russian), Funk. Anal. Prilož. 8, No. 3 (1974) 54-66

[14] Toda, M., Studies on a Nonlinear Lattice, Lecture Notes, Trondheim University, 1974.

[15] Zaharov, V. E., and Faddeev, L. D., "Korteweg-de Vries Equation: A completely Integrable Hamiltonian System," Transl. in Func. Anal. and Its Applic. 5 (1972) 280-287.

Added in February 1975:

A number of recent papers have borne out my premonition (p. 20) that an updating of this lecture would soon be necessary:

P. D. Lax (Comm. Pure Appl. Math. 28, No. 1) has carried out a detailed study of the topology of the set of N-gap potentials, and of the P K de V flows on it. Marčenko (Mat. Sb. 95, No. 3) describes one possible solution of the inverse problem for Hill's equation, and uses this to solve P K de V. I understand that H. P. McKean and P. van Moerbecke have obtained analytical representations of N-gap potentials.

The spectral theory of the periodic Toda lattice is being developed by M. Kac and van Moerbecke (preprint). They have verified that Toda's explicit solutions are indeed 1-gap potentials (cf. the comment in 1.C above).

A number of new finite-dimensional systems admitting a Lax representation have been discovered by J. Moser.

FINITELY MANY MASS POINTS ON THE LINE UNDER THE INFLUENCE OF AN EXPONENTIAL POTENTIAL -- AN INTEGRABLE SYSTEM

Jürgen Moser*

Courant Institute of Mathematical Sciences, NYU, New York 10012

1. Analogue of the Toda Lattice for Finitely Many Mass Points

We consider the analogue of the Toda lattice [8] where only a finite number of mass points are admitted which move freely on the real axis. Denoting the position of the mass points by x_k, $k = 1,\dots,n$, we form the Hamiltonian

$$H = \frac{1}{2} \sum_{k=1}^{n} y_k^2 + \sum_{k=1}^{n-1} e^{(x_k - x_{k+1})} \tag{1.1}$$

with the differential equations

$$(1.2)\qquad \begin{aligned}
\dot{x}_k &= H_{y_k} = y_k , && k = 1,2,\dots,n \\
\dot{y}_k &= -H_{x_k} = e^{x_{k-1}-x_k} - e^{x_k - x_{k+1}} , && k = 2,\dots,n-1 \\
\dot{y}_1 &= -H_{x_1} = - e^{x_1 - x_2} \\
\dot{y}_n &= -H_{x_n} = e^{x_{n-1}-x_n}
\end{aligned}$$

Thus we can write our system (1.2) as

$$(1.2')\qquad x_k = e^{x_{k-1}-x_k} - e^{x_k - x_{k+1}} , \qquad k = 1,\dots,n$$

if we set $e^{x_0 - x_1} = 0$ and $e^{x_n - x_{n+1}} = 0$, that is we have the formal boundary condition

$$(1.3)\qquad x_0 = -\infty , \qquad x_{n+1} = +\infty .$$

It is the aim to study completely the flow determined by this

* This work has been partially supported by the National Science Foundation, Grant GP-42289X.

system of differential equations and relate the solution to the existence of n integrals of the motion. These integrals are essentially the same as those found by Henon [4] and Flaschka [1] for the same system of differential equations (1.2') under periodic boundary conditions, say

$$
(1.3') \qquad \begin{aligned} x_{k+n} &= x_k + 1 \\ y_{k+n} &= y_k \end{aligned} \qquad k = 0, \pm 1, \ldots .
$$

The crucial difference between the two problems is that the boundary condition (1.3') gives rise to a compact energy surface and the solutions are expected to be quasiperiodic, lying on tori, as one is familiar from integrable Hamiltonian systems. If we impose the boundary condition (1.3) instead of (1.3') the energy surface is noncompact, as the particles can run to infinity. In fact, we will show, as is intuitively clear, that for any initial configuration mutual distances between all particles grow indefinitely, i.e. $x_{k-1} - x_k \to \infty$ for $k = 2, \ldots, n$; and they behave asymptotically like free particles depending linearly on time. This suggests the scattering problem: To determine the relation between this asymptotic motion for the past and the future. This can be done explicitly here and one finds that $y_{n-k+1}(+\infty) = y_k(-\infty)$, so that at $t = +\infty$ the first particle has the velocity of the last at $t = -\infty$ etc. as in a familiar experiment of collision of steel balls. Moreover, the phase relation can also be determined explicitly and we will show that

$$
x_{n-k+1}(t) - x_k(-t) - 2y_k^- t \to \sum_{j<k} \log (y_j^- - y_k^-)^2 - \sum_{j>k} \log (y_j^- - y_k^-)^2 ,
$$

where $y_j^- = y_j(-\infty)$ are assumed ordered according to size. Thus the particles behave asymptotically as if they interacted just pairwise! This will be derived in Section 4.

In the limit $t \to +\infty$, the y_k, $k = 1,2,\dots,n$, or their symmetric functions, are t-independent integrals of the motion, and one may ask for integrals of the given system which asymptotically agree with these integrals. This is indeed possible, and Henon's construction of integrals was based on this idea, even though in the periodic case this idea is not really justified and was only a guiding principle for the construction of integrals. For the noncompact case, i.e. boundary condition (1.3), the free system is indeed the limit state and this approach quite natural. On the other hand, the noncompact case is, of course, much less complicated, as the solutions have no recurrence property and the flow has the nature of parallel flow. In fact, we will show that (1.2) can be mapped into the following system of differential equations,

$$\frac{d\lambda_k}{dt} = 0 , \quad \frac{dr_k}{dt} = -\frac{\partial V}{\partial r_k} , \quad k = 1,\dots,n, \tag{1.4}$$

where

$$V = \frac{\sum_{k=1}^{n} \lambda_k r_k^2}{2 \sum_{k=1}^{n} r_k^2} \tag{1.4'}$$

and the variables are restricted to the (2n-1) dimensional domain

$$\lambda_1 < \lambda_2 < \dots < \lambda_n ; \quad \sum_{k=1}^{n} r_k^2 = 1 , \quad r_k > 0 . \tag{1.5}$$

Clearly, the solutions run from the maximum of V at $r_k = \delta_{kn}$ to the minimum of V of $r_k = \delta_{k1}$ as t runs from $-\infty$ to $+\infty$, and $\lambda_1,\lambda_2,\dots,\lambda_n$ are integrals of the motion, while $r_1,\dots,r_{n-1}$ can be viewed as parameters on the surfaces $\lambda_k = \text{const}$. The mapping taking x,y into the variables λ_k, r_k on (1.5) is up to translation of the x_k one to one and will be given explicitly. The inverse mapping illustrates the inverse method of spectral theory.

Thus this note does not claim any new idea and should be considered as providing a simple model illustrating the construction of integrals and its connection with the inverse method of spectral theory in extreme simplicity, yet with all rigor. On the other hand, it leads immediately to an unsolved problem if one wants to carry out this approach for the periodic boundary condition (1.3'). Although the integrals I_k for this problem are well known, no parameters are known on the level surfaces $I_k = c_k$ which determine the x_k (mod 1), y_k uniquely. This is related to the lack of an inverse theory for the Hill's equation $-u'' + q(x)u = \lambda u$, $q(x+1) = q(x)$ under periodic boundary conditions $u(x+1) = u(x)$ where the problem consists in finding a set of quantities which together with the eigenvalues allow one to determine $q(x)$. One can hope to shed some light on this question if one could solve the above finite dimensional problem.

2. Flaschka's Form of the Differential Equation and Asymptotic Behavior

We set, with Flaschka,

$$(2.1) \qquad a_k = \frac{1}{2} e^{(x_k - x_{k+1})/2} , \quad b_k = -\frac{1}{2} y_k$$

so that the differential equations (1.2) go into

$$(2.2) \qquad \begin{aligned} \dot{a}_k &= a_k(b_{k+1} - b_k) , \quad k = 1,2,\ldots,n-1 \\ \dot{b}_k &= 2(a_k^2 - a_{k-1}^2) , \quad k = 1,2,\ldots,n \end{aligned}$$

with the boundary conditions (1.3) being

$$(2.3) \qquad a_0 = 0 , \quad a_n = 0 .$$

Observe that (2.1) provides a transformation of the (x,y) variables into the (a,b)-variables. We identify points (x,y), $(\tilde{x},\tilde{y})$ if $x_k - \tilde{x}_k$ is independent of k, and call the equivalence class a "configuration". It is characterized by the $2n-1$ numbers $x_k - x_n$, $(k = 1,\ldots,n-1$, and y_k, $k = 1,\ldots,n$. Thus (2.1) defines an invertible transformation of the $(2n-1)$-dimensional space of configurations into the domain

$$D = \{a,b \mid a_k > 0 \ , \ k = 1,\ldots,n-1)\} \ ,$$

and it remains to study the flow given by the quadratic differential equation (2.2) in D. The energy is given by

$$H = 4\left\{ \sum_{k=1}^{n-1} a_k^2 + \frac{1}{2} \sum_{k=1}^{n} b_k^2 \right\} . \tag{2.4}$$

We show first that for any solution in D

$$a_k(t) \to 0 \quad \text{for} \quad t \to \pm\infty \quad \text{and} \quad k = 1,\ldots,n-1 \ , \tag{2.5}$$

which amounts to the assertion that $x_{k+1}-x_k \to \infty$ as $t \to \pm\infty$. To prove this we consider the system (2.2) with prescribed $a_0(t), a_n(t) \in L^2(-\infty,+\infty)$ so that $\int_{-\infty}^{+\infty} (a_0^2 + a_n^2)\, dt < \infty$ and prove the

Lemma. For any solution of (2.2) with this modified boundary condition we have

$$\int_{-\infty}^{\infty} (a_1^2 + a_{k-1}^2)\, dt < \infty \ .$$

Proof: Consider the function

$$\phi(t) = b_1 - b_n$$

for which

$$\frac{d\phi}{dt} = \dot{b}_1 - \dot{b}_n = 2(a_0^2 + a_n^2) - 2(a_1^2 + a_{n-1}^2) \ .$$

Thus

$$\psi = \frac{1}{2}\phi - \int_{-\infty}^{t} (a_0^2 + a_n^2)\, dt$$

satisfies

$$\frac{d\psi}{dt} = -\ (a_1^2 + a_{n-1}^2) \ .$$

Since by the energy relation ϕ and hence ψ is bounded, also

$$\int_{-T}^{T} (a_1^2 + a_{n-1}^2)\ dt = \ \psi(-T) - \psi(T)$$

is bounded for $T \to \pm \infty$, proving the lemma.

We can apply this argument, in particular, to $a_0 = a_n = 0$. Applying this lemma to the reduced system where the first and last equations in the first and second line of (2.2) are cancelled we conclude that $\int_{-\infty}^{+\infty} (a_n^2 + a_{n-2}^2)\ dt < \infty$ and inductively that

$$\text{(2.6)} \qquad \sum_{k=1}^{n-1} \int_{-\infty}^{+\infty} a_k^2\ dt < \infty \ .$$

Since on the other hand $|\dot{p}| \le 2 \sum a_k^2 |b_k - b_{k+1}| \le M$ is bounded it follows that $p = \sum_1^{n-1} a_k^2 \to 0$ for $t \to \pm\infty$. Indeed, otherwise there would exist a sequence $|t_k| \to \infty$ with $p(t_k) \ge \delta > 0$. We may assume that the sequence is so selected that $|t_{k+1} - t_k| \ge \delta/M$. Since $p(t) \ge \delta/2$ in the disjoint intervals $|t - t_k| < \frac{1}{2}\frac{\delta}{M}$ it cannot be integrable, contradicting (2.6). This proves (2.5). Moreover, we conclude from (2.2) that b_k tends to a limit $b_k(\infty)$ as $t \to +\infty$.

Flaschka [1,2] noted that the above system (2.2) can be expressed in matrix form

$$\text{(2.7)} \qquad \frac{d}{dt} L = BL - LB$$

where

$$L = \begin{pmatrix} b_1 & a_1 & & & 0 \\ a_1 & b_2 & & & \\ & & \ddots & & \\ & & & b_{n-1} & a_{n-1} \\ 0 & & & a_{n-1} & b_n \end{pmatrix} ; \quad B = \begin{pmatrix} 0 & a_1 & & & 0 \\ -a_1 & 0 & & & \\ & & \ddots & & \\ & & & 0 & a_{n-1} \\ 0 & & & -a_{n-1} & 0 \end{pmatrix}$$

Thus if $U = U(t)$ is the orthogonal matrix satisfying

$$\frac{dU}{dt} = BU \; ; \qquad U(0) = I$$

then by (2.7)

$$\frac{d}{dt} (U^{-1} L\, U) = 0$$

hence

$$U^{-1} L\, U = L(0) \; .$$

Thus, $L(t)$ is similar to $L(0)$ and the eigenvalues λ_k of the Jacobi matrix L, which are real and distinct, are independent of t. This description of the integrals as eigenvalues of a linear operator is due to Lax [5] and Flaschka's derivation was based on his approach.

Thus the characteristic polynomial

$$(2.8) \qquad \Delta_n(\lambda) = \det(\lambda I - L) = \prod_{k=1}^{n} (\lambda - \lambda_k) = \sum_{k=0}^{n} I_k \lambda^{n-k}$$

as well as the coefficients $I_1, \ldots, I_n$ are constants of the motion (2.2). For definiteness we order the eigenvalues according to their size,

$$\lambda_1 < \lambda_2 < \ldots < \lambda_n \quad .$$

Notice that $L(t) \to L(\infty)$ as $t \to +\infty$ where $L(\infty)$ is a diagonal matrix whose diagonal elements must be the eigenvalues λ_k in appropriate order. From

$$\frac{\dot{a}_k}{a_k} \sim b_{k+1}(\infty) - b_k(\infty)$$

and (2.5) we conclude that $b_{k+1}(\infty) < b_k(\infty)$ or

$$b_k(\infty) = \lambda_{n-k+1}$$

i.e.

$$L(\infty) = \mathrm{diag}(\lambda_n, \lambda_{n-1}, \ldots, \lambda_1) .$$

Using that the t-reversing substitution

$$t \to -t \; ; \quad a_k \to a_{n-k} \; ; \quad b_k \to b_{n+1-k}$$

leaves the system invariant, we conclude that

$$L(-\infty) = \mathrm{diag}(\lambda_1, \lambda_2, \ldots, \lambda_n)$$

i.e. $L(\infty)$, $L(-\infty)$ differ just in the order of the diagonal elements. The physical interpretation of this result is: If for $t \to -\infty$ the particles x_k approach the velocities $y_k = -2\lambda_k$ where $y_1 < y_2 < \ldots < y_n$ then for $t \to +\infty$ the particles x_k have the velocities y_{n-k+1} so that the particles exchange their velocities.

This describes the flow for our problem (2.2), (2.3). Still we will find another set of variables, $r_k > 0$, $k = 1,\ldots,n-1$, which together with the λ_k form a set of coordinates, and represent the differential equations in these new variables.

3. Partial Fractions and Continued Fractions

Let

$$R(\lambda) = (\lambda I - L)^{-1}$$

where we suppress the dependence in t. This is an n by n matrix and we single out the element in the last row and last column

(3.1) $R_{nn}(\lambda) = (R(\lambda)e_n, e_n) = f(\lambda)$ where $e_n = (0,0,\ldots,0,1)$,

and $f(\lambda)$ is hereby defined. Since L is symmetric it follows that $f(\lambda)$ is an analytic function for $\mathrm{Im}\,\lambda \neq 0$ and

$$\operatorname{Im} f(\lambda) > 0 \quad \text{for} \quad \operatorname{Im} \lambda > 0 .$$

Moreover, it is rational with simple poles at the eigenvalues λ_k and so admits the partial fraction expansion

$$(3.2) \qquad f(\lambda) = \sum_{k=1}^{n} \frac{r_k^2}{\lambda-\lambda_k} , \qquad r_k > 0 ,$$

with positive residua r_k^2. Moreover, for $|\lambda| \to \infty$ one has $\lambda f(\lambda) \to 1$ and

$$\sum_{k=1}^{n} r_k^2 = 1 .$$

Thus we have a mapping ϕ associating with every point in

$$(3.3) \qquad D = \{a_1,\ldots,a_{n-1},b_1,\ldots,b_n \text{ with } a_k > 0\}$$

a point in

$$(3.4) \quad \Lambda = \{\lambda_1,\ldots,\lambda_n,r_1,\ldots,r_n \text{ with } \lambda_1 < \lambda_2 < \ldots < \lambda_n, \; \sum_{k=1}^{n} r_k^2 = 1, \; r_k > 0\} .$$

We claim that this mapping ϕ: $D \to \Lambda$ is one to one and onto. We will view it as a coordinate transformation and then describe the differential equations in the new variables. The fact that the mapping ϕ has an inverse ϕ^{-1}: $\Lambda \to D$ corresponds to the inverse method of spectral theory, which in the elementary form described here goes back to Stieltjes [3]. It is based on the fact that $f(\lambda)$ admits a continued fraction expansion

$$(3.5) \qquad f(\lambda) = \cfrac{1}{\lambda - b_n - \cfrac{a_{n-1}^2}{\lambda - b_{n-1} \ddots - \cfrac{a_1^2}{\lambda - b_1}}}$$

where the entries a_k, b_k agree precisely with those of L.

To prove this we establish the identity

$$f(\lambda) = \frac{\Delta_{n-1}}{\Delta_n} \tag{3.6}$$

where Δ_n is the characteristic polynomial of $(\lambda I-L)$, see (2.8), and Δ_k the k by k subdeterminant obtained by canceling the last n-k rows and columns of $(\lambda I-L)$. Expanding Δ_k by the last row one finds

$$\Delta_k = (\lambda-b_k)\,\Delta_{k-1} - a_{k-1}^2\,\Delta_{k-2} \tag{3.7}$$

for $k = 3,4,\dots,n$; it holds also for $k=1,2$ if we set

$$\Delta_{-1} = 0\ , \quad \Delta_0 = 1\ .$$

Thus the ratios $s_k = \Delta_k/\Delta_{k-1}$ satisfy the recursion formula

$$s_k = \lambda - b_k - \frac{a_{k-1}^2}{s_{k-1}} \quad \text{for} \quad k = 2,3,\dots,n$$

which leads to a finite continued fraction for $s_n = \Delta_n/\Delta_{n-1} = f^{-1}(\lambda)$.

Thus the representation (3.5) follows from (3.6) which we prove now. For this purpose we compute the last column

$$R\,e_n = z \quad \text{of} \quad R = R(\lambda)\ .$$

We find

$$\left\{\begin{aligned} z_{k+1} &= \frac{\Delta_k}{\Delta_n}\,a_{k+1}\cdots a_{n-1} \quad \text{for} \quad k = 0,1,\dots,n-2 \\ z_n &= \frac{\Delta_{n-1}}{\Delta_n}\ . \end{aligned}\right. \tag{3.8}$$

Indeed z is the solution of

$$(\lambda I-L)\,z = e_n$$

and using the recursion formula one readily verifies (3.8). Thus

$$f(\lambda) = R_{nn} = z_n = \frac{\Delta_{n-1}}{\Delta_n}$$

as we wanted to show.

Thus for a given matrix L we can compute the rational function $f(\lambda)$ which has n simple real poles with positive residua, since $\operatorname{Im} f(\lambda) > 0$ for $\operatorname{Im} \lambda > 0$. Ordering these poles according to size we have defined the mapping ϕ taking D into Λ (see (3.3), (3.4)).

We come to the "inverse problem" which requires that we determine ϕ^{-1}. With any point in Λ we associate $f(\lambda)$ by (3.2). Then $\operatorname{Im} f > 0$ for $\operatorname{Im} \lambda > 0$ and $\lambda f(\Lambda) \to 1$ for $|\lambda| \to \infty$. Thus

$$\frac{1}{f(\lambda)} = \lambda + A - g(\lambda)$$

where A is a real constant and $g(\lambda)$ is a rational function which satisfies

$$\operatorname{Im} g(\lambda) = \operatorname{Im} \lambda + \frac{\operatorname{Im} f}{|f|^2} > 0 \quad \text{for} \quad \operatorname{Im} \lambda > 0 .$$

Thus $g(\lambda)$ has only simple poles on the real axis and their number is n-1. One computes easily $-A = \sum_1 \lambda_k r_k^2$, $\lambda g(\lambda) \to \sum \lambda_k^2 r_k^2 - (\sum \lambda_k r_k^2)^2 > 0$. Thus $g = B\, f_{n-1}$ with $B > 0$, and $\lambda f_{n-1} \to 1$ for $|\lambda| \to \infty$. Thus

$$f(\lambda) = \frac{1}{\lambda + A - Bf_{n-1}}$$

and by induction we get a unique continued fraction of the form (3.5) with $A = -b_n$, $B = a_{n-1}^2 > 0$, etc. This shows that ϕ maps D one to one onto Λ.

Finally we express the differential equation (2.2) in these new variables. For this purpose we deduce from (2.7)

$$\frac{d}{dt} R = R \frac{dL}{dt} R = BR - RB$$

and taking the last element $R_{nn} = f$ in R we find

$$\frac{df}{dt} = (e_n, (BR-RB)e_n) = -2(e_n, Ra_{n-1}e_{n-1}) = -2a_{n-1} R_{n,n-1} .$$

Since $R_{n,n-1}$ agrees with z_{n-1} in (3.8) we obtain

$$\frac{df}{dt} = - 2\, a_{n-1}^2 \frac{\Delta_{n-2}}{\Delta_n} .$$

This formula allows us to determine the desired differential equations. Since we established already that $d\lambda_k/dt = 0$ we have

$$\frac{df}{dt} = \sum_{k=1}^{n} \frac{2r_k \dot{r}_k}{\lambda - \lambda_k}$$

Comparing the residue of the last two expressions we get

$$2r_k\dot{r}_k = - 2\, a_{n-1}^2 \left.\frac{\Delta_{n-2}}{\Delta_n'}\right|_{\lambda=\lambda_k} .$$

By the recursion formula (3.7) we have

$$\Delta_n = (\lambda - b_n)\Delta_{n-1} - a_{n-1}^2 \Delta_{n-2}$$

or, since $\Delta_n(\lambda_k) = 0$,

$$\Delta_{n-2}(\lambda_k) = \frac{\lambda_k - b_n}{a_{n-1}^2} \Delta_{n-1}(\lambda_k) ,$$

hence

$$2r_k\dot{r}_k = - 2(\lambda_k - b_n) \left.\frac{\Delta_{n-1}}{\Delta_n'}\right|_{\lambda=\lambda_k} .$$

A similar comparison of the residua of

$$f(\lambda) = \frac{\Delta_{n-1}}{\Delta_n} = \sum_{k=1}^{n} \frac{r_k^2}{\lambda - \lambda_k}$$

gives

$$\left.\frac{\Delta_{n-1}}{\Delta_n'}\right|_{\lambda=\lambda_k} = r_k^2$$

hence

$$2r_k\dot{r}_k = - 2(\lambda_k - b_n) r_k^2 .$$

Since $\sum_{k=1}^{n} r_k^2 = 1$ we find

$$0 = \sum r_k\dot{r}_k = - \sum \lambda_k r_k^2 + b_n$$

and so, $b_n = \sum \lambda_k r_k^2$, as we had seen before, and

$$\dot{r}_k = -(\lambda_k - \sum \lambda_j r_j^2) r_k$$

which gives the differential equation (1.4) of Section 1. These differential equations represent the vector field along the gradient of the function V(r) (see (1.4')) restricted to the part of the unit sphere lying in the positive quadrant. Thus every solution approaches for $t \to +\infty$ the minimum: $r(t) \to e_1$ and for $t \to -\infty$ the maximum: $r(t) \to e_n$. Of course, it is also possible to give an analytical representation for the solutions, since they are obtained by projecting the linear differential equations $\dot{r}_k = -\lambda_k r_k$ on the unit sphere. Thus we find

$$\begin{cases} \lambda_k(t) = \lambda_k(0) \\ r_k^2(t) = \dfrac{r_k^2(0)\, e^{-2\lambda_k t}}{\sum\limits_{j=1}^{n} r_j^2(0)\, e^{-2\lambda_j t}} . \end{cases}$$

To summarize our result we consider the r_k as homogeneous variables, still positive, and set, accordingly,

$$f(\lambda) = \left(\sum_{k=1}^{n} \frac{r_k^2}{\lambda - \lambda_k} \right) \left(\sum_{k=1}^{n} r_k{}^2 \right)^{-1} . \tag{3.9}$$

From (3.5) and the calculation of continued fractions it is clear that the a_k^2, b_k are rational functions of r_j, λ_j, of degree 0 in the r_j. One verifies that a_k, b_k are of degree 1 in λ_j. Thus we have the following rational transformation

$$(3.10) \qquad \begin{aligned} a_k^2 &= A_k(r,\lambda) , \qquad && k = 1,2,\ldots,n-1 \\ b_k^2 &= B_k(r,\lambda) , \qquad && k = 1,2,\ldots,n \end{aligned}$$

of $\lambda_1 < \ldots < \lambda_n$; $r_k > 0$ into the domain D. If we identify two proportional vectors r the mapping is one to one.

In these homogeneous coordinates r_k the differential equations become linear

$$\frac{d\lambda_k}{dt} = 0 \; ; \qquad \frac{dr_k}{dt} = - \lambda_k r_k \; . \tag{3.11}$$

Thus the solutions of (2.2) can be represented as rational functions of the n constants λ_j and n exponential functions $e^{-\lambda_j t}$.

As we mentioned in Section 1 the flow is particularly simple in this case since no periodic or recurrent solutions are present. In the more interesting case of the periodic boundary condition one has quasiperiodic solutions and the task of finding coordinates for the integral surfaces which are tori, is more difficult. The main problem is the "inverse problem" which consists in recovering L -- which is then a cyclic matrix -- from its eigenvalues and appropriately chosen quantities. This problem seems unsolved as yet.

4. Solution of the Scattering Problem

From the results of Section 2 it follows that the asymptotic behavior of the solutions of our problem (1.2') is given by

$$\begin{aligned} x_k(t) &= \alpha_k^+ t + \beta_k^+ + O(e^{-\delta t}) \\ x_k(-t) &= -\alpha_k^- t + \beta_k^- + O(e^{-\delta t}) \end{aligned} \tag{4.1}$$

for $t \to +\infty$ with some $\delta > 0$. Moreover, we found

$$\alpha_k^+ = \lim_{k \to k + \infty} y_k = - 2\lambda_{n-k+1} \quad \text{and} \quad \alpha_k^- = - 2\lambda_k$$

i.e.

$$\alpha_{n-k+1}^+ = \alpha_k^- \tag{4.2}$$

which expresses that the $(n-k+1)^{st}$ particle has then for $t \to +\infty$ the velocity which the kth particle had in the past.

Our goal is to determine the relation between the phases β_k^+, β_k^- which can be given explicitly too. This remarkable fact is also a consequence of the integrable character of the system and the representation of $e^{x_k - x_{k+1}}$, y_k as rational functions of λ_j, $e^{-\lambda_j t}$ given by (3.10),(3.11). An explicit calculation seems prohibitive; nevertheless the following argument, which uses just rudimentary properties of rational functions will lead to the goal. The result is

$$\beta_{n-k+1}^+ = \beta_k^- + \sum_{j \neq k} \phi_{jk}(\alpha^-) \tag{4.3}$$

where

$$\phi_{jk}(\alpha) = \begin{cases} \log\,(\alpha_j^- - \alpha_k^-)^2 & \text{for} \quad j < k \\ -\log\,(\alpha_j^- - \alpha_k^-)^2 & \text{for} \quad j > k \end{cases} \tag{4.4}$$

For n=2 this amounts to

$$\begin{aligned} \beta_2^+ &= \beta_1^- - \log\,(\alpha_1^- - \alpha_2^-)^2 \\ \beta_1^+ &= \beta_2^- + \log\,(\alpha_1^- - \alpha_2^-)^2 \end{aligned} \tag{4.5}$$

Thus ϕ_{jk} represents the phase shift between two particles with velocities α_j^-, α_k^- at $t = -\infty$. The result (4.3) can therefore be interpreted as follows: The particles are scattered just as if their interaction takes place two at a time! This was suggested to me by M. Kruskal who described an analogous phenomenon for solutions of the Korteweg-de Vries equation (see [6], Theorem 3.7) and by P. D. Lax. This phenomenon which had been discovered by Zakharov et al. (see [6] for references) is obviously intimately related to our result and it is conceivable that one can be derived from the other -- but we have not pursued this point.

We illustrate the statement in Figures 1, 2. Figure 1 illustrates the case n = 2, which is given explicitly in terms of $\cosh\,(\lambda_2 - \lambda_1)t$. The asymptotic behavior can be interpreted as the elastic reflection of two rods of length $\phi_{21} = \log\,(\alpha_2^- - \alpha_1^-)^2$,

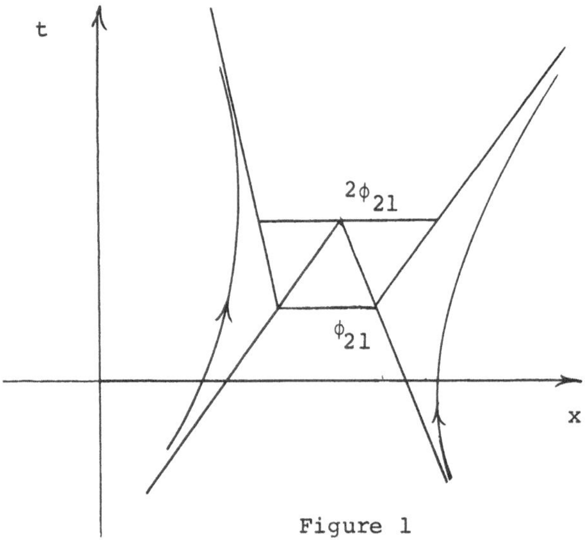

Figure 1

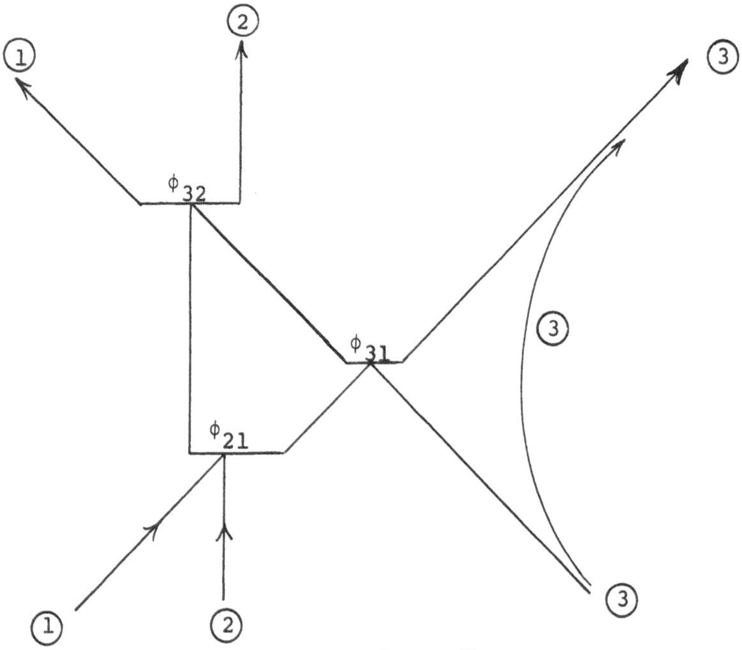

Figure 2

provided this number is positive. For negative values of ϕ_{21} the particles reflect only after passing each other. However, this interpretation is somewhat misleading, especially if $n > 2$, since the length of the rods depends on their velocity, not on the label. We indicate schematically the construction of the scattering for $n = 3$ in Figure 2.

To prove (4.3) first translate it into an asymptotic statement for (2.2). For this purpose we note that on account of the linear t-dependence of the center of mass we have

$$\sum_{k=1}^{n} \beta_k^+ = \sum_{k=1}^{n} \beta_k^-$$

and therefore it suffices to prove (4.3) for the differences $\beta_{k+1}^- - \beta_k^-$, i.e. it suffices to establish

$$\beta_{n-k}^+ - \beta_{n-k+1}^+ = \beta_{k+1}^- - \beta_k^- - \sum_{j\neq k} \phi_{jk} + \sum_{j\neq k+1} \phi_{j,k+1} .$$

Using (2.1), (4.1) this amounts to

$$\lim_{t\to+\infty} a_{n-k}(t)\, a_k(-t)\, e^{2(\lambda_{k+1}-\lambda_k)t} = C_k(\lambda) , \qquad k = 1,2,\ldots,n-1 \tag{4.6}$$

with

$$\log\,(4C_k)^2 = - \sum_{j\neq k} \phi_{j,k} + \sum_{j\neq k+1} \phi_{j,k+1} .$$

Finally, with $\alpha_k^- = -2\lambda_k$ and (4.4) this gives for C_k the expression

$$C_k = \frac{\prod_{j>k} (\lambda_j - \lambda_k)}{\prod_{j<k} (\lambda_k - \lambda_j)} \cdot \frac{\prod_{j<k+1} (\lambda_{k+1} - \lambda_j)}{\prod_{j>k+1} (\lambda_j - \lambda_{k+1})} \tag{4.7}$$

where empty products are to be set equal to 1.

Thus it suffices to prove (4.6) with (4.7). For $n = 2$ this is easily verified. In that case our transformation (3.10) takes the explicit form

$$b_1 = \frac{\lambda_2 r_1^2 + \lambda_1 r_2^2}{r_1^2 + r_2^2}, \qquad b_2 = \frac{\lambda_1 r_1^2 + \lambda_2 r_2^2}{r_1^2 + r_2^2}$$

(4.8)

$$a_1 = \frac{(\lambda_2 - \lambda_1) r_1 r_2}{r_1^2 + r_2^2} = (\lambda_2 - \lambda_1)\left(\frac{r_1}{r_2} + \frac{r_2}{r_1}\right)^{-1}$$

Since $r_k(t) = r_k(0)e^{-\lambda_k t}$ we find

$$\text{(4.9)} \qquad a_1(t)\, a_1(-t)\, e^{2(\lambda_2-\lambda_1)t} \to (\lambda_2-\lambda_1)^2$$

which corresponds to (4.6) for $k = 1$, $n = 2$.

For $n = 3$ one can still, with some effort, verify the above statement by an explicit calculation but for general n this seems a hopeless approach. Therefore we proceed as follows. We know that for every solution

(4.10)

$$a_{n-k}(t) \sim c_{n-k}^{+}\, e^{-(\lambda_{k+1}-\lambda_k)t}$$

$$a_k(-t) \sim c_k^{-}\, e^{-(\lambda_{k+1}-\lambda_k)t} \qquad \text{as} \quad t \to +\infty$$

with positive constants c_{n-k}^{+}, c_k^{-}.

(i) First we establish that c_{n-k}^{+}, c_k^{-} depend real analytically on the initial data $a_j(0)$, $b_j(0)$ of the solution. (ii) Second, we show that

(4.11)

$$c_{n-k}^{+} = \frac{r_{k+1}(0)}{r_k(0)}\, C_{n-k}^{+}(\lambda)$$

$$c_k^{-} = \frac{r_k(0)}{r_{k+1}(0)}\, C_k^{-}(\lambda)$$

with C_{n-k}^{+}, C_k^{-} depending on λ only. This shows that the limit (4.6) is equal to

$$\text{(4.11')} \qquad C_k(\lambda) = C_{n-k}^{+}(\lambda)\, C_k^{-}(\lambda) ,$$

and therefore independent of the initial condition of r_j. This makes the actual determination of $C_k(\lambda)$ easy if we consider various limit situations for the initial conditions, which will be the third step (iii).

To begin with the analytic dependence of the constants c_{n-k}^+, c_k^- on the initial conditions we fix a solution $a_j(t)$, $b_j(t)$ of (2.2) and describe a nearby one by

$$\tilde{a}_j = a_j e^{u_j} , \qquad \tilde{b}_j = b_j + v_j$$

where $u_j(0)$, $v_j(0)$ are small. The differential equations for u,v are then

$$\begin{aligned} \dot{u}_k &= v_{k+1} - v_k \\ \dot{v}_k &= 2\{a_k^2(e^{2u_k} - 1) - a_{k-1}^2(e^{2u_{k-1}} - 1)\} . \end{aligned}$$

The asymptotic behavior of the solutions is given by

$$\begin{aligned} u_k &= (v_{k+1}(\infty) - v_k(\infty))t + \gamma_k + O(e^{-\delta t}) \\ v_k &= v_k(+\infty) + O(e^{-\delta t}) \end{aligned} \qquad \text{for} \quad t \to +\infty \tag{4.12}$$

if the initial data are small enough. It is sufficient to show that $v_k(\infty)$, γ_k depend real analytically on $u_j(0)$, $v_j(0)$ if these are close to zero. For this purpose we permit complex initial values and show that the above asymptotic description holds for a <u>complex</u> neighborhood of the origin. This requires some simple a priori estimates:

Obviously it suffices to establish the analytic dependence on the initial values $u_k(\tau)$, $v_k(\tau)$ for some fixed positive τ. The fixed real solution satisfies an estimate

$$0 < a_k(t) < c_1 e^{-\delta t} \quad \text{for} \quad 0 \leq t < \infty$$

with some positive constants δ, c_1; we may assume $\delta < 1 < c_1$.

With

$$0 < \eta < \frac{\delta}{8}$$

and some τ, to be determined later, we consider complex initial values in

(4.13) $$|u_k(\tau)| < \eta , \qquad |v_k(\tau)| < \eta$$

Let $M(t) = \max_k |v_k(t)|$ and consider this function in an interval $\tau \leq t < \tau'$ in which $M(t) \leq 2\eta$. Then we get from the differential equations for $\tau \leq t < \tau'$

$$|u_k(t)| \leq \eta + 2 \int_\tau^t M(t)\, dt \leq \eta(1 + 4(t-\tau)) .$$

Using the inequality

$$|e^{2u} - 1| \leq 2|u|\, e^{2|u|}$$

and setting $s = t - \tau$ we get for $\tau \leq t < \tau'$

$$|v_k(t)| \leq \eta + 8c_1^2\eta \int_\tau^t e^{-2\delta t}(1 + 4(t-\tau))\, e^{2\eta(1+4(t-\tau))}\, dt$$

hence

$$M(t) \leq \eta\left(1 + c_2\, e^{-2\delta\tau} \int_0^\infty e^{-(2\delta-8\eta)s}\, (1 + 4s)\, ds\right)$$

with $c_2 = 8c_1^2e^2$. Since $2\delta-8\eta > \delta$ we get

$$M(t) \leq \eta\left(1 + c_2\, e^{-2\delta\tau}\, 5\delta^{-2}\right) .$$

Now we fix τ so that

$$5c_2\, e^{-2\delta\tau}\, \delta^{-2} < 1$$

so that

$$M(t) < 2\eta \quad \text{for} \quad \tau \leq t < \tau' .$$

Thus we can take $\tau' = \infty$ and have the estimate

$$|v_k(t)| \le 2\eta \quad \text{for all} \quad t \ge \tau ,$$

and all complex initial data in the polydisk (4.13).

Since $v_k(t)$ depends analytically on those initial data and converges on the real axis for $t \to +\infty$ the limit function $v_k(\infty)$ is analytic in (4.13). From the differential equation we obtain

$$|v_k(t) - v_k(\infty)| \le c_4 \, e^{-\delta t} \quad \text{for} \quad t \ge \tau$$

and

$$u_k(t)-(v_{k+1}(\infty)-v_k(\infty))t = u_k(0) + \int_0^t \left(v_{k+1}(t)-v_{k+1}(\infty)-v_k(t)+v_k(\infty)\right) dt$$

converges for $t \to +\infty$ with a uniform bound. Hence its limit γ_k is analytic in (4.13), completing the proof of (i).

To prove (ii) we use the representation (3.10) of the solutions by

$$a^2_{n-k}(t) = A_{n-k}(r,\lambda) \quad \text{with} \quad r_j = r_j(0)\, e^{-\lambda_j t}, \quad \lambda_j = \lambda_j(0) .$$

Here A_{n-k} is a rational function in r, λ, say,

$$A_{n-k} = \frac{P}{Q}$$

with P, Q being polynomials in r, λ. They are homogeneous in the r_j, both of the same degree, since A_{n-k} is of degree 0. To study its asymptotic behavior for $t \to +\infty$ we assume first that λ_{j+1}/λ_j is sufficiently large for all $j = 1,2,\ldots,n-1$. Then the dominant term in P is the one which comes first in lexicographical ordering of the exponents of r_j. Let $P_0 = \prod_{j=1}^{n} r_j^{p_j}$ be this term in P and $Q_0 \prod_{j=1}^{n} r_j^{q_j}$ the dominant term in Q. Then P_0, Q_0 are polynomials in λ and

$$A_{n-k} \sim \frac{P_0}{Q_0} \prod_{j=1}^{n} r_j^{(p_j-q_j)} .$$

Since on the other hand

$$a_{n-k}^2(t) \sim \text{const. } e^{-2(\lambda_{k+1}-\lambda_k)t}, \quad k=1,\dots,n-1,$$

we conclude that $p_k - q_k = -2$, $p_{k+1} - q_{k+1} = +2$, $p_j = q_j$ otherwise, and

$$A_{n-k} \sim \frac{P_0}{Q_0}\left(\frac{r_{k+1}}{r_k}\right)^2 \quad \text{for} \quad t \to +\infty .$$

Here the coefficient P_0/Q_0 is positive for $\lambda_1 < \lambda_2 < \dots < \lambda_n$. This proves the first line of (4.11) with

$$(c_{n-k}^+)^2 = \frac{P_0}{Q_0}$$

at least for large values of λ_{j+1}/λ_j. Since the coefficient is real analytic for all real λ in $\lambda_1 < \lambda_2 < \dots < \lambda_n$ this equation holds for all those λ.* The second equation of (4.11) follows in just the same manner.

This shows that the coefficient $C_k(\lambda)$ in (4.11') is independent of $r_j(0)$, and that $C_k^2(\lambda)$ is a rational function of λ. We will determine $C_k(\lambda)$ by induction on n. Incidentally, this will show that even $C_k(\lambda)$ is rational. For $n = 2$ the formula (4.7) or equivalently (4.3), (4.4) was verified. We prefer to prove the statement in the form (4.3). From our argument we know that

$$\beta_{n-k+1}^+ = \beta_k^- + \Phi_k(\alpha), \qquad k = 1,2,\dots,$$

where $\Phi_k(\alpha)$ is a real analytic function of $\alpha = (\alpha_1^-,\dots,\alpha_n^-)$, independent of β_j^-. This follows from the fact that $C_k(\lambda)$ is independent of $r_j(0)$, and depends on $\lambda_j = -\frac{1}{2}\alpha_j^-$ only. For n+1

* In general, the limit of such a rational function of exponentials may even be discontinuous, e.g. for

$$\frac{Ar_1r_3 + Br_2^2}{Cr_1r_3 + Dr_2^2} \to \frac{A}{C} \quad \text{for} \quad \frac{\lambda_2 - \lambda_1}{\lambda_3 - \lambda_2} > 1,$$

but

$$\to \frac{B}{D} \quad \text{for} \quad \frac{\lambda_2 - \lambda_1}{\lambda_3 - \lambda_2} < 1 .$$

particles we denote the corresponding function by $\Psi_k(\tilde{\alpha})$, $\tilde{\alpha} = (\alpha_1^-,\ldots,\alpha_{n+1}^-)$. The induction proof requires the verification of

$$\Psi_k(\tilde{\alpha}) = \Phi_k(\alpha) + \phi_{n+1,k} \quad \text{for} \quad k = 1,2,\ldots,n\,. \tag{4.14}$$

The determination of $\Psi_{n+1}(\tilde{\alpha})$ follows then from

$$\sum_{k=1}^{n+1} \Psi_k(\tilde{\alpha}) = 0$$

which is a consequence of the linear t-dependence of the center of mass. Thus it suffices to prove (4.14). Since both sides are independent of the β_j^- we may, and will, choose β_{n+1}^- very large positive, so that the n particles $x_1,x_2,\ldots,x_n$ have already undergone their mutual interaction and are very far apart by the time x_{n+1} interacts with any of them. In other words, when $x_{n+1}(t)$ exerts some force on $x_1,\ldots,x_n$ there are already close to

$$x_k \sim \alpha_k^+ t + \beta_k^+\,, \quad \text{where} \quad \beta_k^+ = \beta_{n-k+1}^- + \Phi_{n-k+1}$$

and t is large positive. Thus the interaction of x_{n+1} with $x_k(t)$, $k \leq n$ takes essentially place pairwise. Since before the interaction with x_{n+1} we have

$$x_{n-k+1} \sim \alpha_k^- t + \beta_k^- + \Phi_k\,, \qquad k = 1,2,\ldots,n,$$

we obtain after interaction

$$x_{n-k+1} \sim \alpha_k^- t + \beta_k^- + \Phi_k + \phi_{n+1,k}$$

which shows that $\Psi_k \sim \Phi_k + \phi_{n+1,k}$ if $\beta_{n+1}^- \to \infty$. But since Ψ_k, Φ_k are independent of β^- the assertions (4.14) follow. The situation is depicted in Figure 3.

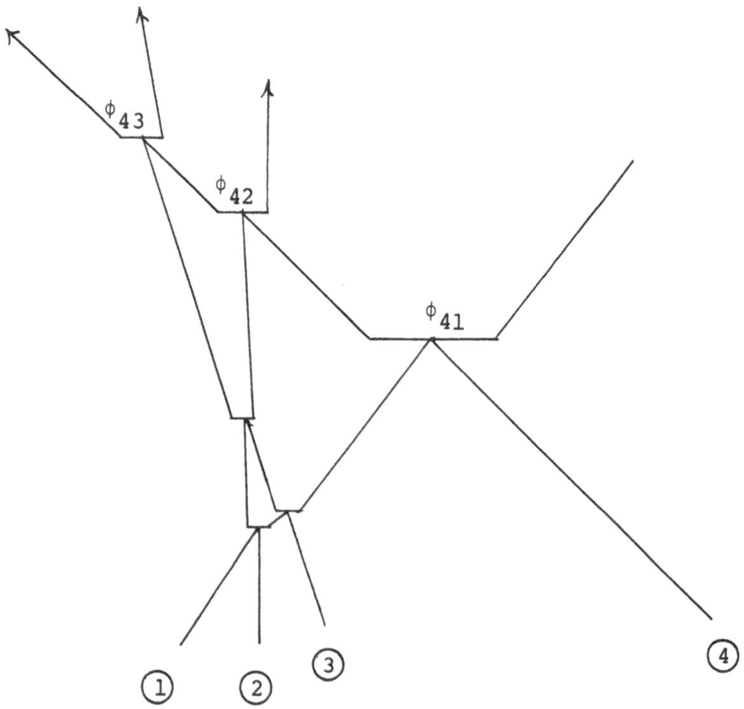

Figure 3

5. Associated Differential Equations

The above Hamiltonian system (1.2) possesses, according to the above, n integrals $I_1, I_2, \ldots, I_n$, which incidentally are polynomials in y_k and $e^{x_k - x_{k+1}}$. One may show, which we will not do here, that these integrals are in involution, i.e. the Poisson bracket for any two of these vanishes. Using the integrals as new Hamiltonians one can introduce n new vector fields which possess the same integrals and commute with each other. This makes the manifolds $I_k = \text{const.}$ into commutative groups. These are well known facts for integrable Hamiltonian systems (see, for example, Appendix 26 of [7]) which we will verify here directly.

As a starting point we take the differential equation (2.7) which represents a deformation of the Jacobi matrix L leaving the spectrum fixed. But there are many such isospectral deformations corresponding to different choices of B. We restrict ourselves to skew symmetric matrices B giving rise to orthogonal similarity transformations. But instead of permitting only one pair of off diagonals we allow several. Let B_p stand for a skew symmetric matrix with p off diagonals above and adjacent to the diagonal. Thus the matrix B defined below (2.7) would be denoted by B_1. We claim that for every p in $1 \leq p < n$ we can find nontrivial matrices B_p such that $\dot{L} = B_p L - L B_p$ defines a meaningful differential equation, that is that the commutator $B_p L - L B_p$ has only one off diagonal above the diagonal, while the others all vanish. We will establish this assertion below but point out first that all these differential equations have the eigenvalues of L as integrals and can therefore be transformed into the variables r_k, λ_k as $\dot{\lambda}_k = 0$, $\dot{r}_k = f_k(\lambda, r)$. For p = 2 one finds the matrix

$$B_2 = \begin{pmatrix} 0 & \beta_1 & \gamma_1 & & & \\ -\beta_1 & 0 & \beta_2 & \gamma_2 & & \\ -\gamma_1 & & & \ddots & & \\ & & \ddots & & \gamma_{n-2} & \\ & \ddots & & & \beta_{n-1} & \\ & & -\gamma_{n-2} & -\beta_{n-1} & 0 & 0 \end{pmatrix}$$

where

$$\text{(5.1)} \qquad \begin{aligned} \beta_k &= (b_k+b_{k+1})a_k , & k &= 1,2,\ldots,n-1 \\ \gamma_k &= a_k\, a_{k+1} , & k &= 1,2,\ldots,n-2 \end{aligned}$$

and the differential equation $\dot{L} = B_2L - L\,B_2$ takes the explicit form

$$\text{(5.2)} \qquad \begin{aligned} \dot{a}_k &= a_k(a_{k+1}^2 - a_{k-1}^2 + b_{k+1}^2 - b_k^2) , & k &\le n-1, \\ \dot{b}_k &= 2\, b_k(a_k^2 - a_{k-1}^2) + 2b_{k+1}a_k^2 - b_{k-1}a_{k-1}^2 , & k &\le n , \end{aligned}$$

where we set $a_0 = 0$, $a_n = 0$. Introducing r, λ again by the transformation (3.10) we find the differential equation

$$\text{(5.3)} \qquad \frac{d\lambda_k}{dt} = 0 , \qquad \frac{dr_k}{dt} = -\lambda_k^2 r_k .$$

We just indicate the calculation. First restricting r_k to $\sum r_k^2 = 1$ we have

$$\text{(5.4)} \qquad \frac{df}{dt} = \sum \frac{2r_k\dot{r}_k}{\lambda - \lambda_k}$$

On the other hand

$$\begin{aligned} \frac{df}{dt} &= (\dot{R}(\lambda)e_n, e_n) = ((B_2R - R\,B_2)e_n, e_n) \\ &= -2(R\,B_2e_n, e_n) = -2(R(\gamma_{n-2}e_{n-2} + \beta_{n-1}e_{n-1}), e_n) \\ &= -2(\gamma_{n-2}R_{n,n-2} + \beta_{n-1}R_{n,n-1}) . \end{aligned}$$

With (5.1) and

$$R_{n,n-1} = \frac{\Delta_{n-2}}{\Delta_n} a_{n-1} , \qquad R_{n,n-2} = \frac{\Delta_{n-3}}{\Delta_n} a_{n-2} a_{n-1}$$

we get

$$\frac{df}{dt} = -2 \frac{a_{n-1}^2}{\Delta_n} (a_{n-2}^2 \Delta_{n-3} + (b_{n-1} + b_n) \Delta_{n-2}) .$$

Using the recursion formulae (3.7) for k = n-1,n we find

$$\frac{df}{dt} = -2(\lambda^2 - b_n^2 - a_{n-1}^2) \frac{\Delta_{n-1}}{\Delta_n} .$$

Comparing the residue of these expressions at λ_k with those of (5.3) we find

$$\dot{r}_k = -(\lambda_k^2 - b_n^2 - a_{n-1}^2) r_k ,$$

or using r_k as homogeneous coordinates

$$\dot{r}_k = -\lambda_k^2 r_k .$$

Thus the solutions of (5.2) can be expressed as rational functions of λ_k and $e^{-\lambda_k^2 t}$ and the asymptotic behavior of its solutions is also completely understood from the results of the previous sect ons.

It is interesting to observe that the differential equations (5.2) possess $b_k = 0$, $k = 1,\dots,n$, as an invariant manifold on which they reduce to

(5.4) $$\dot{a}_k = a_k(a_{k+1}^2 - a_{k-1}^2) , \qquad k = 1,\dots,n-1,$$

which are the deformation equations for a Jacobi matrix L with a zero diagonal.*

To understand which of the solutions of (5.3) corresponds to (5.4) we consider again the continued fraction expansion (3.5), denoting the left-hand side by $f(\lambda,a,b)$. One easily verifies that

* These equations for $n = \infty$ were recently studied by M. Kac and van Moerbeke, according to a letter from M. Kac.

the involution $b_k \to -b_k$, $a_k \to a_k$ gives rise to

$$- f(-\lambda,a,-b) = f(\lambda,a,b) = \sum_{k=1}^{n} \frac{r_k^2}{\lambda-\lambda_k} .$$

Hence, since the eigenvalues λ_k are ordered according to size, the above involution corresponds to

$$\lambda_k \to -\lambda_{n-k+1} , \qquad r_k \to r_{n-k+1} .$$

The fixed points of this involution are the points (a,b) with $b_k = 0$ in the first representation and the points (λ,r) with

(5.5) $$\lambda_k + \lambda_{n-k+1} = 0 , \quad r_k = r_{n-k+1} .$$

This is evident also from the fact that the symmetric Jacobi matrices with zero diagonal have a spectrum symmetric with respect to the origin. Thus the solutions of (5.4) are given by precisely those rational functions in λ_j , $e^{-\lambda_j^2 t}$ for which the λ_j satisfy (5.5)

Using (2.1) it is easy to rewrite the system (5.2) in the variables x_k, y_k and one finds a Hamiltonian system

$$\dot{x}_k = \frac{\partial H_2}{\partial y_k} , \qquad \dot{y}_k = - \frac{\partial H_2}{\partial x_k} , \quad k = 1,2,\ldots,n,$$

with

$$H_2 = - \frac{1}{6} \sum_{k=1}^{n} y_k^3 - \frac{1}{2} \sum_{k=1}^{n} y_k (e^{x_{k-1}-x_k} + e^{x_k-x_{k+1}}) .$$

Again, in the above system one has to set $x_0 = -\infty$, $x_{n+1} = +\infty$. Although this system has no physical interpretation one has a full description of the scattering problem.

If one expresses the above Hamiltonian H_2 in terms of a,b one finds readily

$$H_2 = \frac{4}{3} \operatorname{tr} L^3 = \frac{4}{3} \sum_{k=1}^{n} \lambda_k^3 .$$

Since our original Hamiltonian (1.1) is given by

$$H = 2 \operatorname{tr} L^2 = 2 \sum_{k=1}^{n} \lambda_k^2 ,$$

one can expect that the further differential equations are associated with Hamiltonian proportional to $\operatorname{tr}(L^{p+1})$.

We will not follow this up but conclude with establishing the existence of the matrices B_p for $p = 1,2,\ldots,n-1$. It is convenient to write the matrices as difference operators. Let ξ stand for a double infinite sequence with components ξ_k (k integers) and let σ denote the shift operator

$$(\sigma\,\xi)_k = \xi_{k+1} .$$

We will assume that $\xi_k = 0$ if $k \leq 0$ or $k > n$ and write the matrix L in the form

$$L\,\xi = a(\sigma\,\xi) + b\,\xi + \sigma^{-1}(a\,\xi) .$$

Here a, b stand for sequences with components a_k and $(a\xi)_k = a_k \xi_k$. Thus $\sigma(a\xi) = \sigma(a)\cdot\sigma(\xi)$.

In this notation B_p will be presented by

$$(5.6) \quad B_p\xi = \gamma\sigma^p\xi + \ldots + \beta(\sigma^q\xi) + \ldots - \sigma^{-q}(\beta\,\xi) + \ldots - \sigma^{-p}(\gamma\,\xi) ,$$

where the qth order term indicates a typical term, $1 \leq q < p$.

The commutator $[B_p, L] = B_pL - L\,B_p$ contains σ, σ^{-1} to powers up to $p+1$. In fact, the highest order terms of this commutator are given by

$$[B_p, L] = \{\gamma\sigma^p(a) - a\sigma(\gamma)\}\,\sigma^{p+1} + \ldots$$

and we determine the $\gamma_1, \gamma_2, \ldots, \gamma_{n-1}$ so that

$$(5.7) \qquad (\gamma\sigma^p(a) - a\sigma(\gamma))\xi = 0$$

i.e.

$$\gamma_k a_{k+p} - a_k \gamma_{k+1} = 0 \ , \qquad k = 1,2,\dots n-p-1.$$

This can be satisfied by

$$\gamma_k = a_k a_{k+1} \cdots a_{k+p-1} \ , \qquad k = 1,2,\dots,n-p.$$

Now we proceed inductively and determine the coefficients β of σ^q in (5.6) to remove the terms of order $q+1$ in $[B_p,L]$, decreasing q from $q = p$ to $q = 1$. Analogously to (5.7) this gives an equation of the form

$$(\beta\sigma^q(a) - a\sigma(\beta))\xi = g\cdot\xi$$

where g is a given sequence. In components

$$\beta_k a_{k+q} - \beta_{k+1} a_k = g_k \ , \qquad k = 1,2,\dots n-q-1.$$

These are $n-q-1$ equations for $n-q$ unknowns. The solution is therefore not unique, but if β_1 is fixed arbitrarily, these equations can be solved recursively and uniquely, since $a_k > 0$.

Thus B_p can be so determined that in $[B_p,L]$ all coefficients of σ^{q+1} for $q = 1,2,\dots,p$ vanish. Since $[B_p,L]$ is symmetric it is a Jacobi matrix, giving rise to the desired differential equation

$$\dot{L} = [B_p,L] \ . \tag{5.8}$$

These are clearly the analogues of the higher order Korteweg-de Vries equations.

Finally, it is obvious that the multiplication

$$(r \otimes r)_k = r_k \tilde{r}_k$$

introduces a group structure into the manifolds λ_k = const. making the $n-1$ dimensional manifold of Jacobi matrices L with fixed spectrum into an Abelian group. This group action commutes with the vector field (2.2), and more generally with the vector fields (5.8) for p = 1,2,...n-1.

References

[1] Flaschka, H., The Toda Lattice, I, Phys. Rev. B 9, (1974) 1924-1925.

[2] Flaschka, H., The Toda Lattice, II, Prog. of Theor. Phys. 51 (1974) 703-716.

[3] Gantmacher, F. R., and Krein, M. G., Oszillationsmatrizen, Oszillationskerne und Kleine Schwingungen Mechanischer Systeme, Akad. Verlag, Berlin (1960). See, Anhang II.

[4] Henon, M., to appear in Phys. Rev. B 9, (1974) 1921-1923.

[5] Lax, P. D., Integrals of nonlinear equations of evolution and solitary waves, Comm. Pure Appl. Math. 21 (1968) 467-490.

[6] Gardner, C. S., Greene, J. M., Kruskal, M. D. and Miura, R.M., Korteweg-de Vries Equation and Generalizations VI, Methods for Exact Solutions, Comm. Pure Appl. Math. 27 (1974) 97-133.

[7] Arnold, V. I., and Arez, A., Problèmes Ergodiques de la Mécanique Classique, Gauthiers-Villars, Paris (1967).

[8] Toda, M., Wave propagation in anharmonic lattices, Jour. Phys. Soc. Japan 23 (1967) 501-506.

ON TRAVELING WAVE SOLUTIONS OF NONLINEAR DIFFUSION EQUATIONS

Charles C. Conley*

Department of Mathematics, University of Wisconsin, Madison 53706

Abstract

An existence proof for periodic traveling wave solutions of an equation of Nagumo is outlined. The proof begins with the analysis of a limiting case in which one set of dependent variables moves infinitely fast compared to the remainder. Singular "orbits" are defined for this limiting system and a perturbation argument using isolating blocks allows one to find actual solutions.

The problem discussed here concerns the existence of bounded solutions of a one parameter family of ordinary differential equations. The solutions correspond to traveling wave solutions of a diffusion equation; the parameter is the wave velocity.

The approach here relies on being able to assume that some of the dependent variables change very rapidly compared to the others. One first considers a limiting case where the "slow" variables don't change at all; this has the effect that one studies a lower dimensional system which now depends on more parameters.

Thus the limit case can be described as a flow on a fiber bundle the base space of which is the parameter (slow variable) space and the fiber of which is the fast variable space.

Away from the rest points of this flow, the orbits in the fiber provide a good approximation to solutions of the actual equation. On the rest point set itself, however, one must take account of the equations for the slow variables since their motion

*Sponsored by the U.S. Army under Contract No. DA-31-124-ARO-D-462.

dominates. Taking account of this infinitesimally slow flow on the rest point set, one pieces together curves in the total space near which solutions of the actual system might be found.

These curves are made up of complete orbits in the fiber, which connect rest points, together with pieces of orbits of the infinitesimally slow flow in the rest point set. In order to carry out the perturbation argument the orbits connecting the rest points must do so in some "nondegenerate" way.

Without giving a full grown "theory", these ideas will be illustrated here by a simple example. As a matter of fact, except for some trivial modifications it is the only example known to this author so perhaps a full grown theory isn't of much use at this point.

The example comes from one of the equations discussed by Nagumo, namely

$$u_t = \varepsilon v$$

$$v_t = v_{xx} + f(v) - u .$$

In these equations ε is to be a small positive constant so that u will play the role of the slow variable. The function f is well enough described by the graphs of f and its integral F depicted in Figure 1.

A traveling wave solution of the above equations means a solution which depends only on a variable $\xi = x + \theta t$; θ is called the wave velocity.

Such solutions correspond to solutions of the ordinary differential equations below in which the prime denotes $d/d\xi$.

$$\theta u' = \varepsilon v$$

$$\theta v' = v'' + f(v) - u.$$

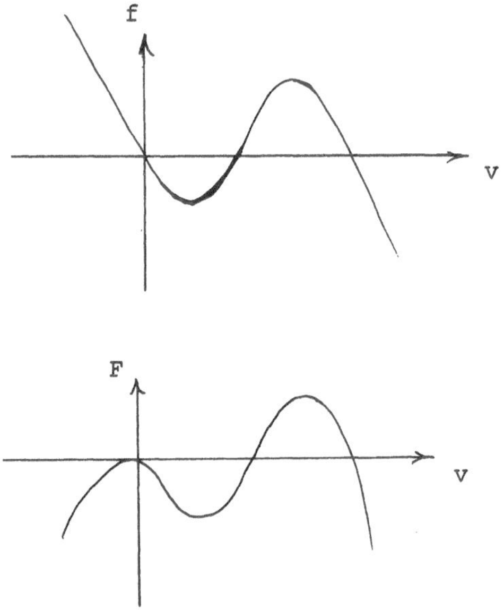

Figure 1

Introducing $\sigma = \varepsilon\,\theta^{-1}$ and $\eta = v'$, these equations can be written as a first order system:

$$u' = \sigma\, v$$

$$v' = \eta$$

$$\eta' = \theta\,\eta - f(v) + u\,.$$

One bounded solution of these equations is the rest point $u = v = \eta = 0$. If $\sigma \neq 0$ $(\varepsilon \neq 0)$ this is the only rest point. Thus the existence of other bounded solutions is not immediately apparent.

Pursuing the scheme outlined, assume now that $\sigma = 0$. The rest point set of this limit equation is $\{(u,v,\eta)\,|\,\eta = 0;\ u = f(v)\}$; the cubic curve in Figure 4 corresponds to this fixed point set. In that figure the vertical u-axis corresponds to the "base space" and the horizontal (v,η) planes to the fibers.

The limit equations will now be discussed for various values of θ. This amounts to describing the planar (v,η)-flow at various u levels. Those depicted are chosen for their later relevance.

The easiest case to treat is that where $\theta = 0$. In this case the equations for v and η form a Hamiltonian system with Hamiltonian function $H = \eta^2/2 + F(v) - uv$. The phase portraits are determined from the graph of F in the bottom picture of Figure 1. In Figure 2 these portraits are drawn for $u = 0$ and for $u = \hat{u}$ (> 0) where $\hat{u}$ is chosen so that the two (local) maxima of $F - uv$ are the same. Observe that in both portraits there are bounded orbits which begin and end at hyperbolic points.

If θ is positive, the system is modified by the addition of a "Negative friction" so that the Hamiltonian function now increases on orbits. One uses this fact to guess at the phase portrait from the aspect of the level curves of the Hamiltonian. For example, consider the case where θ is small and positive, and u is close to $\hat{u}$. It is then not hard to prove that there exist unique values $\bar{u}(\theta)$ and $\underline{u}(\theta)$ with $\bar{u}(\theta) > \hat{u} > \underline{u}(\theta)$ such that for these values of u, there are orbits which run from one hyperbolic rest point to the other. These orbits will play a crucial role. The phase portraits are indicated in Figure 3; the solid lines indicate levels of the Hamiltonian function, and the dashed lines the orbits.

Observe that as θ decreases to zero the values $\bar{u}(\theta)$ and $\underline{u}(\theta)$ go to $\hat{u}$ and the corresponding phase portraits go to the upper one in Figure 2. On the other hand, as θ increases, $\bar{u}(\theta)$ and $\underline{u}(\theta)$ spread apart, the former increasing and the latter decreasing until a value, θ^*, of θ is reached such that $\underline{u}(\theta) = 0$. It will be evident later why this value of θ is singled out.

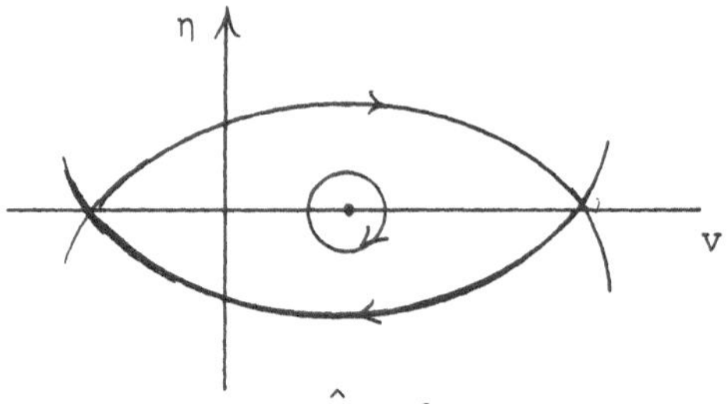

$u = \hat{u} > 0$

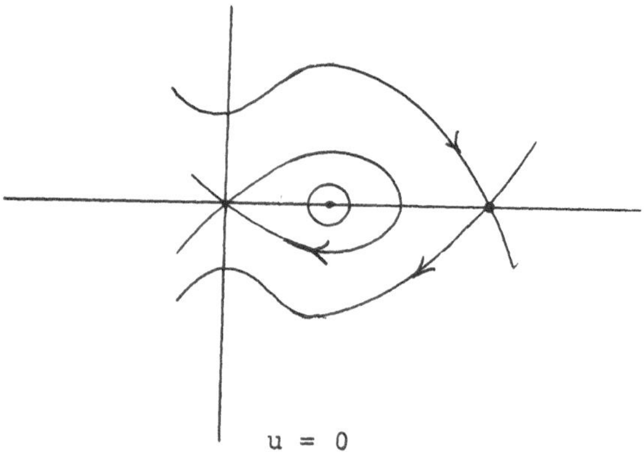

$u = 0$

Figure 2

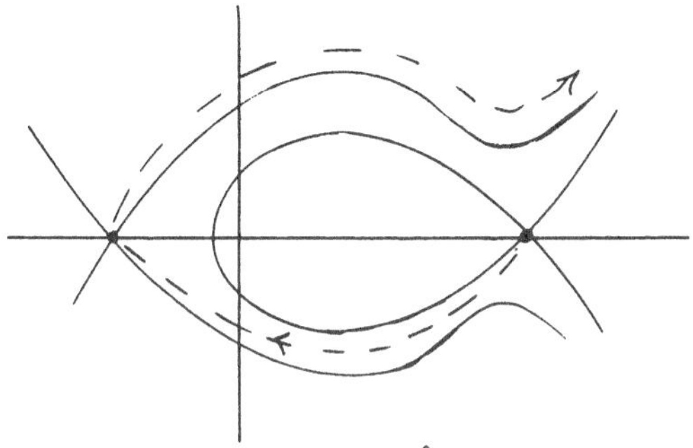

$u = \bar{u}(\theta) > \hat{u}$

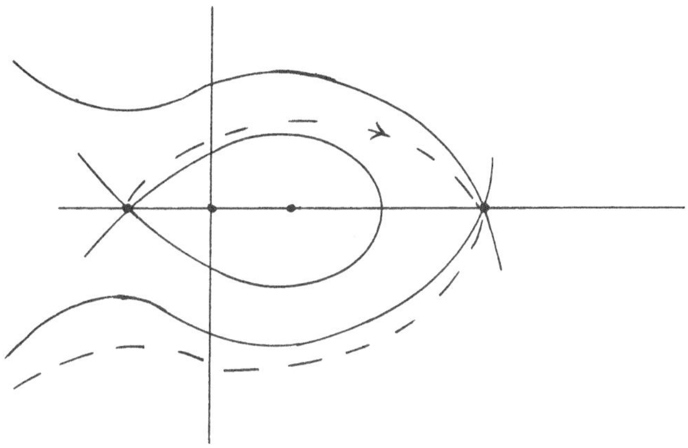

$u = \bar{u}(\theta) < \hat{u}$

<u>Figure 3</u>

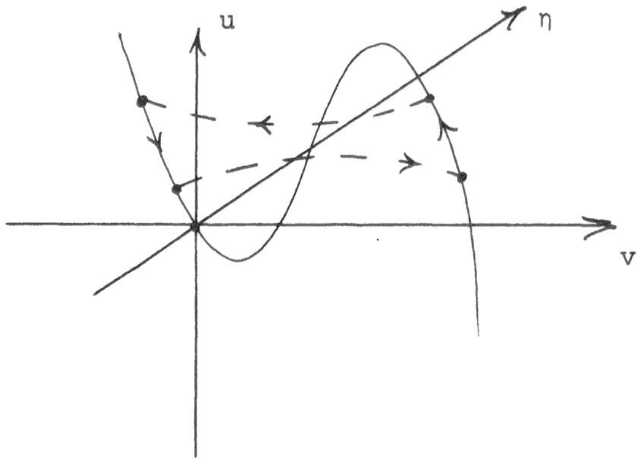

Figure 4. The positive η axis points into the paper.

It is now necessary to view the full three dimensional phase portrait. This is suggested in Figure 4 for a value of θ strictly between 0 and θ^*. In that figure the "cubic" curve indicates the rest point set (when $\sigma = 0$) and the two horizontal orbits are those which connect the hyperbolic points as indicated in Figure 3 (admittedly not much of the phase portrait is drawn, but it is the crucial part).

In this picture one sees a simple closed curve made up of the two "connecting" orbits together with two segments from the cubic curve of rest points. The contention is that for small positive σ, there ought to be a periodic orbit near this curve which follows around it in the direction indicated. Excepting the vertical arcs, the direction indicated is obviously correct (cf. Figure 3). On the arcs, one notes that the equation $\dot{u} = \sigma v$ $(\sigma > 0)$ implies that u must be increasing on the right-hand arc and decreasing on the left-hand arc as indicated. This also explains why the case $\underline{u}(\theta) < 0$ is not considered -- in that case one could not expect to go around the loop as seen from the directions of the infinitesimally slow flow on the rest point set.

The perturbation argument used to prove the existence of the periodic orbit requires also that the connections be in some sense nondegenerate. This will become clear in the following constructions.

Around the right hand vertical arc (or rather a slightly larger one) there is a tube with "square" cross section as indicated in Figure 5. If σ is slightly positive, orbits cross vertically upward through the bottom and top of this tube (since $\dot{u}$ is positive). Also, on viewing the typical cross section of the tube one sees that on the sides labeled 1 and 3, orbits are crossing into the tube while they cross out on the sides labeled 2 and 4. In particular, every boundary point of this tube is a point of strict entrance or strict exit in the sense of Wazewski -- the relevant consequences are that the mapping which assigns to each point its first point of exit is a continuous one as is the corresponding mapping for the backward flow. In particular the topological disk comprised of entrance points (bottom and sides 1 and 3) is carried homeomorphically by the flow to that comprised of exit points (top and sides 2 and 4).

Having understood Figure 5 it is easy to make a similar tube about the left hand vertical arc. It remains to deal with the horizontal orbits.

In order to understand how tubes are to be constructed about these orbits it is necessary to determine how the unstable manifolds of the hyperbolic points behave for values of u close to $\bar{u}(\theta)$ and $\underline{u}(\theta)$. This behavior is depicted in Figure 6 for values of u close to $\bar{u}(\theta)$ and an analogous picture holds for those near $\underline{u}(\theta)$. At this stage the "nondegeneracy" of the connection is visible.

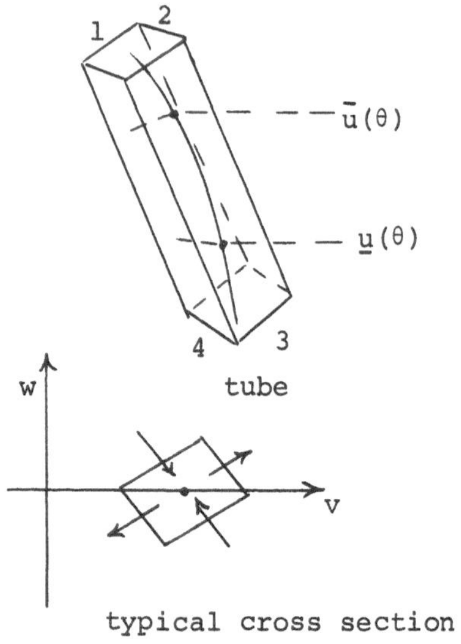

Figure 5

A more precise description is given in Figure 7. There is shown face 4 of the right-hand tube and in it a vertical arc consisting of points on the unstable manifold of the right hand hyperbolic points in each u-level.

The point at level $\bar{u}(\theta)$ is carried to an entering face of the left hand tube and the rest of the arc is carried so that it stretches across this face. By making the vertical tubes small enough one can construct a horizontal tube of orbits as depicted in Figure 8. With a slight modification one can assume that orbits are entering this tube on the sides 3, 3' and 1,1' and leaving it on the sides 2,2' and 4,4'.

This modification has been chosen so that every boundary point of the union of the three tubes is a point of strict entrance or strict exit. The only points that need to be checked are those that lie (also) in the boundary of two of the tubes, for example

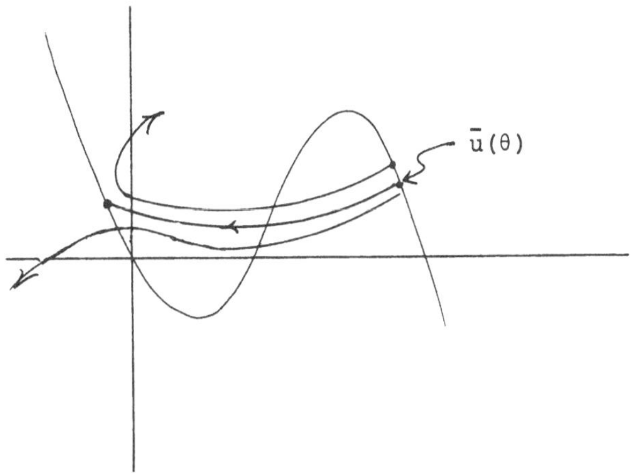

Figure 6

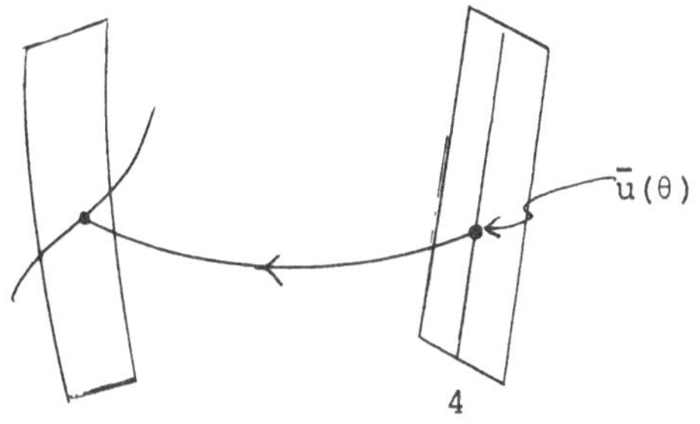

Figure 7

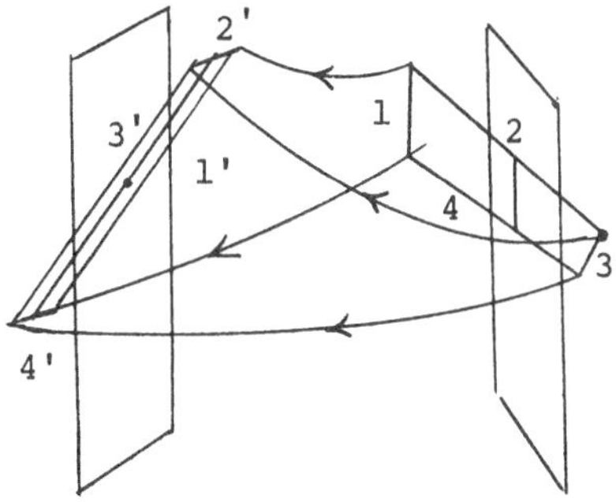

Figure 8

parts of edges 2, 4, 1' and 3' -- there are more.

One now completes the construction of a "cycle" of tubes by building another horizontal one about the other crossing orbit.

Having done this construction one can prove the existence (for small positive σ) of a periodic orbit which lies in this cycle of tubes in a variety of ways. For example: by construction, every boundary point of the union of the tubes is either a point of strict exit or a point of strict entrance (the union is a "block"). It can be verified with pictures that the set of exit points (as well as the set of entrance points) consists of two disjoint annuli. Thus, topologically, the picture is the same as that in Figure 9 where orbits enter on the top and bottom and exit on the cylindrical sides. Furthermore, all the orbits go around the cycle of tubes without "turning around" and there is a local surface of section like that indicated in Figure 9. One can then finish up with an elementary fixed point theorem.

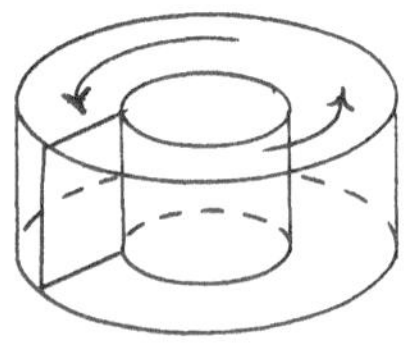

Figure 9

With reference to another article [1] in this collection one can see here "square mappings". The (four) squares can be taken to be the entering sets in each tube and the mappings are those from the entrance to the exit set. In the discussion here, no derivative conditions have been discussed; however, it is easy to conjecture that such conditions are satisfied and that the periodic orbit is the only bounded orbit in the cycle of tubes.

If θ is equal to θ^* the above construction cannot be carried out. In [2] a similar construction is described which shows that for small values of σ (> 0) there exists a value of θ near θ^* such that the equations admit an orbit which begins and ends in the rest point $u = v = \eta = 0$. The existence of such a traveling wave solution also has been proved for the Hodgkin-Huxley equation in [2] as well as in [3] by a somewhat different argument. In [4] the periodic orbits discussed here are treated and in [5] the homoclinic orbit ($\theta = \theta^*$).

A more difficult problem concerning traveling wave solutions of nonlinear diffusion equations concerns their stability as solutions of the partial differential equation. At least with the limited technical knowledge of this author, this question seems more difficult for the equation given than for that obtained by adding a small diffusion (smoothing) term to the u equation: thus

$$u_t = \delta u_{xx} + \varepsilon v$$
$$v_t = v_{xx} + f(v) - u .$$

However, this apparently leads to a more complicated ordinary differential equation for traveling waves. Fortunately the added complication is only apparent (at least if δ is small). The equations in this case can be written as:

$$u' = \xi$$
$$\xi' = \delta^{-1}(\theta\xi - \varepsilon v)$$
$$v' = w$$
$$w' = \theta w - f(v) + u .$$

If δ is small, the ξ variable moves much faster than the others except when $\theta\xi = \varepsilon v$. Treating the other variables as constant for the moment the flow in ξ space is as depicted in Figure 10 where the point is $\xi = \varepsilon\theta^{-1} v$.

Figure 10

One should now imagine that this flow is appended to each point of the cycle of tubes. Then for the full four dimensional flow, the region depicted in Figure 9 is replaced by one which is essentially the product of the region in Figure 9 with that in Figure 10.

The argument for the existence of the periodic orbit then follows just as easily as before. This consideration also carries over to the Hodgkin-Huxley equation.

References

[1] Conley, C., Hyperbolic sets and the shift automorphism, in this volume.

[2] Carpenter, G., Thesis, University of Wisconsin, Madison.

[3] Hastings, S., On traveling wave solutions of the Hodgkin-Huxley equations, (to appear).

[4] Hastings, S., The existence of periodic solutions to Nagumo's equation (to appear), Quarterly Jour. of Math., September.

[5] Hastings, S., The existence of homoclinic orbits for Nagumo's equation, (to appear).

THE EXISTENCE OF HETEROCLINIC ORBITS, AND APPLICATIONS

Charles C. Conley

University of Wisconsin, Madison, Wisconsin 53706

Joel A. Smoller *

University of Michigan, Ann Arbor, Michigan 48104

1. Introduction

Many questions arising in the physical and biological sciences, are concerned with constructing traveling wave solutions of a system of nonlinear partial differential equations. That is, given a system of partial differential equations of the form

(1) $$u_t + f(u)_x = (B(u)u_x)_x ,$$

$-\infty < x < \infty$, $t > 0$, we seek a solution of (1) of the form

(2) $$u = u(\xi) , \qquad \xi = x - ct ,$$

where c is a constant, either given beforehand, or to be determined as part of the problem. Here f is a smooth vector-valued function and $B(u)$ is a positive definite (or semidefinite) matrix-valued function. If we substitute (2) into (1), we are led to a system of ordinary differential equations of the following type:

(3) $$B\dot{u} = \Phi(u) , \qquad \dot{} = \frac{d}{d\xi} .$$

The problem we wish to discuss can now be stated. Suppose that $\Phi(u_0) = \Phi(u_1) = 0$; it is required to find a solution of (3) which connects u_0 and u_1. In our applications, one is given, in addi-

* Research of Conley supported by NSF Research Contract No. NSF GP-28267X2.
Research of Smoller supported by Air Force Office of Scientific Research, Contract Number AFOSR-71-2122.

tion, a "direction" of the orbit, say from u_0 to u_1 (as ξ increases), so that our problem is to integrate (3) subject to the boundary conditions

$$(4) \qquad \lim_{\xi \to -\infty} u(\xi) = u_0 , \qquad \lim_{\xi \to +\infty} u(\xi) = u_1 .$$

Thus, we require that the unstable manifold of u_0 intersects the stable manifold at u_1. If $u_0 \neq u_1$, such a solution is called a <u>heteroclinic orbit</u> while if $u_0 = u_1$, it is called a <u>homoclinic orbit</u>; we are interested here only in heteroclinic orbits.

A moment's reflection shows that the problem (3), (4) can be viewed in a topological framework. Indeed, our approach to the problem is via a modified Morse theory. Thus, given a compact manifold M and a function V(x) defined on M into itself, we consider the equation $\dot{x} = V(x)$. Assuming that V is gradient-like * with respect to a real-valued function, we have the usual Morse inequalities. That is, Morse theory provides us with information on the number of nondegenerate rest points and their indices.** Our method here is to extend the Morse index to more general invariant sets. We shall describe such a general technique which solves the above problem, [4]. Then we shall show how these methods are applied to a specific problem in shock-wave theory, the structure of magnetohydrodynamic shock waves, [5].

---

* See the eighth page of this article for definition.

** The Morse index of such a point is the dimension of the stable manifold at the point; i.e., the number of eigenvalues with negative real parts of the linearized equations at this point.

2. Mathematical Tools: The Index of an Isolated Invariant Set

Let $V(x)$ be a vector field defined in an open subset U of R^n, which defines a flow α; for brevity, we set $\alpha(x,t) = x\cdot t$.*
The main tool in our development is the notion of an isolating block for the flow defined by V, [3]. Thus, let $B \subset U$ be the closure of a bounded open set, and define a subset b^+ of the boundary of B by

$$b^+ = \{x \in \partial B: \exists\ \varepsilon > 0 \text{ with } x\cdot(-\varepsilon,0) \cap B = \phi\} .$$

Hence, if $x \in b^+$, the orbit through x leaves B for a short backwards time (Figure 1).

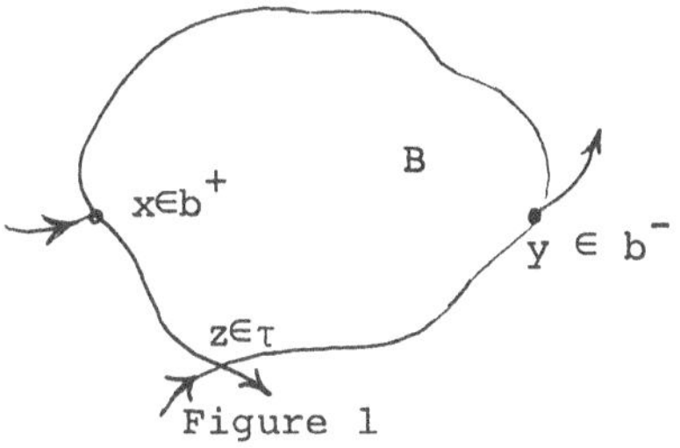

Figure 1

Similarly, let

$$b^- = \{x \in \partial B: \exists\ \varepsilon > 0 \text{ with } x\cdot(0,\varepsilon) \cap B = \phi\} ,$$

and

$$\tau = \{x \in \partial B: V \text{ is tangent to } B \text{ at } x\} .$$

We say that B is an <u>isolating block</u> for the flow if $b^+ \cap b^- = \tau$.
It follows at once that if B is an isolating block, then all the tangencies to B must be external. As a result, the mapping defined by the flow from the set of strict entrance points on ∂B to the set of strict exit points on ∂B, whenever it is defined, is

* $x\cdot t$ is the unique point in phase space lying on the solution curve through x which is t time units away from the initial point x.

continuous, [3]. (This is, perhaps, the key to the usefulness of the notion of an isolating block.) For B an isolating block, let I denote the maximal invariant set * of the flow contained in B. Intuitively, one sees that I is "isolated" from any other invariant set.

We proceed now to describe a somewhat more general approach to these notions. Let X be a compact metric space and consider the complete metric space of continuous functions on $X \times R$ into X with the compact-open topology (uniform convergence on compact subsets). Let F denote the subspace of flows; that is, those functions $\alpha: X \times R \to X$ where $\alpha(x,0) = x$, $\alpha(x,s+t) = \alpha(\alpha(x,s),t)$; we again denote $\alpha(x,t)$ by $x \cdot t$. A compact subset $N \subset X$ is an <u>isolating neighborhood</u> of the flow if whenever $x \in \partial N$, then the orbit through x, $x \cdot R$, is not contained in N. Thus, we see that any invariant set in N lies wholly in the interior of N and doesn't meet the boundary of N. Note that an isolating block is an isolating neighborhood.

Let $\Phi(N) \subset F$ be the set of flows on $X \times R$ for which N is an isolating neighborhood. It is easy to see that $\Phi(N)$ is open in F (if all orbits of $f \in \Phi(N)$ through ∂N leave N, the same is true for nearby flows).

For $f \in \Phi(N)$, let $S(N,f)$ be the maximal invariant set of f contained in N. Then $S(N,f)$ is a compact invariant set contained in the interior of N; such sets are called <u>isolated invariant sets</u>, [3]. We pause here to give some examples. First, consider in the plane, a saddle point p for a flow (Figure 2a). It is easy to see that any sufficiently small disk about p is an isolating neighborhood, and that p is an isolated invariant set. On the

* An invariant set I is a set that gets mapped into itself by the flow; i.e., $I \cdot R = R$.

other hand, if q is a center, in the plane, then it is clear that q is not an isolated invariant set since one cannot find an isolating neighborhood for q (Figure 2b).

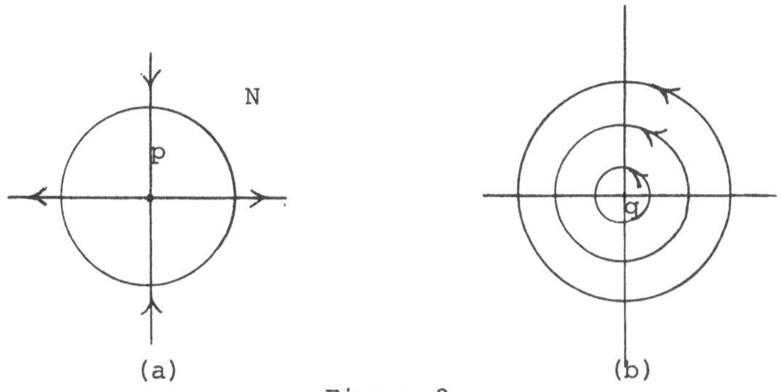

Figure 2

The connection between our two ways of describing isolated invariant sets is given by the following theorem, [3].

Theorem (Conley-Easton). Every isolated invariant set can be realized as the maximal invariant set contained in an isolating block.

We wish now to discuss a notion of "continuation" of isolated invariant sets, with a view to later applications. Thus, let Λ be any connected topological space. By a "family of flows parametrized by Λ," we mean a continuous mapping $\lambda \to f_\lambda$ from Λ into F. Suppose that to each $\lambda \in \Lambda$, there corresponds an isolated invariant set S_λ of f_λ; we say that the sets S_λ are <u>related by continuation</u> if for each $\lambda \in \Lambda$, there exists a compact subset $N_\lambda \in X$ and a neighborhood $U_\lambda \in \Lambda$ of λ such that if $\mu \in U_\lambda$

then N_λ is an isolating neighborhood relative to the flow f_μ, and $S_\mu = S(N_\lambda, f_\mu)$.* We see that two isolated invariant sets are related by continuation if we can go from one to another by a finite sequence $S_{\mu_{i+1}} = S(N_{\mu_i}, F_{\mu_{i+1}})$; this is reminiscent of the usual proofs of the monodromy theorem in complex analysis (draw a picture!).

Our aim is to construct an index for an isolated invariant set which is invariant under continuation. To do this we proceed as follows. For S an isolated invariant set, we use the Conley-Easton theorem to construct an isolating neighborhood about S in the form of an isolating block B. We then define the (homotopy) index $h(S)$ of S as the equivalence class of pairs of pointed spaces

$$h(S) = (B/b^+, B/b^-) \;;$$

the equivalence relation being homotopy equivalence. Of course by B/b^+ we mean the usual quotient space with the quotient topology; that is, the space obtained from B by collapsing b^+ to a point (see any standard topology text). The word "pointed," used in this context is understood to mean that we consider the space B/b^+ together with the distinguished point b^+. B/b^- is defined similarly. We define $h(S) = 0$ if $h(S)$ is the homotopy type of a pair of (pointed) points.

It may strike the reader a bit unusual to define an index as a pair of equivalence classes of topological spaces rather than as

---

* A more general definition, [9], is the following. Let $\mathcal{S} = \{(S,f) : S \text{ is an isolated invariant set for } f \in \mathcal{F}\}$, and for $N \subset X$, let $\sigma_N : \Phi(N) \to \mathcal{S}$ be defined by $\sigma_N = (S(N,f), f)$. We topologize $\mathcal{S}$ by the sets $\sigma_N(V)$, where V is open in $\Phi(N)$, N compact. Then the pairs (S,f) lying in the same component of $\mathcal{S}$ are said to be "related by continuation."

a real number, as is done in the classical cases (e.g. Poincaré or Morse index). However, the connection is given by the following property.

<u>Property 1</u> If x_0 is a nondegenerate rest point for the system $\dot{x} = V(x)$, having Morse index k, then $h(x_0) = (S^k, S^{n-k})$, where S^j denotes a j-dimensional sphere.

Thus, this index carries the same information as the Morse index, whenever the latter is defined.

<u>Property 2</u> $h(S)$ is independent of the isolating block B.

By this we mean that if S is the maximal invariant set contained in two isolating blocks, then the index computed using the first block agrees with that computed from the second block.

Before stating the next property we need a definition. The "wedge" $X \vee Y$ of two pointed spaces (X,x_0), (Y,y_0) is the pointed space obtained by glueing X to Y at their distinguished points (Figure 3). The topology on $X \vee Y$ is the natural one;

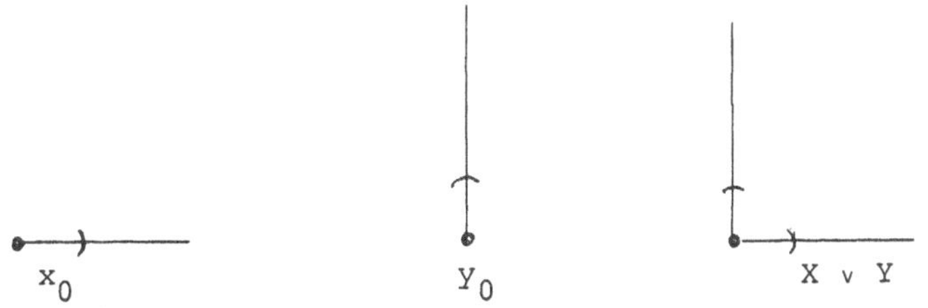

Figure 3

a neighborhood of the distinguished point is made up of unions of neighborhoods of x_0 and y_0 , (Figure 3). Observe that if X is not a point, then $X \vee Y$ is not a point for any pointed space Y. We then have the following "addition" property.

Property 3. If B_1 and B_2 are disjoint isolating blocks with corresponding isolated invariant sets S_1 and S_2 , then $B = B_1 \vee B_2$ is an isolating block with isolated invariant set $S = S_1 \cup S_2$ and $h(S) = h(S_1) \vee h(S_2)$, in the sense that $B/b^{\pm} = B_1/b_1^{\pm} \vee B_2/b_2^{\pm}$.

Property 4. $h(S)$ is invariant under continuation.

The proof of this proposition follows from the remark made earlier that $\Phi(N)$ is open; the proofs of all the above properties are not difficult and can be found in [1], [2], and [9].

Recall that a vector field V is called gradient-like with respect to a real-valued function P if $\langle V, \nabla P\rangle > 0$ except at the rest points of V. That is P increases along (nonconstant) orbits of V; thus P is like a Liapunov function. Observe that gradient-like vector fields cannot be too pathological (and in fact cannot even admit periodic orbits).

Lemma Let V be a gradient-like vector field with respect to P in a region U. Let $B \subset U$ be an isolating block for V containing exactly two nondegenerate rest points x_0, x_1 at least one of which has nonzero index. Suppose that $h(S) = 0$, where S is the isolated invariant set in B. Then there exists an orbit of V connecting x_0 and x_1.

Before giving the proof of this lemma, we remark that it provides the solution for proving the existence of heteroclinic orbits; in particular, it answers a question of Gelfand [7, p. 299].

Proof: It suffices to show that B contains a third orbit γ

different from x_0 and x_1. Since P has a maximum (respectively, minimum) on γ, this maximum (respectively, minimum) must be at an "endpoint" of γ, so that γ connects x_0 to x_1. If no such third orbit exists, then $S = \{x_0\} \vee \{x_1\}$ is the isolated invariant set in B so that $h(S) = h(x_0) \vee h(x_1) = 0$. However this last equation for pointed spaces requires both of $h(x_0)$ and $h(x_1)$ to be zero, contradicting the hypothesis.

3. An Application

We shall consider the so-called structure problem for magneto-hydrodynamic (mhd) shock waves; for other applications, see [4]. The problem is formulated as follows (see [8], and [9]). Let $u = (x_1,x_2,y_1,y_2,V,T) \equiv (x,y,V,T)$ be a point in R^6. Here x_1 and y_1 are mechanical variables (velocity components), x_2 and y_2 are electromagnetic variables (magnetic field components) and V and T denote thermodynamic variables, the specific volume (reciprocal of the density) and temperature, respectively. Let A denote the matrix

$$A = \begin{pmatrix} 1 & -\delta \\ -\delta & V \end{pmatrix}$$

and let Q and P be the real valued functions defined by

$$Q = \frac{1}{2} \langle x,Ax\rangle + \frac{1}{2} \langle y,Ay\rangle + \varepsilon x_2 + \frac{1}{2} V^2 - JV + E - f(V,T) ,$$

$$P = T^{-1} Q ,$$

where δ, ε, J and E are constants. $f(V,T)$ is the Helmholtz free energy function satisfying $f_V = -p$, $f_T = -s$, where p and s denote the pressure and entropy, respectively. Moreover, in terms of these quantities, the internal energy e is given by $e = f + Ts$.

We define the "viscosity" vector λ by $\lambda = (\mu,\nu,\mu_1,k)$, where μ, ν and μ_1 denote the ordinary, second, and magnetic viscosities, respectively, and k is the thermal conductivity. We assume here that all of these "viscosity parameters" are positive.* We define the "viscosity" matrix B by

$$B = B(u,\lambda) = T^{-1} \operatorname{diag} \{\mu,\nu,\mu,\nu,\mu_1,kT^{-1}\} .$$

In terms of this notation, Germain [8] has shown that the equations for the structure of mhd shocks take the form

$$B\,\dot{u} = \nabla P . \tag{5}$$

In [8], Germain also showed that this system admits (at most!) four rest points u_0,u_1,u_2,u_3, all nondegenerate, having Morse indices 0,1,2,3 respectively, and that $P(u_i) < P(u_{i+1})$, $i = 0,1,2$.

The problem is to show that there is an orbit of (5) running from u_0 to u_1, and likewise one from u_2 to u_3. The transition $u_0 \to u_1$ corresponds to the so-called "slow" mhd shock, and that of $u_2 \to u_3$ to the "fast" mhd shock. If such orbits exist, we say that the fast and slow mhd shocks <u>have structure</u>. We shall prove this using the lemma of the last section. Note first that if we differentiate P along orbits of (5), we get

$$\dot{P} = \langle\nabla P,\dot{u}\rangle = \langle\nabla P,B^{-1}\,\nabla P\rangle > 0$$

in $V,T > 0$, except at critical points of P; thus we have

<u>Fact 1</u> The system (5) is gradient-like in $V,T > 0$.

<u>Hypotheses</u>.

(H_a) p,s and e are positive in $V,T > 0$.

(H_b) For fixed $T > 0$, $p(V,T) \to \infty$ as $V \to 0$.

(H_c) Given $V_0,k > 0$, there exists a $T_0 > 0$ such that if $0 < V \le V_0$ and $T \ge T_0$, then $e(V,T) > k$.

(H_d) For $0 < V \le V_0$, $s(V,T) \to 0$ uniformly in V as $T \to 0$.

(H_e) If $p = p(V,s)$, then $p_V < 0$, $p_{VV} > 0$, $p_s > 0$.

* See [6] for the case where we allow some components of λ to equal zero.

Hypothesis (H_d) is called Nernst's third law of thermodynamics. Since there appears to be some controversy about it, we can replace (H_d) by

$(H_{d'})$ If $s = s(V,T)$, then $s_V > 0$, $s_T > 0$ and for <u>fixed</u> V, $s(V,T)$ has a limit independent V as $T \to 0$.

We rewrite (5) as

$$B_0\dot{x} = Ax + \bar{\varepsilon}\,, \qquad B_0\dot{y} = Ay$$

$$(6) \qquad \mu_1\dot{V} = \frac{1}{2}(x_2^2 + y_2^2) + V - J + p(V,T)$$

$$k\dot{T} = -Q + Ts = -(Q+f) + e\,,$$

where $B_0 = \mathrm{diag}\,\{\mu,\nu\}$ and $\bar{\varepsilon} = (0,\varepsilon)^t$. Using (6), we find (see [5]), that at the rest points, the following equations must hold

$$\dot{V} = F_1(V,T) \equiv \frac{1}{2}\varepsilon^2(V-\delta^2)^{-2} + V - J + p(V,T) = 0$$

$$\dot{T} = F_2(V,T) \equiv \frac{1}{2}\varepsilon^2(V-\delta^2)^{-1} - \frac{1}{2}V^2 + JV - E + e(V,T) = 0.$$

Since $\partial F_1/\partial T = p_T > 0$ and $\partial F_2/\partial T = e_T > 0$, we see that $F_1 = 0$ and $F_2 = 0$ are both graphs of functions and using our hypotheses, we can depict them as in Figure 4.

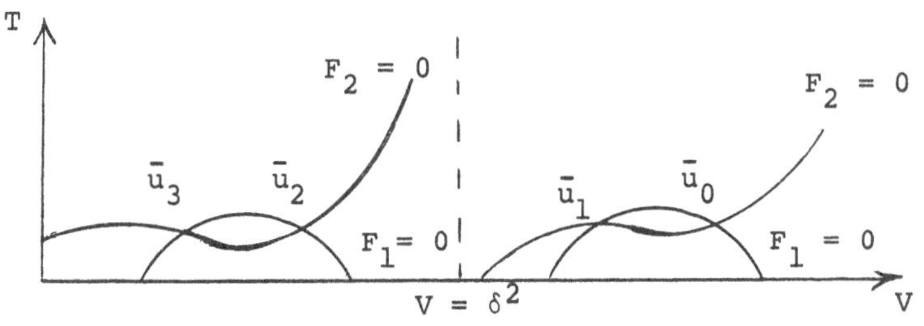

Figure 4 ($\bar{u}_i$ is the projection of u_i onto V-T space).

We remark that as $E \to +\infty$, the curve $F_2(V,T) = 0$ pulls off of the curve $F_1(V,T) = 0$, so that we have:

Fact 2 For sufficiently large E, the system (5) has no rest points.

This fact will be used later to compute the index of our isolated invariant sets.

We now fix the parameter values λ, E and let $S' = S'(\lambda,E)$ denote the set of points lying on complete bounded orbits of (5) lying in $V,T > 0$, corresponding to λ, E.

Lemma. $S'(\lambda,E)$ is bounded for fixed λ, E.

For example, $V \leq J$ follows from the third equation in (6). For, if V were ever larger than J, then $\dot{V} > 0$ at such a point so that the orbit would escape to infinity and not lie in S'. We can also show that $\|x\| + \|y\|$ is bounded in S' (see [5]) and from this it follows that T is bounded on S'. To see this, we use the equation $T = -(Q + f) + e$ and note that $Q + f$ is bounded on S'. But then (H_c) shows that $T \leq$ constant on S'.

Next, let C denote the set of rest points of (5). If $u \in C$, then $\dot{T}(u) = -Q + Ts = 0$ so that at u, $s = T^{-1}Q = P$. Using (H_d), there is a neighborhood U of the interval $[0,T]$ in V-T space and a real number k such that

$$\min s|_C > k > \max s|_U .$$

Let $S = S(\lambda,E) = S'(\lambda,E) \cap \{u\colon P(u) > k\}$. Clearly S is bounded, and it is not too hard to show that S is closed. This is done by showing that S has no limit point on $V = 0$ or $T = 0$ or $P = k$, (see [5] for details).

Thus $S(\lambda,E)$ is an isolated invariant set for each fixed set of parameter values (λ,E). Choose a constant c such that

$P(u_1) < c < P(u_2)$, and set

$$S_{01} = S_{01}(\lambda,E) = S(\lambda,E) \cap \{u: P(u) < c\}$$

$$S_{23} = S_{23}(\lambda,E) = S(\lambda,E) \cap \{u: P(u) > c\} .$$

It is easy to check that S_{01} and S_{23} are disjoint isolated invariant sets. Hence we have:

Fact 3 The pair of nondegenerate rest points $\{u_0,u_1\}$, (respectively $\{u_2,u_3\}$) are contained in an isolating block B_{01} (respectively B_{23}). This isolating block contains no other rest points of (5).

We claim that for fixed λ, the sets $S_{01}(\lambda,E)$ have the same index, independent of E. To see this, note that an isolating block for $S_{01}(\lambda,E)$ is also an isolating block for $S_{01}(\lambda,E\pm\varepsilon)$ if $\varepsilon > 0$ is sufficiently small (recall from Section 2 that $\Phi(N)$ is open in F). Consequently

$$h(S_{01}(\lambda,E\pm\varepsilon)) = h(S_{01}(\lambda,E)) ,$$

and it follows that for any E_1 and E_2 , $h(S_{01}(\lambda,E_1)) = h(S_{01}(\lambda,E_2))$. Using Fact 2, we see that $h(S_{01}(\lambda,E)) = 0$ for all sufficiently large E. Since a similar argument holds for $S_{23}(\lambda,E)$ we have:

Fact 4 $h(S_{01}(\lambda,E)) = h(S_{23}(\lambda,E)) = 0$.

But from Facts 1, 3, and 4, together with our lemma in Section 2, we see that there is an orbit of (5) running from u_0 to u_1 and likewise one running from u_2 to u_3. Thus we have proved:

Theorem. Fast and slow mhd shocks possess structure.

References

[1] Churchill, R. C., Isolated invariant sets in compact metric spaces, J. Diff. Equa. 12, (1972), 330-352.

[2] Conley, C. C., A generalization of the Morse index, Proc. NRL-MRC Conf., ed. by L. Weiss, Academic Press, N.Y., 1971.

[3] Conley, C. C., and R. Easton, Isolated invariant sets and isolating blocks, Trans. Amer. Math. Soc., 158 (1971), 35-61.

[4] Conley, C. C., and J. A. Smoller, Topological methods in the theory of shock waves, Partial Differential Equations, Proc. Symposia in Pure Math., XXIII, Amer. Math. Soc., Providence, (1973).

[5] Conley, C. C., and J. A. Smoller, The structure of magneto-hydrodynamic shock waves, Comm. Pure Appl. Math., 27 (1974), 367-375.

[6] Conley, C. C., and J. A. Smoller, The structure of magneto-hydrodynamic shock waves, II, to appear.

[7] Gelfand, I. M., Some problems in the theory of quasilinear equations, Usp. Mat. Nauk, 14 (1959), 87-158. Eng. Transl. in Amer. Math Soc. Transl. Ser. 2, No. 29 (1963), 295-381.

[8] Germain, P., Contribution à la théorie des ondes de choc en magnetodynamique des fluides, O.N.E.R.A. Publ. No. 97 (1959), Paris.

[9] Kulikovsky, A. G., and G. A. Liubimov, Magnetohydrodynamics, Addison-Wesley, Reading, Mass., 1965.

[10] Montgomery, J. T. Cohomology of isolated invariant sets under perturbation, J. Diff. Equations, 13 (1973), 257-299.

HADAMARD'S GENERALIZATION OF HYPERBOLICITY, WITH APPLICATIONS TO THE HOPF BIFURCATION PROBLEM

Neil Fenichel

Department of Mathematics, University of British Columbia
Vancouver, B. C.

1. Introduction

The simplest example of a dynamical system exhibiting hyperbolic structure is a smooth map $F: R^2 \to R^2$ near a hyperbolic fixed point. In suitable coordinates F takes the form

$$x_1 = \lambda x + \dots , \qquad\qquad y_1 = \mu y + \dots ,$$

where the dots denote terms vanishing to first order, and the hyperbolicity condition is $0 < |\lambda| < 1 < |\mu|$.

The hyperbolicity condition may be expressed in terms of the action of F on squares, as shown in Figure 1. F maps a square in the x-y plane to a rectangle which is elongated in the direction of the y_1 axis and compressed in the direction of the x_1 axis. The hyperbolicity condition also may be expressed in terms of the action of F on cones, as shown in Figures 2 and 3. F maps a cone around the y axis to a longer cone of smaller aperture around the y_1 axis, and F^{-1} maps a cone around the x_1 axis to a longer cone of smaller aperture around the x axis.

The unstable manifold of the origin is defined as the set of points whose orbit under F^{-1} tends to the origin:

$$W^u = \{p \in R^2 : F^{-k}(p) \to 0 \text{ as } k \to \infty\}.$$

It is well known that under our hyperbolicity conditions W^u is a smooth curve tangent to the y-axis. We mention three iteration procedures for construction of this curve. First, let S be any square containing the origin, as shown in Figure 1. The sequence

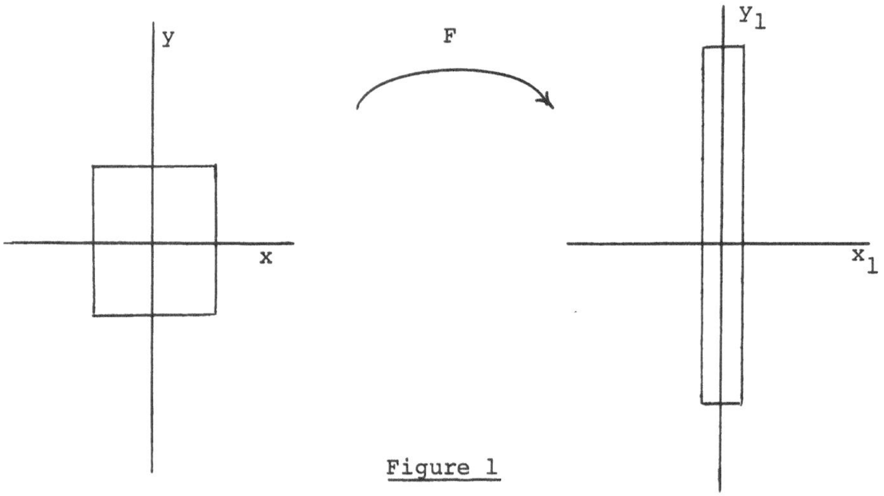

Figure 1

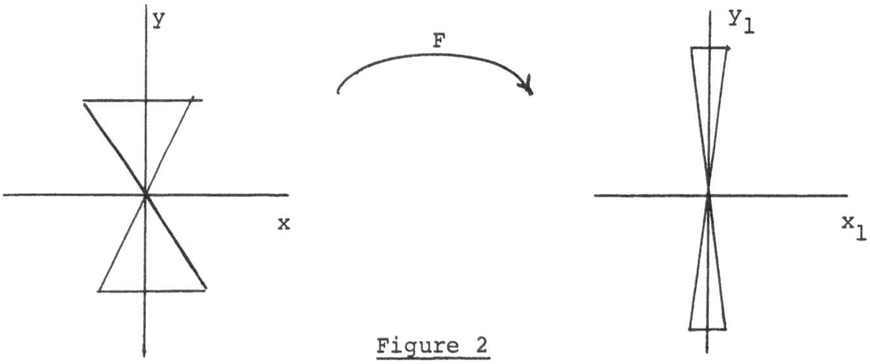

Figure 2

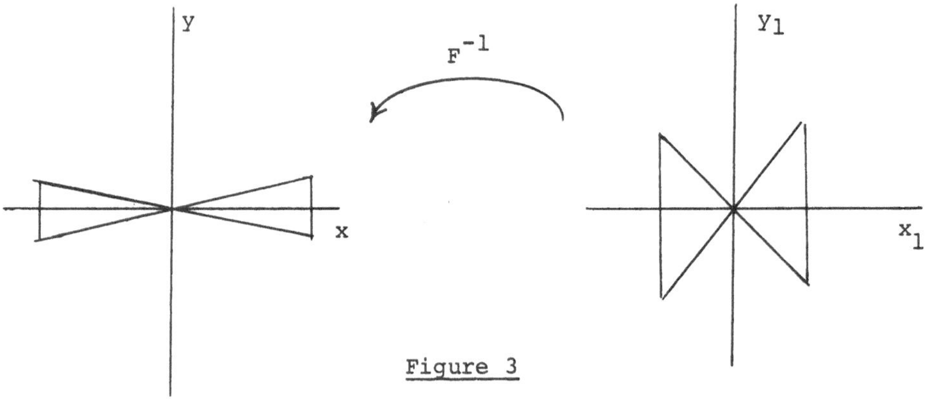

Figure 3

of images, S, $F(S)$, $F^2(S)$, ..., converges to W^u. Second, let C be any cone around the y axis, as shown in Figure 2. The sequence of images C, $F(C)$, $F^2(C)$,..., converges to W^u. Third, let Γ be any smooth curve through the origin, not tangent to the x axis. Then the sequence Γ, $F(\Gamma)$, $F^2(\Gamma)$,..., also converges to W^u. Convergence must be interpreted suitably for each construction.

More than seventy years ago, Hadamard noted that the unstable manifold construction uses only the single cone condition, Figure 2, rather than the double cone condition, Figures 2 and 3. In terms of of the eigenvalues, Hadamard required that $0 < |\lambda| < |\mu|$ and $1 < |\mu|$, rather than $0 < |\lambda| < 1$ and $1 < |\mu|$. Under these hypotheses the rectangles S, $F(S)$, $F^2(S)$, ..., may not converge to a curve. If $|\lambda| > 1$, the widths of successive rectangles actually increase. The cones C, $F(C)$, $F^2(C)$, ..., and the curves Γ, $F(\Gamma)$, $F^2(\Gamma)$,..., do converge, however, to a curve W^E tangent to the y axis. Convergence of these sequences is essentially equivalent to our single cone condition. Thus we have the following theorem of Hadamard.

Theorem 1. Let $F: R^2 \to R^2$ be a diffeomorphism leaving the origin fixed. Let λ and μ be the eigenvalues of the Jacobian matrix, and suppose $|\lambda| < |\mu|$ and $1 < |\mu|$. Then there is a unique curve W^E through the origin, invariant in the sense that $F(W^E)=W^E$, whose tangent at the origin is not an eigenvector for the eigenvalue λ. The tangent of W^E at the origin is an eigenvector for the eigenvalue μ.

2. Remarks.

(1) W^E admits a metric characterization. Let d denote the usual Euclidean distance. Then

$$W^E = \{p \in R^2 : d(F^{-k}(p),0)/\alpha^k \to 0 \text{ as } k \to \infty \text{ for all } \alpha > |\mu|^{-1}\}.$$

(2) Suppose the Jacobian matrix $DF(0)$ is not diagonal, but has the lower triangular form

$$\begin{bmatrix} \lambda & 0 \\ b & \mu \end{bmatrix} .$$

Then the y axis is invariant, so the cone construction for W^E remains essentially unchanged.

(3) Suppose the Jacobian matrix $DF(0)$ is not diagonal, but has the upper triangular form

$$\begin{bmatrix} \lambda & a \\ 0 & \mu \end{bmatrix}$$

Then the x axis is invariant, but if $a \neq 0$ the y axis is not invariant. There is, however, a unique invariant line $\ell: x = my$ which is transversal to the x axis. By a linear change of coordinates we may assume that ℓ is, in fact, the y axis. To see this, note that ℓ is invariant if and only if $DF(0)$ takes any point of the form $\begin{pmatrix} my \\ y \end{pmatrix}$ to a point of the form $\begin{pmatrix} my_1 \\ y_1 \end{pmatrix}$. Equivalently,

$$\lambda my + ay = my_1 , \qquad \mu y = y_1 ,$$

or

$$\lambda m + a = m\mu , \qquad m = a/(\mu-\lambda) .$$

(4) It is convenient to repeat the last construction in a form which generalizes to mappings of R^n, n arbitrary, and which also generalizes to mappings of vector bundles. Suppose λ and μ are invertible square matrices, and every eigenvalue of μ is larger, in modulus, than all eigenvalues of λ. The functional equation for m,

$$\lambda m + a = m\mu ,$$

takes the form

$$m = a\mu^{-1} + \lambda m\mu^{-1} .$$

By repeated substitution of $a\mu^{-1} + \lambda m\mu^{-1}$ for m in the right-hand side of this equation, we find that

$$\begin{aligned} m &= a\mu^{-1} + \lambda m\mu^{-1} \\ &= a\mu^{-1} + \lambda a\mu^{-2} + \lambda^2 m\mu^{-2} \\ &= a\mu^{-1} + \lambda a\mu^{-2} + \lambda^2 a\mu^{-3} + \lambda^3 m\mu^{-3} \\ &\vdots \\ &= \sum_{j=0}^{\infty} \lambda^j a\mu^{-(j+1)} . \end{aligned}$$

We have solved the functional equation for m by summing a geometric series. This is the model for all the contraction mapping proofs we will omit below. Complete proofs are in [1].

(5) Our construction of an invariant complement can be used to prove the following comment made by Ruelle at the Rencontres. Let X be a vector field on a compact manifold M, and let $\phi: M \to (0,\infty)$ be a differentiable function. Let F^t be the flow generated by X, and let G^t be the flow generated by ϕX. Then if F^t is an Anosov flow, G^t also is an Anosov flow.

Let D denote partial differentiation with respect to x, so that DF^t is the Jacobian matrix of the map $x \longmapsto F^t(x)$. We call a subbundle B of the tangent space TM invariant if B is invariant under DF^t for all t. To say that F^t is an Anosov flow means that TM splits as $E^u \oplus (X) \oplus E^s$, where (X) is the one dimensional invariant bundle spanned by X, and E^u and E^s are invariant bundles such that vectors in E^u grow exponentially under DF^t as $t \to \infty$, and vectors in E^s grow exponentially under DF^t as $t \to -\infty$.

The flow G^t satisfies $G^{t+s}(m) = G^t(G^s(m))$. Differentiating this equation at $s = 0$ gives

$$\phi(G^t(m))\ X(G^t(m)) = \phi(m)\ DG^t(m)\ X(m) .$$

Hence (X) is invariant under DG^t for all t, and $DG^t(m)|(X)_m$

is bounded with bounded inverse, uniformly for $m \in M$ and $t \in R$.

Let

$$\tau = \tau(m,t) = \int_0^t (\phi(F^s(m))^{-1} \, ds.$$

Then

$$\frac{\partial \tau}{\partial t}(m,t) = (\phi(F^t(m)))^{-1},$$

and

$$D\tau(m,t) = -\int_0^t (\phi(F^s(m)))^{-2} D\phi(F^s(m)) \, DF^s(m) \, ds .$$

As a function of t, G^τ satisfies

$$\frac{\partial G^{\tau(m,t)}}{\partial t}(m) = \frac{\partial G^\tau(m)}{\partial \tau} \frac{\partial \tau}{\partial t}$$

$$= \phi(G^\tau(m)) \, (\phi(F^t(m)))^{-1} \, X(G^\tau(m)) .$$

F^t also satisfies this equation, with the same initial data, so $F^t(m) = G^{\tau(m,t)}(m)$ for all m and t. Hence

$$DF^t(m) = DG^\tau(m) + \left\{\frac{\partial G^{\tau(m,t)}(m)}{\partial \tau}\right\} D\tau(m,t)$$

$$= DG^\tau(m) + \phi(G^\tau(m)) \, X(G^\tau(m)) \, D\tau(m,t) .$$

From this it follows that $E^u \oplus (X)$ and $E^s \oplus (X)$ are invariant under DG^t for all t, and that vectors in $E^u \oplus (X) \setminus (X)$ grow exponentially under DG^t as $t \to \infty$, and vectors in $E^s \oplus (X) \setminus (X)$ grow exponentially under DG^t as $t \to -\infty$.

It follows that for large t, $DG^t|E^u$ dominates $DG^t|(X)$, and so by the methods of Remark 4, (X) has an invariant complement in $E^u \oplus (X)$. Similarly, for large negative t, $DG^t|E^s$ dominates $DG^t|(X)$, and so (X) has an invariant complement in $E^s \oplus (X)$. Taken with exponential growth under DG^t, this shows that G^t is an Anosov flow.

3. Expanding Structures

We now introduce the notion of invariant set with expanding structure. This will be defined in terms of precise exponential rates, in order to get sharp estimates in Theorem 2. Underlying the definition, however, is Hadamard's original idea. An invariant set has expanding structure if it satisfies a cone condition on each backward orbit. See Figure 4.

Let U and V be open subsets of a manifold M, and let $F: U \to V$ be a diffeomorphism. Suppose $\Lambda \subset U$ is a compact set, invariant in the sense that

(H1) $$\Lambda \subset F(\Lambda) .$$

Suppose also that the tangent space $TM|\Lambda$ splits as $TM|\Lambda = N \oplus E$, and that either N or E is invariant, in the sense that

(H2a) $$N \subset DF\cdot N \quad \text{or}$$

(H2b) $$E \subset DF\cdot E$$

Here DF is the map induced by F on tangent vectors. In coordinates DF acts on a vector $v \in TM|U$ to give $DF(m)\cdot v \in TM|V$.

Let π^E and π^N be the projections corresponding to the splitting $TM|\Lambda = N \oplus E$. For arbitrary $m \in \Lambda$, and arbitrary $v^0 \in N_m$, $w^0 \in E_m$, define

$$m^{-k} = F^{-k}(m)$$

$$v^{-k} = \pi^N DF^{-k}(m)\cdot v^0$$

$$w^{-k} = \pi^E DF^{-k}(m)\cdot w^0 .$$

These definitions are illustrated in Figure 5 for the case in which Λ is an invariant submanifold of U, $N = T\Lambda$, and E is an invariant complement of N in $TM|\Lambda$. Note that $v^{-k} \in N_{m^{-k}}$ and $w^{-k} \in E_{m^{-k}}$.

Choose any Riemannian metric $|\ |$ on TM, and define for all $m \in \Lambda$,

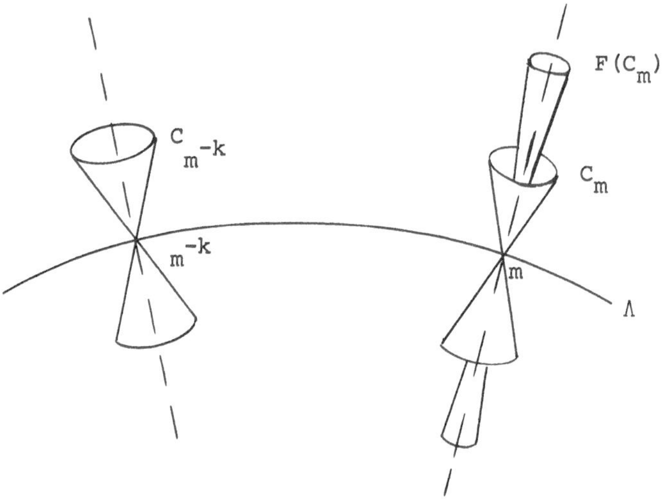

Figure 4

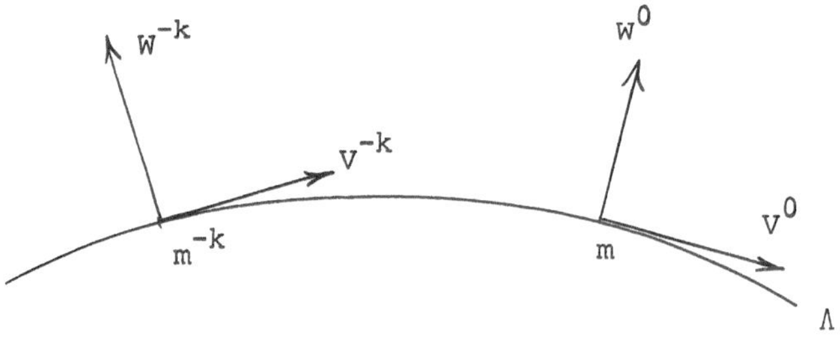

Figure 5

$$\alpha^*(m) = \inf\{\alpha > 0: w^{-k}/\alpha^k \to 0 \text{ as } k \to \infty, \text{ for all } w^0 \in E_m\},$$

$$\rho^*(m) = \inf\{\rho > 0: [|w^{-k}|/|v^{-k}|]/\rho^k \to 0 \text{ as } k \to \infty, \text{ for all } v^0 \in N_m,\ w^0 \in E_m\}.$$

α^* and ρ^* are constant on orbits and independent of the metric. Given N satisfying (H2a), α^* and ρ^* are independent of E. Given E satisfying (H2b), α^* and ρ^* are independent of N. As functions of m, α^* and ρ^* generally are discontinuous. See [1, Lemmas 1 and 2]. Our final hypothesis is

(H3) $\qquad \alpha^*(m) < 1,\ \rho^*(m) < 1$ for all $m \in \Lambda$.

We say that (Λ, E) is an invariant set with expanding structure if (H1), (H2b), and (H3) hold. If (H1), (H2a), and (H3) hold, it follows as in Remark 4 that N has a unique invariant complement E', and that (Λ, E') is an invariant set with expanding structure.

In the example of Hadamard, Λ is the origin, N is the x axis, and E is the y axis. The functions α^* and ρ^* are defined only at the origin, and $\alpha^* = 1/|\mu|$, $\rho^* = |\lambda|/|\mu|$.

The geometric significance of (H3) is that for any continuous family of cones C_m, and for any $m \in \Lambda$, there exists k such that $DF^k: TM_{m^{-k}} \to TM_m$ satisfies a cone condition as shown in Figure 4. Our main technical lemma asserts that there exists K such that $DF^K: TM_{F^{-K}(m)} \to TM_m$ satisfies a cone condition uniformly for $m \in \Lambda$. The precise statement and proof are found in [1, II.B].

Uniformity Lemma. α^* and ρ^* attain their suprema on Λ. Under hypothesis (H3), there exists K such that DF^K satisfies a uniform cone condition.

The following theorem is proved by the iteration procedure outlined above for the proof of Hadamard's theorem. A sharper result is proved in [1, Theorems 1,2,3].

Theorem 2. Let (Λ,E) be an invariant set with expanding structure, and suppose that $\alpha^*(m) < \alpha$ and $\rho^*(m) < \rho$ for all $m \in \Lambda$. Suppose also that $U = V = M$. Let d denote the distance corresponding to any Riemannian metric. Then for each $m \in \Lambda$ there is a manifold $W^E(m)$ diffeomorphic to E_m, such that

(i) $W^E(F^{-1}(m)) = F^{-1}(W^E(m))$.

(ii) $d(F^{-k}(p),F^{-k}(m))/\alpha^k \to 0$ as $k \to \infty$, for all $p \in W^E(m)$.

(iii) $$\frac{d(F^{-k}(p),F^{-k}(m))}{d(F^{-k}(q),F^{-k}(m))}\Big/\rho^k \to 0 \text{ as } k \to \infty,$$
for all $p \in W^E(m)$ and all $q \notin W^E(m)$ such that $d(F^{-k}(q),F^{-k}(m)) \to 0$ as $k \to \infty$.

Example 1. Suppose F is an Anosov diffeomorphism on a manifold M, with splitting $TM = E^u \oplus E^s$. Let $E = E^u$. Then the manifolds W^E are the unstable manifolds of points in M.

Example 1'. Suppose Λ is any hyperbolic invariant set, with splitting $TM|\Lambda = E^u \oplus E^s$. Let $E = E^u$. Then the manifolds W^E again are the unstable manifolds of points. Thus Theorem 2 includes an elementary proof of the unstable manifold theorem of Hirsch and Pugh [4].

Example 2. Let F^t be an Anosov flow on a manifold M, with splitting $TM = E^u \oplus (X) \oplus E^s$. Fix any $t > 0$, and let $F = F^t$, $E = E^u$, and $N = E^s \oplus (X)$. Then the manifolds W^E again are the unstable manifolds of points. Of course, the same construction works if F^t is an arbitrary flow and Λ is a hyperbolic invariant set.

Example 3. A classical result: Let X be a vector field with flow F^t, and let Λ be a closed orbit of X. We say that Λ is asymptotically stable with asymptotic phase if Λ is asymptotically stable and for all p near Λ there exists $m \in \Lambda$ such that $d(F^t(p), F^t(m)) \to 0$ as $t \to \infty$.

Theorem 3. Let Λ be a closed orbit of a vector field X, and suppose that all Floquet multipliers of Λ have modulus less than one. Then Λ is asymptotically stable with asymptotic phase.

Proof: The one dimensional bundle $(X)|\Lambda$ satisfies (H1), (H2a), and (H3) for the flow generated by -X. Hence $(X)|\Lambda$ has an invariant complement E, and (Λ, E) is an invariant set with expanding structure. The manifolds $W^E(m)$ are disjoint and cover a neighborhood of Λ. For each $m \in \Lambda$, $W^E(m)$ is precisely the set of points with asymptotic phase m. See [1, Theorem 4] for details.

Example 4. Let U and V be open subsets of a manifold M, and let $F: U \to V$ be a diffeomorphism. Suppose $\Lambda \subset U$ is a compact manifold, invariant in the sense that $\Lambda = F(\Lambda)$. We say that Λ is asymptotically stable with unique asymptotic phase if Λ is asymptotically stable, and for each p near Λ there exists $m \in \Lambda$ such that

(i) $d(F^{-k}(p), F^{-k}(m)) \to 0$ as $k \to \infty$.

(ii) $\dfrac{d(F^{-k}(p), F^{-k}(m))}{d(F^{-k}(p), F^{-k}(m'))} \to 0$ as $k \to \infty$, for all $m' \in M$, $m' \neq m$.

Theorem 4. Suppose F^{-1} satisfies (H1), (H2a), and (H3), with $N = T\Lambda$, and with E any complement of $T\Lambda$ in $TM|\Lambda$. Then Λ is asymptotically stable with unique asymptotic phase.

Proof: Construct the manifolds $W^E(m)$ as in the previous example. See [1, Theorem 4] for the complete proof of a sharper result.

4. Hopf Bifurcation.

As an application we show how our results give new information about the Hopf bifurcation problem. We treat the case of a diffeomorphism near a fixed point. Similar techniques are applicable to flows near stationary points and to flows near periodic orbits. Details appear in [2]. See also [5] and [6].

Let $F_\mu : R^n \to R^n$ be a family of diffeomorphisms depending on a real parameter μ, and assume that $F_\mu(0) = 0$ for all μ. Suppose that two complex conjugate eigenvalues $\lambda_\mu, \bar{\lambda}_\mu$, of $DF_\mu(0)$ pass to the outside of the unit circle as μ increases through zero, while the remaining eigenvalues remain inside the unit circle. Under a finite number of nonresonance and nondegeneracy conditions, it is possible to transform F_μ to a normal form. In this normal form there is a coefficient β which determines the stability of the origin at $\mu = 0$. In case Re $\beta < 0$ the following structure appears for $\mu > 0$. See Figure 6.

(i) There is a two dimensional manifold with boundary $\bar{D}_\mu = D_\mu \cup \partial D_\mu$ near the plane spanned by the eigenvectors corresponding to λ_μ and $\bar{\lambda}_\mu$. D_μ is invariant in the sense that $F_\mu(\bar{D}_\mu) \subset D_\mu$.

(ii) There is a differentiable circle $\Gamma_\mu \subset D_\mu$, invariant in the sense that $F_\mu(\Gamma_\mu) = \Gamma_\mu$. The radius of Γ_μ is proportional to $\mu^{1/2}$.

(iii) The origin lies in the interior of Γ_μ in D_μ. For $m \neq 0$ in the interior of Γ_μ, $F_\mu^k(m) \to 0$ as $k \to -\infty$ and $F_\mu^k(m) \to \Gamma_\mu$ as $k \to \infty$.

(iv) For m in the exterior of Γ_μ in D_μ, $F_\mu^k(m) \to \Gamma_\mu$ as $k \to \infty$ and $F_\mu^k(m) \notin D_\mu$ for some $k < 0$.

(v) D_μ satisfies the hypotheses of Theorem 4, with suitable modifications to take the boundary into account. Hence D_μ is asymptotically stable with unique asymptotic phase.

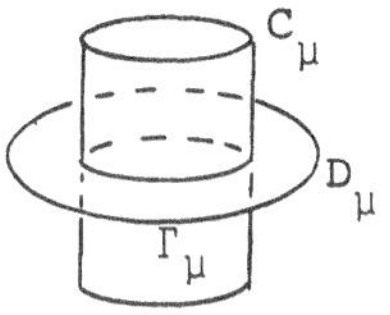

Figure 6

(vi) By (iii) and (v), there is a neighborhood U of D_μ such that for $p \in U - W^s(0)$, $F_\mu^k(p) \to \Gamma_\mu$ as $k \to \infty$. Here $W^s(0)$ denotes the stable manifold of the origin, a manifold of codimension two.

(vii) Γ_μ satisfies the hypotheses of Theorem 4, so Γ_μ is asymptotically stable with unique asymptotic phase. Hence for $p \in U - W^s(0)$, $F_\mu^k(p)$ tends fastest to a unique orbit in Γ_μ.

(viii) Let $W^E(m)$ denote the manifolds of codimension two constructed for D_μ as in (v). Let $C_\mu = \bigsqcup_{m \in \Gamma_\mu} W^E(m)$. Then C_μ is a manifold of codimension 1, a cylinder separating U into an inside and an outside of Γ_μ. C_μ consists of all orbits tending to $D_\mu - \{0\}$ non-tangentially. Our construction shows that C_μ is a C^0 manifold. According to [1, Theorem 6], C_μ is in fact a differentiable manifold.

5. References

[1] Fenichel, N., Asymptotic Stability with Rate Conditions, Indiana Univ. Math. J. 23, 12 (1974) 1109-1137.

[2] Fenichel, N., The Orbit Structure of the Hopf Bifurcation Problem, J. Diff. Equations, 17 (1975) 1-21.

[3] Hadamard, J., Sur l'Iteration et les Solutions Asymptotiques des Equations Differentielles, Bull. de la Soc. Math. de France 29 (1901) 224-228.

[4] Hirsch, M. W. and Pugh, C. C., Stable Manifolds and Hyperbolic Sets, Proc. Symposia in Pure Math., Vol. XIV., Amer. Math. Soc., (1970) 133-163.

[5] Ruelle, D., and Takens, F., On the Nature of Turbulence, Comm. Math. Phys. 20 (1971) 167-192.

[6] Sacker, R., On Invariant Surfaces and Bifurcation of Periodic Solutions of Ordinary Differential Equations, New York Univ., IMM-NYU 333, October 1964.

This research was partially supported by the Army Research Office (Durham) and by the National Science Foundation, Grant GP-43613-X.

HYPERBOLIC SETS AND SHIFT AUTOMORHPISMS

Charles C. Conley*

Department of Mathematics, University of Wisconsin, Madison 53706

Abstract

An elementary discussion of hyperbolic invariant sets is presented and a proof of the theorem that "near any nonperiodic chain recurrent point there lies an embedded shift automorphism" is indicated.

The aim of these remarks is to give a slightly different treatment of compact hyperbolic invariant sets with primary focus on the theorem that near a chain recurrent point which is not periodic, there is an embedded shift automorphism.

1. Notation and Generalities

Let M be a smooth manifold and for $N \subset M$ let TN denote the tangent bundle of M restricted to N. The zero section of TN will be dnoted N_0. On replacing the fibers of TM by their line spaces (i.e., the space R^n by the projective space P^n of lines through the origin), one obtains a projective bundle the restriction of which to N is denoted PN.

Given a diffeomorphism $\phi: M \to M$, let $T\phi$ and $P\phi$ denote the induced bundle homeomorphisms on TM and PM. An invariant set of ϕ means a subset N of M such that $\phi(N) = N$. If N is such a set then TN and PN are respectively $T\phi$- and $P\phi$-invariant sets.

A compact ϕ-invariant set A is called an attractor for ϕ if A admits a compact neighborhood U such that $\bigcap_{n \geq 0} \phi^n(U) = A$. This

*Sponsored by the U.S. Army under Contract No. DA-31-124-ARO-D-462.

is equivalent to saying that A is the largest invariant set contained in some compact neighborhood B with the property that $\phi(B) \subset$ interior of B (this is a "converse Liapounov theorem"). Such a set B will be called an attractor neighborhood for ϕ. Note that if ϕ' is close to ϕ in the compact open topology, then B will also be an attractor neighborhood for ϕ'.

A repeller for ϕ means an attractor for ϕ^{-1}.

Because of the linearity of the fiber maps and the compactness requirement, any attractor of $T\phi$ is contained in the zero section (and so corresponds to an attractor for ϕ). Again because of linearity, any attractor for $P\phi$ meets each fiber of PM in a projective subspace (see [1] for example).

A compact ϕ-invariant set $H \subset M$ is called a hyperbolic invariant set if TH is the sum of two $T\phi$-invariant subbundles T^u and T^s such that H_0 is a repeller for $T\phi|T^u$ and an attractor for $T\phi|T^s$ (Figure 1 illustrates the case where H consists of a point and T^u and T^s have fiber dimension 1). The fiber dimensions of

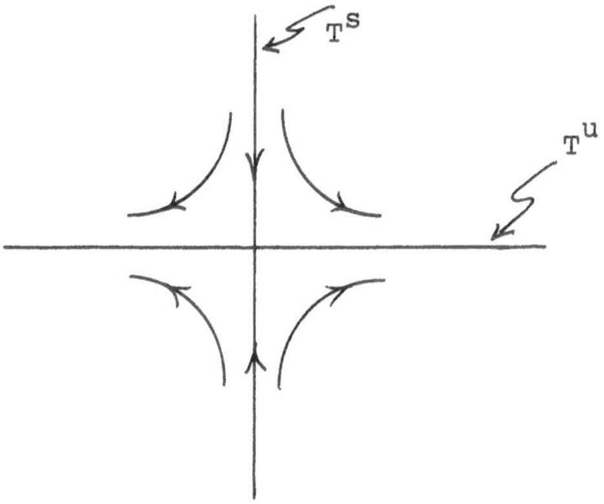

Figure 1

T^u and T^s will be denoted by u and s respectively, note that u+s =n.

Corresponding to T^u and T^s there are $P\phi$-invariant projective sub-bundles, P^u and P^s, of $P\phi$. A brief study of the behavior of lines through the origin in Figure 1 shows that P^u is an attractor, and P^s a repeller, for $P\phi$.

Suppose now that H' is a compact invariant set which is contained in a small neighborhood of H; the claim is that H' must also be hyperbolic. This comes out of the following general considerations. Let I be any compact invariant set for a discrete flow ϕ on a locally compact metric space and suppose $B \subset I$ is an attractor neighborhood for the flow $\phi|I$. Given $\varepsilon > 0$, let $B(\varepsilon)$ be the closed ε-neighborhood of B in the whole space. If ε is small then $B(\varepsilon) \cap I$ will also be an attractor neighborhood of $\phi|I$. Furthermore if ε' is small enough (depending now on ε) and I' is a compact invariant set within ε' of I, then $B' = B(\varepsilon) \cap I'$ is also an attractor neighborhood for $\phi|I'$. One just verifies directly that $\phi(B') \subset$ interior B' relative to I'.

Now if H' is close to H, PH' is close to PH. Given an attractor neighborhood B of P^u for the map $P\phi|PH$ one then finds an attractor neighborhood for $P\phi|PH'$. Since B meets each fiber in a set containing a projective subspace of dimension u-1, B' does also. It follows that the attractor for $P\phi|PH'$ contained in B' is an invariant subbundle with fiber dimension u-1. The complementary repeller is then an invariant subbundle of dimension s-1. The corresponding bundles in TH' give the splitting, and the same style argument shows that H_0' is a repeller for the one and an attractor for the other so that H' is hyperbolic.

2. Mappings of Squares

Let H be a hyperbolic invariant set and let B be an attractor neighborhood for H_0 in T^s. Then the set B^s defined as the convex hull of the symmetrization of B (i.e., the convex hull of $\{\eta | \eta \text{ or } -\eta \in B\}$) is a convex symmetric attractor neighborhood.

In a similar way one finds a convex symmetric repeller neighborhood for H_0 in T^u.

For $x \in H$, let $B^s(x)$ and $B^u(x)$ be the parts of B^s and B^u over x. Let $S(x) = B^u(x) \times B^s(x)$. Under $T\phi$, $B^u(x)$ is carried to a set in the fiber over $\phi(x)$ which contains $B^u(\phi(x))$ in its (relative) interior. Likewise, $B^s(x)$ is carried to the interior of $B^s(\phi(x))$. Figure 2 illustrates the case where $s = u = 1$; the vertical direction corresponds to B^s, the horizontal to B^u.

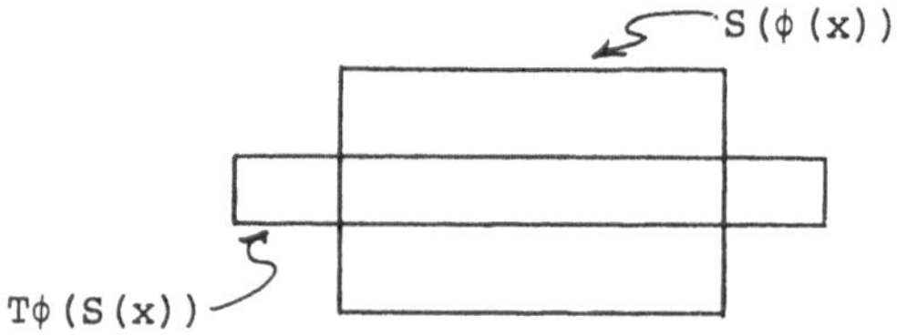

Figure 2

These maps are nondegenerate in a topological sense which can be described in the (1,1) case as follows: any arc connecting the (vertical) sides of the "square" $S(x)$ contains a subarc which is carried to an arc in $S(\phi(x))$ which connects the sides of that square. Observe this property is C^0 stable.

Consider an infinite sequence of squares $S_0, S_1, \ldots$ and maps $\psi_n\colon S_{n-1} \to S_n$ which have this property. The statement is: there exists a continuum in S_0 connecting the top and bottom, the points of which stay in the squares under the maps ψ_n: that is for any $n \geq 1$ and any x in the continuum, $\psi_n \circ \ldots \circ \psi_1(x) \in S_n$. To see this observe that, by induction, any arc in S_0 which connects the sides must contain a subarc which is carried through the squares to an arc in S_n which connects the sides of S_n. Then any such arc contains at least one point which stays in the squares under all iterations and this implies the statement.

If one is given a biinfinite sequence of squares and mappings, a symmetric argument shows there exists a continuum in S_0 which connects the sides, and points of which stay in the squares in the backwards direction. The intersection of these two continua must be nonempty, and points in it stay in the squares in both directions.

A finer statement can be made for C^1 perturbations of the triples $(S(x), S(\phi(x)), T\phi)$. Define a norm on $T\{x\}$ (the fiber of T_M over x) so that $S(x)$ is the unit ball. For $\zeta \in T\{x\}$, let ξ and η denote the components of ζ in T^u and T^s respectively. Call a vector horizontal if $\|\xi\|/\|\eta\| > 1$ and vertical if $\|\xi\|/\|\eta\| < 1$. Then under $T\phi$, horizontal vectors are carried to horizontal ones and their horizontal component is increased in length by a factor greater than 1. From this it follows that the continuum of points which stay in the squares under forward iteration cannot contain two points connected by a horizontal vector -- it is thus a "vertical" arc. Also, using the fact that the cone angle of the cone of vertical vectors is increased by a factor greater than one under $T\phi$, one can show that this vertical arc is smooth (see [2]). Corresponding statements hold for the horizontal continuum and imply in particular that there is only one point which stays in the squares in both directions.

Of course the application to the sequence ... $S(\phi^{-1}(x)), S(x), S(\phi(x))$, ... with maps $T\phi$ is trivial -- the continua are just $B^s(x)$ and $B^u(x)$ and the point is the origin of $T\{x\}$. Now choose an exponential map from T_M to M and identify each $S(x)$ with its image in M. Then provided B^s and B^u are scaled down sufficiently (i.e., replaced by ρB^s and ρB^u for small enough ρ). ϕ itself becomes a C^1 perturbation of $T\phi$. In this way one can prove the local stable manifold theorem for example.

Similarly, if ϕ' is C^1 close to ϕ then the sequence $\ldots S(\phi^{-1}(x)), S(x), S(\phi(x)) \ldots$ with map ϕ' (replacing ϕ) determines a unique point x' in $S(x)$ which stays in the squares in both directions. The mapping $x \to x'$ then defines a conjugacy from $\phi|H$ to $\phi|H'$ where H' is a hyperbolic invariant set of ϕ'.

These things and related ones have been discussed in the literature by several authors including Alekseev, Anosov, Bowen, Fenichel, Franks, Guckenheimer, Hirsch, Moser, Newhouse, Palis, Pugh, Robbin, Robinson, Shub, Sinai, Smale and Walters; Sinai and Bowen have also treated the material in the next section.

3. ε-chains

Given ϕ, an ε, n-chain from x to y means a finite sequence of points $\{x_1,\ldots,x_m\}$ $(m \geq n)$ such that x_1 and x_n are within $\varepsilon/2$ of x and y respectively and $\phi(x_i)$ is within ε of x_{i+1} for $i < m$. A point x is called chain recurrent if for ε, $n > 0$, there is an ε, n-chain from x back to itself.

An ε-pseudo-orbit is a bi-infinite ε-chain, i.e., a bi-infinite sequence of points $\{x_n\}_{-\infty}^{\infty}$ such that $\phi(x_i)$ is within ε of x_{i+1} for all i. Observe that if x is chain recurrent then there is a periodic ε-pseudo orbit near x for any $\varepsilon > 0$; just choose any ε, n-chain from x to x and repeat it forward and back infinitely often.

Now given a hyperbolic invariant set H, and an ε-pseudo-orbit in H ($\{x_n\} \subset H$) the aim is to associate to this pseudo-orbit a bi-infinite sequence of square mappings. In fact the squares $S(x_n)$ (considered to lie in M via the exponential map) can be used -- one has only to check that $S(x_i)$ is carried so that it meets $S(x_{i+1})$ as depicted in Figure 2 wherein $S(x)$ is replaced by $S(x_i)$ and $S(\phi(x))$ by $S(x_{i+1})$.

There is a little to check, because even if x_{i+1} is close to $\phi(x_i)$, the corresponding squares need not be close. However, $T\phi(B^s(x_i))$ is known to be interior to B^s. Thus if it is "translated" by means of a local trivialization to the fiber of T^s over x_{i+1} the image will lie interior to $B^s(x_{i+1})$ at least if x_{i+1} is close enough to $\phi(x_i)$. Using the corresponding statement for $B^u(x_i)$ together with the continuity of the exponential map one finds that the triple $(S(x_i), S(x_{i+1}), \phi)$ is C^0 close to the triple $(S(x_i), tS(\phi(x_i), \phi)$ where $tS(\phi(x_i))$ is the translate (to the fiber over x_{i+1}) of $S(\phi(x_i))$. (To obtain C^1 close one would have to make requirements on the local trivialization and in any case C^1-close is not needed).

Now it is seen that for small ε, an ε-pseudo-orbit determines a unique orbit of ϕ (not necessarily in H). Namely it determines a sequence of square mappings so at least a nonempty set of orbits. But this set can contain only one orbit since invariant sets near H must be hyperbolic.

This now implies that near any chain recurrent point of H there lies a periodic orbit in a hyperbolic set near H. If the chain recurrent point is not itself periodic, there must be infinitely many nearby periodic orbits.

An example is provided by a diffeomorphism of R^2 with a hyperbolic fixed point and a nondegenerate homoclinic point (Figure 3).

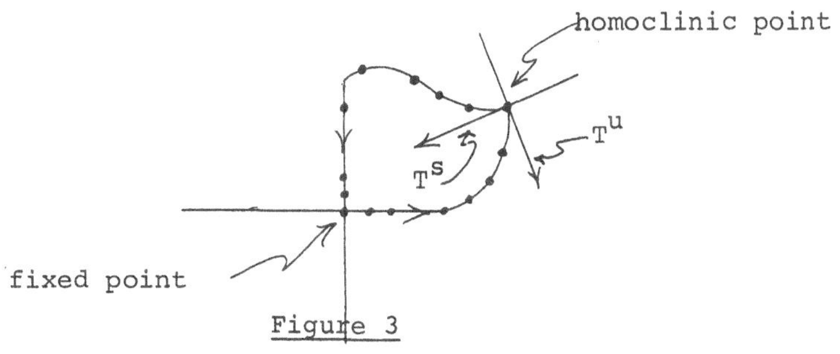

Figure 3

A homoclinic point is one that tends to the hyperbolic point in both time directions; nondegenerate refers to the fact that the stable and unstable manifolds (the curves in Figure 3) cross nontangentially at the homoclinic point. The set H in this example consists of the orbit of the homoclinic point (the countable point set indicated in Figure 3) together with the fixed point. (Of course every point in H is homoclinic to the rest point.) That H is hyperbolic follows from the nondegeneracy of the intersection. Thus T^s at the homoclinic point is spanned by the vector at that point tangent to the stable manifold while T^u is spanned by that tangent to the unstable manifold.

No point of H is periodic except for the hyperbolic fixed point, however every point is chain recurrent. For example, an ε-chain from the homoclinic point x back to itself consists of the (properly ordered) complement of a small enough neighborhood of the rest point.

Following Smale [3], one can see there is a shift automorphism embedded near any chain recurrent point x of H which is not periodic. Thus, near x there must lie two distinct periodic point points x_1 and x_2 each of which determines a bi-infinite sequence of square mappings with S_0 being a square about x. Each of these sequences of square mappings determines a horizontal and a vertical arc in S_0; furthermore the two horizontal arcs cannot intersect each other because points on each tend to the corresponding periodic orbit in backward time -- no point can go to both periodic orbits in backward time. Likewise the two vertical arcs cannot intersect. Thus the situation is as indicated in Figure 4. Now for a sufficiently high n, which is a common period of both periodic orbits, one can describe the set of points in S_0 which stay in the squares under n iterations of one of the sequences, say that corres-

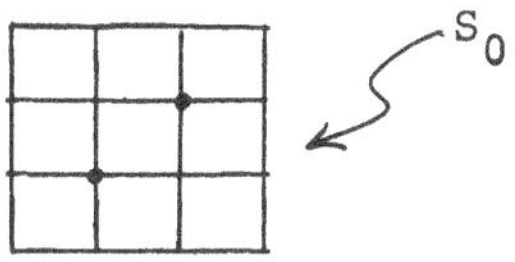

Figure 4

ponding to the upper right-hand periodic orbit. This set consists of a vertical strip about the corresponding vertical arc and is carried by ϕ_n to a horizontal strip about the horizontal arc as indicated in Figure 5. In a similar way the other periodic orbit

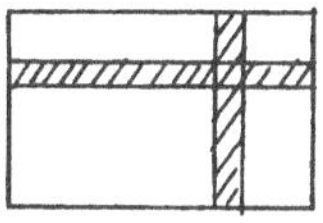

Figure 5

(or rather corresponding sequence of square mappings) determines a vertical strip which is carried to a horizontal strip. If n is large enough, the two vertical strips and likewise the two horizontal ones will be narrow enough that they cannot intersect. Thus one has the situation (depicted in Figure 6) that ϕ^n takes the union of

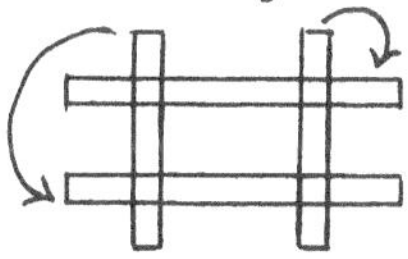

Figure 6

the two vertical strips to the union of the two horizontal ones mapping horizontal and vertical boundary arcs to those of the same type. This mapping has all the essential features of the horseshoe mapping of Smale (cf. [2]) and one concludes that the set of points which stay in the set of Figure 6 under all iterations of $\psi = \phi^n$ is a Cantor set and ϕ restricted to this set is conjugate to the shift automorphism.

References

[1] Selgrade, J., Isolated invariant sets for flows on vector bundles (to appear).

[2] Moser, J., Stable and Random Motions in Dynamical Systems with Special Emphasis on Celestial Mechanics; Annals of Mathematics Studies 77, Princeton University Press 1973.

[3] Smale, S., Diffeomorphisms with many periodic points. Differential and Combinatorial Topology (edited by S. S. Cairns), Princeton University Press (1965) 63-80.

[4] Bowen, R., ω limit sets for axiom A diffeomorphisms (to appear in Jour. Diff. Equa).

[5] Newhouse, S., "Hyperbolic limit sets," Trans. Amer. Math. Soc. 167 (1972) 125-150.

[6] Pugh, C. C., and Shub, M., "The Ω-stability theorem for flows," Inventiones Math. 11 (1970) 150-158.

[7] Shub, M., "Stability and Genericity for Diffeomorphisms," in Dynamical Systems (ed. M. Peixoto) Proccedings of Symposium on Dynamical Systems, Salvador, Bahia, Brazil (1971) Academic Press.

[8] Smale, S., "The Ω-Stability Theorem," in Global Analysis, Proc. Symp. Pure Math. 14 (Providence: A.M.S. 1970) 289-299.

[9] Palis, J., "On Morse-Smale Dynamical Systems," Topology 4 (1969) 385-404.

[10] Anosov, D. V., Geodesic flows on closed Riemannian manifolds with negative curvature, Proc. Steklov Inst. 90 (1967).

[11] Bowen, R., Topological entropy and Axiom A, Proc. Symp. Pure Math. 14 (1970) 23-41.

[12] Bowen, R., Periodic points and measures for axiom A diffeomorphisms, Trans. AMS 154 (1971) 377-397.

[13] Hirsch, M., Palis, J., Pugh, C., and Shub, M., Neighborhoods of hyperbolic sets, Inventiones Math. 9 (1970) 121-134.

[14] Hirsch, M., and Pugh, C., Stable manifolds and hyperbolic sets, Proc. Symp. Pure Math. 14 (1970) 133-163.

[15] Smale, S., Differentiable dynamical systems, Bull. AMS 73 (1967) 747-817.

[16] Walters, P., Anosov diffeomorphisms are topologically stable, Topology 9 (1970) 71-78.

[17] Guckenheimer, J., "Absolutely Ω-stable diffeomorphisms," Topology 11 (1972), 195-197.

[18] Hirsch, M., Palis, J., Pugh, C., and Shub, M., "Neighborhoods of hyperbolic sets," Inventiones Math. 9 (1970) 121-134.

[19] Mane, R., "Persistent manifolds are normally hyperbolic, to appear.

[20] Moser, J., "On a theorem of Anosov," J. Diff. Equa. 5 (1969) 411-440.

[21] Fenichel, N., "Persistence and smoothness of invariant manifolds for flows," Indiana Univ. Math. J. 21 (1971) 193-226.

[22] Franks, J., "Necessary conditions for stability of diffeomorphisms," Trans. AMS 158 (1972) 301-308.

[23] Franks, J., "Differentiably Ω-stable diffeomorphisms," Topology 11 (1972) 107-113.

[24] Franks, J., "Time dependent stable diffeomorphisms," Inventiones Math., 24 (1974), pp. 163-172.

[25] Robbin, J., "A structural stability theorem," Annals Math. 94 (1971) 447-493.

[26] Robinson, R.C., "Structural stability of C^1 flows," to appear Proc. Conf. Applic. Topology and Dynamical Systems, Univ. of Warwick, 1974.

[27] Sacker, R., "A perturbation theorem for invariant manifolds and Hölder continuity," J. Math. Mech. 18 (1969) 705-762.

TRIPLE COLLISION IN NEWTONIAN GRAVITATIONAL SYSTEMS

Richard McGehee*

School of Mathematics, University of Minnesota, Minneapolis, Minnesota

1. Introduction

Consider a system of point masses moving in Euclidean space according to the laws of classical mechanics. We assume only gravitational interactions, so the force between any two particles is proportional to the inverse square of the distance between them. As two or more of the particles approach collision the force becomes infinite and the differential equations describing the motion become undefined.

The behavior of the system near a double collision is well known and will be described below. In previous work the author gave a description of the behavior of orbits near a triple collision [4]. This paper is an expository account of that work.

We shall begin with a general discussion of singularities in celestial mechanics. Then we shall discuss two different approaches to the question of whether orbits can be extended through collisions. Next we shall consider the collinear three-body problem and show that an appropriate transformation allows us to extend the differential equations to include triple collision as an invariant set. Finally, by studying the flow on this invariant set we shall deduce properties of orbits near triple collision. We shall see that orbits cannot be extended through triple collisions. We shall also see that the system can transfer an arbitrary amount of energy from potential to kinetic by passing close to triple collision.

*Sponsored in part by National Science Foundation Grant GP-38955 and in part by the U. S. Army under Contract No. DA-31-124-ARO-D-462.

2. Singularities in Celestial Mechanics

Assume that the system consists of n point masses moving in ν-dimensional Euclidean space. The Newtonian gravitational potential is given by

$$-U(\underset{\sim}{q}) = -\sum_{i<j} \frac{m_i m_j}{\|q_i - q_j\|}$$

where $m_i > 0$ is the mass and $q_i \in R^\nu$ is the position of particle i, $\| \ \|$ denotes the Euclidean norm, and $\underset{\sim}{q} = (q_1,\ldots,q_n) \in (R^\nu)^n$. Let $p_i \in R^\nu$ be the momentum of particle i and write $\underset{\sim}{p} = (p_1,\ldots,p_n)$. Then the kinetic energy is given by

$$T(\underset{\sim}{p}) = \sum_{i=1}^{n} \frac{\|p_i\|^2}{m_i}$$

The Hamiltonian for this system is

$$H(\underset{\sim}{q},\underset{\sim}{p}) = T(\underset{\sim}{p}) - U(\underset{\sim}{q}) \ .$$

The differential equations describing the motion are

$$\text{(2.1)} \qquad \begin{aligned} \dot{\underset{\sim}{q}} &= H_p(\underset{\sim}{q},\underset{\sim}{p}) = \nabla T(\underset{\sim}{p}) \\ \dot{\underset{\sim}{p}} &= -H_q(\underset{\sim}{q},\underset{\sim}{p}) = \nabla U(\underset{\sim}{q}) \ , \end{aligned}$$

where the dot denotes differentiation with respect to time and the subscript on H denotes the gradient with respect to the indicated variable.

We consider the basic question of the existence of solutions of equations (2.1). The singularities of the function U play a role, since the differential equations are not even defined there. These singularities occur whenever two or more of the particles coincide. We shall denote the set of singularities by

$$\Delta = \{\underset{\sim}{q} \in (R^\nu)^n : q_i = q_j \text{ for some } i \neq j\}$$

and refer to it as the set of <u>collision points</u>. Note that U is a real analytic function on $(R^{\nu})^n - \Delta$.

If we start with initial data $(\underset{\sim}{q}(0),\underset{\sim}{p}(0))$, where $\underset{\sim}{q}(0) \notin \Delta$, then standard theorems applied to equations (2.1) give the existence and uniqueness of a solution $(\underset{\sim}{q}(t),\underset{\sim}{p}(t))$, defined on some maximal interval $[0,t^*)$. If $t^* < \infty$, then the solution is said to experience a <u>singularity</u> at t^*.

The singularities of the solutions must correspond in some way to the singularities of the potential. In fact, it is fairly easy to show that $U(\underset{\sim}{q}(t)) \to \infty$ as $t \to t^*$ [9]. In other words, $\underset{\sim}{q}(t)$ must approach Δ as $t \to t^*$.

An important kind of singularity occurs if $\underset{\sim}{q}(t)$ actually approaches a particular point in Δ, i.e.

$$\underset{\sim}{q}(t) \to \underset{\sim}{q}^* \in \Delta \quad \text{as} \quad t \to t^* .$$

Such a singularity is called a <u>collision</u> and has the property that each of the particles approaches a limiting position. If k of the particles coincide in the limit, then the collision is called a <u>k-tuple collision</u>.

The question of whether noncollision singularities exist is open. Work of Painlevé [5], Von Zeipel [13], and Sperling [10] shows that $\underset{\sim}{q}(t) \to \infty$ as $t \to t^*$ for such a singularity. Thus the problem of finding a noncollision singularity is equivalent to that of finding a solution which becomes unbounded in finite time. For $n = 3$ Painlevé [5] proved that all singularities are due to collision. For $n \geq 4$ the question remains open, although Mather and McGehee [3] have recently investigated the system for $n = 4$ with orbits extended through double collisions in a manner made precise below. They found extended solutions which become unbounded in finite time, but the orbits contain an infinite number of double collisions and hence are not examples of noncollision singularities.

Even the singularities due to collision are not fully understood, at least when more than two of the particles are involved. For double collisions, one might expect from physical considerations that a solution can be extended by an elastic bounce. This extension can be mathematically justified in two very different ways and is generally known as "regularizing" the singularity. These two methods will be discussed in the next section.

To describe what is known about collisions of more than two particles we need the following definition [14]: Let $(s_1,\ldots,s_n) \in (R^\nu)^n$ be the position of the particles. We say that the particles are in a central configuration if there exists a $\mu > 0$ so that

$$\nabla_i U(s_1,\ldots,s_n) = -\mu m_i s_i \ , \qquad i = 1,\ldots,n.$$

Here ∇_i denotes the gradient with respect to the ith variable.

This definition is motivated by the existence of the following special solutions of equations (2.1), called homographic solutions [14]. Let $\rho(t) > 0$ satisfy

$$\ddot{\rho}(t) = -\frac{\mu}{\rho(t)^2} \tag{2.2}$$

and let

$$\begin{aligned} q_i(t) &= \rho(t)s_i \ , \\ p_i(t) &= \dot{\rho}(t)m_i s_i \ , \end{aligned} \qquad i = 1,\ldots,n\ .$$

Then $(\underset{\sim}{q}(t),\underset{\sim}{p}(t))$ satisfies equations (2.1). Such solutions always maintain the same configuration of particles; only the scale is changing.

Homographic solutions either begin or end in an n-tuple collision and are thus special cases of collision orbits. Actually, all orbits beginning or ending in n-tuple collision asymptotically approach central configurations. Sundman [12] proved this result for $n = 3$, and Wintner [14] pointed out that Sundman's techniques work for arbitrary n. Little is known about central configurations for $n \geq 4$, although Palmore [6,7] has recently proved that there are only a finite number of distinct ones, at least for most values of the masses.

Not much else is known about collisions except for the case $n = 3$. Siegel [8] has given a complete analytic description of orbits ending in triple collision. One of Siegel's results is that the set of orbits ending in triple collision forms a smooth submanifold of the phase space. Siegel also showed that most solutions cannot be extended through a triple collision. We shall discuss both these results below.

3. Methods of Extending Singular Solutions

The classical approach to celestial mechanics involves expanding solutions in power series as functions of time. The question of whether a solution can be extended through a singularity becomes a question of whether a certain power series has an appropriate analytic continuation. Sundman [11] and Siegel [8] adopted this approach in their investigations of collisions.

A more modern approach is that defined by Easton [1]. Easton considers orbits passing close to the singular one and asks whether it can be extended so that the extension is continuous with respect to initial conditions. For a precise definition we refer to Easton's work [1,2]; for our purposes it will suffice to give four examples illustrating the two methods of regularization. Each of the examples is a vector field on R^2 with a singularity at the origin.

Example 1.

$$\dot{x} = \frac{1}{(x^2+y^2)^{1/3}}$$

$$\dot{y} = 0 .$$

Consider the solution

$$(3.1)\qquad \begin{aligned} x(t) &= -\frac{5}{3}^{3/5}(t^* - t)^{3/5} , \\ y(t) &= 0 . \end{aligned} \qquad t < t^* ,$$

This solution starts at a point on the negative x-axis and approaches the origin as $t \to t^*-$. Since $(t^*-t)^{3/5}$ is a real negative number for $t > t^*$, solution (3.1) can be extended through the singularity by the classical method. The extended solution is given by (3.1) with t arbitrary.

Solution (3.1) can also be extended by Easton's method. The phase portrait for Example 1 is shown in Figure 1. The negative x-axis is an orbit ending at the singularity, while the positive x-axis is an orbit beginning at the singularity. If one connects the two orbits at the origin, making the x-axis a single orbit, then the resulting orbit is continuous with respect to initial conditions. The extension is the same as that given by the classical method.

Example 2.

$$\dot{x} = -\frac{x}{x^2+y^2}$$

$$\dot{y} = \frac{y}{x^2+y^2} .$$

Consider the solution

$$(3.2)\qquad \begin{aligned} x(t) &= -2^{1/2}(t^* - t)^{1/2} , \\ y(t) &= 0 . \end{aligned} \qquad t < t^*,$$

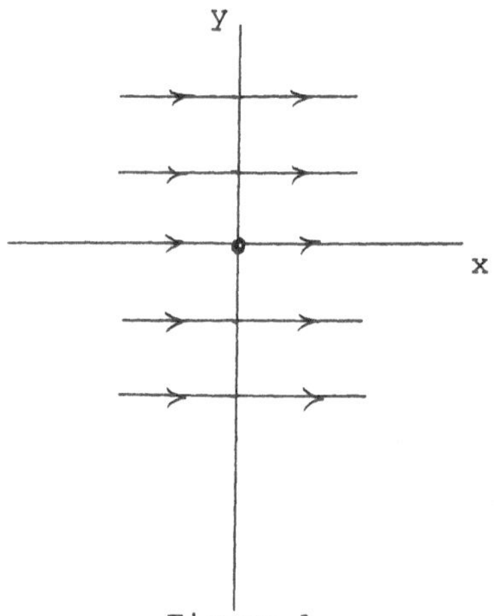

Figure 1

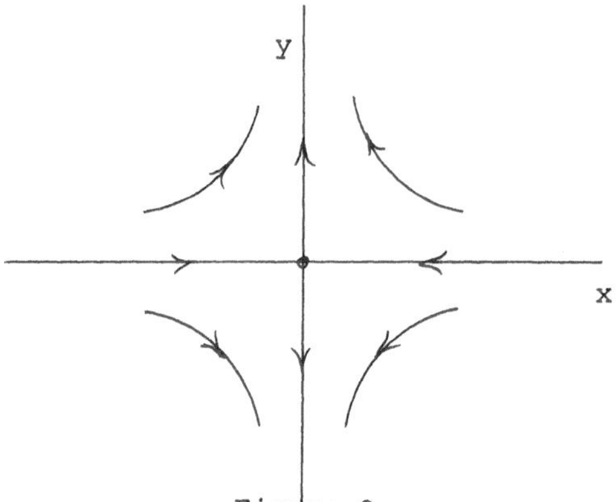

Figure 2

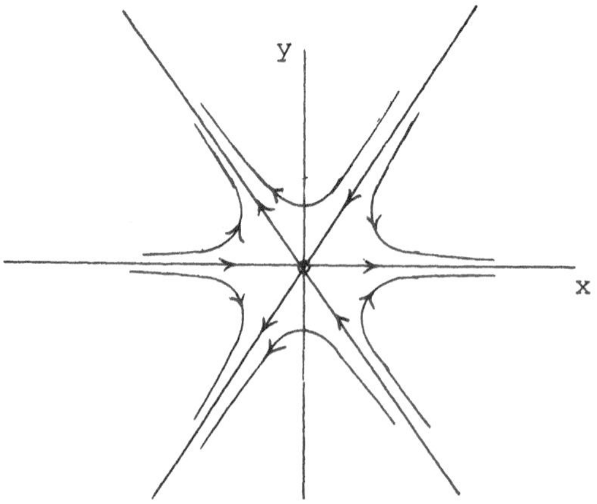

Figure 3

As before, this solution starts on the negative x-axis and approaches the origin as $t \to t^*$. In this case, however, $(t^*-t)^{1/2}$ is imaginary for $t > t^*$, so solution (3.2) cannot be extended by the classical method. Neither can it be extended by Easton's method as can be seen in Figure 2. Orbits starting close to the negative x-axis with positive y end up close to the positive y-axis, while orbits starting with negative y end up close to the negative y-axis. Thus no extension will result in an orbit continuous with respect to initial data.

Example 3.

$$\dot{x} = \frac{1}{(x^2+y^2)^{1/2}}$$

$$\dot{y} = 0 .$$

Equation (3.2) also is a solution for this example. Therefore the solution cannot be extended classically. However, the phase portrait for this example is the same as that for Example 1. (Figure 1). Therefore the solution can be extended by Easton's method.

Example 4.

$$\dot{x} = \frac{x^2-y^2}{(x^2+y^2)^{4/3}}$$

$$\dot{y} = -\frac{xy}{(x^2+y^2)^{4/3}} .$$

Equation (3.1) also gives a solution for this example. It can be extended classically, just as before. However, an argument similar to that given in Example 2 shows that this solution cannot be extended by Easton's method. (Figure 3).

The above examples show that the classical method of regularization and Easton's method are not at all related. The existence or nonexistence of an extension for one method implies nothing about the existence or nonexistence of an extension for the other.

4. Regularization of Collisions

We now ask whether orbits can be extened through collisions. We have two methods from which to choose, but in both cases the results are roughly the same: double collisions can be extended, but triple (and presumably higher order) collisions cannot be extended.

If a double collision involving particles i and j occurs at time t^*, then $q_i(t)-q_j(t)$ can be written as a power series in $(t^*- t)^{1/3}$ [9,14]. The solution can therefore be extended classically to $t > t^*$. The series begins with $(t^*- t)^{2/3}$ and the extension corresponds to an elastic bounce.

If we adopt Easton's viewpoint, we also find that double collisions can be extended [2]. Consider two particles passing close together but not colliding; for example, a comet passing close to the sun. The particles are on highly eccentric orbits and retreat from each other in approximately the opposite direction from which they approach. As we take orbits passing closer and closer to double collision we see that the orbits limit to a collision orbit extended by an elastic bounce. Thus Easton's method produces the same extension as the classical method.

The situation is quite different for higher order collisions. Consider first a homographic solution for which a collision occurs at time t^*. An examination of equation (2.2) reveals that $\rho(t)$ can be written as a power series in $(t^*- t)^{2/3}$. Therefore the homographic solution can be extended classically. However, when Siegel examined triple collision in the three-body problem he found that most orbits ending in triple collision contain irrational powers of $(t^* - t)$ in their expansions. Such orbits cannot be extended classically.

As we have seen in the examples in Section 3, Siegel's conclusion gives us no information about whether orbits can be extended through triple collision by Easton's method. However, we shall see below that orbits passing close to triple collision behave so wildly that it is impossible to extend collision orbits. One expects that the same conclusion holds for higher order collisions.

5. Triple Collision in the Collinear Three-Body Problem

In the remainder of this paper we shall study the behavior of orbits as they pass close to triple collision. We examine the simplest possible system in which a triple collision can occur: the collinear three-body problem.

For this system, $n = 3$, $\nu = 1$, $q_i \in R^1$, and $p_i \in R^1$. For notational purposes we shall consider $\underset{\sim}{q}$ and $\underset{\sim}{p}$ as column vectors in R^3. Let M be the 3×3 diagonal matrix $\mathrm{diag}(m_1, m_2, m_3)$. We can then write the kinetic energy

$$T(\underset{\sim}{p}) = \frac{1}{2} \underset{\sim}{p}^T M^{-1} \underset{\sim}{p}$$

and the differential equation (2.1)

$$(4.1) \qquad \begin{aligned} \dot{\underset{\sim}{q}} &= M^{-1} \underset{\sim}{p} \\ \dot{\underset{\sim}{p}} &= \nabla U(\underset{\sim}{q}) . \end{aligned}$$

We fix the center of mass at the origin. The position $\underset{\sim}{q}$ therefore lies on the plane

$$\underset{\sim}{Q} = \{\underset{\sim}{q} \in R^3 : m_1 q_1 + m_2 q_2 + m_3 q_3 = 0\} ,$$

(Figure 4) and the momentum vector lies on the plane

$$\underset{\sim}{P} = \{\underset{\sim}{p} \in R^3 : p_1 + p_2 + p_3 = 0\} .$$

Equations (4.1) give a vector field on the four-dimensional linear space $\underset{\sim}{Q} \times \underset{\sim}{P}$. Fixing the total energy, we have a vector field on the three-dimensional manifold

$$E(h) = \{(q,p) \in Q \times P\colon H(q,p) = h\} .$$

On $E(h)$ we have the energy relation

$$T(p) - U(q) = h . \tag{4.2}$$

Since the center of mass is fixed at the origin, triple collision can occur only at $q = 0$. Equations (4.1) are of course undefined at $q = 0$. We wish to choose new variables so that the transformed differential equations can be extended to points corresponding to triple collision. Let

$$\begin{aligned} r &= (q^T M q)^{1/2} \\ s &= r^{-1} q . \end{aligned} \tag{4.3}$$

Physically, r^2 is the moment of inertia and s is the configuration of the system of particles. Note that s is confined to the circle

$$S = \{s \in Q\colon s^T M s = 1\} .$$

One can think of (r,s) as polar coordinates on Q (Figure 5). Triple collision occurs at $r = 0$.

Now consider the energy relation (4.2). Since U is homogeneous of degree -1 and T is homogeneous of degree 2 the momentum p behaves like $r^{-1/2}$ near triple collision. We therefore choose a new variable on P:

$$u = r^{1/2} p \in P . \tag{4.4}$$

The energy relation (4.2) then becomes

$$T(u) - U(s) = rh . \tag{4.5}$$

The differential equations in the new coordinates can be computed:

$$\begin{aligned} \dot{r} &= r^{-1/2}\, \underset{\sim}{u}^T \underset{\sim}{s} \\ \dot{\underset{\sim}{s}} &= r^{-3/2} \left[-(\underset{\sim}{u}^T \underset{\sim}{s})\underset{\sim}{s} + M^{-1} \underset{\sim}{u} \right] \\ \dot{\underset{\sim}{u}} &= r^{-3/2} \left[\tfrac{1}{2} (\underset{\sim}{u}^T \underset{\sim}{s})\underset{\sim}{u} + \nabla U(\underset{\sim}{s}) \right] . \end{aligned} \tag{4.6}$$

The energy relation (4.5) defines a new manifold

$$\underset{\sim}{F}(h) = \{(r,\underset{\sim}{s},\underset{\sim}{u}) \in [0,\infty) \times \underset{\sim}{S} \times \underset{\sim}{P}\colon (4.5) \text{ holds}\} .$$

Note that this manifold contains points with $r = 0$ but that the manifold $\underset{\sim}{E}(h)$ contains no points with $\underset{\sim}{q} = 0$. Transformations (4.3) and (4.4) define a diffeomorphism between $\underset{\sim}{E}(h)$ and

$$\underset{\sim}{F}^0(h) = \{(r,\underset{\sim}{s},\underset{\sim}{u}) \in \underset{\sim}{F}(h)\colon r > 0\} .$$

This diffeomorphism carries the vector field (4.1) on $\underset{\sim}{E}(h)$ to the vector field (4.6) on $\underset{\sim}{F}^0(h)$.

Triple collision now corresponds to the manifold

$$\underset{\sim}{F}^c(h) = \underset{\sim}{F}(h) - \underset{\sim}{F}^0(h) = \{(r,\underset{\sim}{s},\underset{\sim}{u}) \in \underset{\sim}{F}(h)\colon r = 0\} .$$

Note that $\underset{\sim}{F}^c(h)$ is two dimensional and is the boundary of $\underset{\sim}{F}(h)$. Letting $r = 0$ in equation (4.5) we have

$$T(\underset{\sim}{u}) - U(\underset{\sim}{s}) = 0 . \tag{4.7}$$

Therefore $\underset{\sim}{F}^c(h)$ is independent of h and we can write

$$\underset{\sim}{F}^c = \{(\underset{\sim}{s},\underset{\sim}{u}) \in \underset{\sim}{S} \times \underset{\sim}{P}\colon (4.7) \text{ holds}\} .$$

We shall refer to $\underset{\sim}{F}^c$ as the triple collision manifold.

The vector field (4.6) is not defined on the triple collision manifold. However, if we scale by the time transformation

$$dt = r^{3/2}\, d\tau \tag{4.8}$$

then the scaled vector field becomes

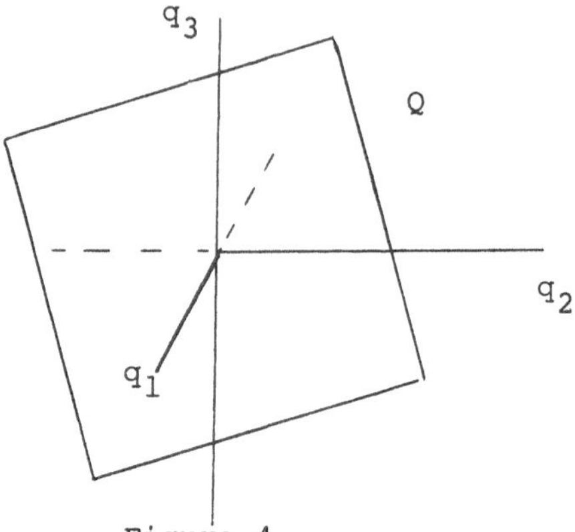

Figure 4

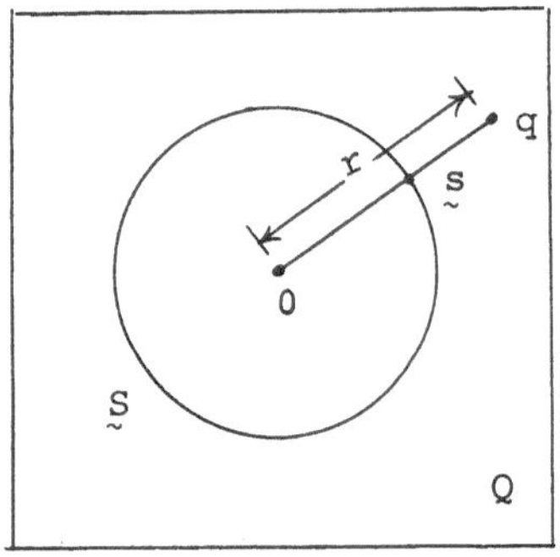

Figure 5

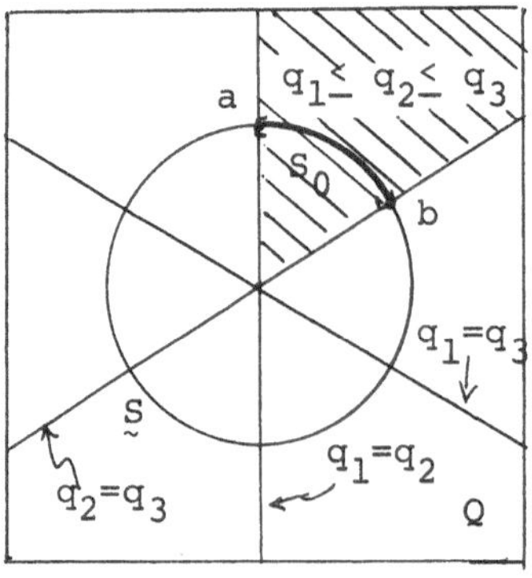

Figure 6

$$\frac{dr}{d\tau} = r\, \underset{\sim}{u}^T \underset{\sim}{s}$$

(4.9)
$$\frac{d\underset{\sim}{s}}{d\tau} = -\,(\underset{\sim}{u}^T \underset{\sim}{s})\underset{\sim}{s} + M^{-1}\underset{\sim}{u}$$

$$\frac{d\underset{\sim}{u}}{d\tau} = \frac{1}{2}\,(\underset{\sim}{u}^T \underset{\sim}{s})\underset{\sim}{u} + \nabla U(\underset{\sim}{s}) \;.$$

This vector field can be extended to $\underset{\sim}{F}^c$ to give a vector field defined on all of $\underset{\sim}{F}(h)$. The triple collision manifold $\underset{\sim}{F}^c$ is an invariant set.

We have now accomplished our first goal. We have pasted a boundary corresponding to triple collision onto the energy manifold. We have scaled the vector field so that it can be extended to the boundary. The scaling acts to slow down orbits ending in triple collision so that they now take an infinite amount of time to reach collision. By studying the flow on the boundary we can draw conclusions about the flow near the boundary and hence about orbits near triple collisions.

6. Coordinates

To facilitate our description of the flow on the triple collision manifold we introduce some new variables.

The order of the particles on the line does not change after a double collision, so we can assume that they satisfy $q_1 \le q_2 \le q_3$. This order determines a sector in $\underset{\sim}{Q}$ and an arc $\underset{\sim}{S}_0$ in $\underset{\sim}{S}$, as shown in Figure 6. The end points of the arc are $\underset{\sim}{a}$ and $\underset{\sim}{b}$, where $\underset{\sim}{a} = (a_1, a_2, a_3)$ is the unique point on $\underset{\sim}{S}$ with $a_1 = a_2 < a_3$ and $\underset{\sim}{b} = (b_1, b_2, b_3)$ is the unique point with $b_1 < b_2 = b_3$. One sees that there is a unique diffeomorphism

$$\sigma\colon [-1,1] \to \underset{\sim}{S}_0$$

with $\sigma(-1) = \underset{\sim}{a}$, $\sigma(1) = \underset{\sim}{b}$, and $\sigma'(s)^T M\, \sigma'(s) = \lambda^{-2}$, $\forall s \in [-1,1]$, where $\lambda > 0$ is a constant depending on the masses. This diffeomorphism defines a parametrization of the possible configurations of particles. The point -1 corresponds to the configuration with particles 1 and 2 coinciding. The point +1 corresponds to the configuration with particles 2 and 3 coinciding. Points in the open interval (-1,1) correspond to configurations with particle 2 between particles 1 and 3. In all of these configurations the order is always $q_1 \leq q_2 \leq q_3$.

We now introduce new variables $s \in [-1,1]$, $v \in R^1$, and $w \in R^1$ as follows:

$$\underset{\sim}{s} = \sigma(s) \tag{6.1}$$

$$\underset{\sim}{u} = vM\,\sigma(s) + w\lambda M\,\sigma'(s) \ .$$

The variable v can be interpreted as the radial component of momentum, while w can be interpreted as the tangential component (Figure 7).

In the new coordinates the energy relation (4.5) becomes

$$\frac{1}{2}(v^2 + w^2) - V(s) = rh \ , \tag{6.2}$$

where $V(s) = U(\sigma(s))$. The differential equations (4.9) can be computed:

$$\begin{aligned} \frac{dr}{d\tau} &= rv \\ \frac{ds}{d\tau} &= \lambda w \\ \frac{dv}{d\tau} &= \frac{1}{2}v^2 + w^2 - V(s) \\ \frac{dw}{d\tau} &= -\frac{1}{2}vw + \lambda V'(s) \ . \end{aligned} \tag{6.3}$$

This computation is straightforward if one recalls the Euler formula

$$\underset{\sim}{q}^T \nabla U(\underset{\sim}{q}) = -U(\underset{\sim}{q}) \ .$$

Equations (6.1) determine a diffeomorphism between $\underset{\sim}{F}(h)$ and

$$\underset{\sim}{G}(h) = \{(r,s,v,w) \in [0,\infty) \times [-1,1] \times R^1 \times R^1 \colon (6.2) \text{ holds}\} .$$

At triple collision $r = 0$, so the energy relation (6.2) becomes

$$\frac{1}{2}(v^2 + w^2) - V(s) = 0 . \tag{6.4}$$

The triple collision manifold $\underset{\sim}{F}^c$ is transformed to the manifold:

$$\underset{\sim}{G}^c = \{(s,v,w) \in [-1,1] \times R^1 \times R^1 \colon (6.4) \text{ holds}\} .$$

This manifold is a cylinder, as shown in Figure 8. The radius of the cylinder goes to infinity as $s \to \pm 1$.

The last three of equations (6.3) give a vector field on $\underset{\sim}{G}^c$. We have not yet regularized double collisions, so orbits flow off the manifold at $s = \pm 1$. However, one can show that a double collision orbit has the property: $s \to \pm 1$, $w \to \pm\infty$, and $v \to v^* < \infty$. The regularization discussed in Section 4 has the effect of pasting an orbit ending in double collision with a certain limiting value of v to the orbit beginning in double collision with the same limiting value of v. For example, the orbit with the property $s \to +1$, $w \to +\infty$, and $v \to v^*$ as $t \to t^*-$ is connected to the orbit with the property $s \to +1$, $w \to -\infty$, and $v \to v^*$ as $t \to t^*+$. These connections can be thought of as pasting the manifold together along two lines at $s = \pm 1$. The actual analytic procedure is to introduce a new variable w' with the property that $w' \to 0$ as an orbit approaches a double collision. The orbit is then extended through $w' = 0$. The details of this procedure are given in [4]. The resulting regularized triple collision manifold is a two-sphere minus four points and is shown in Figure 9.

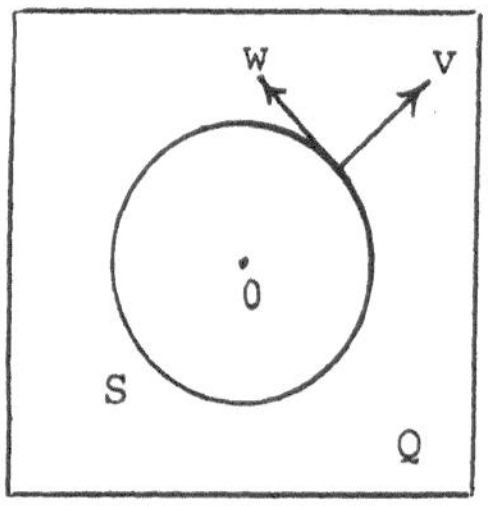

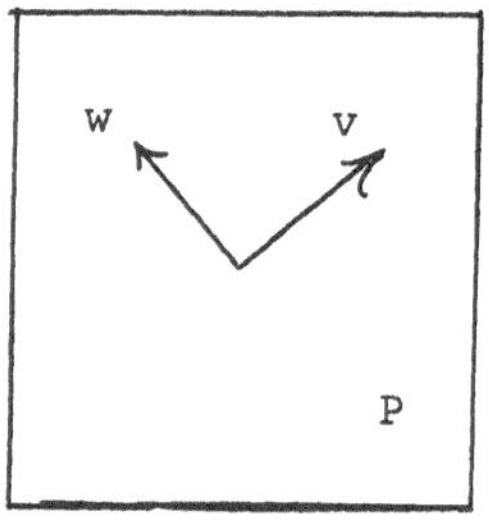

Figure 7

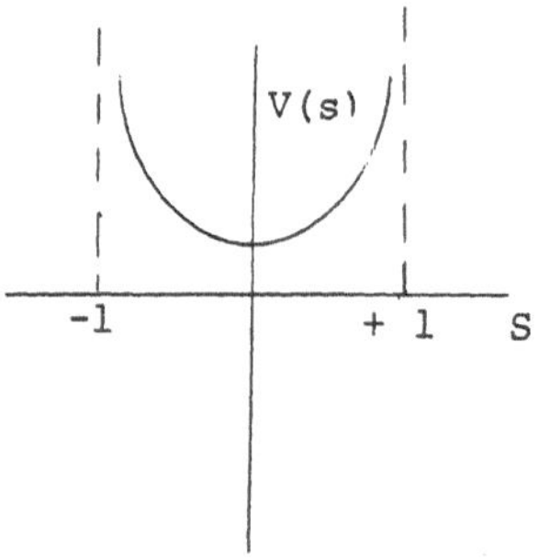

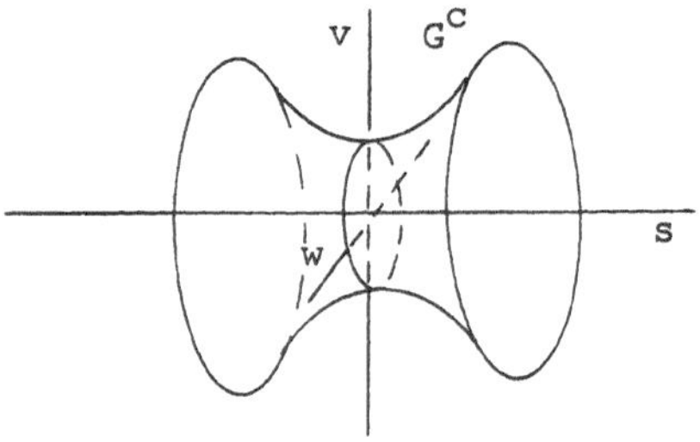

Figure 8

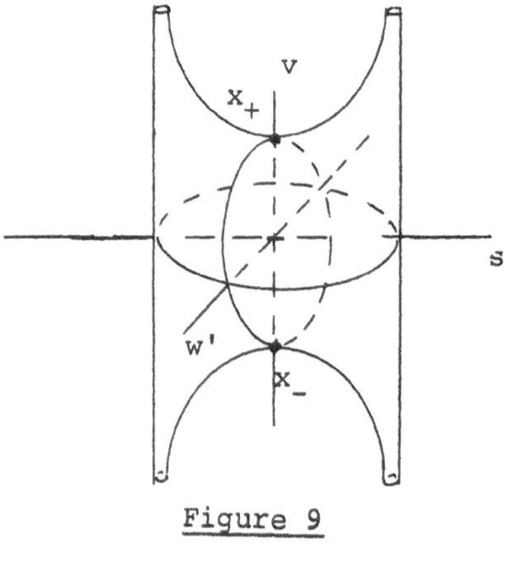

Figure 9

7. The Triple Collision Manifold

We can now give a description of the flow on the triple colli-collision manifold. First we shall find the rest points and show that the flow is gradient-like. Then we shall discuss the results of Sundman and Siegel as they apply to the collinear problem. Finally we shall discuss the behavior of orbits passing close to triple collision.

From equations (6.3) we see that a rest point $(s_0, v_0, w_0) \in \underset{\sim}{G}^c$ must satisfy $w_0 = 0$, $V'(s_0) = 0$, and $v_0^2 = 2V(s_0)$. One can show that $V'(s) = 0$ if and only if $\sigma(s)$ is a central configuration [4]. An appeal to classical results or a direct computation shows that there is only one central configuration with $q_1 < q_2 < q_3$, i.e. V' has a unique zero s_0 on $(-1,1)$. We thus have the existence of exactly two rest points for the flow on $\underset{\sim}{G}^c$. We label these $x_+ = (s_0, +v_0, 0)$ and $x_- = (s_0, -v_0, 0)$ (Figure 9).

Substituting equation (6.4) into the third of equations (6.3) we see that

$$\frac{dv}{d\tau} = \frac{1}{2} w^2$$

on $\underset{\sim}{G}^c$. Therefore v increases along orbits except at the two rest points. Such a function is called a "Liapunov function", and a flow with a Liapunov function is called "gradient-like".

An orbit ending in triple collision must be asymptotic to $\underset{\sim}{G}^c$. It follows from the gradient-like structure of the flow on $\underset{\sim}{G}^c$ that such an orbit must be asymptotic to one of the rest points [4]. Since the rest points correspond to the central configuration, we have established Sundman's result that orbits ending in triple collision must approach a central configuration.

We also see from the gradient-like structure that the two rest points are saddle points for the flow on $\underset{\sim}{G}^c$. In fact, a computation of the eigenvalues of the Jacobian matrix at each of the rest points shows that these points are hyperbolic. The first of equations (6.3) shows that x_- is attracting in the r-direction, while x_+ is repelling in the r-direction. Thus orbits ending in triple collision are asymptotic to x_-, while orbits beginning in triple collision are asymptotic to x_+. The stable manifold theorem applied to each of the rest points then gives Siegel's result for the collinear problem: the set of orbits ending in triple collision forms a real analytic immersed submanifold of codimension 1. (Figure 10).

Now consider an orbit starting near the stable manifold of x_-. This orbit will move toward x_- close to the stable manifold, pass by x_-, and then move away from x_- close to the unstable manifold. The closer it starts to the stable manifold, the longer it will stay close to the unstable manifold. Thus we see that the behavior of the unstable manifold of x_- determines the behavior of orbits passing close to triple collision.

The unstable manifold of x_- has two branches, each consisting of a single orbit. Which branch is followed by a near-collision orbit depends upon on which side of the stable manifold the orbit begins. The gradient-like structure of the flow on $\underset{\sim}{G}^c$ implies that each branch can behave in one of only two possible ways:

(1) it asymptotically approaches x_+ (Figure 11), or

(2) it goes up one of the two arms of $\underset{\sim}{G}^c$ extending to $v = +\infty$ (Figure 12). Of course, the flow on $\underset{\sim}{G}^c$ depends only on the values of the masses. It can be shown that behavior (1) occurs for some masses, while behavior (2) occurs for others [4]. The proof will be indicated below, but first consider the implications of the second behavior.

An orbit passing close to triple collision leaves a neighborhood of $\underset{\sim}{G}^c$ with arbitrarily large v. Recall that v is the radial component of the momentum. An orbit with large v has large kinetic energy. This kinetic energy comes from the loss in potential energy caused by the close proximity of two of the particles. By passing close to triple collision the system transfers an arbitrary amount of energy from potential to kinetic. The system emerges with two of the particles close together and travelling with large velocity in one direction and the third particle travelling with large velocity in the opposite direction.

This arbitrary transfer of energy precludes the possibility of regularizing triple collision in the sense of Easton. An orbit ending in triple collision would have to be connected to a non-existent orbit with infinite kinetic energy. This behavior also forms the basis for the construction of orbits in the four-body problem which become unbounded in finite time [3].

These results are valid only if behavior (2) holds. To see that this behavior actually does occur for some masses, consider the special case when the two outside particles have equal mass ($m_1 = m_3$). In this case reflection through the v-axis is a symmetry for the flow on $\underset{\sim}{G}^c$. Therefore either (a) the unstable manifold of x_- is exactly the stable manifold of x_+ (Figure 11) or (b) one branch of the unstable manifold of x_- goes up one arm of $\underset{\sim}{G}^c$ while the other branch goes up the other arm (Figure 12).

If the central mass m_2 is taken to be small, one can show that the unstable manifold of x_- spirals many times around the v-axis before arriving at x_+ [4]. This spiralling corresponds to the small central particle bouncing back and forth many times as the two outside masses approach and retreat. The amount of spiralling becomes infinite as the central mass approaches zero. Therefore

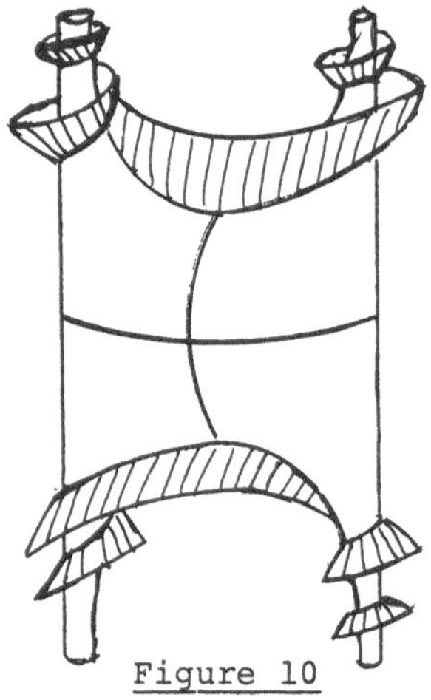

Figure 10

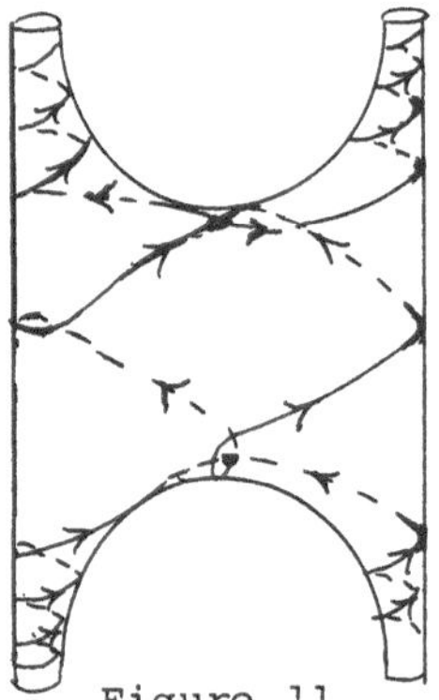

Figure 11

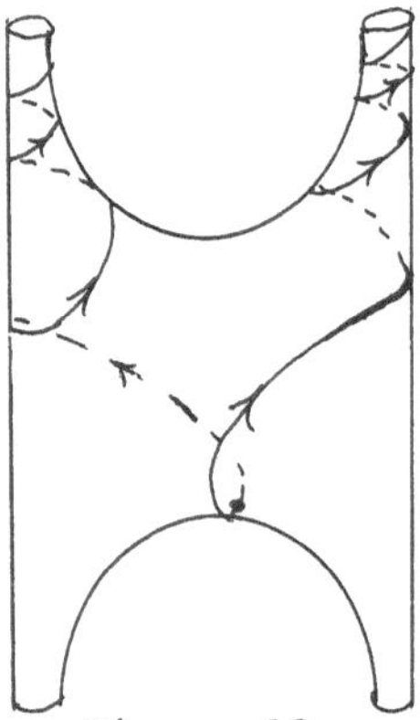

Figure 12

sometimes the unstable manifold of x_- hits x_+ and (a) occurs, while sometimes it misses x_+ and (b) occurs.

Behavior (b) has an interesting feature. One arm of $\underset{\sim}{G}^c$ has values of s close to -1, while the other arm has values of s close to +1. Therefore orbits passing close to triple collision have radically different behavior depending upon on which side of the stable manifold of x_- they begin. On one side orbits emerge with particles 1 and 2 close together, while on the other side they emerge with particles 2 and 3 close together. In both cases there is a large transfer of energy from potential to kinetic.

The set of masses for which behavior (2) occurs is clearly open and, as we have seen, nonempty. It follows from analyticity that this set is also dense. Therefore the arbitrary transfer of energy, and hence the nonregularizability of triple collision, occurs for almost all masses.

8. Other Collisions

The main tools in our study of triple collision in the collinear three-body problem are the transformations given in Section 5. These transformations allow us to study orbits near collision by studying the flow on a ficticious manifold introduced at collision.

Similar transformations can be used to introduce a similar ficticious manifold at n-tuple collision for the n-body problem in R^ν. The flow on this manifold is again gradient-like and rest points again correspond to central configurations. Therefore orbits ending in n-tuple collision must approach central configurations. A careful study of the flow on this manifold might provide interesting new insight into the behavior of orbits passing close to collision. In particular, triple collision in the planar three-body problem would be interesting to study.

The technique used in Section 5 also works for an arbitrary homogeneous potential. This technique may be able to provide new information for classical systems other than gravitational.

References

[1] Easton, R., Regularization of vector fields by surgery, J. Differential Equations 10 (1971) 92-99.

[2] Easton, R., The Topology of the regularization integral surfaces of the 3-body problem, J. Diff. Eqs. 12 (1972) 361-384.

[3] Mather, J., and McGehee, R., Orbits for the collinear four-body problem which become unbounded in finite time, in this work.

[4] McGehee, R., Triple Collision in the collinear three-body problem, Invent. Math., 27, (1974) 191-227.

[5] Painlevé, P., Lecons sur la théorie analytique des équations differentielles, (Stockholm, 1895), A. Hermann, Paris (1897).

[6] Palmore, J., Classifying relative equilibria I, Bull. Amer. Math. Soc. 79, 5 (1973) 904-908.

[7] Palmore, J., Classifying relative equilibria II, preprint.

[8] Siegel, C., Der Dreierstoss, Ann. Math. 42 (1941) 127-168.

[9] Siegel, C., and Moser, J., Lectures on Celeastial Mechanics, Springer Verlag (1971).

[10] Sperling, H., On the real singularities of the n-body problem, J. Reine Angew. Math. 245 (1970) 14-50.

[11] Sundman, K., Mémoire sur la problème des trois corps, Acta Math. 36 (1912) 105-179.

[12] Sundman, K., Nouvelles recherches sur la problème des trois corps, Acta Soc. Sci. Fenn. 35, 9 (1909).

[13] von Zeipel, H., Sur les singularités du problème des n corps, Ark. Mat. Astr. Fys. 4, No. 32 (1908).

[14] Wintner, A., The Analytical Foundations of Celestial Mechanics, Princeton University Press (1941).

SOLUTIONS OF THE COLLINEAR FOUR BODY PROBLEM WHICH BECOME UNBOUNDED IN FINITE TIME

J. N. Mather *

Department of Mathematics, Harvard University, Cambridge, Mass. 02138

R. McGehee **

School of Mathematics, University of Minnesota, Minneapolis, Minn.

In this paper, we prove the existence of solutions of the (Newtonian) collinear four body problem which become unbounded in finite time. We consider four point masses m_1, m_2, m_3, m_4 moving on the line according to the inverse square law

$$\ddot{q}_i = \sum_{j \neq i} m_j \frac{q_j - q_i}{|q_j - q_i|^3},$$

where q_i denotes the position of the ith particle. Binary collisions will be regularized, so that the system is defined and continuous beyond a binary collision, However, triple collisions will be regarded as singularities. This means that if a triple collision occurs at t_0, then $q_i(t)$ is defined only for $t < t_0$. A detailed description of this set-up is given in [1]. Note that $q_1 \leq q_2 \leq q_3 \leq q_4$ for all t.

The time parameter in the regularized system differs from the time parameter in the original system. In this paper, t always means the time parameter for the original system.

We will show the existence of solutions which become unbounded in finite time for special values of the masses. These solutions will be of the following type. There will exist an increasing bounded sequence $t_1, t_2, t_3, \ldots$ of times at which binary collisions

* This research was partially supported by NSF Grant No. GP-43613X and a Sloan Foundation Fellowship.
Currently visiting at Princeton University.
** Partially supported by National Science Foundation Grant GP38955.

occur. As $t \to t_\infty = \lim t_i$, we will have that $q_1 \to -\infty$, $q_3, q_4 \to +\infty$, and q_2 bounces back and forth infinitely often between q_1 and q_3. We will also have that $q_4 - q_3 \to 0$. The loss of potential energy stemming from $q_4 - q_3 \to 0$ provides the kinetic energy which causes the solution to become unbounded in finite time.

1. Statement of Results

By phase space, we will mean the phase space of the collinear four body problem, subject to the conditions $q_1 \leq q_2 \leq q_3 \leq q_4$, where the binary collisions are regularized. Phase space is an 8-manifold. By fixing the center of mass at the origin, we obtain a submanifold of dimension 6. We let Ω be any energy hypersurface of the latter manifold. Then Ω is a 5-manifold.

We let Σ be the subset of Ω consisting of initial conditions such that the corresponding trajectories end in triple collision of the 2nd, 3rd, and 4th particles, i.e., $q_2 = q_3 = q_4$. Theorem 2 below asserts that Σ is a codimension 1 (immersed) manifold.

The following result is the main theorem of this paper.

Theorem 1. There exist positive values m_1, m_2, m_3, m_4 of the masses such that the following holds. Let Γ be any arc in Ω crossing Σ. Then there is an uncountable set A of points on Γ having the following property. Let $x \in A$. There exists $t_0 > 0$ such that the trajectory with initial condition x is defined for $t < t_0$ and satisfies

$$q_1(t) \to -\infty, \quad q_3(t), q_4(t) \to +\infty, \quad \text{as} \quad t \to t_0 .$$

The rest of this paper consists of a proof of this theorem.

If $x \in \Sigma$, let $\tau(x)$ denote the time at which the trajectory with initial condition x ends in triple collision. The proof of Theorem 1 is based on the following theorem.

Theorem 2. Σ is a codimension 1 immersed submanifold of Ω. Furthermore there exist positive values of the masses such that

$$\frac{m_1-m_2}{m_1+m_2} > \frac{m_2}{m_3+m_4} \tag{1}$$

and the following holds.

Let Γ be an arc in Ω crossing Σ. Then there is a subarc Γ_0 of Γ, having one end point x_0 in Σ, and a continuous positive function τ_- defined on Γ_0 such that

(a) $\tau_-(x_0) = \tau(x_0)$

(b) If $x \in \Gamma_0 - \{x_0\}$, then the trajectory with initial condition x is defined for $0 \leq t \leq \tau_-(x)$ and $q_2(\tau_-(x)) = 0$.

(c) $\dot{q}_2(\tau_-(x)) \to -\infty$ as $x \to x_0$.

Theorem 2 will be proved in Section 5.

Now we outline the proof of Theorem 1, assuming Theorem 2. We assume that the masses satisfy (1) and the conclusion of Theorem 2 holds. Let $x \in \Gamma_0$, $x \neq x_0$, and consider the trajectory with initial datum x. At $t = \tau_-(x)$, the second particle crosses the origin, traveling very fast in the negative direction. It then collides with the first particle at a time $\hat{\tau}(x) > \tau_-(x)$. Next, it passes through the origin again, this time in the positive direction, at a time $\tau_+(x) > \hat{\tau}(x)$. It then either collides with the third particle, or the third and fourth particle simutaneously, at a time $\tau'(x) > \tau_+(x)$. In the latter case, we have triple collision, and $x \in \Sigma$.

Let $\tau^*(x)$ be the time of the last collision between the second and third particle before $\tau_-(x)$. Let $N(x)$ be the number of collisions between the third and fourth particle in the time interval $(\tau^*(x),\tau'(x))$. Then $N(x) \to \infty$ as $x \in \Gamma - x_0$ approaches

x_0. Each time the value of N changes, Γ_0 crosses Σ, so we deduce that Γ_0 crosses Σ infinitely often.

Then we may apply Theorem 2 again, this time for Γ_0 in place of Γ. Thus, we find a subarc Γ_1 in the interior of Γ_0, having an endpoint x_1 in Σ such that the conclusions of Theorem 2 hold. Moreover $\tau(x_1) \to \tau'(x_1)$, and $\tau'(x) \to \tau(x_0)$ as $x \in \Gamma_0 - x_0$ approaches x_0, so we may choose x_1 so that $\tau(x_1) \leq \tau(x_0) + \varepsilon$ for $\varepsilon > 0$ arbitrarily small.

Continuing in this way, we choose an arc Γ_2 in the interior of Γ_1 with endpoint $x_2 \in \Sigma$, so the conclusions of Theorem 2 hold, and then an arc Γ_3 in the interior of Γ_2 with end point $x_3 \in \Sigma$, etc. We may make the choices so that x_n converges, and $\tau(x_n)$ converges. Let $x = \lim x_n$ and $t_0 = \lim \tau(x_n)$. Then the trajectory with initial datum x is defined for $t < t_0$ and satisfies the conditions $q_1(t) \to -\infty$, $q_3(t), q_4(t) \to +\infty$ as $t \to t_0$.

To get the uncountable set A of Theorem 1, we need only modify the above proof slightly. Since Γ_0 crosses Σ infinitely often, we may choose two disjoint arcs Γ_1^0 and Γ_1^1 in the interior of Γ_0 which satisfy the conditions we required of Γ_1 before. Likewise we may choose two disjoint arcs Γ_2^{i0} and Γ_2^{i1} in Γ_1^i which satisfy the conditions we required of Γ_2 above. Continuing in this way we get a nested sequence of arcs which converge to a Cantor set. Any initial condition in the Cantor set gives a trajectory of the required type.

We will give a formal proof of Theorem 1 (assuming Theorem 2) in the following sections.

2. Return to Triple Collision

In this section, we will give the details of the first part of the proof of Theorem 2, sketched in the end of the previous section. Our main objective is to show that after the system passes close to triple collision of the 2nd, 3rd, and 4th particles, it can return to triple collision of the same particles.

Definition. By an α-*arc*, we will mean an arc Γ_0 satisfying the conclusions of Theorem 2. We will call x_0 the *terminal point* of Γ_0, and τ_- the τ_--function on Γ_0. We let $\mu(\Gamma_0) = \tau(x_0)$.

Throughout this section, the α-arc Γ_0, the terminal point x_0, and τ_- will be fixed. We let $x \in \Gamma_0 - x_0$. We consider the trajectory with initial condition x. We let

$$\tau^*(x) = \sup\{t: q_2(t) = q_3(t) \text{ and } t < \tau_-(x)\}$$

$$\hat{\tau}(x) = \inf\{t: q_1(t) = q_2(t) \text{ and } t > \tau_-(x)\}$$

$$\tau_+(x) = \inf\{t: q_2(t) = 0 \text{ and } t > \hat{\tau}(x)\}$$

$$\tau'(x) = \inf\{t: q_2(t) = q_3(t) \text{ and } t > \tau_+(x)\}$$

$$V_-(x) = -\dot{q}_2(\tau_-(x))$$

$$V_+(x) = \dot{q}_2(\tau_+(x)) .$$

It is obvious that $\tau^*(x) < \tau_-(x) < \hat{\tau}(x) < \tau_+(x) < \tau'(x)$ if these numbers exist.

For any subset S of $\{1,2,3,4\}$ we define the center of mass

$$q_S = \sum_{i\in S} m_i q_i \Big/ \sum_{i\in S} m_i .$$

We let

$$W_-(x) = \dot{q}_{34}(\tau_-(x))$$
$$W_+(x) = \dot{q}_{34}(\tau_+(x)) .$$

Lemma 1. For x sufficiently close to x_0, we have that $\tau^*(x)$, $\hat{\tau}(x)$, $\tau_+(x)$, $\tau'(x)$ exist. Moreover,

$$V_+ = \frac{m_1-m_2}{m_1+m_2} V_- + o(V_-) \tag{1}$$

$$W_- = \frac{m_2}{m_3+m_4} V_- + o(V_-) \tag{2}$$

$$W_+ = W_- + o(V_-) \tag{3}$$

$$\tau^*,\ \tau_-,\ \hat{\tau},\ \tau_+,\ \tau' \to \tau(x_0) \tag{4}$$

as $x \to x_0$.

Proof: The existence of $\tau^*(x)$ and the fact that $\tau^* \to \tau(x_0)$ are obvious. The existence of τ_- and the fact that $\tau_- \to \tau(x_0)$ is part of Theorem 2. For the proof of the remaining assertions, we use the fact that $V_- \to \infty$ as $x \to x_0$.

We will also need that the change in momentum and energy in a subsystem which is distant from the remaining particles is relatively small. For momentum, this is easy. For $S \subset \{1,2,3,4\}$, we define the momentum of the subsystem S to be

$$p_S = m_S\dot{q}_S = \sum_{i\in S} m_i\dot{q}_i$$

where m_S is the mass of the subsystem S:

$$m_S = \sum_{i\in S} m_i .$$

It follows immediately from the inverse square law that

$$\dot{p}_S = \sum_{i\in S}\sum_{j\notin S} m_i m_j \frac{q_i - q_j}{|q_i-q_j|^3} .$$

It is easily seen that $q_1(\tau_-(x))$, $p_1(\tau_-(x))$, $q_{234}(\tau_-(x))$, and $p_{234}(\tau_-(x))$ converge as $x \to x_0$, since $\tau_-(x) \to \tau(x_0)$ as $x \to x_0$. Furthermore, $q_{34}(\tau_-(x))$ converges as $x \to x_0$, since

$q_2(\tau_-(x)) = 0$. Since $\dot{q}_2(\tau_-(x)) \to -\infty$ as $x \to x_0$, we may assume that the second particle has velocity arbitrarily above escape velocity relative to the third and fourth particles as $x \to x_0$. Since the position and velocity of the first particle are bounded for $x \in \Gamma_0$, we deduce the existence of $\hat{\tau}(x)$ for x sufficiently close to x_0.

To go further, we will need an estimate for the change of internal energy of a subsystem. If $S \subset \{1,2,3,4\}$, we define the internal energy of S to be

$$h_S = \sum_{i\in S} \frac{m_i}{2} (\dot{q}_i - \dot{q}_S)^2 - \sum_{i,j\in S} \frac{m_i m_j}{|q_i - q_j|} .$$

We let $r_{ij} = |q_i - q_j|$, $S' = \{1,2,3,4\} - S$, and $r_{SS'} = \inf \{r_{ij}: i \in S, j \in S'\}$.

<u>Lemma 2</u>. Let $\rho, r^0 > 0$. There exists $C > 0$ such that the following holds for any trajectory. If $h_{12}(t_0) > 0$, $r_{12}(t_0) \leq r^0$, and $r_{SS'}(t_0) \geq \rho$, then

$$\frac{d}{dt} \log h_{12}(t_0) \leq C(h_{12}(t_0))^{-1/2}.$$

This will be proved in Section 3.

By taking x sufficiently close to x_0, we can take V_- arbitrarily large and therefore also $h_{12}(\tau_-)$ arbitrarily large. Let $\varepsilon > 0$, and let $t_1(x) = \min(\tau_-(x)+\varepsilon, \tau_+(x))$ if $\tau_+(x)$ exists and $t_1(x) = \tau_-(x)+\varepsilon$ otherwise. Since $-p_{12}(\tau_-)$ is arbitrarily large, we see that the lemma applies, and by taking x sufficiently close to x_0, we can arrange that the change in $\log h_{12}(t)$ is arbitrarily small in the interval $\tau_- \leq t \leq t_1$. It follows that the second particle must return to the origin in arbitrarily small time, i.e., $\tau_+(x)$ exists, and $\tau_+(x) - \tau_-(x) \to 0$ as $x \to x_0$. (Here, we use $m_1 > m_2$.)

Then $|\log h_{12}(\tau_+) - \log h_{12}(\tau_-)| \to 0$ as $x \to x_0$, $|p_{12}(\tau_+) - p_{12}(\tau_-)| \to 0$ as $x \to x_0$, so we get (1). Since $p_{234}(\tau_-)$ approaches a limit as $x \to x_0$, we get (2). Since $|p_{34}(\tau_-) - p_{34}(\tau_+)| \to 0$, as $x \to x_0$, we get (3). As for (4), we have already shown $\tau^*, \tau_-, \hat{\tau}, \tau_+ \to \tau(x_0)$.

It only remains to show τ' exists and $\tau' \to \tau(x_0)$. However, from the inequality (1) in Section 1, we get that $V_+ > W_+$ for x sufficiently close to x_0, and the second particle must eventually overtake the third and fourth particles, and in arbitrarily short time.

For what we do next, we will need:

<u>Lemma 3</u>.

$$\sup_{\tau_- \le t \le \tau_+} |h_{34}(t) - h_{34}(\tau_-)| \to 0 , \quad \text{as} \quad x \to x_0 .$$

This will be proved in Section 3.

As $x \to x_0$, we have $V_- \to -\infty$. Thus if $C < m_2/2$, the kinetic energy is $> CV_-^2$ for x sufficiently close to x_0. From conservation of energy, and the fact that $q_1(\tau_-), q_2(\tau_-), q_{34}(\tau_-)$ converge to distinct limits as $x \to x_0$, it follows that there exists $C_1 > 0$ such that

$$h_{34}(\tau_-) < -C_1 V_-^2$$

for all x sufficiently near x_0. From Lemma 3, we obtain that if $C_2 < C_1$ then

$$\sup_{\tau_- \le t \le \tau_+} h_{34}(t) < -C_2 V_-^2$$

if x is sufficiently near x_0. Therefore

(5) $$\sup_{\tau_- \le t \le \tau_+} r_{34}(t) \le C_3 V_-^{-2} ,$$

for a suitable $C_3 > 0$ and all x sufficiently near x_0. Since we are considering an inverse square law, there exists $C_4 > 0$ such that

(6) $$\inf_{\tau_- \le t \le \tau_+} \ddot{r}_{34}(t) \le - C_4 V_-^4$$

for all x sufficiently near x_0 (where we make the convention that $\ddot{r}_{34}(t) = -\infty$ if $q_3(t) = q_4(t)$). Combining (5) and (6), we get

Lemma 4. There exists $C_5 > 0$ such that the maximum time between collisions of the third and fourth particles for $\tau_- \le t \le \tau_+$ is $\le C_5 V_-^{-3}$.

On the other hand, it is easily seen that there exists $C_6 > 0$ such that $\tau_+ - \tau_- > C_6 V_-^{-1}$ if x is sufficiently close to x_0. Combining this remark with Lemma 4, we see that the number of collisions between the third and fourth particles in the time interval $[\tau_-, \tau_+]$ goes to ∞ as $x \to x_0$. Therefore, if we let $N(x)$ be the number of collisions between the third and fourth particles in the time interval $[\tau^*, \tau']$, we find

(7) $$N(x) \to \infty, \quad \text{as} \quad x \to x_0.$$

Since the value of the integer valued function $N(x)$ changes only when $x \in \Gamma_0$ crosses Σ, we obtain:

Lemma 5. Γ_0 crosses Σ infinitely often. Moreover, for any $\varepsilon > 0$, we may choose an infinite sequence $\Gamma_1^0, \Gamma_1^1, \Gamma_1^2, \ldots$ of disjoint α-arcs in Γ_0 such that for all i, we have $\mu(\Gamma_1^i) < \tau(x_0) + \varepsilon$.

3. Change of Energy of a Subsystem

In this section, we will prove Lemmas 2 and 3 in Section 2. We will be interested in the change in energy of a two-particle subsystem. Say $S = \{1,2\}$. Then

$$h_S = h_{12} = \frac{m_1 m_2}{2(m_1+m_2)} (\dot{q}_1-\dot{q}_2)^2 - \frac{m_1 m_2}{r_{12}} .,$$

where $r_{ij} = |q_i-q_j|$. Moreover,

$$\dot{h}_{12} = \frac{m_1 m_2}{m_1+m_2} (\dot{q}_1-\dot{q}_2)(\ddot{q}_1-\ddot{q}_2) + \frac{m_1 m_2}{r_{12}^3} (q_1-q_2)(\dot{q}_1-\dot{q}_2) .$$

The inverse square law gives

$$\ddot{q}_1-\ddot{q}_2 = -(m_1+m_2) \frac{q_1-q_2}{r_{12}^3} + R ,$$

where

$$R = \sum_{i \notin S} m_i \left(\frac{q_i-q_1}{r_{i1}^3} - \frac{q_i-q_2}{r_{i2}^3} \right)$$

Then

$$\dot{h}_{12} = \frac{m_1 m_2}{m_1+m_2} (\dot{q}_1-\dot{q}_2) R .$$

Clearly,

$$R = \sum_{i \notin S} m_i \left(\frac{q_2-q_1}{r_{i2}^3} + (q_1-q_i)\left(\frac{1}{r_{i2}^3} - \frac{1}{r_{i1}^3}\right) \right) .$$

If $r_{i1} \leq r_{i2}$, then

$$\left| \frac{q_2-q_1}{r_{i2}^3} + (q_1-q_i)\left(\frac{1}{r_{i2}^3} - \frac{1}{r_{i1}^3}\right) \right| \leq 4 \frac{r_{12}}{r_{SS'}^3} ,$$

where $r_{SS'} = \inf\{r_{ij}: i \in S, j \in S'\}$ and S' denotes the complement of S. This is also true if $r_{i2} \leq r_{i1}$, since the quantity on the left side can be rewritten as

$$\left| \frac{q_i-q_1}{r_{i1}^3} - \frac{q_i-q_2}{r_{i2}^3} \right|$$

which is left unchanged by transposition of {1,2}. Therefore

$$|R| \leq 4\left(\sum_{i\notin S} m_i\right) \frac{r_{12}}{r_{SS'}^{\ 3}}$$

and

(1) $$|\dot{h}_{12}| \leq 4\left(\sum_{i\notin S} m_i\right) |\dot{q}_1-\dot{q}_2| \frac{r_{12}}{r_{SS'}^{\ 3}} .$$

Let $v = |\dot{q}_1-\dot{q}_2|$. If $h_{12} < 0$, we can solve the problem of maximizing vr_{12} subject to the condition

(2) $$h_{12} = \frac{1}{2}\frac{m_1m_2}{m_1+m_2} v^2 - \frac{m_1m_2}{r_{12}} .$$

and we find that the maximum occurs at

$$r_{12} = -\frac{m_1m_2}{2h_{12}}$$

and takes the value

$$vr_{12} = \sqrt{-\frac{m_1m_2(m_1+m_2)}{2h_{12}}}$$

Therefore

(3) $$|\dot{h}_{12}| \leq C(-h_{12})^{-1/2}\, r_{SS'}^{\ -3}$$

where

$$C = 4\left(\sum_{i\notin S} m_i\right) \sqrt{m_1m_2(m_1+m_2)/2} .$$

Obviously, we can apply the above to the subsystem {3,4}, as well as the subsystem {1,2}, and Lemma 3 follows immediately.

Now we consider the case when $h_{12} > 0$. Let $v = |\dot{q}_1-\dot{q}_2|$. It is easily seen that vr_{12} is an increasing function of r_{12} for $r_{12} > 0$ and fixed h_{12}. Therefore, it follows from (1) that

$$|\dot{h}_{12}| \leq \sqrt{C_2h_{12} + C_3}$$

where

$$C_2 = \left(4\left(\sum_{i\notin S} m_i\right) r^0\, \rho^{-3}\right)^2 \frac{2(m_1+m_2)}{m_1m_2}$$

$$C_3 = \left[4\left(\sum_{i \notin S} m_i\right) r^0 \rho^{-3}\right]^2 \frac{2(m_1+m_2)}{r^0} .$$

Lemma 2 follows immediately.

4. Proof of Theorem 1

By Theorem 2, we may choose an α-arc Γ^0 in Γ . Let x_0 be the terminal point of Γ^0. The following assertion is a consequence of Lemma 5 and induction on k. For each i, $0 \leq i \leq k$, and each sequence $(j_1,\ldots,j_i)$ of 0's and 1's, there exists an α-arc $\Gamma^i_{j_1\ldots j_i}$ such that

(a) $$\Gamma^i_{j_1\ldots j_i} \subset \Gamma^{i-1}_{j_1\ldots j_{i-1}} , \quad \Gamma^1_j \subset \Gamma^0$$

(b) for fixed i, the various $\Gamma^i_{j_1\ldots j_i}$ are disjoint, and

(c) $\mu(\Gamma^i_{j_1\ldots j_i}) \leq \tau(x_0) + 2^{-1} + 2^{-2} + \ldots + 2^{-i}$.

Let $\Gamma^i = \cup\, \Gamma^i_{j_1\ldots j_i}$, where the union is taken over all sequences $(j_1,\ldots j_i)$ of 0's and 1's of length i. Let

$$A = \bigcap_i \Gamma^i .$$

Clearly A is uncountable. For $x \in A$, the trajectory with initial datum x must cross the origin infinitely often as $t \to t_0 < \tau(x) + 1$. For such x, the trajectory has the behavior described in Theorem 1.

5. Proof of Theorem 2

Consider the three-particle subsystem {2,3,4}. Using the techniques of [1], we shall show that solutions passing close to triple collision can emerge with particle 2 having arbitrarily large velocity. This behavior is known to occur for certain values of the masses with $m_2 = m_4$. Since $m_2/(m_3+m_4) < 1$, we can choose m_1 so large that (1) holds.

Denote the momentum of particle i by $p_i = m_i\dot{q}_i$ and write $\underset{\sim}{q} = (q_1,q_2,q_3,q_4)$ and $\underset{\sim}{p} = (p_1,p_2,p_3,p_4)$. The Hamiltonian for the system is

$$H(\underset{\sim}{q},\underset{\sim}{p}) = T(\underset{\sim}{p}) - U(\underset{\sim}{q}) ,$$

where

$$T(\underset{\sim}{p}) = \frac{1}{2}\sum_{i=1}^{4} \frac{p_i^2}{m_i}$$

$$U(\underset{\sim}{q}) = \sum_{i<j} \frac{m_i m_j}{|q_i - q_j|} .$$

We consider the system as having two components, (1) the three-particle system {2,3,4}, and (2) the two-particle system composed of particle 1 and the center of mass of subsystem {2,3,4}. Denote this center of mass by

$$\bar{x} = \frac{m_2q_2 + m_3q_3 + m_4q_4}{m_2 + m_3 + m_4}$$

and let

$$\bar{y} = \frac{p_2 + p_3 + p_4}{m_2 + m_3 + m_4} .$$

Define new variables for the subsystem {2,3,4}:

$$x_1 = q_2 - \bar{x} \qquad y_1 = p_2 - m_2\bar{y}$$

$$x_2 = q_3 - \bar{x} \qquad y_2 = p_3 - m_3\bar{y}$$

$$x_3 = q_4 - \bar{x} \qquad y_3 = p_4 - m_4\bar{y} .$$

Note that

$$\underset{\sim}{x} = (x_1,x_2,x_3) \in X = \{\underset{\sim}{x} \in R^3\colon m_1x_1 + m_2x_2 + m_3x_3 = 0\}$$

$$\tilde{y} = (y_1, y_2, y_3) \in Y = \{\tilde{y} \in R^3 : y_1 + y_2 + y_3 = 0\} .$$

Define new variables for the two-particle system:

$$z_1 = q_1 \qquad w_1 = p_1$$
$$z_2 = \bar{x} \qquad w_2 = (m_2+m_3+m_4)\, \bar{y} .$$

Since the center of mass of the whole system is fixed at the origin we have

$$\tilde{z} = (z_1, z_2) \in Z = \{\tilde{z} \in R^2 : m_1 z_1 + (m_2+m_3+m_4) z_2 = 0\}$$
$$\tilde{w} = (w_1, w_2) \in W = \{\tilde{w} \in R^2 : w_1 + w_2 = 0\}.$$

We think of $\tilde{x}, \tilde{y}, \tilde{z}, \tilde{w}$ as column vectors and define two diagonal matrices:

$$M_1 = \text{diag}\,(m_2, m_3, m_4)$$
$$M_2 = \text{diag}\,(m_1, m_2+m_3+m_4) .$$

We can now write the Hamiltonian

$$H(\tilde{x}, \tilde{z}, \tilde{y}, \tilde{w}) = \frac{1}{2} \tilde{y}^T M_1^{-1} \tilde{y} + \frac{1}{2} \tilde{w}^T M_2^{-1} \tilde{w} - U_1(\tilde{x}) - U_2(\tilde{x}, \tilde{z}) ,$$

where

$$U_1(\tilde{x}) = \frac{m_2 m_3}{|x_2 - x_3|} + \frac{m_2 m_4}{|x_2 - x_4|} + \frac{m_3 m_4}{|x_3 - x_4|}$$

$$U_2(\tilde{x}, \tilde{z}) = \frac{m_1 m_2}{|z_1 - z_2 - x_1|} + \frac{m_1 m_3}{|z_1 - z_2 - x_2|} + \frac{m_1 m_4}{|z_1 - z_2 - x_3|} .$$

The equations of motion in these coordinates are

$$\dot{\tilde{x}} = M_1^{-1} \tilde{y}$$
$$\dot{\tilde{y}} = \nabla U_1(\tilde{x}) + \nabla_1 U_2(\tilde{x}, \tilde{z})$$
$$\dot{\tilde{z}} = M_2^{-1} \tilde{w}$$
$$\dot{\tilde{w}} = \nabla_2 U_2(\tilde{x}, \tilde{z}) .$$

Here ∇_1 (∇_2) denotes the gradient with respect to the first (second) variable.

We now introduce the variables used in [1]. Let

$$r = (\underset{\sim}{x}^T M_1 \underset{\sim}{x})^{1/2}$$
$$\underset{\sim}{s} = \underset{\sim}{x}/r$$
$$\underset{\sim}{u} = r^{1/2} \underset{\sim}{y} .$$

Note that r^2 is the moment of inertia of the three-particle system {2,3,4} with respect to its center of mass. Also,

$$\underset{\sim}{s} \in S = \{\underset{\sim}{x} \in X: \underset{\sim}{x}^T M_1 \underset{\sim}{x} = 1\} .$$

Since X is a two-dimensional linear space, S is homeomorphic to a circle. The equations of motion in these variables become

$$\dot{r} = r^{-1/2} \underset{\sim}{u}^T \underset{\sim}{s}$$

$$\dot{\underset{\sim}{s}} = r^{-3/2} [-(\underset{\sim}{u}^T \underset{\sim}{s})\underset{\sim}{s} + M_1^{-1}\underset{\sim}{u}]$$

$$\dot{\underset{\sim}{u}} = r^{-3/2} [\tfrac{1}{2} (\underset{\sim}{u}^T\underset{\sim}{s})\underset{\sim}{u} + \nabla_1 U(\underset{\sim}{s}) + r^2 \nabla_1 U_2(r\underset{\sim}{s},\underset{\sim}{z})]$$

$$\dot{\underset{\sim}{z}} = M_2^{-1} \underset{\sim}{w}$$

$$\dot{\underset{\sim}{w}} = \nabla_2 U_2(r\underset{\sim}{s},\underset{\sim}{z}) .$$

Points on S corrrespond to configurations for the system {2,3,4}. There is an arc of configurations such that $q_2 \leq q_3 \leq q_4$. Let $\sigma: [-1,1] \to S^1$ be a diffeomorphism onto this arc with the property that

$$\sigma'(s)^T M_1 \sigma'(s) = \lambda^{-2} ,$$

where λ is a constant. If we assume that $s = -1$ corresponds to $q_2 = q_3$ and $s = +1$ corresponds to $q_3 = q_4$, then σ is unique and can be written explicitly [1]. Let $\underset{\sim}{z}_0 \in Z$ correspond to a point with $q_1 < 0$ and satisfy $\underset{\sim}{z}_0^T M_2 \underset{\sim}{z}_0 = 1$.

We now introduce new variables s, v, u, z, and w satisfying

$$\underset{\sim}{s} = \sigma(s)$$
$$\underset{\sim}{u} = vM_1\sigma(s) + u\lambda M_1\sigma'(s)$$
$$\underset{\sim}{z} = z\,\underset{\sim}{z}_0$$
$$\underset{\sim}{w} = wM_2\underset{\sim}{z}_0\ .$$

We also define

$$V_1(s) = U_1(\sigma(s))$$
$$V_2(r,s,z) = U_2(r\sigma(s), z\,\underset{\sim}{z}_0)$$

The equations of motion can then be written

$$\text{(1)}\quad \begin{aligned}
\dot{r} &= r^{-1/2}\, v \\
\dot{s} &= r^{-3/2}\, \lambda\, u \\
\dot{v} &= r^{-3/2}[\tfrac{1}{2}\, v^2 + u^2 - V_1(s) + r^2\, \frac{\partial V_2}{\partial r}\, (r,s,z)] \\
\dot{u} &= r^{-3/2}[-\, \tfrac{1}{2}\, vu + \lambda V_1'(s) + \lambda r\, \frac{\partial V_2}{\partial s}\, (r,s,z)] \\
\dot{z} &= w \\
\dot{w} &= \frac{\partial V_2}{\partial z}\, (r,s,z)\ .
\end{aligned}$$

Again as in [1] we make the time transformation

$$dt = r^{3/2}\, dt'$$

so as to slow down orbits ending in triple collision. We then have

$$\text{(2)}\quad \begin{aligned}
\frac{dr}{dt'} &= rv \\
\frac{ds}{dt'} &= \lambda u \\
\frac{dv}{dt'} &= \frac{1}{2}\, v^2 + u^2 - V_1(s) + r^2\, \frac{\partial V_2}{\partial r}\, (r,s,z) \\
\frac{du}{dt'} &= -\, \frac{1}{2}\, vu + \lambda V_1'(s) + \lambda r\, \frac{\partial V_2}{\partial s}\, (r,s,z) \\
\frac{dz}{dt'} &= r^{3/2}\, w \\
\frac{dw}{dt'} &= r^{3/2}\, \frac{\partial V_2}{\partial z}\, (r,s,z)\ .
\end{aligned}$$

We fix the total energy $H(\underset{\sim}{q},\underset{\sim}{p}) = h_0$. We write the total energy as the sum of the energy h_1 of the three-particle system plus the energy h_2 of the two-particle system.

$$h_1 = \frac{1}{2}\underset{\sim}{y}^T M_1^{-1}\underset{\sim}{y} - U_1(\underset{\sim}{x}) = \frac{1}{r}\left[\frac{1}{2}(u^2+v^2) - V_1(s)\right]$$

$$h_2 = \frac{1}{2}\underset{\sim}{w}^T M_2^{-1}\underset{\sim}{w} - U_2(\underset{\sim}{x},\underset{\sim}{z}) = \frac{1}{2}w^2 - V_2(r,s,z)$$

$$h_1 + h_2 = h_0 \ .$$

Note that h_0 is a constant of motion but that h_1 and h_2 are not.

If $z \neq 0$ (i.e. we are not at a quadruple collision) then $V_2(r,s,z)$ is smooth at $r = 0$. Therefore the equations of motion extend to $\{r = 0\}$ where we have a flow on a manifold $\Omega_c \times (0,\infty) \times R^1$ given by

$$(3) \qquad \begin{aligned} \frac{ds}{dt'} &= \lambda u \\ \frac{dv}{dt'} &= \frac{1}{2}v^2 + u^2 - V_1(s) \\ \frac{du}{dt'} &= -\frac{1}{2}vu + \lambda V_1'(s) \\ \frac{dz}{dt'} &= 0 \\ \frac{dw}{dt'} &= 0 \ . \end{aligned}$$

Here

$$(s,v,u) \in \Omega_c = \{\tfrac{1}{2}(u^2+v^2) - V_1(s) = 0\} \ ,$$

$$z \in (0,\infty) \ , \quad w \in R^1.$$

The above flow is the product of the identity flow on $(0,\infty)\times R^1$ with a flow on Ω_c. The flow on Ω_c is exactly the flow studied in [1], i.e., the effect of particle 1 disappears in the limit as we approach triple collision.

Vector fields (2) and (3) still have singularities at $s = \pm 1$. These singularities correspond to double collisions between particles 2 and 3 and between particles 3 and 4. Orbits can be extended through these double collisions by exactly the same transformations used in [1]. We shall not carry out these transformations here, but we shall speak of the flow as though the orbits were extended through double collisions.

The manifold structure of Σ now follows from the computations in [1] and from Theorem 1 of [2]. There is a unique point $\omega \in \Omega_c$ such that an orbit ending in triple collision must be asymptotic to ω. Thus Σ is exactly the stable manifold of the invariant set $\{\omega\} \times (0,\infty) \times R^1$. In [1] it was shown that ω has two attracting directions. Adding the two neutral directions of $(0,\infty) \times R^1$ we have that Σ is four-dimensional (codimension 1).

The unstable manifold of ω in Ω_c consists of two orbits. In [1] it was shown that, for certain values of m_2, m_3 and m_4, with $m_2 = m_4$, v becomes unbounded on these two orbits. Furthermore, on one of the orbits $s \to -1$, while on the other orbit $s \to +1$. As mentioned above, throughout this section we are assuming that m_2, m_3 and m_4 are fixed at values such that the flow on Ω_c satisfies these properties.

We have an arc Γ crossing Σ at x_0. The orbit through x_0 is asymptotic to $\{\omega\} \times (0,\infty) \times R^1$ and hence to a point (ω, z^*, w^*) [3]. The orbit through x_0 takes an infinite time to reach triple collision in the time coordinate t'. In the original time coordinate t, however, the orbit arrives at triple collision in time $\tau(x_0) < \infty$.

Now let $\rho(s,z)$ be such that the position coordinates $(r = \rho(s,z), s, z)$ correspond to the position coordinates $(q_1, q_2 = 0, q_3, q_4)$. Then ρ is a well defined continuous function. Let

$$\Lambda = \{(r,s,u,v,z,w): r = \rho(s,z)\} .$$

We shall now prove Theorem 2, assuming the estimates given in the following lemma. These estimates will be proved at the end of this section.

Lemma 6. Let $(r,s,v,u,z,w)(t')$ be a solution of (2). There exists a $\bar{\mu} > 0$ such that the following holds. Given any $\bar{v} > 0$, there exist $\tilde{\varepsilon} > 0$ and $\tilde{v} > 0$ such that, if

$$|z(0)-z^*| < \tilde{\varepsilon},\ |w(0)-w^*| < \tilde{\varepsilon},\ r(0) < \tilde{\varepsilon} \text{ and } v(0) > \tilde{v} ,$$

then

$$v(t') > \bar{v},\ |h_1(t')| < \bar{\mu},\ r(t') < \bar{\mu}, \text{ and } |w(t')-w^*| < \bar{\mu}$$

until the solution crosses Λ.

Proof of Theorem 2: That Σ is a codimension 1 immersed submanifold and that the masses can be chosen to satisfy (1) has already been proved. We have only to show the existence of Γ_0 and τ_- satisfying (a), (b), and (c).

For $x \in \Gamma$, define $\tau_-(x)$ as the least positive time such that the orbit through x hits Λ at that time. Whenever τ_- is defined, it satisfies (b).

Since Σ has codimension 1, it is locally two-sided. Orbits starting on one side of Σ tend to follow one branch of the unstable manifold of ω, while orbits starting on the other side of Σ tend to follow the other branch. Let Σ_0 be a small open neighborhood of x_0 in Σ with respect to the manifold topology. Pick a subarc $\tilde{\Gamma}$ of Γ such that $\tilde{\Gamma} \cap \Sigma_0 = \{x_0\}$ and $\tilde{\Gamma}$ lies on the side of Σ_0 corresponding to the branch of the unstable manifold with $s \to +1$.

Orbits starting close to the stable manifold of a point tend to follow the unstable manifold of that point arbitrarily far. Therefore, given any $a > 0$, any $\tilde{v} > 0$, and any open neighborhood $\mathcal{U}$

of the closure of the positive orbit through x_0 , there exists a subarc $\hat{\Gamma}$ of $\tilde{\Gamma}$, abutting on x_0, with the following properties: (1) for each $x \in \hat{\Gamma}$, the orbit through x remains in $U \cup \{r < a\}$ until v exceeds $\tilde{v}$, and (2), as $x \to x_0$, the orbit passes arbitrarily close to the point on the unstable manifold of (ω, z^*, w^*) with $v = \tilde{v}$. Note that this point has $z = z^*$ and $w = w^*$. Recall that $dt = r^{3/2}\,dt'$. Therefore in the time coordinate t the time for the orbit to go from $\hat{\Gamma}$ to $v = \tilde{v}$ can be made arbitrarily close to $\tau(x_0)$ by choosing a and U small.

Now let $\bar{\mu}$ be given by Lemma 6. From equations (1) we have

$$\frac{dr}{dt} = r^{-1/2}\, v \geq \bar{\mu}^{-1/2}\, \bar{v}$$

as long as $v > \bar{v}$. Therefore the orbits described by Lemma 6 cross Λ in time less than $\bar{\mu}^{3/2}/\bar{v}$. For any $\bar{v}$, let $\tilde{\varepsilon}$, $\tilde{v}$ be given by Lemma 6. Choose $\hat{\Gamma}$ so small that the orbits starting in $\hat{\Gamma}$ satisfy the hypotheses of that lemma the first time v equals $\tilde{v}$. Then τ_- will be defined on $\hat{\Gamma}$ and $\tau_-(x) \to \tau(x_0)$ as $x \to x_0$. Taking $\Gamma_0 = \hat{\Gamma} \cup \{x_0\}$, we have proved (a) and (b).

To prove (c), we first note that, for $x \in \hat{\Gamma}$, the value of v when the orbit through x intersects Λ goes to infinity as $x \to x_0$. The definition of h_1 gives us

$$\frac{1}{2}\,(u^2+v^2) - V_1(s) = rh_1 \ ,$$

so

$$\frac{1}{2}\, v^2 - \bar{\mu}^2 \leq V_1(s) \ .$$

Therefore $s \to +1$ as $v \to \infty$. From the explicit form of σ given in [1], we can write

$$y_1 = \lambda_1 r^{-1/2}[v \cos \lambda(1-s) + u \sin \lambda(1-s)] \ ,$$

where $\lambda_1 < 0$ is a constant depending only on the masses. Again by the definition of h_1 ,

$$u^2 \le 2V_1(s) + 2\bar{\mu}^2$$

Since $(1-s)V_1(s)$ is bounded as $s \to +1$, $|u \sin \lambda(1-s)|$ is bounded. Therefore $y_1 \to -\infty$ as $v \to \infty$. But

$$\dot{q}_2 = \dot{x}_1 + \dot{z}_2 = y_1/m_1 + w_2/(m_1+m_2+m_3).$$

Since w_2 is bounded by a constant times $|w^*| + \bar{\mu}$, $\dot{q}_2 \to -\infty$ as $v \to \infty$, which establishes (c). Since Λ is a section if $\dot{q}_2 < 0$, τ_- is continuous near x_0. If necessary, choose Γ_0 smaller, so that τ_- is continuous on Γ_0. The proof of Theorem 2 is complete.

We shall prove Lemma 6 by a series of estimates given in Lemma 7 through 10 below. Let $\mathcal{R} = \{(r,s,z): r \le \rho(s,z)\}$.

Lemma 7. There exist positive ε_0 and μ_0 so that

$$|V_2(r,s,z)| < \mu_0 , \quad |\frac{\partial V_2}{\partial r}(r,s,z)| < \mu_0 , \quad \text{and} \quad |\frac{\partial V_2}{\partial z}(r,s,z)| < \mu_0$$

whenever $(r,s,z) \in R$ and $|z-z^*| < \varepsilon_0$.

Proof: V_2 is real analytic in a neighborhood of the compact set $\{(r,s,z^*): 0 \le r \le \rho(s,z^*),\ s \in [-1,]\}$. The estimates follow from continuity.

For Lemmas 8, 9 and 10, let $(r,s,v,u,z,w)(t')$ be a solution of (2) such that $(r,s,z)(t') \in \mathcal{R}$ for $t' \in [0,\bar{t}']$. Let $\rho^* = \max\{\rho(s,z^*): s \in [-1,1]\}$ and fix $\varepsilon^* > 0$.

Lemma 8. There exist $\varepsilon_1 > 0$ and $\mu_1 > 0$ so that

$$|z(\bar{t}') - z(0)| < \mu_1/\hat{v}$$

whenever

$$|w(t')-w^*| < \varepsilon_1 , \ r(t') < \rho^*+\varepsilon^* , \text{ and } v(t') > \hat{v} > 0 ,$$

for $t' \in [0,\bar{t}')$.

Proof: Pick any $\varepsilon_1 > 0$ and let $\mu_1 > \frac{2}{3}(\rho^*+\varepsilon^*)^{3/2}(|w^*|+\varepsilon_1)$. Since $\frac{dr}{dt'} = rv > 0$, we have

$$z(\bar{t}') - z(0) = \int_0^{\bar{t}'} \frac{dz}{dt'}\, dt' = \int_{r(0)}^{r(\bar{t}')} \frac{dz}{dr}\, dr$$

Therefore

$$|z(\bar{t}')-z(0)| \leq \int_{r(0)}^{r(\bar{t}')} r^{1/2} \frac{|w|}{v}\, dr$$

$$\leq \frac{2}{3}(\rho^*+\varepsilon^*)^{3/2}(|w^*|+\varepsilon_1)/\hat{v} < \mu_1/\hat{v} .$$

Lemma 9. There exist positive ε_2, v_2, and μ_2 so that

$$|w(\bar{t}')-w(0)| < \mu_2/\hat{v} \quad ,$$

$$|h(\bar{t}')| < \mu_2 \quad ,$$

$$v(\bar{t}') > v(0) - \rho^* - \varepsilon^* ,$$

whenever $|z(0)-z^*| < \varepsilon_2$ and whenever

$$|w(t')-w^*| < \varepsilon_2 , \quad r(t') < \rho^*+\varepsilon^* , \quad \text{and} \quad v(t') > \hat{v} \geq v_2$$

for $t' \in [0,\bar{t}')$.

Proof: First choose $\varepsilon_2 < \min(\varepsilon_1, \frac{\varepsilon_0}{2})$. Then choose μ_2 so that

$$\mu_2 > \max\left(\frac{2}{3}(\rho^*+\varepsilon^*)^{3/2}\mu_0,\ |h_0| + \frac{1}{2}(|w^*|+\varepsilon_2)^2+\mu_0\right) .$$

Finally choose v_2 so that

$$v_2 > \max\left(2\mu_1/\varepsilon_0,\ \mu_2+(\rho^*+\varepsilon^*)\mu_0\right) .$$

By Lemma 8,

$$|z(t')-z(0)| < \mu_1/v_2 < \varepsilon_0/2 .$$

Therefore $|z(t')-z^*| < \varepsilon_0$ and the estimates of Lemma 7 hold. Hence

$$|w(\bar{t}')-w(0)| \leq \int_{r(0)}^{r(\bar{t}')} r^{1/2} \left|\frac{\partial V_2}{\partial z}(r,s,z)\right|/v \, dr$$

$$\leq \frac{2}{3} (\rho^*+\varepsilon^*)^{3/2} \mu_0/\hat{v} < \mu_2/\hat{v} \qquad ,$$

which establishes the first estimate.

For the second estimate,

$$h_1 = h_0 - h_2 = h_0 - \frac{1}{2} w^2 + V_2(r,s,z)$$

so

$$|h_1(\bar{t}')| \leq |h_0| + \frac{1}{2} (|w^*|+\varepsilon_2)^2 + \mu_0 < \mu_2 \ .$$

For the third estimate, we see from equations (2) that

$$\frac{d(r+v)}{dt'} = \frac{1}{2} u^2 + r[v + h_1 + r \frac{\partial V_2}{\partial r} (r,s,z)] \ .$$

Therefore

$$\frac{d(r+v)}{dt'} > r[v_2 - \mu_2 - (\rho^*+\varepsilon^*)\mu_0] > 0 \ .$$

Hence

$$r(\bar{t}') + v(\bar{t}') > r(0) + v(0) > v(0) \ ,$$

and

$$v(\bar{t}') > v(0) - \rho^* - \varepsilon^* \ ,$$

which completes the proof of Lemma 9.

<u>Lemma 10</u>. Given any positive δ and ν, there exist positive ε_3 and v_3 so that the following estimates hold:

$$|w(\bar{t}')-w^*| < \delta \quad \text{and} \quad v(\bar{t}') > \nu$$

for all $\bar{t}'$ such that $r(t') < \rho^*+\varepsilon^*$ for $t' \in [0,\bar{t}')$ and whenever

$$|z(0)-z^*| < \varepsilon_3 \ , \quad |w(0)-w^*| < \varepsilon_3 \ , \quad \text{and} \quad v(0) > v_3 \ .$$

<u>Proof</u>: We can take $\delta < \varepsilon_2$ and $\nu > v_2$ without losing generality. Choose $\varepsilon_3 < \varepsilon_2$ and $v_3 > \nu + \rho^* + \varepsilon^*$. Also choose v_3 so large

that

$$\mu_2/(v_3-\rho^*-\varepsilon^*) < \delta .$$

We proceed by contradiction. Assume there is a $\hat{t}'$ such that $|w(\hat{t}')-w^*| \geq \delta$ or $v(\hat{t}') \leq \nu$. Choose the least such $\hat{t}'$. Then $|w(t')-w^*| < \delta < \varepsilon_2$ and $v(t') > \nu > v_2$ for $t' \in [0,\hat{t}')$. By Lemma 9,

$$v(\hat{t}') > v_3 - \rho^* - \varepsilon^* > \nu .$$

Therefore we must have $|w(\hat{t}') - w^*| \geq \delta$. But, again by Lemma 9,

$$|w(\hat{t}')-w(0)| < \mu_2/(v_3-\rho^*-\varepsilon^*) < \delta,$$

which contradicts the assumption and proves Lemma 10.

<u>Proof of Lemma 6</u>: The solution crosses Λ when $r = \rho(s,z)$. From the definition of ρ^*, there exists an ε_4 so that, if $r(0) < \varepsilon_4$, $r(\hat{t}') \geq \rho^*+\varepsilon^*$, and $|z(t')-z^*| < \varepsilon_4$, for all $t \in [0,\hat{t}']$, then the solution crosses Λ somewhere in the time interval $(0,\hat{t}')$.

Let $\bar{\mu} > \max(\mu_2,\rho^*+\varepsilon^*)$. Given any $\bar{v}$, pick $\nu > \max(\bar{v},v_2,2\mu_1/\varepsilon_4)$, $\delta < \min(\varepsilon_1,\varepsilon_2,\bar{\mu})$, and let ε_3 and v_3 be given by Lemma 10. Now take $\tilde{\varepsilon} < \min(\varepsilon_2,\varepsilon_3,\varepsilon_4/2)$ and $\tilde{v} > v_3$.

Pick any $\bar{t}'$ so that $r(t') < \rho^*+\varepsilon^*$ for all $t' \in [0,\bar{t}')$. Then, by Lemma 10,

$$|w(t')-w^*| < \delta < \varepsilon_1 \text{ and } v(t') > \nu , \quad \forall\, t' \in [0,\bar{t}'].$$

Therefore, by Lemma 8,

$$|z(t')-z(0)| < \mu_1/\nu < \varepsilon_4/2 .$$

But $|z(0)-z^*| < \tilde{\varepsilon} < \varepsilon_4/2$, so

$$|z(t')-z^*| < \varepsilon_4 , \qquad \forall\, t' \in [0,\bar{t}'].$$

Therefore, for any $\tilde{t}'$ such that the solution does not cross Λ in the interval $[0,\tilde{t}']$, we must have that $(r,s,z)(t') \in R$ and

$$r(t') < \rho^* + \varepsilon^* , \qquad \forall\ t' \in [0,\tilde{t}'].$$

By Lemma 10,

$$|w(t')-w^*| < \delta \quad \text{and} \quad v(t') > \nu , \qquad \forall\ t' \in [0,\tilde{t}'].$$

Since $\nu > \bar{v}$ and $\delta < \bar{\mu}$, we have two of the estimates of Lemma 6. Since $\nu > v_2$, $\delta < \varepsilon_2$, and $\tilde{\varepsilon} < \varepsilon_2$, we can apply Lemma 9 to get

$$|h(t')| < \mu_2 < \bar{\mu} , \qquad \forall\ t' \in [0,\tilde{t}'].$$

Finally, $r(t') < \rho^* + \varepsilon^* < \bar{\mu}$, which completes the proof of Lemma 6.

References

[1] McGehee, R., Triple collision in the collinear three-body problem, Inventiones math., to appear.

[2] Fenichel, N., Persistence and smoothness of invariant manifolds for flows, Indiana Univ. Math. J. 21 (1971) 193-226.

[3] Fenichel, N., Asymptotic stability with rate conditions, Indiana Univ. Math. J., 23 (1974) 1109-1137.

ON OPTIMAL ESTIMATES FOR THE SOLUTIONS OF LINEAR PARTIAL DIFFERENTIAL EQUATIONS OF FIRST ORDER WITH CONSTANT COEFFICIENTS ON THE TORUS

Helmut Rüssmann

1. Introduction

We consider for $n \geq 2$ the linear partial differential equation

(1.1) $$\omega_1 \frac{\partial u}{\partial x_1} + \ldots + \omega_n \frac{\partial u}{\partial x_n} = f(x) ,$$

where $\omega_1, \ldots, \omega_n$ are rationally independent real numbers and $x = (x_1, \ldots, x_n) \mapsto f(x)$ is a complex valued function, which is defined and analytic in a strip

$$|\text{Im } x| = \max_{1 \leq j \leq n} |\text{Im } x_j| < r$$

of positive width $2r$ in the space $\mathbb{C}^n$ and which has period 2π in each variable $x_1, \ldots, x_n$. We consider analytic solutions u of (1.1) of period 2π in $x_1, \ldots, x_n$. In other words, identifying all variables mod 2π we consider the differential equation (1.1) on a complex extension of the torus $\mathbb{R}^n / 2\pi \cdot \mathbb{Z}^n$.

This problem arises for example in the proof of the Kolmogorov-Arnold-Moser theorem on the preservation of invariant tori under small perturbation of the Hamiltonian, where in each step of the Newton iteration process such a differential equation (1.1) has to be solved and the solution properly estimated. In the differentiable case one needs even estimates, which are in some sense optimal, in order to get optimal differentiability conditions. It is the aim of this paper to give such optimal estimates under certain conditions for the vector $\omega = (\omega_1, \ldots, \omega_n)$.

It is easy to construct a formal solution u of (1.1) by

expanding f in a Fourier series

$$f(x) = \sum_{k \in \mathbb{Z}^n} f_k \, e^{i\langle k,x\rangle}$$

where the symbol $\langle\cdot,\cdot\rangle$ is defined by

$$\langle x,y\rangle = x_1 y_1 + x_2 y_2 + \ldots + x_n y_n$$

if

$$x = (x_1,\ldots,x_n) \;, \quad y = (y_1,\ldots,y_n) \in \mathbb{C}^n$$

and

$$f_k = \left(\frac{1}{2\pi}\int_{-\pi}^{\pi}\right)^n f(x)\, e^{-i\langle k,x\rangle}\, dx_1\ldots dx_n \;, \quad k \in \mathbb{Z}^n$$

are the Fourier coefficients of f. Inserting a Fourier series for u with unknown coefficients in (1.1) we get the series

$$(1.2) \qquad u(x) = \sum_{k\neq 0} \frac{f_k}{\langle k,\omega\rangle} e^{i\langle k,x\rangle}$$

provided that the mean value f_0 of f vanishes. The series (1.2) is then uniquely determined up to an arbitrary constant.

The denominators $\langle k,\omega\rangle$ in (1.2) do not vanish because of the rational independence of $\omega_1,\ldots,\omega_n$. However, they come arbitrarily close to zero, so that we have here a problem with "small divisors". If some of the denominators $\langle k,\omega\rangle$ tend to zero too rapidly, one can construct examples for f, such that (1.1) has no solutions on the torus $\mathbb{R}^n/2\pi\cdot\mathbb{Z}^n$ (see Siegel-Moser [1], p. 260).

In order to prevent the denominators from getting small too rapidly, one usually assumes, that the vector $\omega = (\omega_1,\ldots,\omega_n)$ satisfies the inequalities

$$(1.3) \qquad |\langle k,\omega\rangle| \geq \gamma |k|^{-\tau} \;, \qquad k \in \mathbb{Z}^n \setminus \{0\} \;,$$

where by $|\cdot|$ we denote the maximum norm defined by

$$|y| = \max_{1 \le j \le m} |y_j| , \qquad y = (y_1,\dots,y_m, \quad m = 1,2,\dots,$$

and γ, τ are some positive constants. This condition has the advantage of being symmetric, but for estimates involving explicit numerical constants it is not quite adequate, because too many inequalities are superfluous. Therefore we prefer a slightly different condition, which is used in the theory of diophantine approximation. For a proper formulation of this condition we put

$$q = (k_1,\dots,k_{n-1}) , \quad p = k_n$$

and

$$D(q,\omega) = \min_{p \in \mathbb{Z}} |\langle (q,p),\omega\rangle| , \qquad q \in \mathbb{Z}^{n-1} . \tag{1.4}$$

Then we require

$$D(q,\omega) \ge \gamma |q|^{-\tau} , \qquad q \in \mathbb{Z}^{n-1} \setminus \{0\}, \tag{1.5}$$

with positive constants γ and τ. One easily sees that (1.3) follows from (1.5), and conversely (1.5) is a consequence of (1.3), if we replace γ by

$$\gamma \left(\frac{|\omega_n|}{|\omega_1| + \dots + |\omega_n|}\right)^{\tau}$$

in (1.5). But it is just this change of constants depending on τ we want to avoid, since we like to clear the dependence on τ in our estimates. Therefore we assume (1.5) instead of (1.3).

Another formulation of (1.5) is

$$\gamma(\omega,\tau) \equiv \inf_{q \in \mathbb{Z}^{n-1} \setminus \{0\}} \{|q|^{\tau} D(q,\omega)\} > 0 .$$

Here of course the question arises if there exist $\omega \in \mathbb{R}^n$ satisfying (1.5). For an answer of this question, given $\tau > 0$, we define

$$\Omega(\tau) = \{\omega \in \mathbb{R}^n \mid \gamma(\omega,\tau) > 0\} ,$$

that is, $\Omega(\tau)$ is the set of all vectors $\omega = (\omega_1,\ldots,\omega_n)$ satisfying the inequalities (1.5) for a certain $\gamma > 0$. Then there are three cases:

Case I. $0 < \tau < n-1$.

In this case we have $\Omega(\tau) = \emptyset$. This is an immediate consequence of a classical theorem of Dirichlet, which states that the inequality

$$|q|^{n-1} D(q,\omega) < |\omega_n|$$

has infinitely many solutions $q \in \mathbb{Z}^{n-1}$ for every $\omega \in \mathbb{R}^n$. (See Cassels [2], p. 14.)

Case II. $n-1 < \tau$.

In this case the set $\mathbb{R}^n \setminus \Omega(\tau)$ has n-dimensional Lebesgue measure zero; that is, for a given $\tau > n-1$ almost all $\omega \in \mathbb{R}^n$ satisfy (1.5) with some $\gamma = \gamma(\omega) > 0$. According to our remark on the relation between (1.3) and (1.5) this means that for a given $\tau > n-1$ almost all $\omega \in \mathbb{R}^n$ satisfy (1.3) with some $\gamma = \gamma(\omega) > 0$. For a proof see Arnold [3], p. 98.

Case III. $\tau = n-1$.

In this case

(a) $\Omega(n-1)$ has n-dimensional Lebesgue measure zero, but

(b) every open set contains continuum many elements of $\Omega(n-1)$, in fact $\Omega(n-1)$ has Hausdorf dimension n.

Statement (a) has been proved by Khinchine [4], [5] (see also Cassels [1], Chapters V and VII).

Statement (b) has been proved by Schmidt [6] (see also Jarnik [7]).

Actually the above statements have not been formulated for the set $\Omega(n-1)$ but for the set

$$\left\{\left(\frac{\omega_1}{\omega_n},\ldots,\frac{\omega_{n-1}}{\omega_n}\right) \;\middle|\; \omega = (\omega_1,\ldots,\omega_n) \in \Omega(n-1)\right\} \subseteq \mathbb{R}^{n-1}$$

as is usual in the theory of diophantine approximation. However, the transference is obvious.

After this remark on the existence of vectors $\omega \in \mathbb{R}^n$ satisfying the inequalities (1.5) we formulate our main reuslt on best estimates of the solutions of (1.1) in the following

Theorem 1.1. Let $\omega = (\omega_1,\ldots,\omega_n) \in \mathbb{R}^n$ be such that the inequalities (1.5) are valid with some constants $\gamma > 0$ and $\tau \geq n-1$. Furthermore let f be a complex valued function, which is analytic in the strip $|\mathrm{Im}\, x| < r$ of $\mathbb{C}^n$, satisfies the estimate

$$|f(x)| \leq M, \qquad |\mathrm{Im}\, x| < r,$$

with some positive constant M, is of period 2π in each variable $x_1,\ldots,x_n$, and has mean value zero. Then the differential equation (1.1) has a unique solution u, which is analytic in $|\mathrm{Im}\, x| < r$, is of period 2π in $x_1,\ldots,x_n$, and has mean value zero. Moreover this solution possesses the estimate

$$(1.6) \quad |u(x)| \leq \frac{3\pi}{2\gamma}\, 6^{n/2}\, \frac{\sqrt{\tau\, \Gamma(2\tau)}}{(2\delta)^{\tau}}\, M, \qquad |\mathrm{Im}\, x| < r-\delta, \quad 0 < \delta < r.$$

If f is real for real x, then so is u.

Actually we prove a more general theorem (see Theorem 3.1), but here we have the most important case.

For controlling the quality of the estimate (1.6) let us take in (1.1) for f the exponentials $x \longmapsto e^{i\langle k,x\rangle}$, $k \in \mathbb{Z}^n$. Then an easy calculation shows that for these special cases under the assumption (1.5) the estimate

$$(1.7) \quad |u(x)| \leq \frac{c}{\gamma}\left(\frac{\tau}{e\,\delta}\right)^{\tau} M, \qquad |\mathrm{Im}\, x| < r-\delta, \quad 0 < \delta < r,$$

with $c = 1$ can be established. It seems that the best possible estimate in Theorem 1.1 is (1.7) with a constant c depending on n. In (1.6) we have reached this result with respect to δ, but not with respect to τ, if τ tends to infinity. For Stirling's formula gives

$$\Gamma(2\tau) \sim (2\tau)^{2\tau-1/2} e^{-2\tau} \sqrt{2\pi} , \qquad \tau \to \infty ,$$

so (1.6) fails (1.7) by the factor $\tau^{1/4}$. Using the technique developed in this paper and the theory of continued fractions we are able to prove Theorem 1.1 with (1.7) instead of (1.6) in the case $n = 2$. In higher dimensions, however, some facts of the geometry of numbers are needed, which do not seem to be available up to this time.

For applications of Theorem 1.1 to the Kolmogorov-Arnold-Moser theorem, it is clear, one is only inter sted in values of τ near $n-1$, and then the estimate (1.6) is as good as (1.7).

The proof of Theorem 1.1 is divided into several lemmas, which are of interest for themselves. In Section 4 we still consider the case that f is a continuous function, which is analytic only in $x_1,\dots,x_{n-1}$. Then the solution u of (1.1) with mean value zero is analytic in $x_1,\dots,x_{n-1}$ and continuously differentiable in x_n, but the estimate (1.6) remains valid with the only restrictions $x_n \in \mathbb{R}$, $\delta \le 2$ (Theorem 4.2).

By a direct estimate of the series (1.2) using (1.3) or (1.5), one easily proves

$$(1.8) \qquad |u(x)| \le C\, \delta^{-\beta} M , \qquad |\mathrm{Im}\, x| < r-\delta , \quad 0 < \delta < r ,$$

with a positive constant C depending on n and τ and with $\beta = \tau+n$. A better result was first established by Moser [8], who showed (1.8) with $\beta = \tau+1$. Thereafter we obtained in [9] an estimate for sums with small divisors, which allows us to prove

(1.8) with $\beta = \tau$, however only for $\tau > n-1$.

There is quite another way to prove (1.6) than offered in this paper, possibly with a better constant depending on n and τ. In this way, one needs the theory of best approximation of analytic and periodic functions in several complex variables by trigonometric polynomials, then the connection between such functions and almost periodic functions in one variable, and finally a very interesting paper of Bohr [10] on the best estimate of the indefinite integral of trigonometric polynomials.

2. Fourier Series of Analytic Periodic Functions

Let $x = (x_1,\dots,x_n) \longmapsto f(x)$ be a complex valued function, which is defined and analytic in the strip

$$|\mathrm{Im}\, x| = \max_{1\le j\le n} |\mathrm{Im}\, x_j| < r$$

of positive width $2r$ in the space $\mathbb{C}^n$ and has period 2π in each variable $x_1,\dots,x_n$. We denote the set of all such functions by $\mathfrak{f}(r)$. The simplest way of getting an inner product in $\mathfrak{f}(r)$ is to define

$$\langle f,g\rangle = \left(\frac{1}{2\pi}\int_{-\pi}^{\pi}\right)^n f(x)\,\overline{g(x)}\,dx_1\dots dx_n \ , \quad f,g \in \mathfrak{f}(r) \ , \tag{2.1}$$

so that the exponentials

$$x \longmapsto e_k(x) \equiv e^{i\langle k,x\rangle} \ , \qquad k \in \mathbb{Z}^n \ , \tag{2.2}$$

form an orthonormal set in $\mathfrak{f}(r)$ relative to this inner product. Clearly there are other possibilities of defining an inner product, for example

$$\langle f,g\rangle_r = \left(\frac{1}{4\pi r}\int_{-\pi}^{\pi}\int_{-r}^{r}\right)^n f(x+iy)\,\overline{g(x+iy)}\,dx_1 dy_1\dots dx_n dy_n$$

for bounded functions $f,g \in \mathfrak{f}(r)$. But then for normalization the exponentials (2.1) have to be multiplied by factors depending on r, which makes the formulas unnecessarily circumstantial. So we prefer (2.1) and prove some technical lemmas concerning the Fourier coefficients

$$f_k = \langle f, e_k \rangle , \qquad k \in \mathbb{Z}^n ,$$

and the Fourier series

$$\sum_{k \in \mathbb{Z}^n} f_k \, e^{i\langle k,x \rangle}$$

of $f \in \mathfrak{f}(r)$, which are useful in the sequel.

Lemma 2.1. Let f belong to $\mathfrak{f}(r)$ and satisfy the estimate

$$|f(x)| \leq M , \qquad |\mathrm{Im}\, x| < r, \tag{2.3}$$

with some positive constant M. Then we have the inequality

$$\sum_{k \in \mathbb{Z}^n} |f_k|^2 \, e^{2r\|k\|} \leq 2^n M^2 \tag{2.4}$$

where

$$\|k\| = |k_1| + \ldots + |k_n| , \qquad k = (k_1,\ldots,k_n),$$

and f_k are the Fourier coefficients of f.

Proof: For every $a \in \mathbb{R}^n$ with

$$|a| = \max_{1 \leq j \leq n} |a_j| < r$$

the function $x \longmapsto f(x+ia)$ belongs to $\mathfrak{f}(r-|a|)$. Its Fourier coefficients are

$$f_k(a) = \left(\frac{1}{2\pi} \int_{-\pi}^{\pi} \right)^n f(x+ia) \, e^{-i\langle k,x \rangle} \, dx_1 \ldots dx_n , \quad k \in \mathbb{Z}^n,$$

so Bessel's inequality gives

$$\sum_{k \in \mathbb{Z}^n} |f_k(a)|^2 \leq \left(\frac{1}{2\pi} \int_{-\pi}^{\pi} \right)^n |f(x+ia)|^2 \, dx_1 \ldots dx_n ,$$

and therefore we have

$$\sum_{k \in \mathbb{Z}^n} |f_k(a)|^2 \leq M^2 , \qquad |a| < r, \tag{2.5}$$

in view of (2.3).

Now we remark that the function

$$a \longmapsto f_k(a)\, e^{<k,a>} = \left(\frac{1}{2\pi}\int_{-\pi}^{\pi}\right)^n f(x+ia)\, e^{-i<k,x+ia>}\, dx_1 \,..dx_n$$

is independent of a in the cube $|a| < r$. For this cube is connected and

$$\frac{\partial}{\partial a_j}\left\{f_k(a) e^{<k,a>}\right\} = \left(\frac{1}{2\pi}\int_{-\pi}\right)^n i\, \frac{\partial}{\partial x_j}\left\{f(x+ia)\, e^{-i<k,x+ia>}\right\} dx_1..dx_n$$

vanishes as mean value of a partial derivative, $j = 1,\ldots,n$. Hence we have

$$f_k(a)\, e^{<k,a>} = f_k(0) = f_k$$

and consequently

$$\sum_{k\in\mathbb{Z}^n} |f_k|^2\, e^{-2<k,a>} \leq M^2\,, \qquad |a| < r, \tag{2.6}$$

in view of (2.5).

To finish the proof we denote by e_μ, $\mu = 1,\ldots,2^n$, those vectors in $\mathbb{R}^n$ which have components ± 1, and define

$$Q_\mu = \left\{k \in \mathbb{Z}^n \mid <k,e_\mu> = -\|k\|\right\}.$$

Then we have

$$\bigcup_{\mu=1}^{2^n} Q_\mu = \mathbb{Z}^n\,, \tag{2.7}$$

and

$$\sum_{k\in Q_\mu} |f_k|^2\, e^{2s\|k\|} \leq M^2\,, \qquad 0 < s < r\,, \qquad \mu = 1,\ldots,2^n\,,$$

putting $a = s\,e_\mu$ in (2.6). Passing to the limit $s \to r$, we get

$$\sum_{k\in Q_\mu} |f_k|^2\, e^{2r\|k\|} \leq M^2\,, \qquad \mu = 1,\ldots,2^n.$$

Adding these inequalities and using (2.7) we obtain the desired inequality (2.4).

Lemma 2.2. Let ϕ be a real function, which is defined and continuous in the interval $[0,\infty[$ such that

$$0 \leq \phi(s) \leq \phi(t)\ , \qquad 0 \leq s \leq t \tag{2.8}$$

and

$$\lim_{s\to\infty} s^{-1} \log \phi(s) = 0\ . \tag{2.9}$$

Furthermore let $a_0, a_1, a_2, \ldots$ be a sequence of nonnegative real numbers such that

$$\sum_{\mu=0}^{m} a_\mu \leq \phi(m)\ , \qquad m = 0,1,\ldots\ . \tag{2.10}$$

Then for every $\delta > 0$ we have

$$\sum_{\mu=0}^{\infty} a_\mu\, e^{-\delta\mu} \leq \int_0^\infty e^{-s}\, \phi(\tfrac{s}{\delta})\, ds < \infty\ . \tag{2.11}$$

Proof: We put

$$A_\mu = \sum_{\nu=0}^{\mu} a_\nu\ , \quad b_\mu = e^{-\delta\mu}\ , \qquad \mu = 0,1,\ldots\ .$$

Then we get

$$A_\mu \leq \phi(\mu) \qquad \mu = 0,1,\ldots\ . \tag{2.12}$$

Moreover using Abel's partial summation formula

$$\sum_{\mu=0}^{N} a_\mu b_\mu = \sum_{\mu=0}^{N} A_\mu (b_\mu - b_{\mu+1}) + A_N\, b_{N+1}$$

we obtain by (2.8), (2.10) and (2.12)

$$\sum_{\mu=0}^{N} a_\mu b_\mu - A_N b_{N+1} = \sum_{\mu=0}^{N} A_\mu (b_\mu - b_{\mu+1})$$

$$\tag{2.13} \left\{ \begin{aligned} &\leq \sum_{\mu=0}^{N} \phi(\mu)\, (b_\mu - b_{\mu+1}) \\ &\leq \sum_{\mu=0}^{N} \phi(\mu+1)\, (b_\mu - b_{\mu+1}) \end{aligned} \right.$$

$$\leq \sum_{\mu=0}^{N} \phi(\mu+1)\, b_\mu \leq e^{\delta} \sum_{\mu=1}^{\infty} \phi(\mu)\, e^{-\delta\mu}.$$

The infinite series at the end of these estimates converges for every $\delta > 0$, since from (2.9) we have

$$\lim_{\mu\to\infty} \sqrt[\mu]{\phi(\mu)} = 1 ,$$

so that the radius of convergence of the power series

$$\sum_{\mu=1}^{\infty} \phi(\mu)\, z^{\mu}$$

is 1. The convergence of the above series implies that its terms $\phi(\mu)\, e^{-\delta\mu}$ tend to zero, so (2.12) yields

$$\lim_{\mu\to\infty} A_\mu\, b_{\mu+1} = 0 . \tag{2.14}$$

Since ϕ is not decreasing by (2.8) and the function $t \mapsto -e^{\delta t}$ is increasing, the expressions (2.13) form a lower sum and an upper sum of the Stieltjes integral

$$\int_0^{N+1} \phi(t)\, d(-e^{-\delta t}) = \int_0^{(N+1)\delta} e^{-s}\, \phi(\frac{s}{\delta})\, ds .$$

Therefore the above chain of estimates gives

$$\sum_{\mu=0}^{N} a_\mu b_\mu - A_N b_{N+1} \leq \int_0^{(N+1)\delta} e^{-s}\, \phi(\frac{s}{\delta})\, ds$$

$$\leq e^{\delta} \sum_{\mu=1}^{\infty} \phi(\mu)\, e^{-\delta\mu} < \infty.$$

Passing to the limit $N \to \infty$ we get by (2.14) the desired estimate (2.11).

<u>Lemma 2.3</u>. Let ϕ be a real function, which is defined and continuous in the interval $[0,\infty[$ such that

$$0 \leq \phi(s) \leq \phi(t) , \qquad 0 \leq s \leq t ,$$

and

$$\lim_{s\to\infty} s^{-1} \log \phi(s) = 0 .$$

Furthermore let $\{c_k\}_{k\in\mathbb{Z}^n}$ be a sequence of complex numbers such that

$$\sum_{|k|\leq m} |c_k|^2 \leq \phi(m)\ , \qquad m = 0,1,\dots, \tag{2.15}$$

where $\quad |k| = \max_{1\leq j\leq n} |k_j|\ , \qquad k = (k_1,\dots k_n)$.

Finally let f be a function which belongs to $\mathfrak{f}(r)$ and satisfies

$$|f(x)| \leq M\ , \qquad |\mathrm{Im}\ x| < r\ ,$$

for some positive constant M. Then the series

$$F(x) = \sum_{k\in\mathbb{Z}^n} c_k f_k\ e^{i<k,x>}$$

where f_k are the Fourier coefficients of f, converges absolutely and uniformly in every strip $|\mathrm{Im}\ x| < r-\delta$, $0 < \delta < r$, and consequently defines a function of $\mathfrak{f}(r)$. Moreover we have the estimate

$$\left\{\begin{aligned} |F(x)| &\leq \sum_{k\in\mathbb{Z}^n} |c_k f_k\ e^{i<k,x>}| \\ &\leq 2^{n/2}\ M \sqrt{\int_0^\infty e^{-s}\ \phi(\tfrac{s}{2\delta})\ ds} < \infty, \\ &\qquad |\mathrm{Im}\ x| < r-\delta,\quad 0 < \delta < r. \end{aligned}\right. \tag{2.16}$$

<u>Proof</u>: The uniform convergence in $|\mathrm{Im}\ x| < r-\delta$, $0 < \delta < r$ follows below in the proof of (2.16). Then in view of the periodicity of the exponentials F is of period 2π in $x_1,\dots,x_n$ and hence converges uniformly in every compact subset of the strip $|\mathrm{Im}\ x| < r$. Therefore F is analytic in the strip by a well known theorem in function theorey (see Theorem 9.12.1 in Dieudonné [11]), and so belongs to $\mathfrak{f}(r)$.

Using Schwarz' inequality and Lemma 2.1 we get for $|\mathrm{Im}\ x| < r-\delta$,

$$|F(x)| \leq \sum_{k\in\mathbb{Z}^n} |c_k f_k\ e^{i<k,x>}|$$

$$\leq \sum_{k\in\mathbb{Z}^n} |c_k|\ |f_k|\ e^{\|k\|(r-\delta)} \leq \sum_{k\in\mathbb{Z}^n} |f_k|\ e^{\|k\| r}\ |c_k| e^{-|k|\delta}$$

$$\leq \sqrt{\sum_{k\in\mathbb{Z}^n} |f_k|^2\ e^{2r\|k\|}} \sqrt{\sum_{k\in\mathbb{Z}^n} |c_k|^2\ e^{-2\delta|k|}}$$

$$\leq 2^{n/2}\ M \sqrt{\sum_{\mu=0}^{\infty} a_\mu\ e^{-2\delta\mu}}$$

putting

$$a_\mu = \sum_{|k|=\mu} |c_k|^2 , \qquad \mu = 0,1,\ldots .$$

By (2.15) we have

$$\sum_{\mu=0}^{m} a_\mu \leq \phi(m) , \qquad m = 0,1,\ldots .$$

Hence by Lemma 2.2,

$$\sum_{\mu=0}^{\infty} a_\mu\ e^{-2\delta\mu} \leq \int_0^\infty e^{-s}\ \phi\left(\frac{s}{2\delta}\right)\ ds < \infty .$$

This and the above estimate for F obviously yield (2.16).

Lemma 2.4. Let f be a function of $\mathfrak{f}(r)$. Then f can be expanded in its Fourier series:

$$f(x) = \sum_{k\in\mathbb{Z}^n} f_k\ e^{i\langle k,x\rangle} , \qquad |\mathrm{Im}\ x| < r .$$

This series converges absolutely and uniformly in every strip

$$|\mathrm{Im}\ x| < r-\delta , \qquad 0 < \delta < r .$$

Proof: Since f is continuous in $|\mathrm{Im}\ x| < r$ and of period 2π in each variable $x_1,\ldots,x_n$ we have

$$M_\delta \equiv \sup_{|\mathrm{Im}\ x|<r-\delta} |f(x)| < \infty , \qquad 0 < \delta < r.$$

Clearly

$$\sum_{\substack{k\in\mathbb{Z}^n \\ |k|\leq m}} 1 = (2m+1)^n , \qquad m = 0,1,\ldots,$$

and the function $s \mapsto \phi(s) = (2s+1)^n$ fulfills the assumptions in Lemma 2.3. So we can apply Lemma 2.3 with $c_k = 1$, $k \in \mathbb{Z}^n$, to the Fourier series

$$F(x) = \sum_{k\in\mathbb{Z}^n} f_k \, e^{i<k,x>}$$

of f in the strip $|\mathrm{Im}\, x| < \hat{r} = r-\delta$. Therefore this series converges absolutely and uniformly in every strip $|\mathrm{Im}\, x| < \hat{r}-\delta = r-2\delta$, $0 < 2\delta < r$ and represents a function of $\mathfrak{f}(r)$.

Now we still have to prove

$$(2.17) \qquad f(x) = F(x) \, , \qquad x \in \mathbb{R}^n \, ,$$

since by analytic continuation this equation remains valid for all of the strip $|\mathrm{Im}\, x| < r$ (see Theorem 9.4.4 in [8]). But f and F have the same Fourier coefficients, so equation (2.17) holds, because the orthonormal set of the exponentials (2.2)--considered as functions in $\mathbb{R}^n$--is complete, as is well known. (See Courant-Hilbert [12], II, Section 4.)

3. Estimates for Sums Cont ining Small Divisors

If we wish to apply Lemma 2.3 to the formal solution (1.2) of (1.1), we have to find estimates for the su s

$$(3.1) \qquad \sum_{\substack{k \in \mathbb{Z}^n \\ 0 < |k| \le m}} \frac{1}{|<k,\omega>|^2} \, , \qquad m = 1,2,\ldots,$$

provided a condition like (1.5) with positive constants $\gamma > 0$, $\tau \geq n-1$, is fulfilled. In this section we give such estimates, but instead of (1.5) we require more general

$$(3.2) \qquad D(q,\omega) \geq \psi\left(|q|\right) \, , \qquad q \in \mathbb{Z}^{n-1}\setminus\{0\},$$

where ψ is a real function with the following three properties:

$$(3.3) \qquad \left\{ \begin{array}{l} \psi \text{ is defined and continuous on the positive} \\ \text{real axis and has positive values.} \end{array} \right.$$

$$(3.4) \qquad \psi(s)\, s^{n-1} \geq \psi(t)\, t^{n-1}, \qquad 0 < s \leq t,$$

$$(3.5) \qquad \lim_{s \downarrow 0} \psi(s) = \infty.$$

Since we always provide $n \geq 2$, we see from (3.4) that ψ is strictly decreasing and

$$\lim_{s \to \infty} \psi(s) = 0.$$

So we conclude that ψ defines a homeomorphism of the positive real axis onto itself.

Condition (3.4) is reasonable because of the fact mentioned in the introduction that there does not exist any $\omega \in \mathbb{R}^n$ which satisfies (3.2), with

$$\psi(s) = \gamma s^{-\tau}, \qquad \gamma > 0, \quad 0 < \tau < n-1.$$

A function ψ with the properties (3.3), (3.4) and (3.5) we call an approximation function in the sequel.

Now we prove three lemmas, which yield optimal estimates for the sums (3.1). Thereafter we apply Lemma 2.3 to prove a more general version of Theorem 1.1.

<u>Lemma 3.1</u>. Let $n \geq 2$ and $\omega = (\omega_1, \ldots, \omega_n)$ be a vector in $\mathbb{R}^n$, the components of which are rationally independent. Then for $m = 1, 2, \ldots$ we have

$$(3.6) \qquad \sum_{\substack{k \in \mathbb{Z}^n \\ 0<|k|\leq m}} \frac{1}{|\langle k,\omega\rangle|^2} \leq \frac{\pi}{2} \sum_{\substack{q \in \mathbb{Z}^{n-1} \\ 0<|q|\leq m}} \frac{1}{D(q,\omega)^2} + \frac{\pi^2}{3|\omega_n|^2},$$

where $D(q,\omega)$ is defined in (1.4).

<u>Proof:</u> We have

$$\sum_{\substack{k \in \mathbb{Z}^n \\ 0<|k|\leq m}} |\langle k,\omega\rangle|^{-2} = \sum_{\substack{q\in \mathbb{Z}^{n-1} \\ 0<|q|\leq m}} \sum_{\substack{p\in \mathbb{Z} \\ |p|\leq m}} |\langle (q,p),\omega\rangle|^{-2} + \sum_{\substack{p \in \mathbb{Z} \\ 0<|p|\leq m}} (\omega_n\, p)^{-2}$$

$$\leq \sum_{\substack{q \in \mathbb{Z}^{n-1} \\ 0<|q|\leq m}} \sum_{p=-\infty}^{\infty} |<(q,p),\omega>|^{-2} + 2\omega_n^{-2} \sum_{p=1}^{\infty} p^{-2}$$

$$= \sum_{\substack{q \in \mathbb{Z}^{n-1} \\ 0<|q|\leq m}} \omega_n^{-2} \sum_{p=-\infty}^{\infty} \left(\omega_n^{-1} D(q,\omega) + p\right)^{-2}$$

$$+ 2\omega_n^{-2} \sum_{p=1}^{\infty} p^{-2} .$$

Now we use the well known formula

$$\sum_{p=-\infty}^{\infty} \frac{1}{(a+p)^2} = \frac{\pi^2}{\sin^2 \pi a} , \qquad a \in \mathbb{R} \setminus \mathbb{Z} , \tag{3.7}$$

to get

$$\sum_{p=-\infty}^{\infty} \frac{1}{(a+p)^2} \leq \frac{\pi^2}{4a^2} , \qquad 0 < |a| \leq \frac{1}{2} , \tag{3.8}$$

in view of

$$\frac{\sin s}{s} \geq \frac{2}{\pi} , \qquad 0 < s \leq \frac{\pi}{2} . \tag{3.9}$$

From (3.7) we also get the well known equaton

$$\sum_{p=1}^{\infty} \frac{1}{p^2} = \frac{\pi^2}{6} \tag{3.10}$$

by subtracting a^{-2} and passing to the limit $a \to 0$.

As a consequence of the definition (1.4) we have

$$0 < |\omega_n|^{-1} D(q,\omega) < \frac{1}{2} , \qquad q \in \mathbb{Z}^{n-1} \setminus \{0\}, \tag{3.11}$$

so that this expression can be inserted in (3.7) instead of a. Then (3.8) and (3.10) together with the calculation at the beginning of this proof give the desired inequality (3.6).

<u>Lemma 3.2.</u> Let $\omega \in \mathbb{R}^n$ be such that the inequalities (3.2) are satisfied, where $D(q,\omega)$ is defined in (1.4) and ψ is an approximation function. Then for $m = 1,2,\ldots,$ we have

$$(3.12)\qquad \sum_{\substack{q\in\mathbb{Z}^{n-1}\setminus\{0\}\\ 0<|q|\le m}} \frac{1}{D(q,\omega)^2} \le \frac{3^{n+2}}{2\psi(m)}\left(\frac{1}{\psi(m)} - \frac{4}{3|\omega_n|}\right).$$

Proof: We put

$$(3.13)\qquad c_j = \frac{|\omega_n|}{2}\left(\frac{2}{3}\right)^j , \qquad j = 0,1,\ldots,$$

and

$$(3.14)\qquad Q^{(j)} = \{q \in \mathbb{Z}^{n-1} \mid c_{j+1} < D(q,\omega) \le c_j\},\ j = 0,1,\ldots .$$

Then by (3.2) we get

$$\psi(|q|) \le D(q,\omega) \le c_j \le 2c_j , \qquad q \in Q^{(j)} ,$$

and therefore

$$(3.15)\qquad |q| \ge a_j , \qquad q \in Q^{(j)} ,$$

where

$$(3.16)\qquad a_j = \psi^{-1}(2c_j) , \qquad j = 0,1,\ldots,$$

and ψ^{-1} denotes the inverse of ψ.

Using (1.4) and (3.2) we obtain for different $q,q' \in Q^{(j)}$ the relation

$$\begin{aligned}\psi(|q-q'|) \le D(q-q',\omega) &\le |<(q-q',p-p'),\omega>| \\ &\le |<(q,p),\omega>| + |<(q',p'),\omega>| \\ &= D(q,\omega) + D(q',\omega) \le 2c_j ,\end{aligned}$$

where the integers p and p' are uniquely determined by the equality. From this relation we deduce

$$(3.17)\qquad |q-q'| \ge a_j , \qquad q,q' \in Q^{(j)}, q \ne q'.$$

Now we define an integer $r \ge 0$ by

$$(3.18)\qquad c_{r+1} < \psi(m) \le c_r .$$

Such an integer exists, since (3.2), (3.11), (3.13) and the fact that ψ is decreasing imply

(3.19) $\psi(m) \leq \psi(|q|) \leq D(q,\omega) \leq c_0 \, , \quad q \in \mathbb{Z}^{n-1}, \quad 0 < |q| \leq m,$

for $m = 1,2,\ldots$. The inverse ψ^{-1} of ψ is also decreasing, so we get

(3.20) $$a_j \leq m \, , \qquad\qquad 0 \leq j \leq r \, ,$$

by (3.13), (3.16) and (3.18). From (3.18) and (3.19), we obtain

$$c_{r+1} < D(q,\omega) \leq c_0 \, , \qquad\qquad 0 < |q| \leq m,$$

and hence by (3.14)

(3.21) $$\left\{ \begin{aligned} \sum_{\substack{q \in \mathbb{Z}^{n-1} \\ 0<|q|\leq m}} D(q,\omega)^{-2} &= \sum_{j=0}^{r} \sum_{\substack{q \in Q^{(j)} \\ |q|\leq m}} D(q,\omega)^{-2} \\ &\leq \sum_{j=0}^{r} c_{j+1}^{-2} \, Z_j \, , \end{aligned} \right.$$

where

$$Z_j = \sum_{\substack{q \in Q^{(j)} \\ |q| \leq m}} 1 \, , \qquad\qquad 0 \leq j \leq r \, .$$

For estimating Z_j we consider every point $q \in Q^{(j)}$ as center of an open cube W_q of length a_j parallel to the coordinate axes. All these cubes are disjoint, because we get for a point

$$x \in W_q \cap W_{q'} \, , \qquad q,q' \in Q^{(j)}, \quad q \neq q',$$

from (3.17) the contradiction

$$a_j \leq |q-q'| \leq |x-q| + |x-q'| < \frac{a_j}{2} + \frac{a_j}{2} \leq a_j \, .$$

Those cubes W_q with centers satisfying

$$|q| \leq m \, , \quad q \in Q^{(j)}$$

lie in the cube

$$\left\{ x \;\middle|\; |x| \leq m + \frac{a_j}{2} \right\}$$

of volume $(2m+a_j)^{n-1}$. The number of these disjoint cubes is Z_j. Because every such cube has volume a_j^{n-1}, we get

$$Z_j \leq \left(\frac{2m}{a_j} + 1\right)^{n-1} \leq \left(\frac{3m}{a_j}\right)^{n-1}$$

where in the last inequality we have used (3.20). Inserting this estimate for Z_j in (3.21) we obtain

$$(3.22) \qquad \sum_{\substack{q \in \mathbb{Z}^{n-1} \\ 0<|q|\leq m}} D(q,\omega)^{-2} \leq (3m)^{n-1} \sum_{j=0}^{r} \frac{1}{c_{j+1}^2 \, a_j^{n-1}} .$$

Now we recall property (3.4) of ψ and the definitions of c_j and a_j in (3.13) and (3.16). Then from (3.20) it is clear that

$$c_{j+1} \, a_j^{n-1} = \frac{1}{3} \psi(a_j) \, a_j^{n-1} \geq \frac{1}{3} \psi(m) \, m^{n-1} ,$$

hence (3.22) yields

$$\sum_{\substack{q\in\mathbb{Z}^{n-1} \\ 0<|q|\leq m}} D(q,\omega)^{-2} \leq \frac{3^n}{\psi(m)} \sum_{j=0}^{r} c_{j+1}^{-1}$$

$$= \frac{3^n}{\psi(m)} \frac{6}{|\omega_n|} \left[\left(\frac{3}{2}\right)^{r+1} - 1\right] = \frac{3^{n+2}}{2\psi(m)} \left[\frac{1}{c_r} - \frac{4}{3|\omega_n|}\right] ,$$

so that with (3.18) the desired estimate (3.12) follows.

<u>Lemma 3.3</u>. Let $\omega = (\omega_1,\dots,\omega_n) \in \mathbb{R}^n$ satisfy the inequalities (3.2), where $D(q,\omega)$ is defined in (1.4) and ψ is an approximation function. Then for $m = 1,2,\dots,$ we have

$$(3.23) \qquad \sum_{\substack{k \in \mathbb{Z}^n \\ 0<|k|\leq m}} \frac{1}{|\langle k,\omega\rangle|^2} \leq \frac{\pi^2}{8} \frac{3^{n+2}}{\psi(m)^2}.$$

<u>Proof</u>: Since ψ is in Lemma 3.2 an approximate function too, we can take (3.13) and (3.19) to obtain

$$(3.24) \qquad \psi(m) \leq \frac{|\omega_n|}{2} , \qquad m = 1,2,\dots,$$

and therefore

$$\frac{1}{|\omega_n|^2} \leq \frac{1}{|\omega_n| \, \psi(m)} , \qquad m = 1,2,\dots,$$

This yields (3.23) by comparing Lemma 3.1 and Lemma 3.2.

<u>Theorem 3.1</u>. Let $\omega = (\omega_1,\dots,\omega_n) \in \mathbb{R}^n$ satisfy the inequalities

$$D(q,\omega) \geq \psi(|q|) , \qquad q \in \mathbb{Z}^{n-1} \setminus \{0\}$$

where

$$D(q,\omega) = \min_{p\in\mathbb{Z}} |<(q,p),\omega>| , \qquad q \in \mathbb{Z}^{n-1}$$

and ψ is an approximation function as defined in (3.3), (3.4) and (3.5), which moreover fulfills the condition

$$\lim_{s\to\infty} \frac{1}{s} \log \frac{1}{\psi(s)} = 0 . \tag{3.25}$$

Furthermore let f be a complex valued function, which is analytic in the strip

$$|\text{Im } x| = \max_{1\leq j\leq n} |\text{Im } x_j| < r$$

of $\mathbb{C}^n$, satisfies the estimate

$$|f(x)| \leq M , \qquad |\text{Im } x| < r ,$$

is of period 2π in each variable $x_1,\dots,x_n$, and has mean value zero. Then the differential equation

$$\omega_1 \frac{\partial u}{\partial x_1} + \dots + \omega_n \frac{\partial u}{\partial x_n} = f(x) \tag{3.26}$$

has a unique solution u, which is analytic in $|\text{Im } x| < r$, is of period 2π in $x_1,\dots,x_n$ and has mean value zero. Moreover, this solution possesses for $|\text{Im } x| < r-\delta$, $0 < \delta < r$ the estimate

$$|u(x)| \leq \frac{3\pi}{2\sqrt{2}} 6^{n/2} M \sqrt{\int_0^\infty e^{-s} \psi\left(\frac{s}{2\delta}\right)^{-2} ds} < \infty. \tag{3.27}$$

If f is real for real x, then so is u.

<u>Proof</u>: We define

$$c_0 = 0 , \quad c_k = \frac{1}{i<k,\omega>} , \qquad k = \mathbb{Z}^n \setminus \{0\},$$

and

$$\phi(s) = \begin{cases} \dfrac{\pi^2}{8} \dfrac{3^{n+2}}{\psi(s)^2} & \text{for } s > 0 \\ 0 & \text{for } s = 0 . \end{cases}$$

Then all assumptions of Lemma 2.3 are fulfilled including the inequality (2.15), which follows from Lemma 3.3. So we may conclude that the series

$$u(x) = \sum_{k \neq 0} \frac{f_k}{i\langle k,\omega\rangle} e^{i\langle k,x\rangle} \tag{3.28}$$

converges absolutely and uniformly in every compact subset of the strip $|\mathrm{Im}\, x| < r$ and therefore defines a function of $\mathfrak{G}(r)$. The uniform convergence moreover allows differentiation and integration term by term in $|\mathrm{Im}\, x| < r$ (see Theorem 9.12.1 in [11]) in order to verify that u is the unique solution of (3.26) with mean value zero. Here of course we also use Lemma 2.4. The estimate (3.27) follows from (2.16) for $F = u$ and our definition of ϕ.

If f is real for real x, we get for the Fourier coefficients of f the equalities

$$\bar{f}_k = f_{-k} , \qquad k \in \mathbb{Z}^n ,$$

and therefore

$$\overline{u(x)} = u(x) , \qquad x \in \mathbb{R}^n ,$$

from (3.28).

<u>Proof of Theorem 1.1</u>: We have only to verify that

$$s \mapsto \psi(s) = \gamma s^{-\tau} , \qquad \gamma > 0,\ \tau \geq n-1,$$

defines an approximation function and fulfills (3.25), which is obvious. Then Theorem 3.1 yields Theorem 1.1.

4. <u>Continuity in One Variable</u>

In order to mark out one variable, for which only continuity is required we change our notation and put

$$y = (y_1,\dots,y_{n-1}) = (x_1,\dots,x_{n-1}) , \qquad z = x_n .$$

Then (1.1) takes the form

$$(4.1) \qquad \omega_1 \frac{\partial u}{\partial y_1} + \ldots + \omega_{n-1} \frac{\partial u}{\partial y_{n-1}} + \omega_n \frac{\partial u}{\partial z} = f(y,z) \ .$$

We assume that f satisfies the following three conditions:

(i) f is a complex valued function defined and continuous in the strip

$$\left\{ y \in \mathbb{C}^{n-1} \;\middle|\; |\mathrm{Im}\, y| < r \right\} \times \left\{ z \in \mathbb{R} \right\}$$

of $\mathbb{C}^n$, where r is a positive number and $|\cdot|$ denotes the maximum norm as always in this paper.

(ii) f is analytic in the variables $y_1, \ldots, y_{n-1}$.

(iii) f has period 2π in all variables $y_1, \ldots, y_{n-1}$, z, and the mean value

$$\left(\frac{1}{2\pi} \int_{-\pi}^{\pi} \right)^n f(y,z) \, dy_1 \ldots dy_{n-1} dz$$

of f vanishes.

The collection of all such functions we denote by $\mathcal{O}\!\mathcal{J}(r)$.

We have

<u>Theorem 4.1.</u> Let $\omega = (\omega_1, \ldots, \omega_n) \in \mathbb{R}^n$ satisfy the inequalities

$$(4.2) \qquad D(q,\omega) \geq \psi(|q|) \ , \qquad q \in \mathbb{Z}^{n-1} \setminus \{0\},$$

where

$$D(q,\omega) = \min_{p \in \mathbb{Z}} |\langle (q,p), \omega \rangle| \ , \qquad q \in \mathbb{Z}^{n-1} \ ,$$

and ψ is an approximation function as defined in (3.3), (3.4) and (3.5), which moreover fulfills the condition

$$\lim_{s \to \infty} \frac{1}{s} \log \frac{1}{\psi(s)} = 0 \ .$$

Furthermore let f belong to $\mathcal{O}\!\mathcal{J}(r)$ and satisfy the estimate

$$(4.3) \qquad |f(y,z)| \leq M \ , \qquad |\mathrm{Im}\, y| < r, \quad z \in \mathbb{R},$$

with some constant $M > 0$. Then the differential equation (4.1) has a unique solution u, which belongs to $\mathcal{O}\!\mathcal{J}(r)$ and is differntiable with respect to z. Moreover this solution possesses for

$$|\mathrm{Im}\, y| < r-\delta\ , \quad z \in \mathbb{R}\ , \qquad 0<\delta<r,\ \delta \le 2,$$

the estimate

(4.4) $$|u(y,z)| \le \frac{3\pi}{2\sqrt{2}}\, 6^{n/2}\, M \sqrt{\int_0^\infty e^{-s}\, \psi\left(\frac{s}{2\delta}\right)^{-2} ds} < \infty\ .$$

If f is real for real y, z, then so is u.

<u>Proof</u>: First we expand f for fixed z in a Fourier series with respect to y according to Lemma 2.4:

$$f(y,z) = \sum_{q\in\mathbb{Z}^{n-1}} f_q(z)\, e^{i\langle q,y\rangle}\ .$$

Here $\langle q,y\rangle = q_1y_1 + \ldots + q_{n-1}y_{n-1}$, and the Fourier coefficients f_q of f are given by

$$f_q(z) = \left(\frac{1}{2\pi}\int_{-\pi}^{\pi}\right)^{n-1} f(y,z)\, e^{-i\langle q,y\rangle}\, dy_1\ldots dy_{n-1}\ ,\quad q\in\mathbb{Z}^{n-1},$$

and hence are continuous 2π-periodic functions of z. Moreover by Lemma 2.1 we have the estimate

(4.5) $$|f_q(z)| \le 2^{(n-1)/2}\, M\, e^{-r|q|}\ , \qquad z\in\mathbb{R}\ ,\ q\in\mathbb{Z}^{n-1}.$$

Now we consider the series

(4.6) $$U(y,z) = \sum_{q\in\mathbb{Z}^{n-1}} c_q\, f_q(z)\, e^{i\langle q,y\rangle}\ ,$$

where we define

(4.7) $$c_0 = 0\ ,\qquad c_q = \frac{\pi}{i\omega_n \sin \frac{\pi}{\omega_n}\langle q,\hat{\omega}\rangle}\ , \qquad q \in \mathbb{Z}^{n-1}\setminus\{0\},$$

and

$$\hat{\omega} = (\omega_1,\ldots,\omega_{n-1})\ .$$

Then we get by (3.9) and (3.11)

(4.8) $$|c_q| = \frac{\pi}{|\omega_n|\, \sin \frac{\pi}{|\omega_n|}\, D(q,\omega)} \le \frac{\pi}{2D(q,\omega)}\ ,\quad q\in\mathbb{Z}^{n-1}\setminus\{0\},$$

and therefore

$$\sum_{|q|\leq m} |c_q|^2 \leq \phi(m) , \qquad m = 0,1,\ldots$$

by Lemma 3.2, if we define

$$(4.9) \qquad \phi(s) = \begin{cases} \max\left[0, \frac{\pi^2}{8}\frac{3^{n+2}}{\psi(s)}\left(\frac{1}{\psi(s)} - \frac{4}{3|\omega_n|}\right)\right] & \text{for } s > 0 \\ 0 & \text{for } s = 0 \end{cases}$$

According to the assumptions on ψ the function ϕ fulfills all conditions required in Lemma 2.3. Furthermore the other conditions of Lemma 2.3 are satisfied. So we can apply this lemma to the series $U(y,z)$ for n-1 instead of n, and obtain from (2.6) the estimate

$$(4.10) \qquad \begin{cases} |U(y,z)| \leq 2^{(n-1)/2} M \sqrt{\int_0^\infty e^{-s}\, \phi(\frac{s}{2\delta})\, ds} < \infty \\ |\mathrm{Im}\, y| < r-\delta , \quad 0 < \delta < r , \quad z \in \mathbb{R} . \end{cases}$$

Lemma 2.3 supplies uniform convergence of $U(y,z)$ with respect to y, while the uniform convergence in all variables is not immediately clear. But this follows from Lemma 2.2, if we put

$$a_\mu = \sum_{|q|=\mu} |c_q| , \qquad \mu = 0,1,\ldots .$$

For then we have in view of (4.2), (4.8) and the decrease of ψ for $m = 1,2,\ldots,$

$$\sum_{\mu=0}^{m} a_\mu \leq \frac{\pi}{2\psi(m)} \sum_{|q|\leq m} 1 = \frac{\pi}{2}\frac{(2m+1)^{n-1}}{\psi(m)}$$

where

$$s \longmapsto \begin{cases} \frac{\pi}{2}\frac{(2s+1)^{n-1}}{\psi(s)} & \text{for } s > 0 \\ 0 & \text{for } s = 0 \end{cases}$$

has all properties of the function ϕ in Lemma 2.2. So applying (2.11) and (4.5) we get for $|\mathrm{Im}\, y| < r-\delta$, $0 < \delta < r$, $z \in \mathbb{R}$ the estimate

$$\sum_{q\in\mathbb{Z}^{n-1}} |c_q\, f_q(z)\, e^{i<q,y>}| \leq 2^{(n-1)/2} M \sum_{q\in\mathbb{Z}^{n-1}} |c_q|\, e^{-\delta|q|} =$$

$$= 2^{(n-1)/2} M \sum_{\mu=0}^{\infty} a_\mu e^{-\delta\mu} < \infty .$$

Therefore the series $U(y,z)$ represents a continuous function in $|\mathrm{Im}\, y| < r$, $z \in R$, which is analytic in $y_1,\dots,y_{n-1}$ and has period 2π in all variables.

We now define

$$(4.11)\quad \begin{cases} u(y,z) = \dfrac{1}{2\pi} \displaystyle\int_{-\pi}^{\pi} \left\{ \dfrac{s-z}{\omega_n} f_0(s+\pi) + \mathcal{U}(y+(s-z)\dfrac{\hat{\omega}}{\omega_n}, s+\pi) \right\} ds \\ \quad = \dfrac{1}{2\pi} \displaystyle\int_{-\pi}^{\pi} \left\{ \dfrac{s}{\omega_n} f_0(s+z+\pi) + \mathcal{U}(y+ s\dfrac{\hat{\omega}}{\omega_n}, s+z+\pi) \right\} ds . \end{cases}$$

Using the definition (4.6), (4.7) of $U(y,z)$ one easily verifies that u is differentiable with respect to z and is a solution of (4.1) belonging to $\mathcal{O}\!\!\mathcal{J}(r)$. For the proof of u being a solution the first expression in (4.11) is adequate. The second expression shows that the mean value of u vanishes. In both cases one must remember that the mean value of f and hence the mean value of f_0 vanishes.

The uniqueness of the solution of (4.1) with vanishing mean value is a consequence of the compeleteness of the orthonormal set (2.2) and the fact that the Fourier series of such a solution is uniquely determined by the Fourier series of f (see (1.2)). These Fourier series for f and u also show that u is real for real y, z, if this is true for f.

It remains to prove (4.4). The relations (4.3), (4.10) and (4.11) yield for $|\mathrm{Im}\, y| < r-\delta$, $0 < \delta < r$, $z \in \mathbb{R}$ the estimate

$$|u(y,z)| \leq \frac{\pi}{2|\omega_n|} M + 2^{(n-1)/2} M \sqrt{\int_0^\infty e^{-s} \phi(\tfrac{s}{2\delta})\, ds}$$

$$\leq \frac{\pi}{2} M \sqrt{\frac{2}{|\omega_n|^2} + \frac{2^{n+2}}{\pi^2} \int_0^\infty e^{-s} \phi(\tfrac{s}{2\delta})\, ds} < \infty.$$

Therefore applying (3.24) and (4.9), we get

$$\frac{2}{|\omega_n|^2} + \frac{2^{n+2}}{\pi^2}\int_0^\infty e^{-s}\,\phi\left(\frac{s}{2\delta}\right)\,ds$$

$$\leq \frac{1}{\psi(1)|\omega_n|} - 6^{n+1}\int_{2\delta}^\infty \frac{e^{-s}\,ds}{\psi(\frac{s}{2\delta})|\omega_n|} + \frac{9}{2}\,6^n\int_0^\infty \frac{e^{-s}\,ds}{\psi(\frac{s}{2\delta})^2}$$

$$\leq \frac{9}{2}\,6^n\int_0^\infty \frac{e^{-s}\,ds}{\psi(\frac{s}{2\delta})^2} < \infty$$

and hence (4.4), because ψ is decreasing, and we consequently have

$$\frac{1}{\psi(1)|\omega_n|} - 6^{n+1}\int_{2\delta}^\infty \frac{e^{-s}\,ds}{\psi(\frac{s}{2\delta})|\omega_n|} \leq \frac{1}{\psi(1)|\omega_n|}\left(1-6^{n+1}\int_{2\delta}^\infty e^{-s}\,ds\right) < 0$$

for $\delta \leq 2$ and for $n \geq 2$, as always provided.

The analog to Theorem 1.1 is

<u>Theorem 4.2</u>. Let $\omega = (\omega_1,\ldots,\omega_n) \in \mathbb{R}^n$ satisfy the inequalities (1.5) with some constants $\gamma > 0$, $\tau \geq n-1$. Furthermore let f belong to $\mathfrak{O}(r)$ and fulfill the estimate

$$|f(y,z)| \leq M\,, \qquad |\mathrm{Im}\,y| < r,\ z \in \mathbb{R},$$

with some constant $M > 0$. Then the differential equation (4.1) has a unique solution u, which belongs to $\mathfrak{O}(r)$ and is differentiable with respect to z. Moreover this solution possesses the estimate

$$(4.12)\qquad |u(y,z)| \leq \frac{3\pi}{2\gamma}\,6^{n/2}\,\frac{\sqrt{\tau\,\Gamma(2\tau)}}{(2\delta)^\tau}\,M\,, \qquad \begin{array}{l}|\mathrm{Im}\,y| < r-\delta,\quad 0 < \delta < r,\\ \delta \leq 2,\quad z \in \mathbb{R}\,.\end{array}$$

If f is real for real y,z, then so is u.

<u>Proof</u>: Since $s \mapsto \psi(s) = \gamma s^{-\tau}$ defines an approximation function for $\gamma > 0$, $\tau \geq n-1$, which has the property $\lim_{s\to\infty}\frac{1}{s}\log\frac{1}{\psi(s)} = 0$, Theorem 4.2 follows from Theorem 4.1.

<u>Remark</u>. An estimate for $\partial/\partial z\ u$ is easily obtained from (4.1), (4.12) and Cauchy's estimate for analytic derivatives.

References

[1] Siegel, C. L., Moser, J. K., Lectures on Celestial Mechanics, Springer, 1971.

[2] Cassels, J. W. S., An Introduction to Diophantine Approximation, Cambridge University Press, 1957.

[3] Arnold, V. I., Small denominators and problems of stability of motion in classical and celestial mechanics, Russian Math. Surveys 18, 85-192 (1963).

[4] Khintchine, A., Zur metrischen Theorie der diophantischen Approximationen, Math. Z. 24, (1926) 706-714.

[5] Khintchine, A., Über eine Klasse linearer diophantischer Approximationen, Rendic. Circ. Mat. Palermo 50, 170-195 (1926).

[6] Schmidt, W. M., Badly approximable systems of linear forms, J. Number Theory 1, 139-154 (1969).

[7] Jarnik, V., Über die simultanen diophantischen Approximationen, Math. Z. 33, 505-543 (1931).

[8] Moser, J., A rapidly convergent iteration method and nonlinear differential equations II, Scuola norm. sup., Pisa, 1966.

[9] Rüssmann, H., Über invariante Kurven differenzierbarer Abbildungen eines Kreisringes, Nachr. Akad. Wiss. Göttingen II, Math. Phys. Kl,, 67-105 (1970).

[10] Bohr, H., Collected Mathematical Works, Vol. II, C 36, Dansk Matematisk Forening, København (1952).

[11] Dieudonné, J., Foundations of Modern Analysis, Academic Press, New York and London (1960).

[12] Courant, R., Hilbert, D., Methods of Mathematical Physics, Vol. I, Interscience-Wiley, New York.

Lecture Notes in Physics

Bisher erschienen/Already published

Vol. 1: J. C. Erdmann, Wärmeleitung in Kristallen, theoretische Grundlagen und fortgeschrittenene experimentelle Methoden. 1969. DM 22,–

Vol. 2: K. Hepp, Théorie de la renormalisation. 1969. DM 20,–

Vol. 3: A. Martin, Scattering Theory: Unitarity, Analyticity and Crossing. 1969. DM 18,–

Vol. 4: G. Ludwig, Deutung des Begriffs physikalische Theorie und axiomatische Grundlegung der Hilbertraumstruktur der Quantenmechanik durch Hauptsätze des Messens. 1970. Vergriffen.

Vol. 5: M. Schaaf, The Reduction of the Product of Two Irreducible Unitary Representations of the Proper Orthochronous Quantummechanical Poincaré Group. 1970. DM 18,–

Vol. 6: Group Representations in Mathematics and Physics. Edited by V. Bargmann. 1970. DM 27,–

Vol. 7: R. Balescu, J. L. Lebowitz, I. Prigogine, P. Résibois, Z. W. Salsburg, Lectures in Statistical Physics. 1971. DM 20,–

Vol. 8: Proceedings of the Second International Conference on Numerical Methods in Fluid Dynamics. Edited by M. Holt. 1971. Out of print.

Vol. 9: D. W. Robinson, The Thermodynamic Pressure in Quantum Statistical Mechanics. 1971. DM 18,–

Vol. 10: J. M. Stewart, Non-Equilibrium Relativistic Kinetic Theory. 1971. DM 18,–

Vol. 11: O. Steinmann, Perturbation Expansions in Axiomatic Field Theory. 1971. DM 18,–

Vol. 12: Statistical Models and Turbulence. Edited by M. Rosenblatt and C. Van Atta. 1972. Out of print.

Vol. 13: M. Ryan, Hamiltonian Cosmology. 1972. DM 20,–

Vol. 14: Methods of Local and Global Differential Geometry in General Relativity. Edited by D. Farnsworth, J. Fink, J. Porter and A. Thompson. 1972. DM 20,–

Vol. 15: M. Fierz, Vorlesungen zur Entwicklungsgeschichte der Mechanik. 1972. DM 18,–

Vol. 16: H.-O. Georgii, Phasenübergang 1. Art bei Gittergasmodellen. 1972. DM 20,–

Vol. 17: Strong Interaction Physics. Edited by W. Rühl and A. Vancura. 1973. DM 32,–

Vol. 18: Proceedings of the Third International Conference on Numerical Methods in Fluid Mechanics, Vol. I. Edited by H. Cabannes and R. Temam. 1973. DM 20,–

Vol. 19: Proceedings of the Third International Conference on Numerical Methods in Fluid Mechanics, Vol. II. Edited by H. Cabannes and R. Temam. 1973. DM 29,–

Vol. 20: Statistical Mechanics and Mathematical Problems. Edited by A. Lenard. 1973. DM 24,–

Vol. 21: Optimization and Stability Problems in Continuum Mechanics. Edited by P. K. C. Wang. 1973. DM 18,–

Vol. 22: Proceedings of the Europhysics Study Conference on Intermediate Processes in Nuclear Reactions. Edited by N. Cindro, P. Kulišić and Th. Mayer-Kuckuk. 1973. DM 29,–

Vol. 23: Nuclear Structure Physics. Proceedings of the Minerva Symposium on Physics. Edited by U. Smilansky, I. Talmi, and H. A. Weidenmüller. 1973. DM 29,–

Vol. 24: R. F. Snipes, Statistical Mechanical Theory of the Electrolytic Transport of Non-electrolytes. 1973. DM 22,–

Vol. 25: Constructive Quantum Field Theory. The 1973 "Ettore Majorana" International School of Mathematical Physics. Edited by G. Velo and A. Wightman. 1973. DM 29,–

Vol. 26: A. Hubert, Theorie der Domänenwände in geordneten Medien. 1974. DM 28,–

Vol. 27: R. Kh. Zeytounian, Notes sur les Ecoulements Rotationnels de Fluides Parfaits. 1974. DM 28,–

Vol. 28: Lectures in Statistical Physics. Edited by W. C. Schieve and J. S. Turner. 1974. DM 24,–

Vol. 29: Foundations of Quantum Mechanics and Ordered Linear Spaces. Advanced Study Institute Held in Marburg 1973. Edited by A. Hartkämper and H. Neumann. 1974. DM 26,–

Vol. 30: Polarization Nuclear Physics. Proceedings of a Meeting held at Ebermannstadt October 1–5, 1973. Edited by D. Fick. 1974. DM 24,–

Vol. 31: Transport Phenomena. Sitges International School of Statistical Mechanics, June 1974. Edited by G. Kirczenow and J. Marro. DM 39,–

Vol. 32: Particles, Quantum Fields and Statistical Mechanics. Proceedings of the 1973 Summer Institute in Theoretical Physics held at the Centro de Investigacion y de Estudios Avanzados del IPN – Mexico City. Edited by M. Alexanian and A. Zepeda. 1975. DM 18,–

Vol. 33: Classical and Quantum Mechanical Aspects of Heavy Ion Collisions. Symposium held at the Max-Planck-Institut für Kernphysik, Heidelberg, Germany, October 2–5, 1974. Edited by H. L. Harney, P. Braun-Munzinger and C. K. Gelbke. 1975. DM 28,–

Vol. 34: One-Dimensional Conductors, GPS Summer School Proceedings, 1974. Edited by H. G. Schuster. 1975. DM 32,–

Vol. 35: Proceedings of the Fourth International Conference on Numerical Methods in Fluid Dynamics. June 24–28, 1974, University of Colorado. Edited by R. D. Richtmyer. 1975. DM 37,–

Vol. 36: R. Gatignol, Théorie Cinétique des Gaz à Répartition Discrète de Vitesses. 1975. DM 23,–

Vol. 37: Trends in Elementary Particle Theory. Proceedings 1974. Edited by H. Rollnik and K. Dietz. 1975. DM 37,–

Vol. 12: Statistical Models and Turbulence. Edited by C. Van Atta and M. Rosenblatt. Reprint of the First Edition 1975. DM 28,–

Vol. 38: Dynamical Systems, Theory and Applications. Proceedings 1974. Edited by J. Moser. 1975. DM 46,–